CHEMISTRY OF THE ENVIRONMENT

ACADEMIC PRESS RAPID MANUSCRIPT REPRODUCTION

CHEMISTRY OF THE ENVIRONMENT

R. A. Bailey
H. M. Clark
J. P. Ferris
S. Krause
R. L. Strong

Department of Chemistry
Rensselaer Polytechnic Institute
Troy, New York

ACADEMIC PRESS New York San Francisco London 1978
A Subsidiary of Harcourt Brace Jovanovich, Publishers

ACADEMIC PRESS, INC.
111 Fifth Avenue, New York, New York 10003

United Kingdom Edition published by
ACADEMIC PRESS, INC. (LONDON) LTD.
24/28 Oval Road, London NW1 7DX

Library of Congress Cataloging in Publication Data

Main entry under title:

Introduction to chemistry of the environment.

Includes bibliographies.
1. Environmental chemistry. I. Bailey, Ronald Albert, Date
QD31.2.I57 540 78–11293
ISBN 0–12–073050–2

PRINTED IN THE UNITED STATES OF AMERICA

80 81 82 9 8 7 6 5 4 3 2

CONTENTS

PREFACE

This book was developed from the notes for a course in "Chemistry of the Environment" given by the authors over the past several years to juniors, seniors, and graduate students in Science and Engineering at Rensselaer Polytechnic Institute. While in its present form this book may serve as a text for a course of this sort, it also can be a valuable background source for those who wish to broaden their understanding of the chemical processes important in the environment, or who require a more thorough grounding in particular aspects of environmental chemistry. It should be useful to the chemist who finds that his or her conventional courses have not dealt adequately with environmental topics or to the nonchemist who needs to expand his knowledge in this area. The nonchemist, however, will need some knowledge of basic chemistry, including elementary organic chemistry, in order to profit from some areas of this book. The objective is to take the reader from a basic level of chemical knowledge to the point where he can deal with advanced monographs and the research literature in environmental chemistry.

The topics covered reflect to some extent the interests of the authors, but we have emphasized those subjects that seem to be of greatest environmental importance. Thus, we have dealt extensively with petroleum, chlorinated hydrocarbons, pesticides, heavy metals, nuclear chemistry, and atmospheric chemistry. We have also dealt with topics such as atmospheric circulation, which are important to understanding environmental processes. Both natural processes and technological processes that impinge on the environment are included. Obviously, not all important topics have been discussed, but the principles included here should be widely applicable to a general understanding of environmental problems. While specific environmental problems may vary, the basic chemistry of environment processes do not change. For example, the chemical and biochemical processes for the environmental degradation of crude oil or a plastic cup are the same. Knowledge of the chemistry as presented here will be applicable both to existing problems and to new problems as they arise.

A bibliography of items ranging from the very general to the very specific is given at the end of each chapter for those who wish to pursue a topic in depth. These references are not intended to provide complete coverage of a subject, but they will provide a significant introduction to the literature of many environmental problems.

Anyone dealing with the broad aspects of the environment must expand his or her knowledge into many fields of science and technology. One of the problems encountered while reading the literature in all these areas arises from the terminology and traditionally established units in each of the fields. We have chosen, as a practical matter, to use the units that are normally encountered in the literature of each field. In the interest of uniformity, however, we have very often expressed a given value for some property or process in both conventional and SI units.

We are indebted to many of our students for helpful suggestions during the development of this book. We also thank our colleagues, Professor Harry Herbrandson in particular, for valuable comments and suggestions.

1
INTRODUCTION

1.1 GENERAL

Man evolved on earth with its atmosphere, land and water systems, and types of climate in such a way that he can cope reasonably well with this particular environment. Being intelligent and inquisitive, mankind has not only investigated the environment extensively, but has done many things to change it. Other living things also change their environment; the roots of large trees can crack rocks, the herds of elephants in present-day African game parks are uprooting the trees and turning forests into grasslands, but no other living thing can change its environment in so many ways or as rapidly as man. The possibility exists that man can change his environment into one that he cannot live in, just as the African elephant may be doing on a smaller scale. The elephants cannot learn to stop uprooting the trees they need for survival, but we should be capable of learning about our environment and about the problems we ourselves create.

One of the vital problems (among the many) that man must solve is how to continue his technologically based civilization without at the same time damaging irreversibly the environment in which he evolved and which supports his life. This environment is complex; the interrelations of its component parts are subtle and sometimes unexpected, and stress in one area may have far-reaching effects. This environment is also finite (hence, we have the expression "spaceship earth"). That being so, there is necessarily a limit to all things that can be tolerated before significant environmental changes take place. Modern society places extreme stress on the environment, and many environmental problems come from prior failure to understand what these limits are.

In this book, we shall use the term environment to refer to the atmosphere-water-earth surroundings that make life possible; basically this is a physicochemical system. Man's total environment consists also of cultural and esthetic components, which we shall not consider here. Neither shall we deal extensively with ecology--that is, with the interaction of living things with the environment--although some of the effects of the physiochemical system on life, and conversely, of living things on the physicochemical environment, will be included.

An immediate association with "environment" is "pollution." What constitutes a pollutant is not easily defined. One tends to associate the term with man-made materials entering the environment with harmful effects especially on living organisms--for example, SO_2 from combustion of sulfur-containing fuels, or hydrocarbons that contribute to smog. However, these and other "pollutants" would be present even in the absence of man, sometimes in considerable amounts. We shall consider a "pollutant" to be any substance not normally present, or which is present in larger concentrations than normal. Although any discussion of the environment must consider pollution, this will not be our primary aim. Rather, we shall be concerned with some of the more important chemical principles that govern the behavior of the physicochemical environment and the interactions of its various facets.

Two human activities that are strongly connected with environmental chemistry in general and pollution in particular are waste disposal and energy production. A few aspects of the chemistry of these activities will be considered later. Some general comments are given to serve as an introduction to the scope of the problem.

1.2 WASTE DISPOSAL

The environmental system represents a very large amount of material: 5×10^{18} kgm of air and 1.5×10^{18} m^3 of water. In principle, comparatively large amounts of other materials can be dispersed in these, and consequently both the atmosphere and water bodies have long been used for disposal of wastes. This is sometimes done directly, and sometimes after partial degradation such as by incineration. In part, such disposal is based on the principle of dilution; when the waste material is sufficiently dilute, it is not noticeable and perhaps not even detectable. A variety of chemical and biological reactions may also intervene to change such input into normal environmental constituents e.g., the chemical degradation of organic materials to CO_2 and H_2O.

Such disposal-by-dilution methods are not necessarily harmful in principle, but several factors must be considered in practice. The first is the problem of mixing. While the eventual dilution of a given material may be acceptable, local concentrations may be quite high since mixing is far from instantaneous. As the disposal of wastes from more and more sources becomes necessary because of population or industrial growth in a particular area, it becomes more difficult to avoid undesirable local excess either continuously or as natural mixing conditions go through inefficient periods. A second factor of long-term importance is the rate of degradation of waste material. It is obvious that if the rate of removal of any substance from a given sector of the environment remains slower than its rate of input, then the concentration of that material eventually will build up to an undesirable level. This is the case with many modern chemical products, including some pesticides and plastics that are only very slowly broken down into normal environmental components or rendered inaccessible by being trapped in sediments. Global contamination by such materials results from the dilution and mixing processes, and concentration levels increase while input continues. The examples of DDT and PCBs, discussed in Chapter 7, are well known. Other examples, such as the occurrence of pellets of plastics in significant amounts even in the open ocean, may ultimately be of equal concern.

A third problem in disposal by dilution is the question of what concentration is necessary before harmful effects are produced. This is often difficult to answer, both because effects may be slow to develop and noticeable only statistically on large samples, and because in many cases concentrations less than the part per million range have to be considered. This is very low indeed, and often methods of detection and measurement become limiting. However, various natural processes of reconcentrating materials may bring levels well above those expected on the basis of uniform mixing of even small portions of the environmental sink available (e.g., DDT and mercury in the food chain). Finally, synergistic effects may come into play--that is, two or more substances that exist at concentrations that separately present no problems, together may be much more deleterious.

Disposal by burying is a process that does not depend on dilution. In general, however, it ultimately depends either on degradation to harmless materials, or alternatively on the hope that the material stays where it is put. Such considerations are particularly important with highly toxic or radioactive waste materials. Leaching of harmful components from such disposal sites and consequent contamination of ground water or extraction from soils and concentration in plants are obvious concerns with such disposal methods.

Disposal of insoluble materials in the oceans also depends upon the principle of isolation of the waste. Solid materials may be attacked by sea water and harmful components leached out into the water. Barrels or other containers for liquid wastes may be corroded and leak. Contamination of the biological environment is a possible result. Ocean currents may play a role in distributing such wastes in unforeseen ways, as is the case with New York City sewage sludge that has been dumped in the Atlantic Ocean.

It is clear that understanding of a disposal problem involves knowledge of physical processes such as mixing, chemical processes that are involved in degradation reactions, chemical reactions of the waste and its products with various aspects of the environment, biological processes, meteorological processes, geology, oceanography, etc. Scientific understanding in all of these fields and the interactions among them are necessary to understand not only waste disposal but environmental problems generally. In this book we shall discuss some of the chemistry involved.

1.3 PROBLEMS ASSOCIATED WITH ENERGY PRODUCTION

Energy production is a major activity of modern man that gives rise to many environmental problems. Some of these problems are associated with the acquisition of the energy source itself (e.g., acid mine drainage, oil spills), others with the energy production step (e.g., combustion by-products such as CO, SO_2, partially burned hydrocarbons, nitrogen oxides), while still others arise from disposal of wastes (e.g., nuclear fission products). In any device depending on conversion of heat to mechanical energy, (e.g., a turbine) the fraction of the heat that can be so converted depends on the difference between the temperature at which the heat enters, and that at which it leaves the device. Practical restrictions on these temperatures limit efficiency, and also result in waste heat that must be dissipated, because it is not practical to make the exhaust temperature equal to ambient temperatures. This leads to the problem of thermal pollution. Such efficiency considerations also lead to the desire for ever higher input temperatures (i.e., the temperature in the combustion chamber). This in itself can produce secondary pollution problems, such as enhanced generation of nitrogen oxides. Some aspects of the energy problem are discussed in Chapter 2.

1.4 THE SCOPE OF CHEMISTRY OF THE ENVIRONMENT

Almost everything that happens in the world around us could come under the general heading "Chemistry of the Environment." Chemical reactions of all kinds occur continuously in the atmosphere, in oceans, lakes, and rivers, in all living things, and even underneath the earth's crust. These reactions take place quite independent of man's activities. The latter serve to complicate an already complex subject.

In order to understand environmental problems, we must have knowledge not only of what materials are being deliberately or inadvertently released into the environment, but also what processes they then undergo. More than this, we need to understand the general principles underlying these processes so that reasonable predictions can be made about the effects to be expected from new but related substances. We must also understand the principles that underlie natural environmental processes in order to anticipate interferences from man's activities. Since this book is chemistry-oriented, the chemical principles underlying environmental processes are emphasized. However, knowledge in biological, meteorological, oceanographic, and other fields is equally important to the overall understanding of the environment. Indeed, although it is convenient to segment topics for study purposes, Commoner's first law of the environment should always be kept in mind: "Everything is related to everything else."

2
ENERGY

2.1 INTRODUCTION

Figure 2.1 shows the annual energy flow pattern in the United States for a recent year. Changes in this picture have been relatively minor, except that our oil imports and our total energy consumption have increased. Many observations important to the discussion in this chapter can be made using Fig. 2.1. From the left-hand side, which shows the energy sources, the enormous preponderance of fossil fuels (oil, coal, and natural gas) in our energy picture is obvious. These energy sources are limited in amount and irreplaceable, except over huge stretches of geologic time. This alone makes it important for us to discover how to use these energy sources most efficiently. Furthermore, some of these energy sources, especially petroleum, are valuable sources of chemicals in addition to their value as fuels.

With regard to efficiency, it is apparent from the right-hand side of Fig. 2.1 that roughly 50% of the total energy input becomes "rejected energy." This rejected energy and its frequent inevitability will be considered below; it is a consequence of one of the major laws of science, specifically, of the second law of thermodynamics (Section 2.2.1). Whether this heat is used further, i.e., to heat a dwelling or workplace or to provide heat for a high-temperature chemical reaction, or whether it is discarded depends on circumstances. If we ever wish to use energy with efficiency, it will be necessary to find ways of turning much of the present rejected energy into useful heat. It has been suggested, for example, that the waste heat from nuclear power plants might be used to heat large enclosed areas in order to use them as greenhouses, i.e., in order to grow tropical fruits such as bananas in Vermont!

Present day use of electrical energy is particularly inefficient, because rejected or waste heat appears both during the generation of the electricity and during its use for residential, commercial, and industrial purposes. It should become possible to generate electrical energy with much lower heat losses in the future, but this will only be possible when more of our electrical energy is generated without the use of

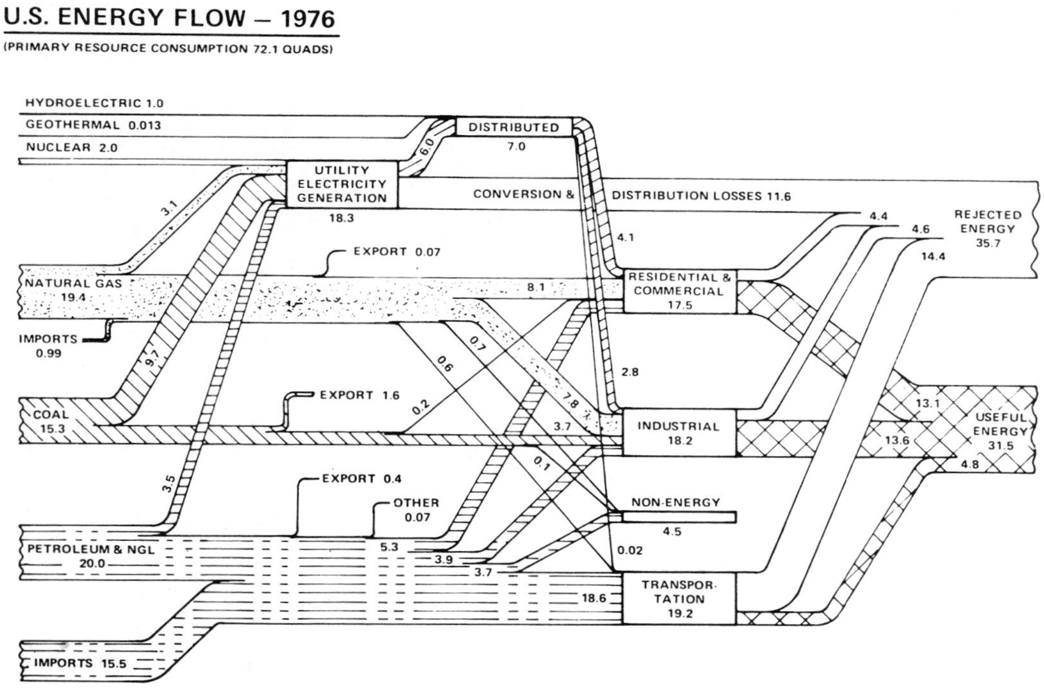

Fig. 2.1. *The energy flow pattern in the U.S. in 1976. The units are in terms of Q (quads) (1Q = 10^{15} Btu = 1.055×10^{18} J). From W. J. Ramsey, U.S. Energy Flow in 1976, Lawrence Livermore Laboratory, March 24, 1977.*

so-called heat engines (see Section 2.2.1).

2.2 THERMODYNAMIC CONSIDERATIONS

2.2.1 Laws of Thermodynamics

Energy flow and energy use are governed by the law of conservation of energy. This law appears sometimes as the first law of thermodynamics:

$$\Delta U = Q + W \tag{2.1}$$

where ΔU is the change in internal energy of the system being studied when heat Q flows in or out, and work W is done on or by the system. The system being studied can be any part of the universe--a person , a city, a forest, a gasoline engine, a nuclear power plant--within certain limits that will not be discussed further here.[1] When ΔU is positive, the internal energy of the system is increasing; when it is negative, the internal energy of the system is decreasing. When Q is positive, heat is being absorbed by the system; when it is negative, heat is being evolved by the system. When W is positive, work is being done *on* the system and when it is negative, work is being done *by* the system.

We shall be discussing the heat released during various chemical reactions, for example, the combustion of coal or oil. At constant pressure,

$$Q = \Delta H \tag{2.2}$$

where ΔH is the enthalpy change during the reaction. Tabulations of ΔH for various reactions are common, especially for combustion reactions. If ΔH is positive, heat is absorbed during the reaction, and the reaction is called endothermic; if ΔH is negative, heat is given off during the reaction, and the reaction is exothermic. The first law of thermodynamics [Eq. (2.1)] is followed during all reactions. The first law of thermodynamics, by itself, is a statement that energy can neither be created or destroyed, and allows us to make certain instructive calculations. For example, let us consider the fate of the energy released when gasoline is burned in an average automobile engine while the car is cruising down a highway at about 50 mi/hour (Table 2.1). About 75% of the energy appears as heat; only the final 25% of the energy in the table is used to move the automobile. Autombiles are thus a

[1] *For rigorous application of the laws of thermodynamics to a system, it must have either isothermal or adiabatic boundaries at any time.*

Table 2.1

Energy Released When Burning Hydrocarbons in an Average Automobile Engine

Fate of energy	*% of Energy*
Exhaust heat losses	*40*
Radiator and engine heat losses	*35*
Wind resistance and road friction	*20*
Transmission, generator, etc.	*5*
TOTAL	*100*

particularly inefficient way of using the internal (chemical) energy stored in gasoline.

In order to discuss the inevitable unavailability of some of the heat to do useful work under various circumstances, we must use the second law of thermodynamics. The first law, looked at in a somewhat different way than just Eq. (2.1), states that any process is possible as long as energy is neither created nor destroyed while this process is going on. However, the second law essentially states that not all processes in which energy is conserved are possible, and, like most scientific laws, it allows one to make quantitative calculations about such processes.

The second law of thermodynamics is a quantitative statement about the well-known fact that most occurrences in this world are irreversible. For example, a hot cup of coffee inevitably cools to room temperature; a cold cup of coffee has never been observed to extract heat from a cold room in order to become a hot cup of coffee. Surprisingly, the transfer of heat from the room to the cup of coffee in order to heat it above room temperature would not violate the first law of thermodynamics even though this kind of heat transfer never occurs in real life. One statement of the second law, however, is very specific in this regard: No process is possible whose sole effect is to remove heat from an object at a low temperature and to supply this heat to an object at higher temperature. This is a very definite but qualitative statement; in order to make the second law quanitative, the concept of entropy S is required.

The most rigorous definition of entropy is in terms of the number of states that are possible for a system under the conditions of temperature, pressure, composition, etc., that we wish to study:

$$S = k \ \ln \Omega \tag{2.3}$$

where k is the Boltzmann constant and Ω is the number of states possible for the system. The state of the system involves the position and energy of every atom and molecule in the system. The more states that are possible for a system, the less we know about which "state" the system is actually in; therefore, we can think of entropy as a measure of our ignorance. For the present, it may be more useful to think of entropy in a less precise way, as a measure of the disorder in a system. A system with many possible states is usually changing from state to state rapidly and this can be considered as a sort of disorder. A gas, for example, has higher entropy than a crystalline solid, partly because the molecules in the gas can have many positions and velocities that are constantly changing when the gas is at some fixed temperature and pressure. The molecules in a crystalline solid, however, are fixed in space in a crystal lattice although they vibrate about their fixed positions; the crystal has fewer accessible states than the gas and therefore has lower entropy. One may also consider that the crystal is more ordered than the gas and therefore has lower entropy.

During discussions of heat and energy, a definition of entropy in terms of heat is particularly useful. The entropy change of a system during some process, ΔS, may be defined as

$$\Delta S = \int dQ_{\text{rev}}/T \tag{2.4}$$

or, if the temperature during the process is constant:

$$\Delta S = Q_{\text{rev}}/T \qquad (T \text{ constant}) \tag{2.5}$$

where Q_{rev} refers to the heat that would have been absorbed or given off during the process if it had proceeded in a reversible manner. It has already been stated that most real occurrences or processes are irreversible. A reversible process is in a special category of "just barely possible" processes that can be reversed by changing very slightly (infinitesimally) the conditions under which the process is occurring. (Conditions include such things as temperature and pressure.)

In terms of entropy changes, the second law of thermodynamics states that no process will ever occur in which the entropy of the universe decreases, i.e.,

$$\Delta S_{universe} \geq 0 \qquad (2.6)$$

where $\Delta S_{universe}$ is the change in entropy of the system and all its surroundings, during any possible process. One example is the well-known observation that a hot cup of coffee in a cold room will eventually cool down to room temperature. This is a general observation; any hot object will lose heat to cold objects so that the temperature of the hot object will decrease and the temperature of the cold objects will increase until the temperatures of all the objects are the same. (The word "object" includes the air in the room in which a cup of hot coffee is cooling. If the room is large, the increase in temperature of the air may be negligible.)

It can be shown that the entropy of the universe always increases during irreversible heat flow. Let us consider a (large) cold object at temperature T to which a quantity of heat Q has flowed from a hot object at temperature $T + \Delta T$. We shall use Eq. (2.5) to calculate the entropy change of these two objects during this heat flow; in this process, $Q = Q_{rev}$, so that Eq. (2.5) can be used directly (For a given temperature change, and all other conditions being constant, the amount of heat transferred to or from an object on heating or cooling is independent of how the change occurs, i.e., whether it is reversible or irreversible.) We must note that the hot object is *losing* heat, so that $Q_{hot\ obj} < 0$, and that $Q_{hot\ obj} = -Q_{cold\ obj}$.

$$\Delta S_{universe} = \Delta S_{hot\ object} + \Delta S_{cold\ object}$$

$$\Delta S_{universe} = \frac{Q_{hot\ object}}{T + \Delta T} + \frac{Q_{cold\ object}}{T}$$

$$\Delta S_{universe} = -\frac{Q_{cold\ object}}{T + \Delta T} + \frac{Q_{cold\ object}}{T}$$

$$\Delta S_{universe} = \frac{(\Delta T)\ Q_{cold\ object}}{T\ (T + \Delta T)} > 0$$

If we assumed that the cold object lost heat to the hot object, then $Q_{cold\ object} < 0$, and $\Delta S_{universe} < 0$. This process would violate the second law of thermodynamics.

Equation (2.6) is a very general statement about processes that are allowed in this universe. It turns out that all real or irreversible processes follow the criterion

$$\Delta S_{\text{universe}} > 0 \tag{2.7}$$

while the criterion

$$\Delta S_{\text{universe}} = 0 \tag{2.8}$$

denotes what happens during reversible processes. Equations (2.4), (2.5), and (2.8) are all exact equations written in terms of reversible processes. In contrast, Eqs. (2.6) and (2.7), written for irreversible processes, are qualitative rather than exact quantitative expressions. It is therefore not surprising that many calculations have been made using reversible processes as examples. The results of such calculations are usually limiting statements; in the next section we shall come across the "maximum efficiency of a heat engine," calculated for a hypothetical reversible heat engine during whose operation Eq. (2.8) would be strictly observed.

2.2.2 Thermodynamic Efficiency of Energy Conversion Processes

In many uses of energy (Fig. 2.1), such as industrial and transportation uses, some kind of work is done, and we think of some kind of machine or engine. If we consider an automobile engine again, we find that chemical energy from the combustion of gasoline appears as heat energy at about 1000 K in the combustion chamber above the pistons. Some of this heat energy appears as work as the pistons move and drive the transmission, generator, etc., through a complex series of linkages, and much heat energy is "rejected," some at the temperature of exhaust, and some at the temperature of the cooling system. We have already noted (Table 2.1) that only 25% of the energy input appears as useful work during the operation of the car, and we shall now consider the upper limit to the percentage of energy input that may appear as useful work.

Such an upper limit may be calculated by assuming that the engine operates in a cyclic way as a *reversible* engine operating between a high temperature T_1 (in this case, 1000 K), and a lower temperature T_2 (in this case not lower than 350 K, somewhat below the boiling point of water). Work is done *by* the engine as it cools from T_1 to T_2, and some work is done *on* the engine elsewhere in the cycle. The type of reversible cycle considered in the calculation is called a Carnot cycle, and is discussed at length in many physical chemistry textbooks. The efficiency of such a Carnot cycle or engine is defined as the net work ouput divided by the heat input at the high temperature and, if the engine is reversible:

$$\text{eff}_{\text{rev Carnot engine}} = (T_1 - T_2)/T_1 \tag{2.9a}$$

$$\text{eff}_{\text{real}} < \text{eff}_{\text{reversible}} \tag{2.9b}$$

where T_1 and T_2 must be in degrees absolute. Thus, the efficiency of a reversible engine depends on the temperature between which it operates; it becomes more efficient as the temperatures T_1 and T_2 move farther apart. Since, for most purposes, room temperature is usually the lowest practical value for T_2, much effort has been expended to make T_1 as high as possible for various engines. For example, superheated steam has been used in steam engines (high pressures are necessary for this) rather than ordinary steam.

An automobile engine does not operate in a Carnot cycle, but in an Otto cycle,[2] in which more than two temperatures need to be considered. The maximum efficiency of a (reversible) Otto engine can be expressed as

$$\text{eff}_{\text{rev Otto engine}} = (1 - 1/r^{\gamma-1}) \tag{2.10}$$

where r is the compression ratio of the engine and γ is the ratio of heat capacity at constant pressure to heat capacity at constant volume of the gas in the engine ($\gamma \approx 1.4$ for air). For an automobile engine with a compression ratio $r = 9$, and using $\gamma = 1.4$, the maximum efficiency would be 0.58, or 58%. Considered as a Carnot engine, the automobile, assumed to be operating between $T_1 \approx 1000$ K and $T_2 \approx 350$ K, would have an efficiency, if reversible, of 0.65, i.e., 65% of the energy of combustion of the gasoline would be converted into useful work. The actual efficiency in the average car, 25%, is very far from the maximum possible efficiency calculated from Eq. (2.9a) or (2.10). Not all engines operate as far from their maximum efficiencies as gasoline engines in cars; the best of the engines used in steam electric power plants (steam generated by burning fossil fuels) have actual efficiencies of 40%, in contrast to maximum theoretical efficiencies of about 60%.

The important thing to note here is that there *is* a theoretical maximum efficiency for engines that do useful work [Eq. (2.9a) or (2.10)], and that real engines generally have less than two-thirds their maximum possible efficiency. There is always a lot of waste heat or "rejected" energy when using such an engine. Some of this waste heat is

[2] *See M. W. Zemansky, "Heat and thermodynamics 5th ed. McGraw-Hill, New York, 1968.*

inevitable, but some is caused by the design of the engine.

Energy conversion processes that do not involve heat engines may have no theoretical limitations to the efficiency of the conversion (except limitations caused by the law of conservation of energy). For example, the direct conversion of solar energy to electrical energy (see Section 2.5) and conversion of chemical energy directly to electrical energy in fuel cells are such processes. It would be very useful if we could generate more of our electricity by such processes.

As already mentioned, the waste heat from our various engines is usually rejected into the environment and there becomes "thermal pollution." We shall see in Chapter 4 that cities are always warmer than their surrounding countrysides partly because of the rejected heat from all of the machines and engines used in the city. Large electric power plants, whether of the steam or nuclear type, reject their waste heat into a very small volume of the environment. When these power plants are water-cooled, the streams and lakes into which this water is usually released may become much warmer and change the environment for aquatic life. Large populations of fish may be killed, or, alternatively, attracted by the new, warmer environment.

2.2.3 Units of Energy and Power

In this book, we shall use joules as units of energy as much as possible since we are using SI units. However, so many other units of energy are in common use that some conversion factors will be given here.

One joule equals 10^7 ergs; 9.480×10^{-4} Btu; 0.7376 ft-lb; 0.2389 cal; 2.778×10^{-4} W-hr.

The unit of power, energy per unit time, in SI units is the watt. Some conversion factors for the watt are given.

One watt equals 3.4129 Btu/hr; 107 ergs/sec; 44.27 ft-lb/min; 1.341×10^{-3} HP; 0.001 kW.

Large amounts of energy are often stated in terms such as 1 *Q* eqals 10^{18} Btu; 2.93×10^{14} kW-hr; 2.52×10^{17} kcal.

Sometimes energy is presented in terms of various units labeled "thermal" or "electrical," especially in relation to various fuels used for electric power plants. The word thermal is often used in connection with the amount of energy that is released by a certain fuel, for example, the heat released during combustion of coal. The word electrical is used for the actual electrical energy produced in an electric power plant. It takes more than 1 kW-hr thermal from the combustion of coal or from any other fuel to produce 1 kW-hr electrical. This is partly because of the limitations imposed by the second law of thermodynamics [such as Eq. (2.9a)] and partly because no real process is as efficient as a hypothetical reversible cycle.

2.3 FOSSIL FUELS

2.3.1 Introduction

Fossil fuels in the form of coal, petroleum, and natural gas now account for 90% of all the energy used in the U.S. (Fig. 2.1). The growth in the use of fossil fuels as an energy source has been very rapid as evidenced by the fact that wood supplied 90% and coal only 10% of the energy used in the U.S. in 1850. The proprortion of energy supplied by coal and wood was equal in 1885. Undoubtedly this change from wood to coal reflected the greater ease in obtaining large amounts of coal required for the new industries developing in the U.S. However, some of the incentive for the change must have also been the energy crisis resulting from cutting all the readily available trees in the northeastern U.S. The supply of fossil fuels is limited and at the present rate of consumption they can only serve as an energy source for a short period in the earth's history (Fig. 2.2).

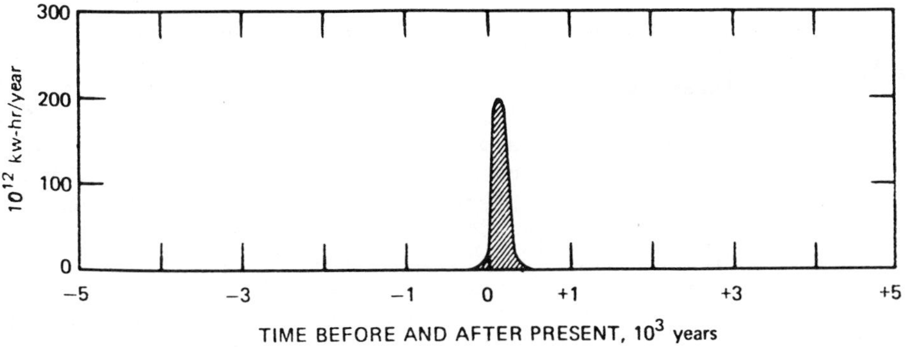

Fig. 2.2. Complete cycle of world consumption of fossil fuels on a time scale of 5000 years before and after the present. Reprinted from M. K. Hubbert, 1972, "Man's conquest of energy. Its ecological and human consequences," A. B. Kline, Jr., Editor in "The environmental and ecological forum," 1970-1971, p. 1-50. U.S. Atomic Energy Commission, Office of Information Services 1972. Publication 71D 25857, U. S. Department of Commerce, Springfield, Virginia 22151.

2.3.2 Liquid Hydrocarbon Fuels

(a) Petroleum

Oil is a convenient energy source because it can be obtained cheaply, it is easy to transport, and, when burned properly, it produces low levels of environmental pollutants. The technology for the separation and conversion of crude oil into a variety of hydrocarbon fractions is well worked out (see Chapter 5). Unfortunately the tremendous use of petroleum in the U.S. has resulted in the consumption of over 50% of our known reserves and, as a consequence, the rate of U.S. crude oil production is decreasing. In 1976, we imported 40% of our crude oil from Northern Africa, the Persian Gulf, and South Africa. The dependence on petroleum as an energy source is even more severe when we consider that according to some estimates 50% of the oil reserves in the world will have been consumed by the year 2000 at the current rate of use. Other energy sources must be found to take the place of petroleum. The consumption of crude oil results not only in the loss of an energy source but also in the loss of the multitude of useful products that are manufactured from petroleum compounds (see Chapters 5-6-7 and 12).

(b) Tar Sands and Oil Shales

Products very similar to those obtained from crude oil may be obtained from tar sands and oil shales. The potential for petroleum production from these cources is estimated to be equal to that of all the crude oil reserves in the world.

Tar sands are a mixture of 85% sand and 15% hydrocarbons. The hydrocarbons have the consistency of paving asphalt. There are large deposits of this black sand in Alberta, Canada and smaller ones in the western U.S. Some of the tar sands can be obtained using open-pit mining techniques, but most of the deposits are so deep that other methods must be used. The approach looked on with favor at the present time is to warm the tar by injecting steam into the bed. Then a hot sodium hydroxide solution will be added and a free-flowing oil-water emulsion is formed which is brought to the surface. The tar can then be separated, cracked, and refined to give the usual petroleum products (see Chapter 5).

The largest oil shale deposits in the world are located in the western U.S. in the states of Utah, Colorado, and Wyoming. The organic portion of shale oil is a highly cross-linked material called kerogen. Consequently, the shale must be heated to crack this hydrocarbon so that the organic products may be removed from the shale. When shale is heated, gaseous hydrocarbons, an oil and a cokelike residue, are obtained. The oil can be processed in the same fashion as natural petroleum (Chapter 5) while the coke is used an an energy source to thermally crack more shale rock.

Several environmental problems arise from the utilization of oil shale. First the oil shale must be mined either by strip mining or by underground mining techniques. Strip mining poses the environmental problem of restoring the land after the mining is completed. A second problem is the absence of the large sources of water in the western U.S. needed for processing of the oil shale and for land reclamation after the strip mining is completed. Water pollution is a real possibility if limited local sources are used. The greatest environmental problem is the disposal of the spent shale after the hydrocarbons are removed. The problem is acute because processed shale occupies a larger volume than the material mined originally. It has been estimated that a plant producing 50,000 barrels of oil a day would also produce 70,000 cubic yards of spent shale per day--enough to make a pile 3 ft x 3 ft x 40 mi. The successful utilization of oil shale as a major source of petroleum hinges to a large extent on the development of a method of disposing of the spent shale in an environmentally acceptable fashion, as well as on economic factors.

(c) Synthetic Petroleum Products from Coal

The conversion of coal to petroleum has been investigated sporadically for about 40 years. Germany obtained some of its petroleum by this means during World War II. South Africa is currently preparing some of its petroleum in this way so as to reduce its dependence on imported oil. At present the U.S. is only testing potential coal liquification techniques at the pilot plant level.

The currently used processes of coal liquefaction are inherently inefficient because they involve first the breaking of carbon-carbon bonds of coal (Chapter 5), and then their catalytic reformation to form hydrocarbons. The high cost of this process relative to the cost of pumping crude oil from the ground has impeded the serious development of coal liquefaction in the U.S. The pilot plants currently in operation in the U.S. are designed to produce a low-sulfur fuel suitable for use in power plants. The sulfur in the coal is converted to hydrogen sulfide which is easily removed before the fuel is burned. This circumvents the use of scrubbers to remove oxides of sulfur from the emissions when this fuel is used in power plants (see Section 2.3.2).

An alternative conversion of coal to methanol may be economically more attractive than its conversion to petroleum. In this relatively simply process the coal is oxidized to carbon monoxide, which is then hydrogenated to methanol. It has been suggested that methanol may serve as a partial or total replacement for gasoline as a low pollution fuel for automobiles. It could also replace hydrocarbons as the fuel for turbines and other machines.

2.3.3 Coal

About 20% of the energy produced in the U.S. is generated from coal. The coal deposits in this country are sufficiently large to supply all its energy needs for the next 100 years. The rate of coal production will have to be increased by at least 50% if coal is to be used to make up the energy deficit resulting from our decreasing supply of domestic oil.

Several severe environmental problems may result from using greater amounts of coal in energy production. One major problem is the release of oxides of sulfur when coal is burnt. Bituminous coal from the eastern and midcontinent U.S. usually has a high sulfur content (3-6%). The Clean Air Act of 1976 standards for the emissions of oxides of sulfur can be met only if the sulfur content is 1% or less. About half the sulfur in coal is in form of pyrite (FeS_2), which can be removed by mechanical cleaning, but the remainder is covalently bound to the carbon and can only be removed by chemical reaction. If this high sulfur coal is used, then the sulfur oxides (mainly SO_2) must be removed from the stack gases after the coal is burnt or the coal must be chemically changed (e.g., coal gasification or liquification) in such a way that the covalently bound sulfur can be removed.

The removal of SO_2 from stack gases has been studied for 40 years but the technology that has been developed is still controversial. The method that is generally considered to be the most efficient is the use of limestone to absorb the SO_2. The main drawback is that up to 350 lb of limestone are required to absorb the SO_2 emissions from a ton of high sulfur bituminous coal. As a consequence there are 8 to 9 ft^3 of sludge that must be disposed of per ton of coal processed. The sludge from a power plant in Pennsylvania will, over the next 25 years, be disposed of by filling a valley which is 5 mi long to a depth of 400 ft.

An alternative possibility is the use of low sulfur coal (1% or less) from the Rocky Mountain states. The use of this resource also raises several environmental problems. A major concern is that this coal can be obtained most economically by strip mining techniques. This process scars the countryside and the arid western land may prove difficult to reclaim after the mining is completed. Since much of the coal is on land owned by the government, the national policy on the utilization and reclamation of this land will significantly affect the degree to which this coal is utilized. A second problem is the added expense in terms of energy and dollars of shipping this coal to the large population centers in the Midwest and East. It may prove less expensive to remove the sulfur from eastern coal than to ship low sulfur coal from the West. A third problem is that western coal has only 50-75% the heat value of eastern coal. The advantage of the low sulfur content is thus offset by the larger amounts of coal that must be used to produce the same amount of energy. As a consequence, the actual emissions of sulfur oxides per Btu of energy released approach those of eastern bituminous coal.

2.3.4 Natural and Synthetic Gas

Natural gas is an ideal fuel. It is virtually pollution-free, it is easy to transport by pipeline, and it requires little or no processing after it comes from the gas well. Natural gas consists mainly of methane together with some other low molecular weight hydrocarbons. In 1976 it supplied about 30% of the energy requirements of the U.S., but since it is currently found in association with petroleum, its supply from these sources will drop sharply in the next 50 years.

It is possible to prepare synthetic gas mixtures from coal. One type, called substitute natural gas, consists mainly of methane and can be used as a direct replacement for natural gas. It is prepared by the hydrogenation of a lower energy gas (low-Btu gas) that is obtained by the controlled oxidation of coal in the presence of water vapor (Chapter 5). The low Btu gas consists of methane, CO, CO_2, H_2, and N_2. As the low-Btu gas is produced at the site of a power plant, it can be used directly as a fuel. Because of its low energy content, it is not economically feasible to ship this low energy gas by pipeline.

The conversion of coal to methane appears to be one of the most promising ways for the utilization of the energy in the large coal reserves in the U.S. It is particularly attractive from an environmental point of view because the sulfur is converted to H_2S, which is readily removed and converted to sulfur. The sulfur is easily stored and can be utilized for the production of chemicals. A second advantage is the possibility of obtaining the low-Btu gas by the underground gasification of coal. This process circumvents the hazards of underground coal mining and the environmental problems of strip mining. Underground gasification would also recover the large volumes of methane that are usually associated with some coal deposits. This methane constitutes one of the greatest hazards of underground coal mining so it is flushed out of the coal mine with ventilating air and lost. It should be noted that the technology of underground coal gasification has not been completely worked out. Most of the pilot plants currently testing coal gasification processes use mined coal.

2.4 NUCLEAR ENERGY

2.4.1 General Aspects

The role of nuclear energy in the history of our planet begins with the formation of the earth about 4.5 billion years ago. Since then, the earth's environment has been continuously regulated by the release of nuclear energy in the sun and in the earth itself. Thus, solar energy, which is critical for life on the earth and which provides us with other useful sources of energy such as fossil fuels, wind, and ocean thermal gradients, stems from nuclear energy. Radioactive decay of some of the earth's primordial elements, a spontaneous process for releasing nuclear energy, has been a continuous although slowly diminishing source of internal energy.

In general, the materials that can be used as sources of nuclear energy are now too highly dispersed on the earth to be harnessed in situ for useful purposes. Geothermal sources may, perhaps, be considered as a means for tapping some of the nuclear energy released within the earth.

There are three types of exothermic nuclear reactions that we recognize today as useful sources of energy. These are fusion, fission, and radioactive decay.

Although fusion, a process in which light nuclei react to form heavier nuclei, occurs in the sun and much is known about the process, man has not yet learned how to use it on the earth as a controlled source of energy. Ironically, man has learned how to use it as a military explosive weapon, the H-bomb, capable of destruction almost beyond the imagination. Controlled thermal fusion has yet to be demonstrated in the laboratory. Fusion power reactors are, therefore, in a preliminary conceptual stage.

As is well known, the technology for the large scale release of fission energy was developed for military applications. Today three decades after the release of the first atom bomb, about 8% of the electricity generated in the U.S.A. is produced in nuclear power plants in which the energy is obtained from the fission reaction. In addition, fission provides the power for the propulsion of many naval vessels.

Radioactive power sources, commonly called isotopic cources, are useful when the need is for a low-power source that is physically small, has a long life, and does not require frequent maintenance. Examples of such power sources are those used for heart pacemakers and for powering apparatus and instruments used for experiments in outer space.

The fundamental nuclear chemical aspects of fusion, fission, and isotopic power sources are discussed later in Chapter 17. In this chapter we shall deal only with two of these, namely, fission, which is now used in some of the large central station electric power plants, and fusion, which has the potential for such use in the future. At the present, nuclear (fission) power plants are substituting for oil-fired power plants. A greater fraction of the supply of petroleum is thereby made available for the production of petrochemicals and the fuels used in most of our current means of transportation.

2.4.2 *Nuclear Fission Power Plants*

(a) Types of Fuel

The current, first generation, commercial nuclear power plants in the U.S. use $^{235}_{92}U$ as the primary fuel and most are light water reactors (LWR) that use H_2O to (1) remove energy from the neutrons released in fission, i.e., thermalize the neutrons, (2) remove the heat released in the fission process, and (3) provide steam to drive the turbines for generating electricity. As the water thermalizes the neutrons needed to maintain criticality (the condition in which there is a self-sustaining fission chain reaction in the reactor), it captures some of them. In order to compensate for this, the concentration of $^{235}_{92}U$ in the fuel (a mixture of $^{235}_{92}U$ and $^{238}_{92}U$ with negligible amounts of $^{234}_{92}U$) must be increased above the 0.720 atom % present in normal uranium. Additionally, there must be some $^{235}_{92}U$ which can be "burned" in the reactor for power generation before the concentration falls to the minimum level needed to just maintain criticality. Typically, the $^{235}_{92}U$ content is enriched to about 2-4 %. Enrichment, of course, adds to the cost of the reactor fuel.

A very small fraction of the $^{238}_{92}U$ undergoes fission in an LWR. A larger but still small fraction is converted to $^{239}_{94}Pu$, a fissile isotope (one that can undergo fission following the capture of a thermalized neutron) of plutonium that also contributes to the power output of the reactor.

As a nuclear power plant operates, the $^{235}_{92}U$ content of the fuel (as sealed metal tubes or rods containing UO_2) decreases, and the content of fission products, the nuclides formed in fission, increases. The fission products capture some of the neutrons needed to maintain the fission chain reaction. They can also cause undesirable physical distortion of the fuel rods.

The spent fuel must be replaced, therefore, by new fuel long before the $^{235}_{92}U$ content drops to the minimum level for criticality for fresh fuel. When the spent fuel is reprocessed, the $^{235}_{92}U$ and the $^{239}_{94}Pu$ are separated from each other and from the fission products. The latter mixture of radionuclides together with radiosotopes of actinide elements produced in the reactor constitute the waste that must be stored and prevented from entering the environment.

The $^{239}_{94}Pu$ can be used instead of $^{235}_{92}U$ to fuel a different type of power reactor--a fast reactor in which the neutrons are not thermalized. If properly designed, the reactor should be able to produce more $^{239}_{94}Pu$ from $^{238}_{92}U$ than what it burns. Several prototype power reactors of this kind, known as breeder reactors, have been built abroad; one is under construction in the U.S., but its future is in doubt.

A third nuclear fuel will very likely become increasingly important. It is $^{233}_{92}U$, a fissile isotope of uranium that can be made from $^{232}_{90}Th$ in a nuclear reactor. Thorium-232 is the predominant isotope of thorium in nature. According to calculation, it should be possible to design a thermal breeder reactor using $^{233}_{92}U$ and $^{232}_{90}Th$. As for the case of $^{239}_{94}Pu$, another fissile nuclide such as $^{235}_{92}U$ or $^{239}_{94}Pu$ would be needed to build up an initial supply of $^{233}_{92}U$.

(b) The Role of Nuclear Fission Power Plants in the Future

There is a limit to the amount of $^{235}_{92}U$ available on the earth for use in the types of nuclear power plants that are now in operation, being constructed, or scheduled to be constructed. Estimates of the reserves of uranium ore vary, but on the basis of the currently projected growth of the nuclear power industry, $^{235}_{92}U$ could be in short supply by the year 2000. Many factors can cause errors in predictions of the time when $^{235}_{92}U$ (or, for that matter, the various fossil fuel supplies) will be exhausted For example, at any given time, the incentive to prospect for new deposits varies with the price that the users are willing to pay in accordance with the law of supply and demand. Thus, the estimated supply may refer to the amount of material known to exist at a concentration that will bring a specified price. We can never be sure when all the deposits of a given material have been discovered. In addition, the oceans, which constitute an untapped source of many materials at low concentration, are seldom included as sources of supply. For materials used to provide energy, sources such as the oceans are not useful unless the energy that can be obtained from a given quantity of materia exceeds that required to produce it. In other words, a source of material becomes useful if the value of that material makes it economically feasible to use existing technology to extract the material from its source in such a way as to hold the impact on the environment to within acceptable limits.

Predictions of the time when $^{235}_{92}U$ will be depleted necessarily take into account the projected rate of consumption. This rate is less today than that which was projected for today ten years ago. The cost of construction of both fossil plants and nuclear plants has increased along with the cost of financing new plant construction. In several instances, therefore, construction of proposed power plants of either type has been postponed or even canceled. Construction costs for nuclear plants have increased more rapidly than those for fossil plants partly because of an increase in the time required for construction (now about ten years) and partly because of the addition of components such as cooling towers to reduce the amount of heat rejected into the cooling water leaving the plant. Increased public concern for the environmental impact and for the safety of operation of nuclear power plants has made more complex and time-consuming both the site selection and the design and construction of a nuclear power plant. Difficulties in siting have also led to cancellation of plans for plant construction. In addition, planning for new plants has been slowed by public concern about the disposal of radioactive waste and about the safety aspects of the plants.

Nuclear power plants are not economically competitive in areas of the U.S. where gas or oil or coal are readily available. The cost of electricity generated from these fossil fuels is very sensitive to the cost of these fuels. The cost of nuclear power, however, is comparatively insensitive to the cost of uranium. In principle, then, one can expect to utilize low grade uranium ores before the supply of $^{235}_{92}U$ is considered to be exhausted.

If the $^{238}_{92}U$ is converted to $^{239}_{94}Pu$ in breeder reactors, the supply of uranium ore can be stretched out to the order of hundreds of years. Thus, on the basis of economics alone, coal and plutonium would seem to be the two fuels that could together meet the energy needs for a few hundred years. A full cost-benefit analysis requires consideration of a number of factors in addition to the economic ones. The impact of energy production on human health, whether it be obvious or subtle, short-term or long-term, can no longer be overlooked regardless of the source of energy. Thus, in the case of coal there is the major problem of atmospheric pollution and there are additional problems associated with the several aspects of coal mining. Similarly, questions have been raised about the safety of fast breeder reactors, about the risks associated with the health hazards of plutonium, and about the feasibility of adequate nuclear safeguards, i.e., prevention of the diversion of plutonium by one means or another from its intended peaceful use as a reactor fuel to its potentially disastrous use in nuclear weapons. These are difficult questions to answer. They are acting as a caution light on the road to a plutonium-based nuclear power industry. In any event, one cannot turn back the clock to the prenuclear era and neither the termination of the development of breeder reactors in the U.S. nor the mere signing of agreements limiting the use of plutonium to the generation of power will, of themselves, prevent either the development of breeder reactors elsewhere or the irresponsible use of plutonium.

The use of $^{232}_{90}Th$ by conversion to $^{233}_{92}U$ would add a few hundred years to the future of nuclear power. Nuclear safeguards would still be a problem and the economic aspects may be less favorable than for $^{239}_{94}Pu$ breeders, but disenchantment with the latter, because of potential health and safety problems, could shift development emphasis to $^{233}_{92}U$ breeders.

2.4.3 Nuclear Fusion Power Plants

(a) Types of Fuel

Given sufficient kinetic energy, two deuterium ($^{2}_{1}H$) nuclei will react (D-D reactions) to form $^{3}_{2}He$ and a neutron or $^{3}_{1}H$ (tritium) and $^{1}_{1}H$ with the release of energy. Similarly, the nuclei of $^{2}_{1}H$ and $^{3}_{1}H$ will react (D-T reaction) exothermically to form $^{4}_{2}He$ and a neutron. These fusion reactions in which a product heavier than either reactant is produced can be studied on a small scale by means of a particle accelerator without recovery of the energy released. In order to obtain useful energy from these reactions, it will be necessary to arrange their occurring on a large scale in a device which provides the means for controlling the reaction rate and removing the energy released. If the energy removed is in excess of that needed to operate the device, the excess can be used for some useful purpose such as the generation of electricity.

Several methods for maintaining a fusion reaction on a large scale are under investigation (see Section 17.8.12). It is hoped that one or more of these can be used to develop fusion power reactors. Problems associated with the materials of construction of large fusion power reactors will be severe. One problem will be damage caused by radiation.

Fusion power reactors will not be free of radioactivity. Those using tritium as a fuel will contain relatively large amounts of this radioisotope of hydrogen. In addition, for any fusion reactor neutrons released in the fusion reaction will induce radioactivity in the materials of construction. On the basis of preliminary estimates, it is believed that there will be less radioactivity associated with fusion reactors than with fission reactors of comparable power output. However, the radioactive species and the quantities produced will vary with the as yet unselected materials of construction. One of the proposed reactors is a hybrid that combines fusion and fission and would, therefore, contain in addition the fission products and plutonium that are characterisitic by-products of reactors now in service.

(b) The role of Nuclear Fusion Power Plants in the Future

Relative to the projected energy requirements for the future, there is a virtually unlimited supply of deuterium on the earth. Normal hydrogen contains 0.015 atom % deuterium and relatively inexpensive methods have been developed for obtaining deuterium with high isotopic purity.

Tritium, on the other hand, is not available in useful amounts in nature. It will be necessary to produce it in the fusion reactor, probably by the reaction of neutrons with lithium present as the metal or as one or more compounds. There is some disagreement as to whether or not the supply of lithium will limit the use of fusion reactors in the future. Actually, this may not be an important question. It seems likely that tritium will be used only in the first generation reactors to demonstrate the feasibility of fusion power because the D-T reaction has a higher probability than the D-D reactions for occurring at a given temperature, and it provides a higher energy release per reaction than deuterium alone.

Until the first prototype fusion reactor is built, one cannot assess the environmental impact of such reactors in terms of the release of radioactivity during normal operation, the release of radioactivity in the event of an accident, the disposal of radioactive waste, and the discharge of unused heat into the surroundings. It appears unlikely that the testing of a fusion power reactor will take place before the end of this century. Assuming that development of the necessary technology is just a matter of time, the overall risk-benefit analysis will eventually include the evaluation of the magnitude of two additional potential problems that are analogous to two mentioned for fission breeder reactors. One is the safety aspects of reactor control. At present it is believed the challenge will be to prevent a fusion reactor from shutting itself down rather than developing uncontrolled excess power. The second potential problem is safeguards for the fuel, which can also serve as one of the ingredients of an H-bomb.

2.5 SOLAR ENERGY

2.5.1 Introduction

Solar energy is the most abundant form of energy available on earth. Certainly there is an ample supply of solar radiation at the earth's surface to meet any conceivable energy need of man in the foreseeable future. It is the only continuously reliable (on a long-term basis) and renewable source of energy, and its usage, at least now, does not present significant pollution and waste disposable problems.

Why, then, is direct utilization of solar radiation not the major source of power in a country as highly developed technologically as the U.S? The major problem is one of economics: At the present time direct solar-generated power cannot be delivered cost competitively with that from conventional fossil fuels (coal, oil, and gas), from nuclear fission, or from indirect solar energy utilization (such as hydroelectric and wind power). However, recent cost increases of gas and oil have made solar space heating and cooling of homes and industrial facilities economically feasible, and commercial development of this aspect of solar energy utilization is now well underway. Furthermore, unlike solar energy, the other energy sources such as fossil fuels are not being replenished as they are used, and they are not inexhaustible. Recent recognition of the importance of solar energy utilization in the long-range energy needs of the U.S. is reflected in the Energy Research and Development Administration (ERDA) budgets over the past several years: In fiscal 1975, $42,000,000 (3.5%) was authorized, while in 1977 the authorization was $290,000,000 (13.2%).

There are two major problems associated with efficient usage of solar radiation -- its diffuseness and its intermittency. We shall see in Chapter 3 that the power from the sun outside the earth's atmosphere is about 2 cal/cm^2 min, or 1400 W/m^2. Only about half of this reaches the surface of the earth, however, and the amount varies considerably depending on weather conditions, time of year, and geographical location. The average annual solar radiation impinging on the U.S., factored over day and night conditions and the four seasons, ranges from about 260 W/m^2 in the southwest to less than 150 W/m^2 in the northeast and far northwest. Average power consumption in the U.S. (10^{13} m^2 surface area) in 1976 is estimated to be about 3×10^{12} W, so that solar radiation falling on 0.2% of its surface area could supply present energy usage in this country. In terms of total surface area this is very small, but it is of the same order of magnitude as the roof areas of all man-made structures now standing, and therefore building of radiation collectors would represent a massive construction effort. In addition, these structures may in themselves contribute to local ecological disturbances if required to be located near major population centers. The intermittent nature of solar energy (day, night, cloudy conditions) necessitates energy storage capabilities. In fact, energy storage and portability are major problems as all alternative energy sources to gas and oil are explored.

There are two qualitatively different ways to collect the radiation from the sun. In the first, the solar energy may be collected simply as heat, then used in a variety of thermal devices such as home and industrial heaters or coolers, thermal motors, and water evaporation. In the second, the energy may be collected as quanta of energy in such processes as photochemical or photoelectric conversions for direct energy storage or utilization. The first way leads to degradation of visible light quanta, ranging in energies from approximately 170 kJ/mole to 300 kJ/mole (see Section 9.2.3) into thermal quanta of about 4 kJ/mole; the second method makes use of the photons of light from the sun in their original high-energy form. Many schemes are being proposed and explored using both ways of solar energy utilization in attempts to develop economically viable alternatives to fossil and nuclear fuels.

2.5.2 Physical Methods of Utilization of Solar Energy

Solar heaters to provide domestic hot water for homes and residential buildings (hotels and dormitories) have been commercially available in countries such as Japan and Israel for many years. In addition to this mode, probably the most developed and economically feasible usage of solar energy at the present time is in the heating of buildings, particularly with the continuing rise of fossil fuel costs. A typical procedure is to use a thermal collector to absorb the solar energy, which is transferred by heat pipes or fluids directly for low temperature ($<100^{o}C$) heating or for storage. This thermal collector may consist of simply a black radiation absorber with a transparent cover of glass or plastic that is opaque to infrared radiation. Facilities for storage may be insulated tanks for fluids, pebble beds for gases, or materials that undergo a chemical phase change near room temperature; an example of such a material is $Na_2SO_4 \cdot 10H_2O$, which dehydrates endothermally at $32.3^{o}C$.

Solar air conditioning, although not as developed as solar heating, has the advantage that it is most needed on hot sunny days when solar radiation is also plentiful. Several types of solar cooling systems are being explored. Some use flat-plate heat collectors of the type also used for solar heating, with the heat, however, being used to vaporize a fluid in an expansion refrigeration unit similar to gas-fired refrigerators. Other units use the vapors mechanically to drive a turbine air conditioner. Another technique for solar cooling is to use a desiccant to remove water vapor from the air; this dried air is then cooled by further evaporation of water with the desiccant regenerated by solar heating. Probably solar air conditioning will become economically feasible initially when used in conjunction with solar heating units, since much of the collecting and recycling equipment will be common to both system. Even greater economies will be realized if the solar units can also be used to supply domestic hot water to the buildings being heated or cooled.

The solar generation of electric power through heat has been extensively studied. Techniques are similar to those used in conventional power plants with the exception that solar radiation absorption rather than gas, coal, oil, or nuclear-fired furnaces is the source of heat to generate steam; again, cost is a major factor, and at the present time it is estimated that solar thermal generation of electrical power is approximately four times as costly as generation by fossil fuel combustion. A promising design is to focus sunlight from a large sun-following bank of reflectors onto a high-temperature collector or receiver, resulting in temperatures as high as 1000°C. Direct sunlight must be used, because indirect light cannot be focussed, so that these units at least initially will be feasible in the U.S. only in the "sunbelt" region. Again, efficient energy storage becomes a major design factor. Alternately, relatively small turbine generators, using vapors generated by heat collectors and operating at low (less than 200°C) temperatures, located near the sites of electric usage may be the most feasible technique. Other systems such as thermoionic (thermal ionization) devices have been proposed, but these are quite limited because of the high temperatures required.

2.5.3 Photoelectric and Photochemical Utilization of Solar Energy

Direct conversion of solar energy to electricity with photovoltaic cells is well known. Such devices are used, for example, in photographic light exposure meters and to control street lights. These cells are extremely inefficient, although new semiconductor materials and developing techniques in mass fabrication of solar cells may lead to a competitive system. Silicon photovoltaic cells with efficiencies of approximately 15% have been used very effectively in recent years in systems where cost is only a minor consideration -- long-duration space missions, for example. Although silicon is one of the most plentiful elements on earth, the high purity, perfect crystallinity, and special doping techniques required to fabricate it into an effective semiconductor photovoltaic device make it very expensive. It is estimated, for example, that at present the electrical cost per kilowatt from a photovoltaic source is from 50 to 100 times that from a conventional fuel power plant. It should be pointed out however, that photovoltaic devices will probably be almost as efficient for collecting solar radiation in small units as in large arrays. This means that they could be deployed on individual homes or buildings and thus not require construction of large collecting reflectors or banks. Again it should be pointed out that electrical storage becomes a major factor in solar energy utilization of this type.

Another means of direct conversion of solar to electrical energy is by photogalvanic cells. This type of cell is analogous to a thermodynamic galvanic cell and is covered in detail in Section 11.2.4.

Solar radiation may be absorbed directly to stimulate photochemical reactions that produce energy-rich products for storage and/or transporation as fuel. A major advantage of photochemical transformation over solar thermal devices is that a temperature gradient is not required, and therefore the thermodynamic second-law restriction does not apply in the production of these fuels. Suitable solar energy-storage photochemical reactions have been quite difficult to find, however, even though the necessary properties of such a system have been recognized for some years. (These properties are discussed in Section 11.2.1.)

Photochemical utilization of solar energy is taking place all around us in the process known as photosynthesis. Details of this will be covered in Section 11.3. It is worth emphasizing here, however, that this remarkable biological process that converts sunlight into organic matter meets virtually all of the criteria for a practical solar energy conversion and storage system. Overall efficiency is quite low, however, and competition for usable land between production of food and any large-scale production of "energy" crops could become a major factor in future consideration. Photosynthesis does provide a very useful model process, though, and research is underway in several laboratories to develop a "synthetic leaf" employing chlorophyll in a photoelectric device for practical solar energy conversion.

Photodissociation of water into molecular hydrogen and oxygen is an extremely important nonbiological process to consider for potential solar energy utilization. Hydrogen may very well be the most important synthetic fuel of the future, and certainly the inexhaustible supply of the two reactants--

sunlight and water--makes this reaction very attractive. Unfortunately water does not absorb visible light, and therefore direct photodissociation cannot occur in the biosphere. Attempts are being made to devise suitable photocatalysts (sensitizer dyes) operating in closed-cycle processes to affect the dissociation. Another approach to solar-induced water dissociation is by photoelectrolysis. Light is absorbed by an electrode that excites electrons to higher energy levels, followed by charge transfer to the electrolyte solution and subsequent reduction of hydrogen or oxidation of oxygen; although the lifetime of excited electrons in metals is too short for the transfer to occur, it is long enough to take place readily from semiconducting electrodes. A major problem is long-term stability of these electrodes under atmospheric conditions.

2.6 GEOTHERMAL ENERGY

Hot springs and geysers occur in many parts of the world and are visible evidence for reserves of thermal energy present in the earth. They have been used for many years as energy sources in a few areas, for example, Iceland, Italy, and New Zealand, and a large source is under development in California. One estimate suggests that there is as much energy available in accessible thermal reservoirs world-wide as in all of the coal deposits. Development of such sources is technologically feasible and in appropriate locations they may be expected to contribute significantly to future energy supplies. They may, of course, produce hot water for heating purposes on a local basis, but their most important potential is for the production of electricity.

Although steam and hot water issue naturally from geothermal sources, practical utilization of them involves drilling wells into the heated strata. Three types of systems are possible.

(a) Dry steam, that is, steam accompanied by very little water, is the simplest system, and is produced by some reservoirs. The steam can be used directly to power turbines for the generation of electric power.

(b) Wet steam, which is steam mixed with a considerable amount of hot liquid water, is the most common geothermal energy source. Indeed, the liquid usually predominates in amount. These waters, which have been in contact with hot rock under pressure, contain large concentrations of dissolved minerals that give rise to severe problems of corrosion and scale deposition from precipitation of dissolved substances. Three approaches to the use of such energy sources are:

(1) separation of the liquid from the steam, and use of only the latter for power generation; (2) use of turbines capable of operating with a two phase mixture; (3) use of the mixture or the hot water to vaporize a second liquid such as isobutane or a freon (see Section 6.1) to operate a turbine. In the last case, a closed cycle is required for the working fluid; the latter must be condensed and reused.

(c) Hot, dry rock is a potential energy source, although thermal reserves of this sort are not in use at present. For use, holes would be drilled into the porous or fractured rock, water pumped into some of them, and steam collected from others.

The environmental effects of geothermal power plants have been discussed for one particular example by Axtman.[3] Many of the problems would be expected to be common to geothermal sources in general. Most obvious is the hot water waste from condensed steam, geothermal water, and cooling water, although some hot water would be released naturally, whether used or not, from most reservoirs. Since the temperature of geothermal water and steam is comparatively low (of the order of 100 to 400°C), the thermal efficiency of power generation from these sources is low in comparison to a modern fossil-fueled plant and a large fraction of the thermal energy of such a system is waste. In addition to the thermal pollution problem, the high mineral content of geothermal waters can be a hazard, particularly since significant amounts of toxic materials such as arsenic and mercury may be present. This, of course, depends on the composition of the rocks at the source. It has been suggested that injection of waste geothermal waters back into the system could avoid the pollution problem and also prolong the useful life of the source.

A second pollution potential arises from gases released with the steam and water. The CO_2 released may exceed that from an equivalent fossil-fueled plant, and hydrogen sulfide is a common constituent. Small amounts of many other gases may be released, depending on the source, but one of major concern is radioactive radon, which occurs in some geothermal sources. Although there are other possible problems, such as earth subsidence, for the most part the environmental problems associated with geothermal energy do not seem to be great.

2.7 WIND AND WATER ENERGY SOURCES

The winds and waters of the earth offer abundant energy, if it could be harnessed economically. Ultimately, all such energy is derived from solar radiation; indeed, it is a form of

[3] *R. C. Axtman, "Environmental impact of a geothermal power plant," Science 187, 795 (1975).*

solar energy and is often classed as such. It is renewable, and since few, if any, chemical reactions are involved in its use, it is free of most of the chemical pollution problems associated with fossil, nuclear, and even geothermal sources.

Wind power has been one of man's earliest forms of energy. Windmills have been used to pump water and perform other kinds of mechanical work for centuries. In the first half of this century many wind-driven electrical generators were in use in rural areas, and in Denmark 200,000 kW of power were generated by wind in 1908. Most of these generators had less than 5 kW capacity, but modern interest is in large units in excess of 100 kW. Although wind power is free, and produces no environmental problems (except possible esthetic ones), it is intermittent and must be used in conjunction with energy storage facilities or as means of reducing fuel consumption when used to supplement conventional generators. Furthermore, the capital costs of large wind-powered generators are very high. Nevertheless, they may contribute significant power in suitable areas, and it has been estimated that 5-10% of the electrical energy needs of the northeastern U.S. could be met by wind power by the end of the century.

Water power was also an early energy source for mankind, and machinery driven by the power of water wheels was a major component of the industrial revolution. Hydroelectric power generation is the cheapest source of electrical power in regions with suitable rivers. In most developed areas such sources are now highly utilized. Water power is pollution-free, although environmental effects from the construction of dams and reservoirs associated with its use may be very serious, and indeed limit the total utilization of the potential power available from this source. In addition, most reservoirs will fill with sediment in two or three hundred years, and some much more rapidly.

Energy is also available from tidal power, particularly in bays where the tidal rise is large. The tide is used to fill an empty reservoir, closed by a dam; the outflow powers a turbine. Such a system is in operation on the Rance estuary in France. Tidal power is environmentally clean except for local topographic effects. The total power potentially available from this source is only a small part of the overall energy needs, but it can be significant in appropriate localities. Methods have also been proposed for capturing the energy of ocean waves. Although there is a great deal of energy in this source, it is diffuse.

In addition to these energy sources, which are sources of kinetic energy, the earth's water systems offer other possibilities for the generation of useful power. One of these is the energy of mixing between salt and fresh waters. One means whereby this might be tapped is a battery in which salt and fresh water are separated by ion exchange membranes. Useful electrical energy may be generated from the potential that is developed across these membranes. Large-scale power generation using river and ocean water would require huge batteries, and many problems of membrane construction and fouling by the natural waters would have to be solved.

Another energy source with vast potential arises from the temperature difference between ocean surface waters and deep waters in the tropics. This difference is the order of 20-25°C, small compared to conventional thermal energy systems, but large enough to provide a practical power output. Indeed, a system utilizing this ocean thermal gradient was tested as early as the 1920s.

The most obvious system, one where the warm surface water is used to vaporize a volatile fluid, which in turn drives a turbine and is then condensed by the cold water, suffers from the poor heat transfer inherent in the small temperature differences available. Water vapor generated from the warm water and condensed by the cold water (in an evacuated system) can be used to drive a turbine, but this must be designed to operate under a low pressure and must be large. A more recent suggestion is to allow the water vapor to raise liquid water in a tube, analogous to the well-known air-lift pump, and to use the energy of the water as it falls back to drive a turbine. It is probable that such systems could generate electricity economically, although problems of condenser fouling, plant siting, and power transmission would be important. Waste waters would be discharged at the depth appropriate to their temperatures, and little environmental impact would be expected.

2.8 ENERGY STORAGE

An efficient means of electrical energy storage is important if intermittent energy sources are used (wind, solar), or for "load leveling" so that energy can be generated in periods of low demand and stored to supplement that generated during peak demand. Pumped storage, in which power is used to pump water to an elevated reservoir and later recovered when the water is allowed to flow back down through a turbine, is one means of accomplishing this. Regrettably, pumped storage reservoirs often entail undesirable changes to the landscape.

On a smaller scale, fly-wheel storage can be used to recover and store energy otherwise wasted in vehicular use. Rechargeable batteries (e.g., the lead-acid battery) provide another means of storage, although those available at present are not economical for large-scale storage use. Some of the newer battery systems, such as the high temperature sodium-sulfur or lithium-sulfur systems may eventually be developed to the point where large battery installations can economically be used to store energy in power generation.

Battery-powered vehicles are attractive both to reduce pollution from internal combustion engines and to supplement or replace liquid fuels as petroleum supplies dwindle. Current battery technology is marginally adequate for short-range, low-performance vehicles such as those used for urban delivery and transportation uses, employing the familiar lead-acid systems. Higher performance will require batteries that can store more electricity per unit weight (specific energy capacity), and deliver more power per unit weight (specific power) than is presently possible with the lead-acid or other proven batteries. Specific energy capacity depends on the energy per unit mass of the electroactive material and the electrolyte and structural components of the cell. The power output depends on the rate at which the stored chemical energy can be released as electrical energy, and on the internal losses through IR drop and polarization effects that make the potential with current flow smaller than the thermodynamic values. Here is a primary reason for interest in new systems such as sodium-sulfur, lithium-sulfur, and lithium-chlorine cells, even though some operate only at high temperatures and involve hazardous materials. As a comparison, the lead-acid system has a specific energy capacity of 20 W-hr/lb if delivering power at 1 W/lb, but to deliver a specific power of 100 W/lb, the recoverable specific energy drops to 1 W-hr/lb. The Na-S cell, on the other hand, may theoretically have a specific power of 100 W/lb at a specific energy of 100 W-hr/lb. That is, the specific power is approaching that of the internal combustion engine, 200-300 W/lb.

Storage of energy by conversion to a suitable fuel is another approach that is particularly appropriate in connection with the use of hydrogen, e.g., photodissociation of water. Hydrogen has many advantages as a fuel, such as freedom from pollution and inexhaustible supply if it is obtained from water. The only combustion product is water, along with some nitrogen oxides if burned in air.

The energy released on burning hydrogen is high, 68.3 kcal/mole (285.8 kJ/mole) for the reaction

$$H_2\ (g) + \frac{1}{2} O_2\ (g) \rightarrow H_2O\ (\ell) \tag{2.11}$$

This corresponds to 142.9 kJ/gm. In comparison, the energy released on combustion of a mole of methane is 212.8 kcal, (890.4 kJ/mole), but this is only 55.6 kJ/gm. (Other hydrocarbons give very nearly the same energy per gram.) However, to supply hydrogen from water requires an energy input at least as great as that which can be recovered on combustion, hence our inclusion of hydrogen as an energy storage system. It may be considered equally as a means of transporting energy, since as a gas it could be transported by pipeline. (At one time, much of the artificially produced gas used for household purposes contained close to 50% hydrogen.) Alternate suggestions are for its use as a liquid, although this requires cryogenic containers since at atmospheric pressure liquid H_2 boils near 20 K. Storage as a solid metallic hydride may also prove to be feasible; some metals will reversibly form a hydride at moderate pressures to provide a high hydrogen density in the solid.

The most obvious means of generating hydrogen in this context is through electrolysis of water, but unfortunately this turns out to be a very uneconomical procedure. A number of thermochemical cycles have been proposed that would utilize the energy from conventional fuels, nuclear reactors, or solar installations to yield hydrogen more efficiently than by the electrolysis procedure after first producing electricity. One example consist of the following steps:

$$3FeCl_2 + 4H_2O \xrightarrow{450°C} Fe_3O_4 + 5HCl + H_2 \qquad \Delta H = +75 \text{ kcal/mole} \tag{2.12}$$

$$Fe_3O_4 + 8HCl \xrightarrow[\text{temperature}]{\text{ambient}} FeCl_2 + 2FeCl_3 + 4H_2O \qquad \Delta H = -50 \text{ kcal/mole} \tag{2.13}$$

$$FeCl_3 \xrightarrow{300°C} FeCl_2 + \frac{1}{2} Cl_2 \qquad \Delta H = +28 \text{ kcal/mole} \tag{2.14}$$

$$Cl_2 + Mg(OH)_2 \xrightarrow[\text{temperature}]{\text{ambient}} MgCl_2 + \frac{1}{2} O_2 + H_2O \qquad \Delta H = -38 \text{ kcal/mole} \tag{2.15}$$

$$MgCl_2 + 2H_2O \xrightarrow{350°C} Mg(OH)_2 + 2HCl \qquad \Delta H = +4 \text{ kcal/mole} \tag{2.16}$$

The overal reaction is simply

$$H_2O \longrightarrow H_2 + \frac{1}{2} O_2 \tag{2.17}$$

Practical use of such reactions depends on how efficiently material handling and separations can be achieved; and it is by no means certain that this or any of the other cycles so far proposed will be useable in a practical system.

2.9 GENERATION OF ELECTRICITY FROM FUELS

All important electric power generation from fossil fuels (and also from nuclear fuels) takes place by allowing the heat from burning fuel to vaporize a liquid, which in turn drives a mechanical device such as a turbine to operate a generator. The heat is rejected at some lower temperature determined by the water used for cooling the working fluid. The Carnot cycle efficiency limitations discussed previously and other losses result in the waste of much of the energy of the fuel. A number of direct heat-to-electricity conversion processes have been proposed to eliminate the inefficiencies associated with the mechanical steps, but they do not affect the heat cycle losses.

Three methods of converting heat directly to electricity are the thermoelectric, thermionic, and magnetohydrodynamic conversions. Thermoelectric conversion operates on the familiar principle of the thermocouple; if two different conductors are made to form two junctions that are maintained at different temperatures, a potential difference and a flow of electricity is set up between them. The thermionic process involves the emission of electrons from a heated cathode. These electrons are collected by an anode at lower temperature, and electricity can flow in the external circuit between the two. In a magnetohydrodynamic generator, the fuel is used to heat and ionize a gas to produce a plasma. Passage of the plasma through a magnetic field generates an electrical potential and current at right angle to the flow of plasma. None of these has yet proven practical for commerical use.

Direct conversion of chemical to electrical energy would also eliminate the Carnot cycle restriction, and may be done in fuel cells. A hydrogen fuel cell, for example, would involve the reaction

$$2H_2 \rightarrow 4H^+ + 4e^- \qquad (2.18)$$

at one electrode, and the reaction

$$O_2 + 4H^+ + 4e^- \rightarrow 2H_2O \qquad (2.19)$$

at the other. The overall reaction is the burning of hydrogen with oxygen to produce water, but if the two processes of reactions (2.18) and (2.19) take place at the surfaces of separate electrodes connected by an external conductor, the flow of electrons necessary for the reactions can be used to do work. This work is equal to that required to transport the required number of electrons through the potential difference E of the cell. On a molar basis, this work is nFE, where n is the number of the electrons transferred in the reaction

(4 in the above reaction), and F is the value of the Faraday constant. If the process is carried out reversibly, the total energy available is equal to the free energy change of the process, that is,

$$\Delta G = -nFE \tag{2.20}$$

Under these conditions, all of the chemical energy is available to do work, and the theoretical efficiency is 100%.

In order to obtain energy at a practical rate from an electrochemical cell, the cell would not in fact operate at equilibrium, and some energy would be lost. The enthalpy change of the reaction is therefore a more realistic quantity on which to base an efficiency comparison, and the maximum intrinsic efficiency of an electrochemical energy-producing process is

$$\varepsilon = \Delta G/\Delta H \tag{2.21}$$

In most cases of interest, the calculated efficiency is greater than 80% (e.g., it is 83% for the H_2-O_2 cell). Some reactions have efficiencies greater than 100% on the basis; this is so if the entropy change is positive, making $|\Delta G|>|\Delta H|$. Such a cell would cool unless heat were provided to it from the surroundings, and in this case both the energy of the fuel and the additional heat energy provided to the cell are converted to electricity.

Operating fuel cells employing a variety of fuels have been constructed. In addition to the hydrogen-oxygen cell referred to, hydrocarbons and alcohols with oxygen or air have been widely investigated. The products in this case are H_2O and CO_2. However, except in specialized applications such as space vehicles, the practical application of fuel cells has not generally been realized because of the small currents that can be transferred per unit area by most electrode materials. In many cases, expensive electrocatalytic materials such as platinum have been employed to obtain satisfactory high current densities at normal operating temperatures. High temperature cells avoid this problem but materials become more expensive and corrosion becomes a major problem. Much research has been carried out on fuel cells over the past quarter century, but the problems above have not yet been overcome to the extent that practical large-scale power generation is possible in spite of the efficiency of fuel conversion theoretically possible.

2.10 THE ENERGY FUTURE

The energy "crisis" that began a few years ago is real. Its solution is not simple, and is further confused by many conflicting claims and arguments; far too many discussions of alternative energy sources are based more on advocacy of personal prejudices than on objective science. We are in no position to resolve these arguments here, but some comments may be pertinent within the general framework of chemistry and the environment. Much of the confusion with regard to future energy supplies come from two causes. The first is the estimate of a given resource. Generally, estimates of proven resources--those already discovered--may be fairly reliable, but estimates of total resources, including those yet to be discovered, are not. Such estimates vary widely. In addition, estimates of the length of time a given resource will last are critically dependent on the estimated rate of use. For example, if a fossil fuel reserve adequate for 1000 years at present levels of use were found, an annual 3.5% increase in consumption would reduce its lifetime to only 104 years. It is not seriously expected that a thousand year supply of any fossil fuel remains to be discovered, and there is no real question that the use of petroleum and natural gas as large-scale energy sources must end in a comparatively short time, 20 to 40 years is a reasonable estimate. Coal resources are larger, and will last several hundred years at present energy production rates, and coal utilization has a proven technology. However, coal utilization involves serious problems of transportation, air pollution, and damage to the land through strip mining and acid mine waste (Section 14.5.3). Coal gasification and liquification processes, which seem mandatory if coal is to be used for household heating and for vehicle power, are technologically sound but require large capital expenditures. Capital cost and waste disposal problems similarly affect oil shale and tar sand use.

This introduces the second cause for confusion as to the most likely future energy source: economic estimates. These also vary widely for a given source, and are sensitive to many influences and assumptions. Obviously, use of an energy source is absolutely uneconomical when the energy expenditure required to collect it and to extract the energy from it equals the energy so obtained, but the practical limit will be reached at some earlier point. For many sources, the capital costs needed to construct the necessary plants are huge, even though operating costs may be low.

Nuclear energy is a proven power source, but uranium supplies, as fossil fuel supplies, are limited. Breeder reactors are technologically feasible, but safety and waste-disposal problems raise major doubts as to whether the public will accept nuclear fission power generation on the scale necessary to be the primary energy source in the near future. Fusion reactors are unproven, and while this process is sufficiently promising to warrant intensive development, any policy based on the assumption at this time that future energy problems will be solved by fusion represents a dangerous gamble.

Solar energy can, with present technology, supplement our energy supply if used for heating and similar processes. However, it cannot supply the main power requirements of most of the world without significant economic and technological breakthroughs in energy storage and conversion. Such breakthroughs are not guaranteed, but this source must be researched intensively.

Wind, water, ocean, and geothermal sources are either too limited to be major energy sources, or involve major collecting and distribution problems that make their widespread use very uneconomical. This may not be true on a local basis, and tidal and geothermal power plants are quite practical in certain locations. Wind power may also be an economical source for supplemental power, but energy storage capabilities are needed if it is to be a primary source.

The most promising approach to energy requirements for the near future at least appears to be a utilization of mixed sources to supplement fossil fuels. As petroleum becomes more costly, its use can be conserved by tapping those solar, geothermal, wind, or other sources that are most economic. Large scale use of coal is necessary to free petroleum for vehicular and chemical uses, and to "buy time." Nuclear fission, if present concerns about safety and waste disposal can be satisfied, may contribute considerably. Energy conservation is an absolute necessity. In the long run, fusion and solar energy may provide an abundance of energy at reasonable cost, but it is equally possible that they may not. The lack of an energy policy in the U.S. over the past years is highly regrettable in view of the magnitude of the problem to be faced.

BIBLIOGRAPHY

M. Fogiel, Ed., "Modern Energy Technology," Vols. I and II, Research and Education Association, New York, 1975. (These volumes give detailed discussions of most important energy sources.)

L.C. Ruedisili and M.W. Firebaugh, Eds., "Perspectives on Energy" Oxford University Press, New York, 1975. (This book contains chapters on virtually all of the topics discussed in this chapter, and others.)

J. M. Hollander and M. K. Simmons, Eds., "Annual Review of Energy," Vol. 1, Annual Reviews, Inc., Palo Alto, California, 1976. (A variety of energy sources and associated problems are discussed.)

J. M. Hollander, M. K. Simmons and D. O. Wood, Eds., "Annual Review of Energy," Vol. 2. Annual Reviews, Inc., Palo Alto, California, 1977.

W. Hafele, A systems approach to energy, *Amer. Sci. 62,* 438 (1976).

E.T. Hayes, Energy implications of materials processing, *Science 191,* 661 (1976).

H. Brown, Energy in our future, *Annu. Rev. Energy 1,* 1 (1976).

E. H. Thorndike, "Energy and the Environment," Addison-Wesley, Reading, Massachusetts, 1976.

S.S. Penner and L. Icerman, Eds., "Energy," Vol. 1. "Demands, Resources, Impact, Technology, and Policy," Vol. 2. "Non-Nuclear Technologies," Vol. 3. "Nuclear Energy and Energy Policies." Addison-Wesley, Reading, Massachusetts, 1974, 1975, 1976.

B. Bolin, The carbon cycle, in "Chemistry in the Environment, Readings from Scientific American." W. H. Freeman, San Francisco, California, 1973, p. 53 (From *Sci. Amer.,* Sept. 1970.)

R. R. Berg, J. C. Calhoun, Jr., and R. L. Whiting, Prognosis for expanded U.S. production of crude oil, *Science 185,* 331 (1974).

L. P. Lessing, Coal, in "Chemistry in the Environment, Readings from Scientific American." W. H. Freeman, San Francisco, California, 1973, p. 159. (From *Sci. Amer.,* July, 1955.)

T. H. Maugh, Gasification: A rediscovered source of clean fuel, *Science 178,* 44 (1972).

E. F. Osborn, Coal and the present energy situation, *Science 183,* 477 (1974).

A. L. Hammond, Coal research (I): Is the program moving ahead, *Science 193,* 665 (1976).

A. L. Hammond, Coal research (II): Gasification faces an uncertain future, *Science 193,* 750 (1976).

A. L. Hammond, Coal research (III): Liquefaction has far to go, *Science 193,* 873 (1976).

A. L. Hammond, Coal research (IV): Direct combustion lags its potential, *Science 194,* 172 (1976).

J. Walsh, Problems of expanding coal production, *Science 185,* 336 (1974).

A. M. Squires, Clean fuels from coal gasification, *Science 185,* 340 (1974).

J. T. Dunham, C. Rampacek, and T. A. Henrie, High-sulfur coal for generating electricity, *Science 185,* 346 (1974).

H. Perry, The gasification of coal, *Sci. Amer. 230,* (3), 19 (1974).

Coal conversion activities picking up, *Chem. Eng. News,* December 1, 1975, p. 24.

O. H. Hammond and R. E. Baron, Synthetic fuels: Prices, prospects and prior art, *Amer. Sci. 64,* 407 (1976).

N. deNevers, Tar sands and oil shales, in "Chemistry in the Environment, Readings from Scientific American," W. H. Freeman, San Francisco, California, 1973, p. 169.

"High recovery proves out for shale oil process," *Chem. Eng. News*, July 1, 1974, p. 15.

W. D. Metz, Oil shale: A huge resource of low-grade fuel, *Science 184,* 1271 (1974).

T. F. Yen, Ed., "Shale oil, tar sands, and related fuel sources." Advances in Chemistry Series #151, p. 184. American Chemical Society, Washington, D. C., 1976.

Prospects for substitute fuels look poor, *Chem. Eng. News,* October 18, 1976, p. 36.

T. B. Reed and R. M. Lerner, Methanol: A versatile fuel for immediate use, *Science 182,* 1299 (1973).

E. E. Wigg, Methanol as a gasoline extender: A critique, *Science 186,* 785 (1974).

G. A. Mills and B. M. Harney, Methanol--the "new fuel" from coal, *Chem. Technol. 4,* 26, (1974).

E. Zebroski and M. Levenson, The nuclear fuel cycle, *Annu. Rev. Energy 1,* 101, (1976).

R. F. Post, Nuclear fusion, *Annu. Rev. Energy 1,* 213, (1976).

F. Daniels, Direct use of the sun's energy, *Amer. Sci. 55,* 15 (1967).

J. A. Duffie and W. A. Beckman, Solar heating and cooling, *Science 191,* 143 (1976).

M. Wolf, Solar energy utilization by physical methods, *Science 184,* 382 (1974).

M. Calvin, Solar energy by photosynthesis, *Science 184,* 375 (1974).

V. Balzani, L. Moggi, M. F. Manfrin, F. Bolletta, and M. Gleria, Solar energy conversion by water photodissociation, *Science 189,* 852 (1975).

B. Chalmers, The photovoltaic generation of electricity, *Sci. Amer. 235,* 34 (October, 1976).

M. D. Archer, Electrochemical aspects of solar energy conversion, *J. Appl. Electrochem. 5,* 17 (1975).

R. R. Hautala, J. Little and E. Sweet, The use of functionalized polymers as photosensitizers in an energy storage reaction, *Solar Energy 19,* 503 (1977).

M. Calvin, Photosynthesis as a resource for energy and materials, *Photochem. Photobiol. 23,* 425 (1976).

D. E. White, "Geothermal energy." U.S. Geological Survey Circular 519, 1965.

L.J.P. Muffler and D.E. White, Energy, in "Perspectives on Energy", (L.C. Ruedisili and M.W. Firebaugh, eds.), p. 352. Oxford University Press, New York, 1975.

H.E. Klei and F. Maslan, Capital and electric production costs for geothermal power plants, *Energy Sources 2,* 331 (1976).

D.H. Cortez, B. Holt, and A.J.L. Hutchinson, Advanced binary cycles for geothermal power generation, *Energy Sources 1,* 73 (1974).

J.P. LeBoff, Wind power feasibility, *Energy Sources 2,* 361 (1976).

W.D. Metz, Wind energy: Large and small systems competing, *Science 197,* 971 (1977).

M.K. Hubbert, Tidal power, in "Perspectives on Energy" (L.C. Ruedisili and M.W. Firebaugh, eds.), p. 359. Oxford University Press, New York, 1975.

W.E. Heronemus, Wind power: A near-term partial solution to the energy crisis, in "Persepctives on Energy" (L.C. Ruedisili and M.W. Firebaugh, eds.), p. 364. Oxford University Press, New York, 1975.

E.J. Beck, Ocean thermal gradient hydraulic power plant, *Science 189,* 293 (1975).

W.D. Metz, Ocean thermal energy; the biggest gamble in solar power, *Science 198,* 178 (1977).

J.N. Weinstein and F.B. Leitz, Electric power from differences in salinity: The dialytic battery, *Science 191,* 557 (1976).

B.H. Clampitt and F.E. Kiviat, Energy recovery from saline water by means of electrochemical cells, *Science 194,* 719 (1975).

I. Fells, Energy - transmission, storage and management, *Chem. Britain 13,* 222 (1977).

D.W. Rakenhorst, Use of flywheels for energy storage, *Energy Systems Policy 2,* 251 (1975).

L.W. Jones, Liquid hydrogen as a fuel for the future, *Science 1974,* 367 (1971).

W.E. Winsche, K.C. Hoffman, and F.J. Salzano, Hydrogen, its future role in the nation's economy, *Science 180,* 1325 (1973).

D.P. Gregory, Electrochemistry and the hydrogen economy, in "Modern Aspects of Electrochemistry," No. 10 (J.O'M. Bockris and B.E. Conroy, eds.). Plenum Press, New York, 1975.

K.V. Kordesch, Power sources for electric vehicles, in "Modern Aspects of Electrochemistry," No. 10 (J.O'M. Bockris and B.E. Conroy, eds.), Chapter 7. Plenum Press, New York, 1975.

R.H. Wentorf, Jr., and R.E. Hanneman, Thermochemical hydrogen generation, *Science 185,* 311 (1974); see also R. Shinnar, *Science 188,* 1036 (1975); M.A. Soliman, W.L. Conger, K.E. Cox, and R.H. Carty, *Science 188,* 1037 (1975); and R.H. Wentorf, Jr. and R.E. Hanneman, *Science 188,* 1038 (1975).

E.A. Fletcher and R.L. Moen, Hydrogen and oxygen from water, *Science 197,* 1050 (1977).

C. Marchetti, The hydrogen economy and the chemist, *Chem. Britain 13,* 219 (1977).

3
ATMOSPHERIC COMPOSITION AND BEHAVIOR

3.1 INTRODUCTION

In recent years, there have been many magazine articles and discussions about the possible effects of various atmospheric pollutants on the earth's climate. On the one hand, it is said that the release of carbon dioxide when we burn fossil fuels has led to a warming trend in the earth's climate since about 1900. On the other hand, it is also said that the release of dust, smoke, and other particulate matter into the atmosphere has led to a cooling trend which, since 1960, has overwhelmed the warming trend. Therefore, at the present time, the earth's climate is presumably getting colder.

These statements should lead to a number of questions in the minds of scientifically inclined readers.

(a) How do we know that the earth's climate became warmer early in the twentieth century and is now becoming cooler? What criteria are used to define these trends? After all, the climate in any one area on the earth's surface is quite variable from year to year. How do we take account of this variability?

(b) Assuming that the criteria for defining the stated changes in the earth's climate are sound, how do these twentieth century climate changes compare with other variations in climate noted in historic but generally preindustrial times? What variations in climate occurred on earth in prehistoric times, even before man evolved to his present state? In other words, is there a possibility that the climate changes noticed in this century are due, at least partially, to natural causes?

(c) What is there about an increase in the carbon dioxide content of the atmosphere that leads to a warming trend in the earth's climate? Why should an increase in the dust content of the atmosphere lead to a cooling trend in the earth's climate? What regulates the earth's climate in general? In other words, what do we know about heat input and heat ouput from the surface of the earth, i.e., the earth's heat balance?

(d) In addition to carbon dioxide and dust, what are the constituents of the earth's atmosphere at the present time? Have these constituents always been the same, and present with the same percentage composition, since the earth was formed? If not, what sorts of evolution have been hypothesized for the earth's atmosphere?

(e) What variations exist in the earth's atmosphere as we go from the surface of the earth to the limits of the atmosphere? This question is closely related to the one about the earth's heat balance.

(f) What do we know about the general circulation of the atmosphere? After all, carbon dioxide and dust particles are released into the atmosphere at particular locations on the earth's surface. How far do they spread through the atmosphere, both horizontally and vertically?

(g) Although the carbon dioxide and dust particles released into the atmosphere by man's efforts are said to affect the total, or macroclimate of the earth, do not man's activities also affect the earth's microclimate? What do we know about the special climates of cities, as opposed to their surrounding countrysides? How do the pollutants produced by man's activities in cities become trapped in the atmosphere above these cities, to become hazards to man's health?

These next two chapters will attempt to answer these questions, though not necessarily in the order in which they have been presented. The answers to some of these questions are complex and, in some cases, not particularly well understood. It will thus be seen that questions and answers concerning the atmosphere and man's effect on the atmosphere and climate are not simple and need a great deal of further study, preferably before irreparable harm is done.

3.2 GASEOUS CONSTITUENTS OF THE ATMOSPHERE

Table 3-1 shows the constituents of clean, dry air near sea level. Usually, the atmosphere also contains water vapor and dust, but these occur in amounts that vary widely from place to place and from time to time. Carbon dioxide, the fourth constituent on the list, has been increasing in concentration during this century, mostly through burning of fossil fuels such as coal and petroleum. The concentrations given in Table 3-1 are approximately for 1976.

A number of nitrogen and sulfur compounds included in the table are naturally occurring compounds as well as man-made pollutants. Naturally occurring nitrogen oxides are produced by the action of lightning discharges on the nitrogen and oxygen of the atmosphere, while the sulfur compounds arise in part from volcanic action. Methane generally comes from decaying organic matter and ammonia, H_2S and N_2O from the decomposition of amino acids by bacteria.

Table 3-1
Constituents of Clean Dry Air Near Sea Level

Component	*Volume (%)*
N_2	78.1
O_2	20.9
Ar	0.934
CO_2	0.0330
Ne	0.00182
He	0.00052
CH_4	0.00016
Kr	0.00011
H_2	0.00005
N_2O	0.000028
CO	0.00001
Xe	9×10^{-6}
O_3	4×10^{-6}
NO_2	2×10^{-6}
NH_3	6×10^{-7}
SO_2	2×10^{-7}
CH_3Cl	5×10^{-8}
C_2H_4	2×10^{-8}
CCl_4	1×10^{-8}
CCl_3F	1×10^{-8}
CCl_2F_2	1×10^{-8}

The concentration of ozone, given in Table 3-1 at sea level, varies a great deal with altitude and somewhat with latitude and will be discussed in greater detail later. Ozone is a very reactive gas, but so is oxygen, so that the high concentration of oxygen in our atmosphere should be very surprising to a chemist. The main reason for our lack of surprise is probably our knowledge that the process of photosynthesis in green plants supplies vast amounts of oxygen to the atmosphere every day (although this may not be the only source). It seems reasonable to suppose that this makes up for the equally vast amounts of oxygen used up by respiration of plants and animals, by weathering of rocks, by burning of fossil fuels, and by other oxidative processes that take place in the atmosphere. This balance, however, requires the presence of green plants that were not present in the very distant past. The various theories that describe the earth as evolving from a primordial, lifeless state to its present condition, and the physical evidence that supports these theories, strongly suggest that oxygen was not present on the primordial earth for any length of time.

Carbon monoxide as well as CO_2 is released to the atmosphere by the burning of fossil fuels, and also by volcanic and biological activity. It is known, however, that carbon monoxide is not accumulating in the atmosphere, and this has led to much recent interest in the origin and fate of atmospheric CO. To the surprise of many people, it has been found that over 90% of the carbon monoxide entering the atmosphere each year comes from natural processes.[1] Estimates of carbon monoxide entering the atmosphere each year are as follows: 3 billion tons from the oxidation of methane arising from the decay of organic matter; 100 million tons from the biosynthesis and degradation of chlorophyll; 400 million tons from the oceans and other natural sources; 270 million tons from man's activities. These figures indicate that CO must be removed from the atmosphere with great efficiency. Although CO can be oxidized by oxygen to CO_2, this process takes place too slowly in the atmosphere to account for the removal. There is evidence that fungi and some bacteria in the soil can utilize CO, converting it to CO_2; these organisms evolved long before man began to contribute carbon monoxide to the atmosphere. Atmospheric carbon monoxide is therefore not one of mankind's long-term problems, but, in the short term it can be very important because of its toxic nature. Although the worldwide concentration of CO is only between 0.1 and 0.5

[1] *T.H. Maugh II, Science 177, 338 (1972).*

parts per million, its concentration in the air over large cities can be fifty to one hundred times as great.

Sulfur and nitrogen compounds in the atmosphere were mentioned briefly in connection with Table 3-1 and will be discussed at greater length in Chapter 14. Recently, there have been some distressing observations made in connection with the sulfur and nitrogen compounds released into the atmosphere by the burning of fossil fuels and by various industrial processes. Many sulfur compounds, after undergoing further oxidation in the atmosphere (see Chapter 9), come back to earth in the form of sulfuric acid dissolved in rainwater. Nitrogen oxides also form acidic solutions in rainwater. Normal rainwater, which always contains CO_2, has a pH of 5.7. In the last 30 years, however, rains and snows of much lower pH, i.e., higher acidity, have been reported from many parts of the world. In New England, specifically at Hubbard Brook, New Hampshire, the average pH of these forms of precipitation during 1970-1971 was 4.03, with occasional pH values reported as low as 2.1. New England is downwind of most U.S. industry, i.e., the Chicago to New York City area. In Scandinavia, which is downwind from England and from West Germany's heavily industrialized Ruhr Valley, rain with a pH as low as 2.8 has been reported. Even in South America's Amazon Basin, pH values as low as 4 have been recorded. Acid rains in Europe have apparently led to rapid weathering of ancient structures and sculptures in recent decades; some of these structures had stood virtually unchanged for hundreds and, in some cases, thousands of years, such as the Parthenon in Athens. In addition, low pH rains and brooks weather rocks much more rapidly than normal waters with a higher pH, and can affect biological systems adversely.

3.3 HISTORY OF THE ATMOSPHERE

3.3.1 Evidence and General Theory

Unfortunately, there is no agreement in the literature either on the composition or on the time of origin of the earth's original atmosphere. Evidence for the composition of the atmosphere at previous times comes from a study of objects that were formed in contact with the atmosphere at those times. These objects may have changed since, but they still provide us with the clues as to the conditions existing when they were formed.

The earliest such objects in existence are sedimentary rocks about 3×10^9 years old. It is believed that these rocks could not have been formed without previous atmospheric weathering or in the absence of large amounts of liquid water.

Therefore, water was probably present in the primitive atmosphere. On the other hand, the compositions of these and later rocks indicate that the atmosphere could not have contained much oxygen before 1.8×10^9 years ago. Also, the relative rarity of carbonate rocks such as limestone and dolomite among the oldest sedimentary rocks seems to indicate that bases such as ammonia could not have been abundant in the primitive atmosphere. The presence of ammonia would have increased the pH of the primitive ocean and thus favored an abundance of carbonate rocks (see Chapter 14). Furthermore, methane was probably not present in the primitive atmosphere either, since it has been surmised that the dissociation of methane in an atmosphere lacking oxygen should have led to large deposits of carbon that have not been found. Because of this geologic evidence and in spite of experiments that have been performed to show that an atmosphere dominated by CH_4 and NH_3 would have been the most suitable environment for the evolution of complex organic molecules, in this chapter we shall assume the absence of CH_4 and NH_3 from the primitive atmosphere, at least in large amounts. There is another school of thought, however, which feels that CH_4 and NH_3 must have been present in the primitive atmosphere, and that insufficient evidence for their absence has been found. Discussion of these conflicting viewpoints is beyond the scope of this book.

A combination of the geologic record and the present composition of the atmosphere, along with some reasonable conjectures, leads to the conclusion that the primitive atmosphere probably contained H_2O, CO_2, N_2, CO, SO_2, HCl, and a few other trace gases of volcanic origin. Volcanic gases contain H_2O, CO_2, N_2, SO_2, H_2S, S, HCl, H_2, CH_4, CO, NH_3, and HF. Of these H_2O is the most abundant, followed by CO_2 and H_2S. One hypothesis states that a major degassing episode in the earth's history may have been initiated by the earth's capture of the moon between 3.5 and 4.5×10^9 years ago. The earth has been degassing at somewhat varying rates ever since, generally by means of hot springs and other types of volcanic action.

The earth at present is emitting mostly water vapor and carbon dioxide, some nitrogen and other gases, and no oxygen. These data must be reconciled with the composition of the atmosphere as given in Table 3-1. Calculations[2] indicate that the total quantity of water vapor released by the earth over all time is about 3×10^5 gm/cm^2 averaged over all the earth's surface. Most of this water vapor has condensed and is presently in lakes and oceans. Some of it, present in the upper atmosphere, is assumed to have been photodissociated into oxygen and hydrogen. The oxygen now present

[2]F.S. Johnson in S. F. Singer, ed., "Global Effects of Environmental Pollution." Springer-Verlag, New York, 1970.

in the atmosphere has probably arisen from photodissociation of water vapor and from photosynthesis in green plants after the latter had evolved.

3.3.2 Carbon Dioxide

Carbon dioxide has been emitted to the extent of about 5×10^4 gm/cm^2 of the earth's surface over all time. Most of this CO_2 has dissolved in the oceans and much has precipitated from the oceans as calcium carbonate. Only about 0.45 gm/cm^2 CO_2 remains in the atmosphere, while about 27 gm/cm^2 averaged over all the earth's surface is dissolved in the oceans. There is 60 times as much CO_2 dissolved in the oceans as there is free in the atmosphere, but most of the carbon dioxide released by the degassing of the earth is now tied up in geologic deposits, i.e., carbonate rocks.

There are complex equilibria, to be discussed in Chapter 14, between the atmospheric CO_2 and that dissolved in the oceans. Recent experiments have shown that it takes about 470 days to dissolve half the CO_2 placed in contact with turbulent seawater. Complete equilibrium takes much longer, probably five to ten years, but, given sufficient time, the oceans and rocks act as an enormous sink for any CO_2 that is added to the atmosphere. That is, most of the CO_2 added to the atmosphere eventually ends up in the oceans and in geologic deposits. However, we are now adding CO_2 to the atmosphere by burning fossil fuels in the amount of approximately 10^{10} metric tons per year. Some of this will remain in the atmosphere even when the atmosphere-ocean equilibrium has been achieved. One estimate of the amount of all recoverable fossil fuels is 1.6×10^{13} metric tons. This would allow 10^{10} metric tons to be burned each year for many more years. The concentration of carbon dioxide in the atmosphere in the nineteenth century was about 290 ppm. while in 1976, as shown in Table 3-1, it was about 330 ppm. It is difficult to calculate how much of the added CO_2 has remained in the atmosphere, but the increased concentration in this century is well documented and will be discussed later, together with its possible effects on climate. Much carbon dioxide is used by green plants in photosynthesis, but this carbon dioxide is returned to the atmosphere within one part in 10^4 by decay and oxidation in living things.

Production of CO_2 from fossil fuels uses oxygen from the atmosphere. However, if all easily recoverable fossil fuels were burned at once, there would be no noticeable effects on the concentration of atmospheric oxygen. For example, if the 1.6×10^{13} metric tons of recoverable fossil fuels were pure carbon and were burned completely, 4.3×10^{13} metric tons of oxygen, or roughly only 10% of the atmospheric oxygen (a total of 10^{21} gm or 5×10^{14} metric tons from Table 3-2) would be used up.

3.3.3 Nitrogen

Compared with CO_2, relatively small amounts of nitrogen have been released by degassing of the earth, only about 10^3 gm/cm^2 of the earth's surface. But nitrogen is chemically inert, and at least 90% of this amount is still present in the atmosphere. The nitrogen no longer present in the atmosphere is mostly present in geologic deposits; the tiny fraction (10^{-8}) of atmospheric nitrogen that is "fixed" each year is soon returned to the atmosphere when plants and animals decay. The nitrogen cycle is discussed more fully in Chapter 14.

3.3.4 Oxygen

As stated above, atmospheric oxygen does not arise directly out of degassing of the earth, but comes from photodissociation of water vapor and photosynthesis. The relative importance of these two processes is in dispute, with estimates ranging from 100% photodissociation to 100% photosynthesis. In the present discussion, it is immaterial where all the oxygen now present in the atmosphere originated. It is interesting to note, however, that calculations exist that show that both processes were probably necessary to account for all the oxygen now present in the atmosphere, in living things, and in oxidized rocks and minerals.

In the photodissociation of water vapor

$$H_2O\ (g) \xrightarrow{h\nu} H_2\ (g) + \frac{1}{2}\, O_2\ (g) \qquad (3\text{-}1)$$

hydrogen gas is produced as well as oxygen gas. Hydrogen molecules, being much less massive than other atmospheric gas molecules, have a higher speed than these molecules at any temperature anywhere in the atmosphere, and many have a velocity greater than the escape velocity from earth's gravitational field. It has been calculated that hydrogen escapes at the rate of approximately 10^8 atoms/sec/cm^2 of earth's surface. Over all time, about 10^{17} sec, then, 10^{25} atoms/cm^2, about 17 gm/cm^2, have escaped, which accounts for about 150 gm/cm^2 of dissociated water vapor. This number corresponds to about 0.05% of the total water vapor released during the degassing of the earth. This is possible, although there is controversy as to whether enough water vapor gets transported to those upper layers of the atmosphere where photodissociation can take place (see Chapter 10).

At what rate did oxygen accumulate in the atmosphere? There is some geologic evidence that the first anaerobic forms of life originated about 3×10^9 years ago. These life forms may have released some oxygen into the atmosphere, adding to that produced by the photodissociation of water vapor. About 1% of the present amount of oxygen had accumulated in the atmosphere 6×10^8 years ago. At this point, respiration could begin in living things and, more important, ozone formation began in the upper atmosphere, as will be discussed in another section of this chapter. The atmospheric ozone screens certain ultraviolet wavelengths completely from the surface of the earth at present. When the oxygen level had reached 1% of its present value, screening of these harmful wavelengths was sufficient such that a variety of life forms became possible in the oceans. By the start of the Silurian era, 4.2×10^8 years ago, the earth's oxygen level had presumably reached about 10% of its present level, and sufficient ozone was present in the upper atmosphere to allow life to evolve on land. Most of the oxygen that has been produced in all this time has been used up in the oxidation of rocks and other surface materials on earth. However, more oxygen is still being produced, both by photodissociation and by photosynthesis. Table 3-2 summarizes the amounts of the four principal gases introduced into the atmosphere since the formation of the earth, and the amounts remaining in the atmosphere today.

3.4 PARTICULATE CONSTITUENTS OF THE ATMOSPHERE

One constituent of the atmosphere that was not included in Table 3-1 was "particulate matter," dust, or, in general, nongaseous matter. We are speaking primarily about particles that are 10^{-7} to 10^{-2} cm in radius; that is, from approximately molecular dimensions to sizes that settle fairly rapidly. Particulate matter in the atmosphere can be natural or man-made. Table 3-3 gives some estimates of the tonnage of particles with radii less than 2×10^{-3} cm (20 μm) emitted to the atmosphere each year.

Table 3-3 indicates that man-made particles constitute between 5 and 45% of the total appearing in the atmosphere annually at the present time. These estimates cover a large range because some occurrences, like volcanic action, are inherently variable and because some of the other sources of

Table 3-2

Atmospheric Gases: Historical Summary[a]

	Total amount introduced into the atmosphere		Amount remaining in atmosphere		Remaining in atmosphere in percent
Gas	(gm)	(gm/cm^2)	(gm)	(gm/cm^2)	(%)
H_2O	1.5×10^{24}	3×10^{5}	10^{19}	2	7×10^{-4}
CO_2	2.5×10^{23}	5×10^{4}	2.25×10^{18}	0.44	9×10^{-4}
N_2	5×10^{21}	10^{3}	4×10^{21}	8×10^{2}	80
O_2	1.5×10^{22}	3×10^{3}	10^{21}	2×10^{2}	7

[a] *There is sixty times as much CO_2, but only 1% as much N_2 and O_2 dissolved in the oceans as there is in the atmosphere.*

Table 3-3.

Estimate of Particles Emitted into or Formed in the Atmosphere Each Year[a]

Particle sources	Quantity (megatons/year)
Natural	
Soil and rock debris	100 - 500
Forest fires and slash-burning	3 - 150
Sea salt	300
Volcanic action	25 - 150
Gaseous emissions:	
Sulfate from H_2S	130 - 200
NH_4^+ salts from NH_3	80 - 270
Nitrate from NO_x	60 - 430
Hydrocarbons from plants	75 - 200
Natural particle subtotal	773 - 2200
Man-made	
Direct emissions, smoke, etc.	10 - 90
Gaseous emissions:	
Sulfate from SO_2	130 - 200
Nitrate from NO_x	30 - 35
Hydrocarbons	15 - 90
Man-made particle subtotal	185 - 415
Grand total	958 - 2615

[a]Reprinted from "inadvertant Climate Modification," Report of the Study of Man's Impact on Climate (SMIC), by permission of the MIT Press, Cambridge, Massachusetts, 1971.

particulate matter are very difficult to estimate.

Most of these particles eventually leave the atmosphere and return to the earth's surface. It is estimated that approximately 70% of the particulate matter that enters the atmosphere is eventually rained back out, probably because the particles act as condensation nuclei for the water droplets that form rain clouds. Large particles can settle out of the atmosphere by sedimentation in the earth's gravitational field. The particles are accelerated by the gravitational force but retarded by friction with the gas molecules of the air, thus reaching a terminal velocity downward which is related to the radius of each particle as shown in Table 3-4.

Table 3-4
Sedimentation Velocity of Particles with Density 1 gm/cm³ in Still Air at 0°C and 1 atm Pressure[a]

Radius of particle μ*m*	*Sedimentation of velocity (cm/sec)*
<0.1	negligible
0.1 (10^{-5} cm)	8×10^{-5}
1.0	4×10^{-3}
10.0	0.3
100.0 (10^{-2} cm)	25

[a]*Reprinted from "Inadvertant Climate Modification, Report of the Study of Man's Impact on Climate (SMIC)" by permission of the MIT Press, Cambridge, Massachusetts, 1971.*

The smallest particles, many with radii from 0.1 to 1.0 μm, have a maximum concentration about 18 km above the earth's surface. Some of these arise from volcanic eruptions while others are particles of ammonium sulfate formed when SO_2 in the atmosphere is oxidized and the resulting H_2SO_4 neutralized by ammonia.

The mean residence time of particulate matter in the lower atmosphere is between 3 and 22 days. In the stratosphere it is 6 months to 5 years, in the mesosphere from 5 to 10 years. (The meaning of "stratosphere" and "mesosphere" will be explained in the next section.) At any rate, the higher in the atmosphere, i.e., the farther above the surface of the earth the particles are introduced, the longer it takes them to come down.

Particles with radii from 0.02 to 10 um may contribute to the turbidity of the atmosphere. We shall see later that the turbidity of the atmosphere is an important determinant of the earth's climate. Upward trends in the turbidity of the atmosphere have been monitored over the past three decades both in urban locations such as Washington, D.C. and in the mountains, such as Davos, Switzerland.

3.5 EXTENT OF THE ATMOSPHERE AND ITS TEMPERATURE AND PRESSURE PROFILE

There are about 5.1×10^{18} kg of atmosphere distributed out over the 5.1×10^{14} m^2 of the earth's surface. This means that atmospheric pressure at sea level is about $10^4 kg/m^2$ or 10^3 gm/cm^2. This is approximately equivalent to one standard atmosphere of pressure (1013 mbar). Most of the atmosphere is fairly close to the surface of the earth. Fifty percent of the atmosphere is within 5 km (3 mi) of sea level, i.e., at a height of somewhat more than 5 km above sea level, atmospheric pressure is 0.5 atm. There are quite a few mountains that are higher than this. One such mountain is Mt. Kilimanjaro in Tanzania which extends just under 6 km above sea level; mountains such as Mt. Whitney and the Matterhorn are just under 5 km in altitude.

Ninety percent of the atmosphere is within 12 km of sea level. If these 12 km are compared to the radius of the earth, 6370 km, it is seen that 90% of the atmosphere exists in an extremely thin layer covering the earth's surface. The total extent of the atmosphere is much harder to specify, since there is no definite upper boundary. For most purposes, one can say that the atmosphere ends between 200 and 400 km above the earth's surface, but on occasion it is necessary to consider a few hundred or a thousand additional kilometers.

Nitrogen, oxygen, and carbon dioxide are reasonably well distributed at all altitudes in the atmosphere. Most of the water vapor and water droplets occur at altitudes below about 14 km, while most of the ozone is localized in the vicinity of 25 km above the earth's surface. There is, however, some water vapor at altitudes above 14 km, about 3 ppm in the stratosphere.

The exact manner in which temperature and pressure change with altitude in the atmosphere depends both on latitude and on the season of the year. Very close to the surface of the earth, temperature and pressure are extremely variable, even beyond the latitude and seasonal variations. For this reason, various national and international groups have proposed "standard atmospheres" at various times. For illustrative purposes (see Table 3-5), we shall use the U.S. Standard Atmosphere, 1962, which depicts idealized year-round mean conditions for middle latitudes such as 45° North, for a range of solar activity that falls between sunspot minimum and sunspot maximum. Standard information for other latitudes and for various times of year can be found in the bibliography at the end of this chapter. Table 3-5 shows that the pressure drops off continuously with increasing altitude, but the temperature goes through some strange gyrations. The temperature of the atmosphere decreases from sea level up to 12 km, remains constant up to 20 km, increases from 20 to 50 km, decreases again from 50 to 80 km, remains constant to 90 km, and then increases asymptotically to about 1500 K above 300 km. These variations are probably less confusing as shown in Fig. 3-1, which also shows the change in pressure with altitude. Note that the temperature scale in Fig. 3-1 is logarithmic, so that the temperature variations in the first 100 km of the atmosphere can be seen better.

Starting at the earth's surface, at sea level, the temperature of the standard atmosphere decreases steadily in the troposphere (sometimes referred to as the lower atmosphere) up to the tropopause. We shall see in Chapter 4 how temperature inversions, i.e., regions where the temperature increases with increasing altitude, can sometimes occur in the troposphere. The troposphere contains most of the mass of the atmosphere, it is the only continuously turbulent part of the atmosphere, and it contains most of our weather systems.

Atmospheric pressure drops off logarithmically with altitude near the earth's surface, but more slowly at higher elevations. At this point, it might be well to ask whether the idea of temperature is still valid at elevations above 350 km where the pressure is less than 10^{-7} mbar. It is a principle of thermodynamics that temperature only has meaning for a system in equilibrium. There must be enough collisions among the molecules that are present to allow a Maxwellian distribution of velocities to be attained among them. We know that the mean free path of a typical molecule in air is about 10^{-5} cm at one atmosphere pressure and the surface temperature of the earth (288 K). At 10^{-7} mbar and 1500 K, the mean free path is about 25m, and it probably takes some time to achieve equilibrium after a particular molecule gains or loses energy. Therefore, it is best to consider the temperature at altitudes above 300 km as a mean kinetic temperature, having to do with the average kinetic energies of the molecules present, whether the distribution of velocities is Maxwellian, i.e., the equilibrium distribution, or not.

Table 3-5
Temperature and Pressure of the Standard Atmosphere[a]

Altitude (km)	*P (mbars)*	*T (K)*
0	1013	288 (15°C)
1	899	282
2	795	275 (2°C)
5	540	256
10	265	223 (-50°C)
12	194	217
15	121	217 (-56°C)
20	55	217
25	25	222
30	12	227
40	3	250
50	0.8	271 (-2°C)
60	0.2	256
70	0.06	220
80	0.01	180
90	0.002	180 (-93°C)
100	0.0003	210
110	7×10^{-5}	257
120	3×10^{-5}	350 (77°C)
130	1×10^{-5}	534
140	7×10^{-6}	714
150	5×10^{-6}	893 (620°C)
175	2×10^{-6}	1130
200	1×10^{-6}	1236
300	2×10^{-7}	1432
400	4×10^{-8}	1487
500	1×10^{-8}	1500 (1227°C)

[a]*Adapted from "U.S. Standard Atmosphere," U.S. Government Printing Office, 1962.*

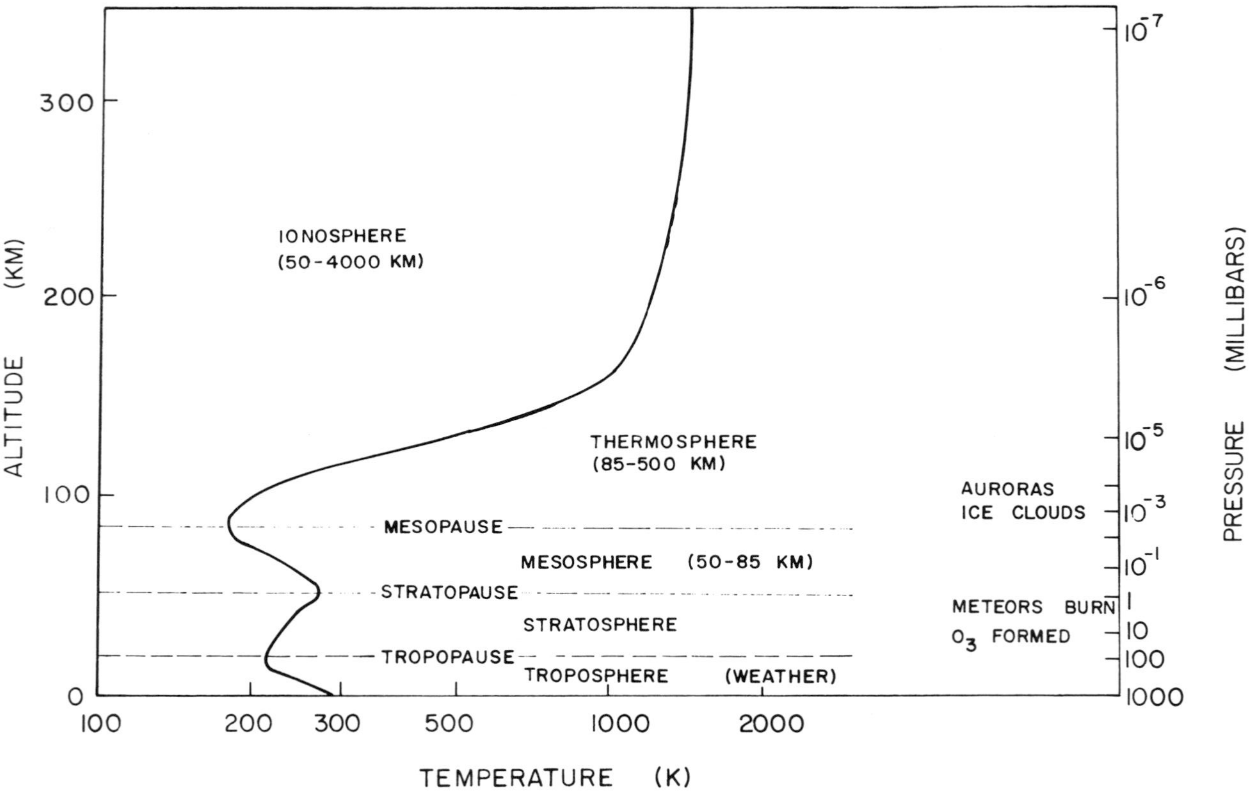

Figure 3-1. Temperature and pressure profile of the U.S. Standard Atmosphere.

Wherever the temperature rises with an increase in altitude in the atmosphere, there must be one or more solar energy absorbing or exothermic chemical reactions involved. Without such processes, the temperature of the atmosphere would vary smoothly from warm at the earth's surface to very cold, close to absolute zero, at very high altitudes. Radiant energy from the sun is absorbed in various photochemical processes both in the stratosphere and in the thermosphere (see Fig. 3-1).

The photochemical processes of the thermosphere will be fully discussed in Chapter 10. At this point, let us simply note that atmospheric oxygen and nitrogen in the thermosphere above 100 km absorb ultraviolet radiation having wavelengths below 180 nm (1800Å). The reactions occurring at these high altitudes absorb virtually all the high-energy electromagnetic radiation that reaches the earth's atmosphere. This circumstance is very fortunate for two reasons. First, this high-energy radiation would initiate chemical reactions in all complex molecules, so that life would be impossible if this radiation reached the earth's surface. Second, the large variations in the intensity of the sun's radiation at these short wavelengths have very little effect on the troposphere where our weather originates, since all these wavelengths are removed at much higher altitudes.

Radiation with wavelengths between 180 and 290 nm would also be destructive to life, but these wavelengths are absorbed in the stratosphere by oxygen and ozone. The maximum absorption of ozone is at 255 nm. The major reactions in the stratosphere, much simplified from the more complete discussion to be found in Chapter 10 are

$$O_2 \rightarrow O + O \tag{3-2}$$

$$O_2 + O + M \rightarrow O_3 + M^* \tag{3-3}$$

$$O_3 \rightarrow O_2 + O \tag{3-4}$$

$$O + O_3 \rightarrow 2O_2 + \text{kinetic energy} \tag{3-5}$$

Reaction (3-2) also occurs at higher altitudes, but does not lead to reaction (3-3) because the pressure is too low to allow trimolecular collisions to occur with any reasonable frequency. The third atom or molecule M is needed for reaction (3-3) so that both energy and momentum can be conserved in the reaction. The symbol M^* refers to the fact that M increases its energy while O_2 is combining with O to form O_3.

Ozone is formed down to about 35 km altitude, below which the ultraviolet radiation necessary for its formation (λ < 240 nm) has been fully absorbed. Most of the ozone in the atmosphere, however, exists at altitudes between 15 and 35 km, so that mixing processes must occur in the stratosphere as well as in the troposphere. Also peculiar is the fact that ozone concentrations are low over the equator, where, on the average, most of the sun's radiations impinges, and high near the poles, at latitudes greater than 50°, where less radiation comes in from the sun. Consequently, some mechanism for transport of ozone from the equator to the poles must exist.

Solar radiation with wavelengths greater than 290 nm (2900 Å) is transmitted to the lower stratosphere, the troposphere, and the earth's surface. We shall see that this remaining radiation, both visible and infrared, comprises more than 97% of the total energy from the sun.

3.6 GENERAL CIRCULATION OF THE ATMOSPHERE

Only very general patterns of atmospheric movement are within the scope of this chapter, where they are pertinent since we are interested in discussing general movements of pollutants and other constituents of the atmosphere. Atmospheric circulation will be considered in an extremely simplified manner. Books on meteorology, such as the ones given as references at the end of this chapter, may be consulted for further details.

There are two major reasons why there exists a general circulation of the atmosphere. First of all, the equatorial regions on earth receive much more solar energy all year round than the polar regions. Hot air therefore tends to move from the equator to the poles, while cold air tends to move from the poles toward the equator. As we shall see, the surface of the earth absorbs a much larger proportion of the solar radiation than the atmosphere above it. Thus, in the troposphere, the air is heated by conduction, convection, and radiation from the ground. There is also some heating of the atmosphere when water vapor, recently evaporated from the earth's surface, condenses to form clouds. Air near the equator is heated to a much greater extent than air near the poles.

If we recall that hot air is less dense than cold air and tends to rise above it (this will be discussed in Chapter 4 in connection with temperature inversions), we shall expect hot air to rise near the equator and flow toward the poles at some elevation above the surface of the earth. The colder air from the polar regions is then expected to flow toward the equator nearer to the surface of the earth as shown in Fig. 3-2. This idea was first formulated by G. Hadley in 1735. Hadley's picture of atmospheric circulation also took account of the earth's rotation, but for the moment we should note the following major oversimplifications in his model. First, no account was taken of unequal heating of land and sea, and, second, no account was taken of variations in solar radiation at various latitudes at different times of the year.

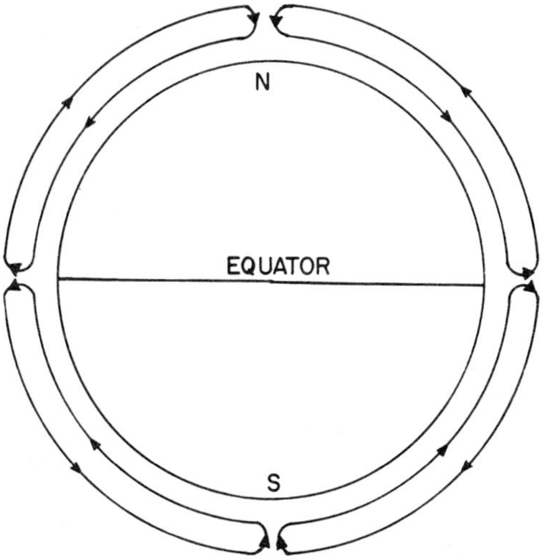

Figure 3-2. General circulation of the atmosphere as proposed by Hadley, 1735. These were the simplest forms of "Hadley cells," with hot air rising near the equator and moving toward the poles in the upper troposphere. Cold air, in this model, moves from the poles toward the equator near the surface of the earth.

What happens when the rotation of the earth from west to east is taken into account? If we think of the earth as rotating around an axis that goes through the north and south poles, we note that the surface of the earth at the poles does not rotate, while the surface of the earth at the equator rotates very quickly. The equatorial circumference of the earth is about 40,000 km and one rotation occurs every 24 h, so that every point on the earth's equator is constantly moving at a velocity of 28 km/min in contrast to the stationary points at the poles. In the same way, the air above the poles does not rotate while the air above the equator rotates with the earth, and at the same velocity. Between the equator and the poles, points on the earth's surface and in the atmosphere above these points rotate at intermediate speeds, between zero and 28 km/min.

When a parcel of air moves from one of the poles toward the equator, it is moving from an area where little or no rotation is taking place to an area where the earth is rotating quickly beneath it. This parcel of air tends to remain motionless with respect to the rotating earth, that is, the earth is rotating away from this parcel of air, moving faster than the parcel of air from west to east. With respect to an observer who is rotating with the earth, this parcel of air seems to be moving backwards, i.e., from east to west. In the northern hemisphere, where the parcel of cold air near the surface of the earth is moving north to south because of unequal heating, the overall result is to give this parcel of air a northeast to southwest motion with respect to someone on the earth's surface. We should therefore expect surface winds in the northern hemisphere to blow generally from the northeast. Similar reasoning would lead us to expect surface winds in the southern hemisphere to blow from the southeast.

A simplified version of the actual prevailing wind patterns on the earth's surface is shown in Fig. 3-3. These prevailing surface wind patterns are most noticeable over the oceans and were of great practical importance in the days of large sailing ships, where the major wind systems received their names. The doldrums near the equator and the horse latitudes were much feared by sailors, because a ship could remain becalmed there for weeks at a time. The various wind belts have a tendency to shift north and south at times, and the winds sometimes reverse because of topographical features, weather patterns, or simply the fact that some of these wind systems are more variable than others. The trade winds and the prevailing westerlies are fairly reliable, at least over the larger oceans, but the polar easterlies barely exist. The prevailing westerlies in both hemispheres seem to be blowing in the wrong direction, at least according to Hadley's original ideas, from the southwest instead of the northeast in the northern hemisphere and from the northwest instead of the southeast in the southern hemisphere. Air is therefore flowing from the equator to the poles along the surface of the earth in these midlatitude regions, and is moving from a more rapidly rotating region of the earth's surface to a more slowly rotating region. Thus, these portions of the atmosphere are moving west to east faster than their destination and the winds, which are a manifestation of this moving air, have a westerly component.

In summary, therefore, near the equator and near the poles, cold air is moving toward the equator near the surface of the earth with an apparent easterly (east to west) component, while, in the midlatitudes, warm air is moving from the equator toward the poles along the surface of the earth with a westerly (west to east) component. These easterly and westerly components of winds moving between portions of the earth that rotate at different speeds are manifestations of the Coriolis effect.

The three main prevailing wind directions in each hemisphere have been explained in terms of a three-cell model (Fig. 3-4) as distinguished from Hadley's original one-cell model (Fig. 3-2). This model has warm air rising at the equator and sinking near the poles just like the original Hadley model. The air circulation systems closest to the equator and to the poles move in the expected directions,

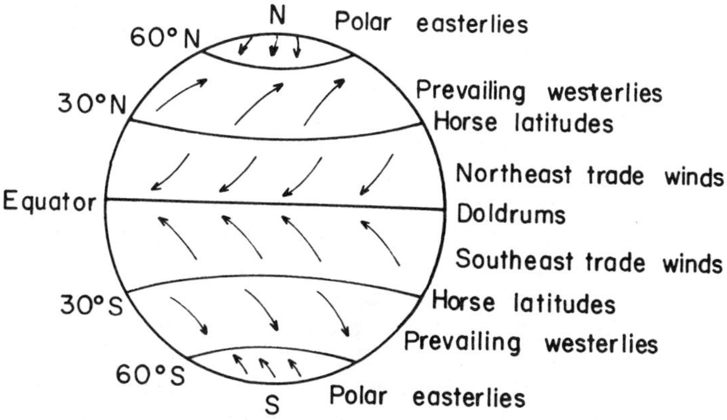

Figure 3-3. An extremely simplified version of prevailing wind patterns on the earth's surface. Polar easterlies, especially, are not to be taken very seriously.

with cold air traveling toward the equator near the surface of the earth, resulting in easterly winds. At latitudes 30°N and 30°S, these cells have downward moving air, while at 60°N and 60°S, they have upward moving air. The midlatitude cells, in order to preserve these wind directions, have warm air moving from the equator toward the poles near the earth's surface, resulting in westerlies. If this three-cell model, designed by Rossby in 1941, seems very hindsighted and artificial, let us note that any model that has only easterly winds all over the globe is impossible. These winds would tend to slow down the earth's west-to-east rotation. In fact, easterly and

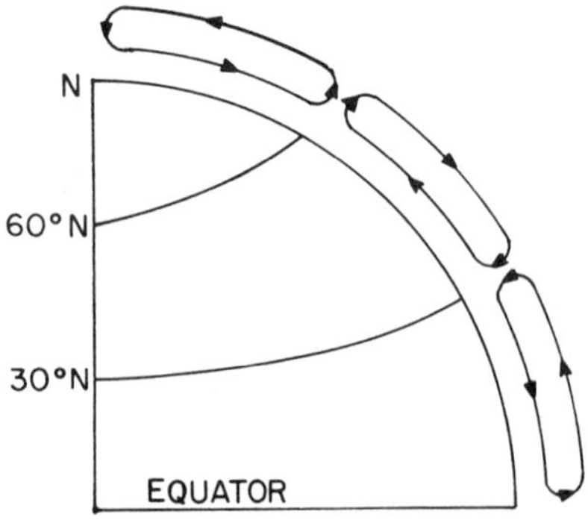

Figure 3-4. Three-cell model of general atmospheric circulation in the northern hemisphere.

westerly winds must balance on the average, since the atmosphere as a whole rotates with the earth. This fact was ignored in the single cell model.

The scheme shown in Fig. 3-4, although it explains the major prevailing wind systems on the earth's surface reasonably well, has been found inadequate to explain numerous phenomena including the so-called jet-stream in the upper troposphere (west to east), the zone of low and high pressure with winds circulating in opposite directions that constitute our weather systems and the general west-to-east movement of these weather systems. It turns out that a good deal of energy is transmitted from the equatorial regions to the polar regions by these cyclone and anticylone weather systems. Further explanations are outside the scope of this chapter. Let us just note here that experiments with model liquid systems, i.e., suspensions of metals in liquids under conditions that simulate conditions in the earth's atmosphere have shown, under certain conditions, flow patterns that correspond to the type of air circulation in cyclones and anticyclones and the slow drift of these flow patterns in the direction of motion of the fastest moving portion of liquid. The experiment is carried out as follows: The liquid suspension is placed between two cyclinders; the inner cylinder is cooled and stationary while the outer cyclinder is heated and rotating. The metal particles in the suspension can be seen in motion back and forth between the stationary cold cylinder wall and the rotating warm cylinder wall. Eddies sometimes form, and these can be seen moving in the direction of rotation of the outer cylinder, but much more slowly. Figure 3-5 is taken from a report of one of these experiments, so that the patterns

Figure 3-5. Photographs illustrating three typical top-surface flow patterns of free thermal convection in a wall-heated rotating fluid annulus; inner wall at 16.3°C; outer wall at 25.8°C; working fluid water. Note: The thick white streak in the lower left quadrant of each picture is the outline of a wire well above the surface of the fluid. Reprinted by permission of the Royal Meterological Society, London, from R. Hide, "Some laboratory experiments on free thermal convection in a rotating fluid subject to a horizontal temperature gradient and their relation to the theory of the global atmospheric circulation," in G. A. Corby, ed., The Global Circulation of the Atmosphere, 1969.

that are observed can be seen.[3] Both the steady waves and the irregular patters in Fig. 3-5 look somewhat like diagrams of the upper atmosphere jet stream seen with the North Pole at the center. The jet stream, observed in this way, has lobes going alternately north and south in which air travels generally west to east, in the direction of earth's rotation. The jet stream moves in the middle latitudes of the hemisphere.

In the U.S., which is in the northern midlatitude zone, everything contributes toward a general west-to-east movement of the atmosphere. This is the region of prevailing westerlies, the jet stream moves west to east, and all major weather systems move west to east. This is why so many sulfur and nitrogen compounds from industrially polluted air seem to end up in rainfall over New England (eastern U.S.), as mentioned earlier. It is probably the jet stream that contributes most to the west-east motion of polluted air over the continental U.S.

[3] *Similar experiments can be observed at the end of the film "Planetary Circulation" produced by Modern Learning Aids in 1967.*

BIBLIOGRAPHY

General

"Inadvertent Climate Modification. Report of the Study of Man's Impact on Climate" (SMIC). The MIT Press, Cambridge, Massachusetts, 1971.

W. H. Matthews, W. W. Kellogg, and G. D. Robinson, Eds., "Man's Impact on the Climate." The MIT Press, Cambridge, Massachusetts, 1971.

W. J. Humphreys, "Physics of the Air," 3rd ed. McGraw-Hill, New York, 1940. Reprinted by Dover, New York, 1964.

G. J. Kukla and H. J. Kukla, Increased surface albedo in the northern hemisphere, *Science 183,* 709 (1974).

T. L. Brown, "Energy and the Environment." Charles E. Merill Publishing Co., Columbus, Ohio, 1971.

A. Turk, J. Turk, and J. T. Wittes, "Ecology, Pollution, Environment." W. B. Saunders Co., Philadelphia, Pennsylvania, 1972.

I. J. Winn (Ed.), "Basic Issues in Environment. Studies in Quiet Desperation." Charles E. Merill Publishing Co., Columbus, Ohio, 1972.

"Chemistry in the Environment, Readings from Scientific American." W. H. Freeman, San Francisco, California, 1973.

"U. S. Standard Atmosphere," 1962; and "U. S. Standard Atmosphere Supplements," 1966. U. S. Government Printing Office, Washington.

Evolution of the Atmosphere

K. A. Kvenvolden, Ed., "Geochemistry and the Origin of Life." Halsted Press, New York, 1975.

S. I. Rasool, D. M. Hunten, and W. M. Kaula, What the exploration of Mars tells us about Earth, *Physics Today,* p. 23, July 1977.

P. Enos, Photosynthesis and atmospheric oxygen, *Science 180,* 515 (1973).

P. E. Cloud, Jr., Atmospheric and hydrospheric evolution on the primitive earth, *Science 160,* 729 (1968).

R. T. Brinkmann, Dissociation of water vapor and evolution of oxygen in the terrestrial atmosphere, *J. Geophys. Research 74,* 5355 (1969).

L. Van Valen, The history and stability of atmospheric oxygen, *Science 171,* 439 (1971).

W. E. McGovern, The primitive Earth: Thermal models of the upper atmosphere for a methane-dominated environment, *J. Atmos. Sci. 26,* 623 (1969).

R. C. Robbins, L. A. Cavanaugh, and L. J. Sales, Analysis of ancient atmospheres, *J. Geophys. Res. 78,* 5341 (1973).

S. W. Fox and K. Dose, "Molecular Evolution and the Origin of Life," Chapter 3. W. H. Freeman, San Francisco, California, 1972.

Meteorology

S. Petterssen, "Introduction to Meteorology." McGraw=Hill, New York, 1941.

R. G. Barry and R. J. Chorley, "Atmosphere, Weather, and Climate." Holt, Rinehart and Winston, Inc., New York, 1970.

G. A. Corby (Ed.), "The Global Circulation of the Atmosphere." Royal Meteorological Society, London, 1969

E. Palmer and C. W. Newton, "Atmospheric Circulation Systems. Their Structure and Physical Interpretation." Academic Press, New York, 1969.

N. Calder, "The Weather Machine." Viking Press, New York, 1974.

W.S. Cleveland, B. Kleiner, J. E. McRae, and J.L. Warner, Photochemical air pollution: Transport from the New York City area into Connecticut and Massachusetts, *Science 191,* 179 (1976).

Pollution

S. F. Singer, (Ed.), "Global Effects of Environmental Pollution." Springer-Verlag, New York, 1970.

J.O. Frohliger and R. Kane, Precipitation: Its acidic nature, *Science 189,* 455 (1975).

Scientists puzzle over acid rain, *Chem. Eng. News,* June 9, 1975, pp. 19-20.

J.N. Gallowey, G.E. Likens, and E.S. Edgerton, Acid precipitation in the northeastern United States: pH and acidity, *Science 194,* 722 (1976).

G.E. Likens and F. H. Bormann, Acidity in rainwater: Has an explanation been presented? *Science 188,* 957 (1975).

G. E. Likens and F. H. Bormann, Acid rain: A serious regional environmental problem, *Science 184,* 1176 (1974).

G.E. Likens, Acid precipitation, *Chem. Eng. News,* November 22, 1976, p. 29.

N.M. Johnson, R.C. Reynolds, and G.E. Likens, Atmospheric sulfur: Its effects on the chemical weathering of New England, *Science 177,* 514 (1972).

P. V. Hobbs, H. Harrison, and E. Robinson, Atmospheric effects of pollutants, *Science 183,* 909 (1974).

R. K. Swartman, V. Ha, M. Julien, and D. J. Whitney, The solar era. Part 5--The pollution of our solar energy, *Mech. Eng.*, December 1972, pp. 23-26.

L. J. Battan, "The Unclean Sky. A Meteorologist Looks at Air Pollution." Doubleday, Garden City, New York, 1966.

E. Robinson and R. C. Robbins, "SRI Project PR-6755, Sources, Abundance, and Fate of Gaseous Atmospheric Pollutants." Stanford Research Institute, Menlo Park, California, 1968.

H. L. Green and W. R. Lane, "Particulate Clouds: Dusts, Smokes, and Mists," 2nd ed. E & F. N. Spon, Ltd., London, 1964.

A. C. Stern, H. C. Wohlers, R. W. Boubel, and W. P. Lowry, "Fundamentals of Air Pollution." Academic Press, New York, 1973.

H. S. Stoker and S. L. Seager, "Environmental Chemistry. Air and Water Pollution.: Scott, Foresman, and Co., Glenview, Illinois, 1972.

G.B. Lubkin, Fluorocarbons and the stratosphere, *Physics Today,* October, p. 34, 1975.

Atmospheric Components

J. London and J. Kelley, Global trends in total atmospheric ozone, *Science 184,* 987 (1974).

J.G. Zeikers and J. C. Ward, Methane formation in living trees: A microbial origin, *Science 184,* 1181 (1974).

R. E. Stoiber and A. Jepsen, Sulfur dioxide contributions to the atmosphere by volcanoes, *Science 182,* 577 (1973).

N. M. Johnson, R. C. Reynolds, and G. E. Likens, Atmospheric sulfur: Its effect on the chemical weathering of New England, *Science 177,* 514 (1972).

Isotopic study confirms CO sources, *Chem. Eng. News,* July 3, 1972, p. 2.

T. H. Maugh II, Carbon monoxide: Natural sources dwarf man's output, *Science 177,* 338 (1972).

H. W. Ellsaesser, Turbidity of the atmosphere: Source of its background variation with the season, *Science 176,* 814 (1972).

K. Heidel, Turbidity trends at Tuscon, Arizona, *Science 177,* 882 (1972).

H. U. Duetsch, The ozone distribution in the atmosphere, *Can. J. Chem. 52,* 1491 (1974).

A. L. Hammond and T. H. Maugh II, Stratospheric pollution: Multiple threats to earth's ozone, *Science 186,* 335 (1974).

4
ENERGY AND CLIMATE

4.1 ENERGY BALANCE OF THE EARTH

The discussion on the general circulation of the atmosphere in Chapter 3 presupposed that most of the heating of the earth's surface comes from solar radiation. The total solar energy reaching the surface of the earth each year is about 2×10^{21} kJ (5×10^{20} kcal). Heat generated by radioactive processes in the earth and conduction from the core contribute 8×10^{17} kJ (2×10^{17} kcal), and man's activities contribute about 4×10^{17} kJ (10^{17} kcal) per year. This means that less than 0.1% of the total energy reaching the earth's surface each year comes from processes other than direct solar radiation. For the purposes of this chapter, therefore, we may treat the energy input to the earth as if it all came from the sun.

The sun is almost a black body radiator (with superimposed line spectra); a perfect black body is one that absorbs all radiation hitting it and hence is also a perfect emitter. The energy density (energy per unit volume) per unit wavelength increment emitted by a black body is given by Planck's black body equation:

$$E_\lambda = 8\pi hc/\lambda^5[\exp(hc/\lambda kT) - 1] \qquad (4\text{-}1)$$

where k (=1.38×10^{-23} J/molecule · K) is Boltzmann's constant or the gas constant per molecule; h is Planck's constant, 6.63×10^{-34} J/sec; c is the speed of light in vacuum, 3.00×10^8 m/sec; λ is the wavelength of interest in meters; and T is the absolute temperature.

Since the sun is emitting radiation mostly from its surface, and since we shall also wish to consider black body radiation from the surface of the earth, we shall rewrite Eq. (4-1) in the form

$$I(\lambda) = C_1/\lambda^5[\exp(C_2/\lambda T) - 1] \qquad (4\text{-}2)$$

where I (λ) is the radiation intensity emitted by each square meter of surface of the black body at wavelength λ, T is the absolute temperature, $C_1 = 3.74 \times 10^{-16}$ W m^2, and $C_2 = 1.438 \times 10^{-2}$ m deg. The total intensity of radiation emitted by a black body at any temperature is given by the Stefan-Boltzmann law,

$$I = \sigma T^4 \tag{4-3}$$

where $\sigma = 5.672 \times 10^{-8}$ W/m^2 deg^4 $= 8.22 \times 10^{-11}$ cal/cm^2 min deg^4.

Equation (4-2) is plotted in Fig. 4-1 as a function of wavelength for two temperatures. It is seen that there is a wavelength of maximum emission λ_{max}, which shifts to lower wavelengths as the temperature is increased. Most commonly the wavelength of maximum black body emission is encountered in the infrared (for example, the 3300 K curve in Fig. 4-1 corresponds approximately to the output of a 200 W tungsten filament lamp). However, at 6000 K (roughly the black body temperature of the sun), λ_{max} is in the visible region (480 nm). Ninety percent of the solar radiation is in the visible

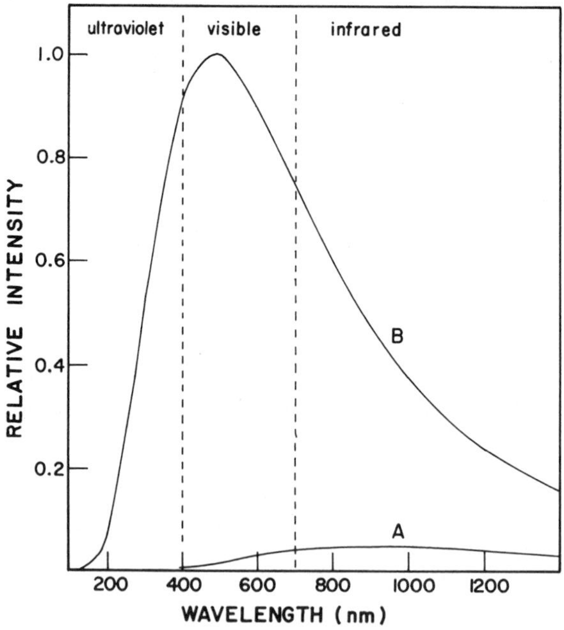

FIGURE 4-1. Distribution of energy from a black body radiator. (A) 3300 K, (B) 6000 K.

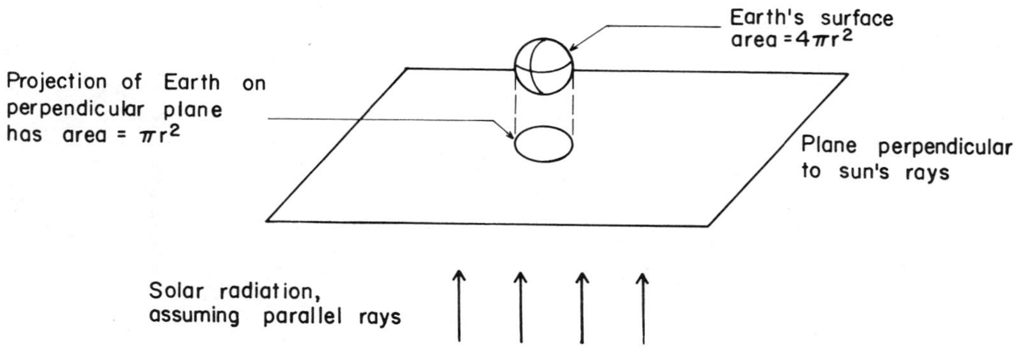

FIGURE 4-2. A comparison of earth's surface area with the area of earth's projection on a plane perpendicular to the sun's rays.

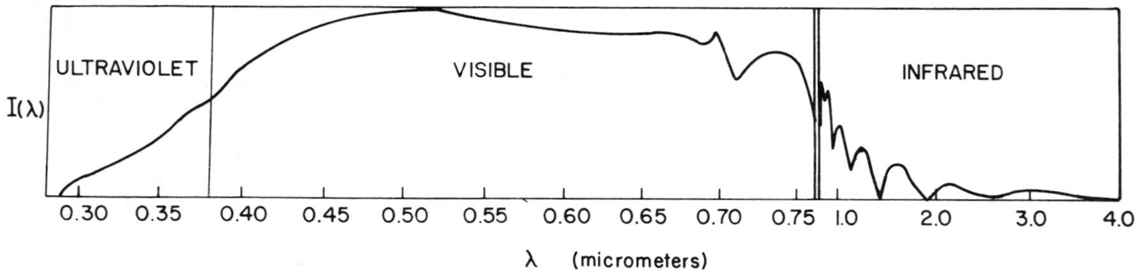

FIGURE 4-3. Intensity of solar radiation versus wavelength for radiation that reaches the surface of the earth.

and infrared, from 0.4 to 4.0 μm. This is just as well, since the radiation intensity of the sun varies somewhat with time in the ultraviolet region of the spectrum and becomes extremely variable in the X-ray region of the spectrum. The wavelength at which maximum emission of radiation occurs at any temperature is given by Wien's displacement law,

$$\lambda_m T = 2897.8 \ (\mu m \ K) \tag{4-4}$$

Solar radiation is emitted in all directions from the sun, and very little of this reaches earth. In fact, earth is so far away from the sun that it picks up only 2×10^{-9} of the total solar energy output. At a distance from the sun equal to the average radius of the earth's orbit, the solar energy passing any surface perpendicular to the solar radiation beam is 1.95 cal/cm^2 min or 1.36 kW/m^2; this is called the solar constant. Since the earth does not consist of a plane surface perpendicular to the path of the solar radiation and since the earth presents a hemispherical surface toward the sun, a recalculation of the solar constant must be made for radiation falling on this hemispherical surface. Actually, since we wish to use an average radiation intensity *averaged over the total surface of the earth,* we must compare the surface area of the whole earth, $4\pi r^2$, where r is the radius of the earth, with the area that the earth projects perpendicular to the sun's rays, πr^2 (see Fig. 4-2). Thus, the earth's surface has four times the area of the circle it projects on a plane perpendicular to the sun's rays, if these rays are assumed parallel at such large distances from the sun. Therefore, the solar radiation that comes in toward the earth's surface is 1/4 × the solar constant, or approximately 0.49 cal/cm^2 min, i.e., 0.34 kW/m^2, averaged over the whole surface. These numbers indicate the total amount of solar radiation coming toward the surface of the earth; some of this radiation is absorbed in the atmosphere and some is reflected.

The solar radiation that actually penetrates to the surface of the earth below the atmosphere no longer has the spectral distribution of black body radiation from a body at 6000 K (Fig. 4-3). Fig. 4-3 is taken from a chart issued by the General Electric Corporation; it shows that no solar radiation with wavelength below 0.28 μm (2800 Å) reaches the surface of the earth. We already knew this, from the discussion of absorption by oxygen in the thermosphere and by ozone and oxygen in the stratosphere. The absorption of solar infrared radiation has not been discussed previously, but a large portion of the sun's infrared radiation is absorbed by water vapor and carbon dioxide in the atmosphere. It can be seen in Fig. 4-3 that this absorption is complete at some wavelengths.

The absorption of solar infrared radiation by various constituents of the atmosphere is not very important because comparatively little of the solar radiation is in the infrared (Fig. 4-1). These absorptions become much more important when we consider radiation emitted by the earth. It should be fairly obvious that the earth is emitting radiation, because, on the average, the earth is in thermal equilibrium. It is in the path of 2×10^{21} kJ of solar radiation each year, and it is necessary that 2×10^{21} kJ per year leave the earth. If less energy than that leaves the earth, it will become hotter. Some of the solar energy, as we shall see, is immediately reflected, but some is absorbed and must be reemitted.

The average temperature T of the earth's surface, taken at any one time over the whole surface of the earth, is close to 14°C (287 K). Even though the temperature varies with time and place, the earth must radiate much like a black body at 14°C. Figure 4-4 shows the intensity of emission $I(\lambda)$ for black bodies at 250 and 300 K. The wavelength for maximum radiation intensity is about 10 μm in this temperature range; i.e., in the infrared region of the spectrum.

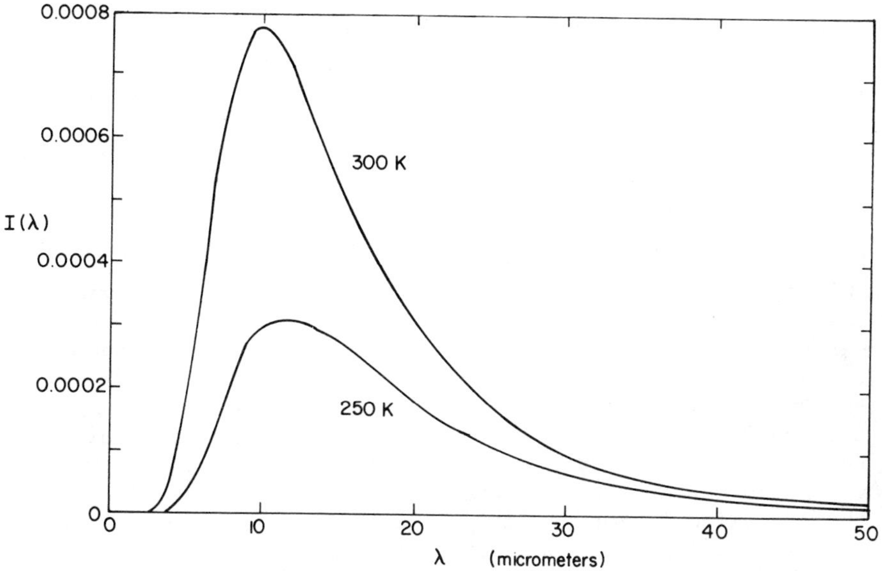

FIGURE 4-4. Radiation intensity versus wavelength for black bodies at 250 K and at 300 K. $I(\lambda)$ *has units of ergs/cm² sec.*

We should be able to find the average value of the earth's temperature by calculation, i.e., by balancing incoming and outgoing radiation at the earth's surface. We shall call this mean radiation temperature T_r. Let us assume that energy in = energy out, ignoring radioactivity and man's activities. We know from satellite measurements that the earth reflects about 30% of the total incoming solar radiation, so that only 70% is absorbed. Therefore, $I_{in} = I_{out} = 0.70 \times 0.49$ cal/cm^2 min. Using the Stefan-Boltzmann Law [Eq. (4-3)],

$$T_r = \left[\frac{0.7 \times 0.49 \text{ cal/cm}^2 \text{ min}}{8.22 \times 10^{-11} \text{ cal/cm}^2 \text{ min deg}^4}\right]^{1/4}$$

and $T_r = 254$ K = $-19°$C. Therefore, in order to satisfy the law of conservation of energy, earth *must* be radiating energy into space at an average of $-19°$C. Therefore, the earth *cannot* be radiating energy out into space directly from its surface, at least on the average. Average temperatures in the neighborhood of $-19°$C are found in only two places from which radiation could logically occur: near the tropopause and in the region of the mesopause in the upper atmosphere. Since radiation cannot occur in the atmosphere unless molecules are present to radiate, it is probable that most of the earth's radiation into space occurs from the atmosphere near the tropopause, probably the upper regions of the troposphere. This leaves us with two main questions: Why does the earth radiate to space from the atmosphere and not from the surface; and why is the surface of the earth hotter than it should be, considering the energy balance calculations? That is, what traps heat between the surface of the earth and the tropopause?

Both of these questions have to do with the absorption of infrared radiation by atmospheric water vapor, carbon dioxide, and small amounts of other gases. Figure 4-4 shows that the earth's radiation is all in the infrared, whether we consider radiation from the surface (black body at 14°C = 287 K) or from the atmosphere (black body at $-19°$C = 254 K). The earth's surface is radiating at its average temperature, and this radiation is absorbed almost completely by atmospheric water vapor and carbon dioxide except in a few wavelength regions, the so-called "atmospheric windows." The main atmospheric window is from 8.5-11 μm, with partially opaque "windows" from 7-8.5 μm and from 11-14 μm. The region below 7 and above 14 μm is completely opaque to infrared radiation. When the sky is clear, terrestrial radiation in the wavelength region of the "windows" escapes directly to space, outside the atmosphere. Clouds, when present, reflect this radiation back to the earth's surface and it does not escape. Hence, on clear nights, especially in the winter, we have the phenomenon of "radiation cooling" of the earth's surface through the atmospheric "windows."

The infrared radiation that is absorbed in the troposphere and in the stratosphere would warm these regions of the atmosphere substantially if energy were not also radiated from these regions. This is the energy radiated at an average temperature of -19°C (this average includes the higher temperatures at which radiation from the earth's surface escapes through the atmospheric windows) that leaves the earth. Thus, the atmosphere intercepts infrared radiation from the earth's surface and then reradiates the energy both back toward the earth's surface, thus warming it, and upward to space, thus cooling the earth so that the constant solar radiation does not slowly increase the earth's temperature. The earth's surface can therefore be much warmer than the average radiation temperature of the earth. This is usually called the "greenhouse effect" because a related phenomenon is observed in a closed greenhouse. The glass or plastic windows of the greenhouse allow the visible solar radiation to enter the inside, but the outgoing infrared radiation from the floor is absorbed or reflected by the same windows. In contrast to the "greenhouse effect" in the atmosphere, however, energy retention in the greenhouse is caused mostly by lack of convection, i.e., lack of mixing of the interior air with the surrounding atmosphere. The same effect occurs in a closed automobile that is left out in the sun. Both the greenhouse and the car interiors may thus become considerably warmer than the temperature of the surrounding atmosphere.

A greenhouse coefficient may be defined as the ratio of the surface temperature to the absolute radiation temperature of the whole system. For the earth, this coefficient is 287 K/254 K = 1.13, while for the planet Venus it appears to be 700 K/265 K = 2.60! Venus thus has a much larger greenhouse effect than earth; this is presently believed to be caused by sulfuric acid in the Venusian atmosphere. Earth has the possibility of varying its greenhouse effect by means of changes in the carbon dioxide content of the atmosphere, since this compound is a prime absorber of the infrared radiation emitted from the earth's surface.

The mean temperature of the earth's surface is certainly affected by the magnitude of the greenhouse effect, but it is also affected by the percentage of incoming solar radiation that is reflected back out to space, that is, by the earth's albedo. It has already been stated that approximately 30% of the incoming radiation is immediately reflected. This is an average value; Table 4-1 shows the albedo of various major features on earth. Major changes in cloud cover, snow and ice, field and forest, and so on, would change the average albedo of the earth. For example, a decrease in snow and ice cover would decrease the earth's albedo, thus leading to absorption of more solar radiation, a higher surface temperature for the earth.

Other factors concerning the energy balance of the earth's surface are also important. We have just seen that 30% of the 0.49 cal/cm^2 min that come toward the earth's surface from the sun is immediately reflected with no change. Calculations indicate that water vapor and clouds in the troposphere absorb about 15% of the remaining solar radiation, while ozone and other ultraviolet absorbers remove an additional 3% of the total. This leaves about 52% of the incoming radiation to be absorbed by the earth's surface, about 0.25 cal/cm^2 min. Radiation from the earth's surface (at 14°C) is about 0.55 cal/cm^2 min, which is over twice the incoming solar energy. Other types of energy transfer also occur, as shown in Table 4-2. Quantities with positive signs in Table 4-2 are energies absorbed by the earth's surface, while quantities with negative signs are energies released by the earth's surface. The contribution from back radiation from the atmosphere may be noted.

TABLE 4-1

Albedo of Various Atmospheric and Surface Features of the Earth

Albedo (% reflected)	*Reflected from*
30	*Total: Earth and atmosphere*
50 - 80	*Clouds*
50 - 80	*Snow and ice*
28	*Deserts*
15 - 30	*Grass*
20 - 25	*Dry, plowed fields*
18 - 24	*Tropical forests*
18	*Deciduous forests in summer*
14 - 18	*Savannahs and semideserts*
7 - 23	*Oceans*
7 - 20	*Bare ground*
3 - 15	*Green fields*

TABLE 4-2

Energy Balance of the Earth's Surface

Type of energy	*Quantity (cal/cm^2 min)*
Solar radiation (absorbed)	*+0.25*
Long wavelength radiation (emitted)	*-0.55*
Back radiation from atmosphere (absorbed)	*+0.45*
Heat transfer to atmosphere by conduction	*-0.05*
Heat transfer to atmosphere by evaporation	*-0.10*
Net energy absorbed	*0*

4.2 CLIMATE HISTORY OF THE EARTH

Before trying to decide whether man is actually succeeding in changing the earth's climate at present, we must consider what sorts of climate changes occurred before man evolved, and what climate changes occurred before the major industrialization of the twentieth century. It is, unfortunately, beyond the scope of this chapter to discuss the various methods of calculating average temperatures at different times in the past; these methods are discussed in some of the references at the end of the chapter.

Figure 4-5 shows a plot of mean temperature between 40°N and 90°N versus a nonlinear function of the time from 500,000,000 B.C. to 1950 A.D. The time scale is arranged so that recent years, for which more reliable data are available, have been emphasized. Very recent times are shown in more detail in Fig. 4-6. Glacial periods occurred between 10,000 and 100,000 B.C. (Wisconsin glacial), 300,000 and 400,000 B.C. (Illinoian glacial), around 600,000 B.C. (Kansan glacial), and just before 1,000,000 B.C. (Nebraskan glacial). It looks, from Fig. 4-5, as though there have also been major glacial episodes around 200,000 B.C. and 500,000,000 B.C. In general, the average temperatures before 1,000,000 B.C. were warmer than those that have occurred since that time. Since 1,000,000 B.C., the interglacial periods, although warmer than either the glacial periods or the present, were much cooler than the eras before that time. Notice that the difference in average temperatures between glacial and interglacial periods was only about 5°C, that is, comparatively small differences in average temperature have large effects on the total climate. Notice also that variations up to 2°C have

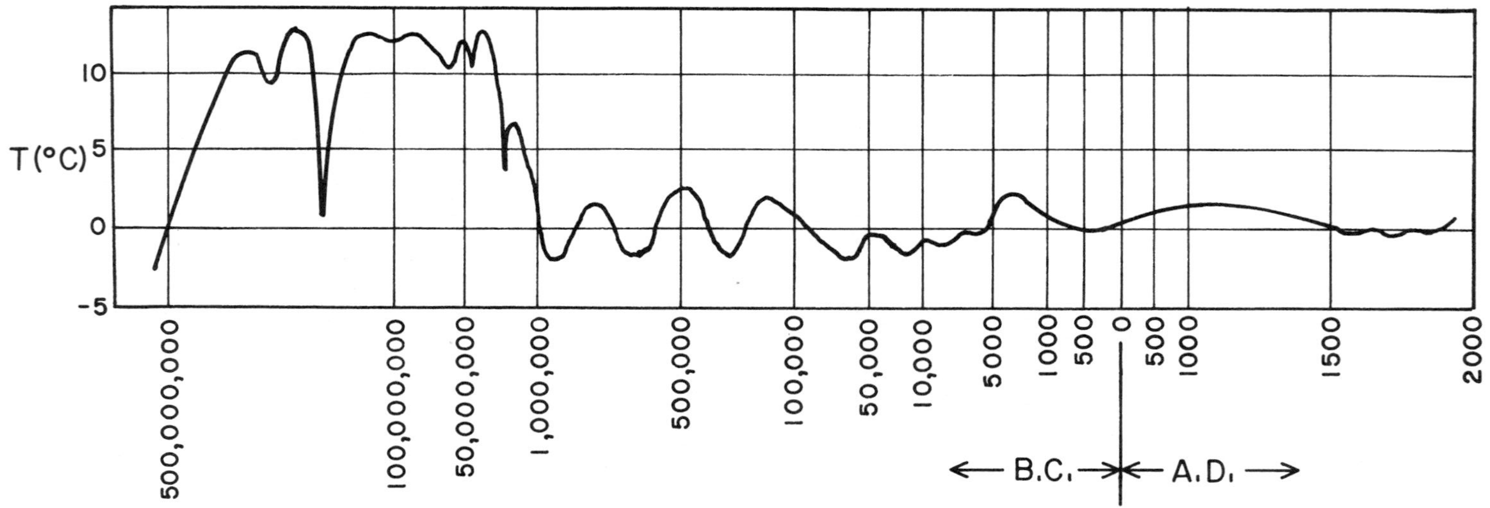

FIGURE 4-5. Temperature variations in the northern hemisphere (40°N - 90°N) as a function of time. Reprinted, by permission of Wiley, New York, from J. E. Oliver, "Climate and man's environment," 1973.

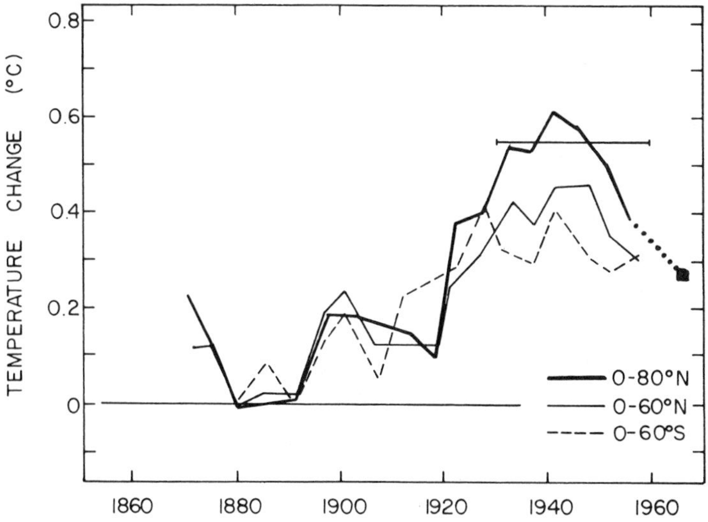

FIGURE 4-6. Mean annual temperature for various latitude bands, 1870 to 1967. Reprinted, by permission of the MIT Press, from "Inadvertent Climate Modification, Report of the Study of Man's Impact on Climate (SMIC), 1971.

occurred since the year 1000 A.D. This means that average temperatures in the northern hemisphere have varied appreciably before man could have had any effect.

Figure 4-5 deals with very long-term variations in average temperature. If we are interested in climate changes possibly attributable to man's intervention, we shall have to deal with much shorter-term fluctuations. Let us, however, start with the period from 4000 to 3000 B.C. where Fig. 4-5 shows a temperature maximum, the so-called "climatic optimum" (optimum for Scandinavia, at least, where the term probably originated). During this time many glaciers had melted, and sea level was much higher than it is today. Major civilizations existed in Egypt. Between 3000 and 750 B.C the world generally become cooler, glaciers advanced, and sea level dropped. Canals were constructed in Egypt to take care of problems associated with the general drop in water levels. Another warming trend culminated in the period 800-1200 A.D. the "little climatic optimum," the warmest climate in several thousand years in the northern hemiphere. Glaciers retreated so far that the Vikings were able to travel to and settle in Greenland and Iceland. Southern Greenland had average temperatures 2-4°C above present temperatures while Europe had about 1°C higher average temperatures than at present. Between 1200 and 1450 A.D., there was a cooling trend that made it harder for ships to travel between Greenland and Iceland and made Greenland cool enough so that the Viking settlements there were abandoned. A warming trend between 1450 and 1550 A.D. was followed by the "little ice age" which did not end in some areas until 1850 A.D. Glaciers advanced and winters were very cold in this period. Since 1850, there has been a distinct warming trend, with precipitous retreats of glaciers, which may have ended around 1940. The retreat of the glaciers is easy enough to document, but these very short-term recent temperature trends are very hard to evaluate. First of all, while glaciers are retreating in one part of the world, they often are advancing elsewhere. What is the average trend? Furthermore, climate and temperatures are variable in any one place from year to year, and it is not easy to find a general trend among fluctuations. The problem in the evaluation of short-term trends consists of both place-to-place and year-to-year variability in climate. The place-to-place variability can be taken care of by averaging temperatures over a large enough area; that is why we see averages over half the Northern Hemisphere, as in Fig. 4-5. The year-to-year variability can be taken care of by using so-called moving averages over 5-year, 20-year, or 30-year time spans. A 20-year temperature average is taken from 1870 to 1890, for example, and attributed to 1880, the central year of the average. Using 30-year moving averages, the temperature increased 0.2°C in the tropics and 0.4°C in the north temperate zone between the period 1890-1919 and the period 1920-1949. These averages, however, still do not tell the whole story; in this period, they represent an increase in the number of warm months during each year, not an overall increase in actual temperature.

Figure 4-6 shows area-average mean annual temperatures for various latitude bands. These mean annual temperatures show a warming trend from 1880 to 1940, followed by a cooling trend that seems to be continuing at present in the northern hemisphere. Since 1964, however, there has been a warming trend in the southern hemisphere as shown by 5-year moving averages in various latitude bands. The warming trend seems to be greatest in the higher latitudes, in the 65°S-90°S latitude band, as contrasted with the 45°S-60°S latitude band. Note that the variations shown in Fig. 4-6 are never larger than 0.6°C, but, as we have seen, it takes only about 5°C to change from a glacial to an interglacial period. The greatest variability in average temperature appears generally to be near the poles, not near the equator. In Fig. 4-6, the average that includes the northernmost latitudes, 60°N-30°N, exhibits the greastest variability in temperature.

4.3 CAUSES OF GLOBAL CLIMATE CHANGES

Climate is mainly effected by wind patterns over the earth's surface, and these changes in wind patterns are generally caused by changes in the earth's energy balance. The average surface temperature can be calculated in principle by employing the concepts discussed in Section 4.1. The starting point, as in Section 4.1, is $I_{in} = I_{out}$, where $I_{in} = (S/4)(1-\alpha)$ and $I_{out} = I_0 - \Delta I$. In these expressions S is the solar constant; α is the earth's albedo; $I_0 = \sigma T^4$, the intensity of radiation emitted at the earth's surface; T is the mean surface temperature of the earth; and ΔI is a term involving the greenhouse effect. Substituting for I_0 and solving for T,

$$T = [S(1 - \alpha) + (\Delta I/\sigma)]^{1/4} \qquad (4\text{-}5)$$

If we take T, the average surface temperature of the earth, as an indicator of climate, then Eq. (4-5) allows us to determine the main factors that affect climate and even the magnitudes of possible effects. It can be seen that the solar constant, the earth's albedo, and the greenhouse effect all help to determine T.

It has always been tempting to attribute major climatic variations, like those between glacial and interglacial periods, to variations in the solar constant. Evidence for this has been extremely inconclusive. The sun is quite variable in its emission of X rays and even some ultraviolet radiation, but the emission of visible and infrared radiation seems to be quite steady. Some extreme variations in sunspot activity may, however, be correlated with peculiarities of climate. The coldest part of the "little ice age" appears to be concurrent with a 70-year minimum in sunspot activity during the late seventeenth and early eighteenth centuries, the so-called "Maunder minimum."[1] There is some evidence that the solar constant decreases by about 2% in a period of very low sunspot activity.[2]

It can be shown, using Eq. (4-5), that an 8% change in the solar constant could change the average surface temperature of the earth by 3°C. The values of S, α, and σ have been given in this chapter, and the present value of ΔI can

[1] *J. A. Eddy, Science 192, 1189 (1976).*
[2] *S. H. Schneider and C. Mass, Science 190, 741 (1975).*

be calculated from σ, the measured mean surface temperature of the earth T, and the mean radiation temperature of the earth T_r, as found in Section 4.1:

$$\Delta I = \sigma \left[T^4 - T_r^4 \right] = 0.215 \qquad (4\text{-}6)$$

Attempts are now being made to monitor the solar constant, so that its effect on climate can be properly evaluated in the future.

Small changes in the greenhouse effect ΔI can have appreciable effects on climate. It is known that there has been a significant increase in the carbon dioxide content of the atmosphere since the beginning of the industrial age. In the past 40 years, the carbon dioxide content of the atmosphere has increased by about 10%. Careful monitoring of carbon dioxide has been recent; Fig. 4-7 shows data obtained at Mauna Loa, Hawaii, between 1958 and 1970. Mauna Loa is situated at a latitude approximately 19°N. Note that there are seasonal variations in the carbon dioxide content of the atmosphere over Mauna Loa; less CO_2 is found in summer, the growing season in the Northern Hemisphere. Hawaii may seem rather far south to feel the effects of this; more northerly areas like Scandinavia and northern Alaska show seasonal variations that have twice the amplitude of those over Mauna Loa. Figure 4-8 shows changes of mean annual carbon dioxide concentrations in other locations; the concentrations are a bit different, but the trends are the same. It has been fashionable, lately, to attribute the worldwide warming trend from the end of the nineteenth century to about 1940 to the increasing carbon dioxide content of the atmosphere which increased the greenhouse effect. Increased burning of fossil fuels is held responsible for this increase in carbon dioxide.

In the period 1958-1968, it has been calculated that enough fossil fuels were burned to increase the carbon dioxide content of the atmosphere by 1.24 parts per million (ppm) per year. The actual increase in CO_2 content was 0.64 ppm/year. This is an indication of the enormous buffering effect of the earth's oceans on the CO_2 concentration in the atmosphere. This buffering effect was discussed in Chapter 3 and is considered further in Chapter 14. It is somewhat puzzling to note that the CO_2 content of the atmosphere increased at a slightly lower rate between 1962 and 1965, 0.46 ppm/year, than between 1958 and 1968 as a whole, although there was no decrease or even leveling off of the use of fossil fuels in those three years. Since 1969, the increase in carbon dioxide content of the atmosphere has accelerated to over 1 ppm/year. A projection has been made for the year 2000 A.D.

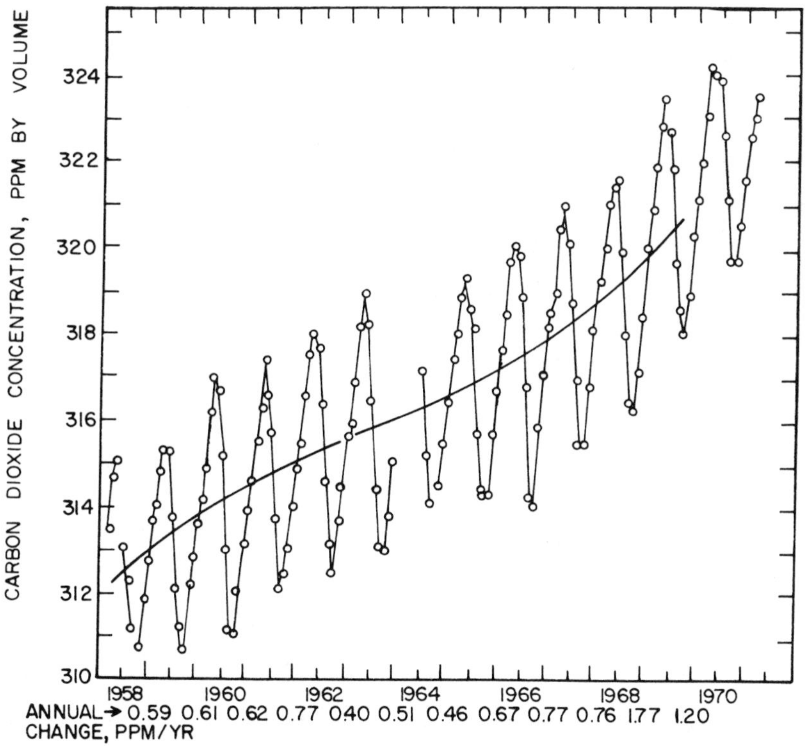

FIGURE 4-7. Mean monthly values of CO_2 concentration at Mauna Loa, Hawaii, from 1958 to 1971. Reprinted by permission of the MIT Press, from "Inadvertent Climate Modification, Report of the Study of Man's Impact on Climate (SMIC), 1971."

assuming increased consumption of fossil fuels each year with 50% of the resulting CO_2 remaining in the atmosphere, that predicts 375 ppm CO_2 in the atmosphere by the year 2000. This value, if it is in fact realized, should significantly increase the greenhouse effect in the earth's atmosphere.

Changes in the earth's albedo may also be implicated in climate changes. Although large-scale cultivation, irrigation, damming of rivers to form lakes, and cutting down of forests all result in changes in the earth's albedo, these causes have heretofore been considered minor. Volcanic and other dusts in the atmosphere have been most investigated.

Figure 4-9 shows an index of volcanic activity on earth between the years 1500 and the present. *Some* of this volcanic activity was followed by worldwide cooling. For example, the "little ice age," 1550 to 1850 A.D., occurred during a

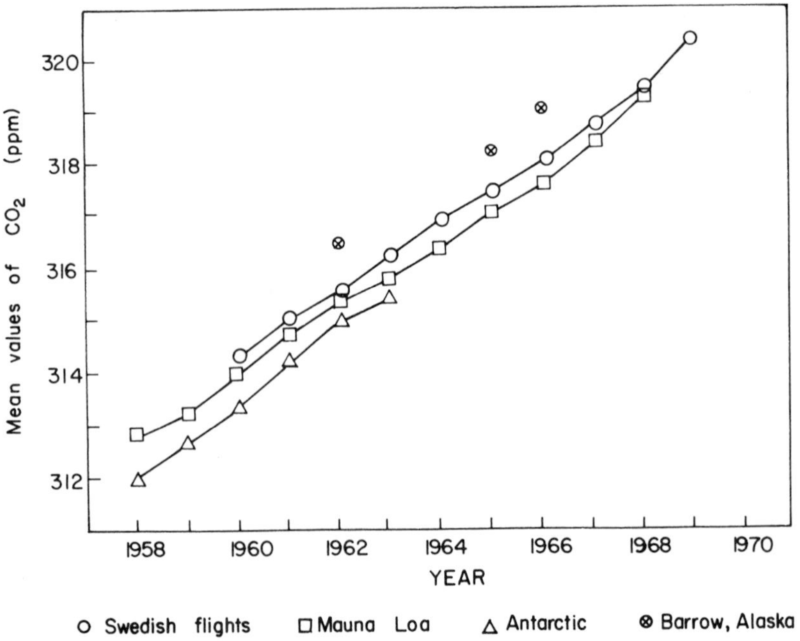

FIGURE 4-8. Annual mean values of CO_2 at various locations. Reprinted, by permission of the MIT Press, from "Man's Impact on Climate," W. H. Matthews, W. W. Kellogg, and G. D. Robinson, Eds., 1971.

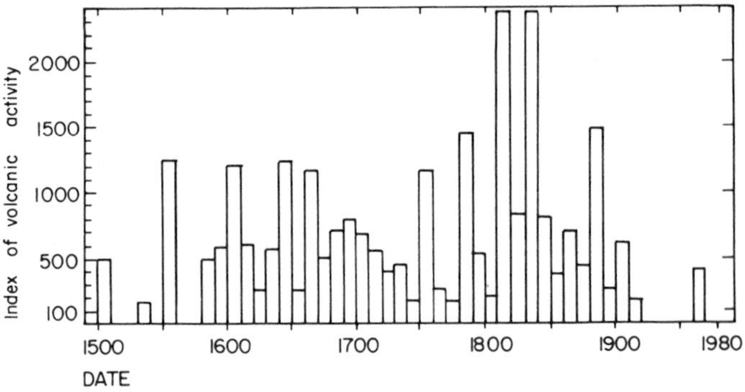

FIGURE 4-9. An index of volcanic activity, in arbitrary units, that is proportional to the relative amount of material injected into the stratosphere in various 10-year periods. Reprinted, by permission of the MIT Press, from "Inadvertent Climate Modification Report of the Study of Man's Impact on Climate (SMIC), 1971.)

period of fairly high volcanic activity as well as low sunspot activity. Some of the highest peaks on Fig. 4-9 are actually tremendous single volcanic explosions that were followed by noticeably cool years. The eruption of Asama in 1783 in Japan (the largest activity peak between 1700 and 1800 in Fig. 4-9) was followed by three cool years, 1783-1785. The year 1816, called the year without summer (in New England, snow fell in June and there was frost in July), followed the eruption of Tamboro, Sumbawa, in the Dutch East Indies in 1815. This eruption killed 5600 people and darkened the sky for days as far as 300 mi away.

We have seen earlier in this chapter that modern man is contributing a large amount of particulate matter to the atmosphere. If this particulate matter, like that emitted by volcanoes, can contribute to a cooling of the earth, presumably by increasing the earth's albedo, then we may have explained the earth's cooling trend from 1940 to 1965, and the further cooling of the northern hemisphere. At any rate, this trend has been attributed to an emission of particulate matter by man's activities that has been great enough to reverse the effects from his simultaneous CO_2 emissions since that time. Whether these interpretations are true is a moot point. We have definitely been increasing both the carbon dioxide and the particle content of the atmosphere and these changes must have some influence on the greenhouse effect and the albedo which have a further effect on climate. It is possible that man's efforts are simply superimposed on natural trends that we do not yet understand. If so, we may be lucky. It could also be that the whole set of man-made processes is self-limiting, at least as far as climate is concerned. For example, suppose that an increase in CO_2 content of the atmosphere caused an increase in temperature because of an increased greenhouse effect, which then caused an increase in evaporation of water from the oceans to form more cloud cover, which then reflected more of the sun's radiation to lower the temperature back to its original value. It is an interesting exercise to dream up these self-limiting processes. The earth-atmosphere-ocean system is so complex that it is easy to imagine such self-limiting or even oscillating processes, but very difficult to develop models that produce results in which one has confidence. Since there are so many unknown factors, it would probably be well if man could leave the composition of the atmosphere unchanged!

4.4 SMALL-SCALE CLIMATE

As we have seen, it is very difficult to ascertain what effects man's activities may have had or will have on large-scale climate. However, man's effects on small-scale climate are well known and worth discussing for that reason. Every town, every reservoir, every plowed field changes the local climate to some extent. In this brief discussion of local climate, we shall confine ourselves to a consideration of urban climates, as distinguished from the surrounding countryside.

On the average, urban climates are warmer and have more precipitation than the surrounding countryside. Table 4-3 shows a comparison of an average city with the surrounding countryside for a number of climate elements. Note that a city is dirtier, cloudier, foggier, has more drizzly rain (precipitation days less than 5 mm total), more snow, less sunshine, less wind, and higher temperatures than the surrounding countryside. Wind speed is lowered in the city, except near very tall buildings, where appreciably higher wind speeds have been found on occasion.

The differences in temperature between a city and its surroundings can be much more variable than the Table 4-3 averages might imply. In London on May 14, 1959, for example, the minimum temperature in the city's center was 52°F (11°C) and 40°F (4°C) at its edges. Taking mean annual temperatures for London, however, in the period 1931-1960, we obtain 11.0°C (51.8°F) at the city's center, 10.3°C (50.5°F) in the suburbs, and 9.6°C (49.2°F) in the adjacent countryside.

Urban heat is generally assumed to arise from three causes. The most obvious source is the heat from combustion and general energy use in the city. A major cause of the warmer city nights is the release of heat stored in structures, sidewalks, streets, etc., in the day time. A minor contributor to the warmth of cities is the back reflection and reradiation of heat from pollutants in the atmosphere above the city.

In the absence of winds, the pollutants in the atmosphere above a city often occupy a so-called "urban dome" (Fig. 4-10), a sheath of air around the city in which pollutants are reasonably well mixed and which may be situated under a temperature inversion. The edge of this urban dome can often be distinguished from airplanes or from mountains near the city. Looking down on the urban dome of Los Angeles is something like observing a large pot of yellow pea soup! Urban domes have varying heights depending on conditions; an urban dome above a subarctic city in the

TABLE 4-3
Microclimate of Urban Areas Compared with Surrounding Countryside[a]

Climate element	Comparison with countryside
Particulate matter	10 times more
Emitted gases	5-25 times more
Cloud cover	5-10% more
Fog, winter	2 times more
Fog, summer	30% more
Total precipitation	5-10% more
Days with less than 5 mm	10% more
Snowfall	5% more
Relative humidity, winter	2% less
Relative humidity, summer	8% less
UV radiation, winter	30% less
UV radiation, summer	5% less
Sunshine duration	5-15% less
Annual mean temperature	0.5° to 1.0°C higher
Winter minimum temperature (average)	1° to 2°C higher
Annual mean wind speed	20 to 30% less

[a]*Reprinted from "Inadvertant Climate Modification, Report of the Study of Man's Impact on Climate (SMIC)," by permission of the MIT Press, Cambridge, Massachusetts, 1971.*

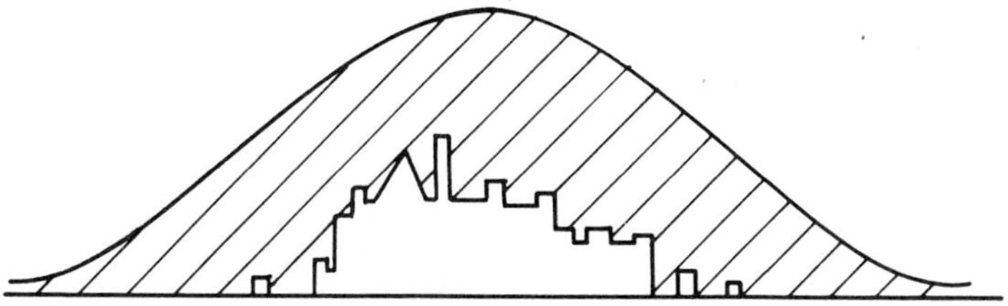

FIGURE 4-10. The urban dome.

winter may be only 100 m thick, while the dome over an oasis in the subtropics may be several kilometers thick. As we have noted, the urban dome tends to be warm; heat from combustion may supply up to 200 W/m^2, as in Moscow in the winter. When the wind blows, the urban dome becomes an urban plume, moving heat and pollutants downwind (Fig. 4-11).

The existence of a visible urban dome implies a different atmospheric composition over the city than in the surrounding country. The air over a city contains smoke, dust, SO_2, and so on, in much higher concentrations than country air. The particles in the air above the city increase the albedo above the city, thus reducing the intensity of the incoming solar radiation. In some cities, on bad days, the intensity of the incoming solar radiation may be reduced up to 50%. This means that cities tend to have cooler days than the surrounding country; these cooler days do not, however, make up for the warmer nights. The added particulate matter over cities also increases the frequency of fogs since the particles can act as condensation nuclei for water vapor.

The lower wind velocity in cities probably is caused by many obstacles (buildings) that are in the path of any wind in a city. The lack of standing water and the fast drainage of precipitation into storm drains result in a decrease of local evaporation and thus in the observed decrease of humidity in cities. However, when a city contains many cooling towers and there is generally a high water vapor output from burning of fossil fuels and industry, this, together with the extra condensation nuclei produced by the city, may cause an increased frequency of showers and thunderstorms downwind from the city and a weak drizzle within the city.

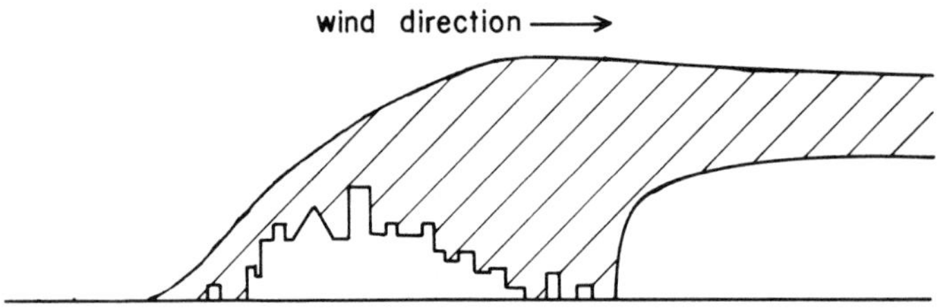

FIGURE 4-11. The urban plume.

4.5 TEMPERATURE INVERSIONS

When pollutants are trapped for a long time over a city, it is usually because of a temperature inversion above the city. This may result in accumulations that can reach lethal proportions, at least for the sick and elderly. Temperature inversions occur when air at some elevation in the troposphere ceases to decrease smoothly in temperature with increasing altitude as is normal near the surface of the earth (see Fig. 3-1). At a temperature inversion, the air temperature increases with increasing altitude for a distance; at higher altitudes, the normal decrease in air temperature with increasing altitude begins again and continues up to the tropopause. Temperature inversions are diagrammed on Fig. 4-12 and 4-13, to be explained below.

In order to discuss why a temperature inversion traps pollutants, it is necessary to know what happens in the absence of such an inversion. In the daytime, the ground is heated by solar radiation, and the air near the ground is heated by conduction from the ground. The air, if we assume that it is an ideal gas, obeys the equation of state for ideal gases:

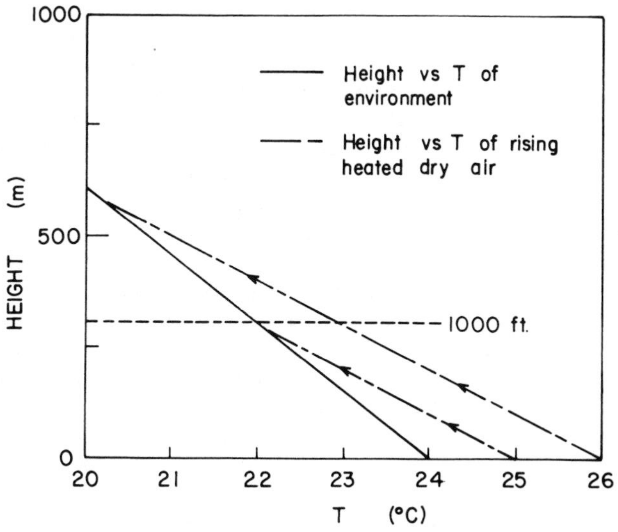

FIGURE 4-12. The fate of dry air heated by the earth's surface when the temperature lapse rate of the environment is normal.

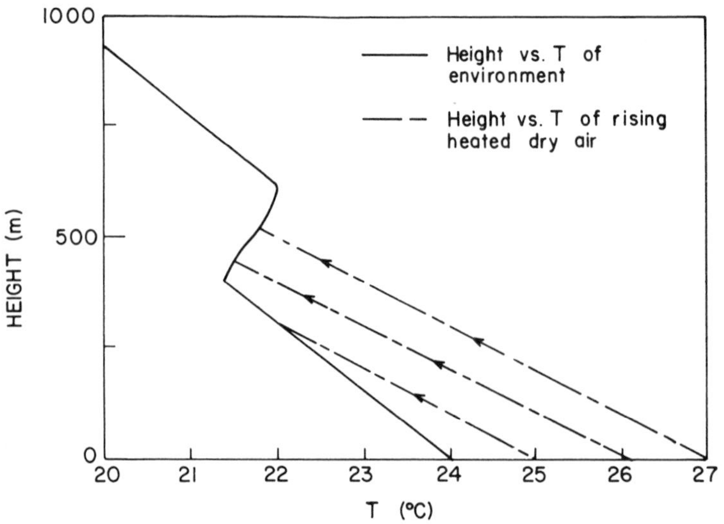

FIGURE 4-13. How a temperature inversion traps heated air and its associated pollutants.

$$PV = nRT \tag{4-7}$$

where P is the pressure, V is the volume of n moles of gas, R is the gas constant, and T is the absolute temperature. For our heated air, the pressure P of the atmosphere above it remains constant, but T increases. If we consider a "packet" of n moles of this air, and rewrite Eq. (4-7) as $V = nRT/P$, then we see that V must increase, and the density, n/V, decrease on heating. Consequently, this packet of air will rise until it reaches a place in the troposphere where the surrounding air has the same density; that is, the same temperature and pressure. Both the temperature and pressure of the rising air packet change continuously as it moves to regions of lower pressure and therefore expands. The expansion can be assumed to be adiabatic; no heat flows into or out of the packet during the process. Derivations of the equations for temperature and pressure changes in reversible adiabatic expansion of an ideal gas are given in most physical chemistry and many general chemistry textbooks. The result is

$$P_1/P_2 = \left[T_1/T_2\right]^{C_p/R} \tag{4-8}$$

where C_p is the heat capacity at constant pressure of the gas, and subscripts 1 and 2 refer to properties of the gas before and after expansion, respectively. Qualitatively, this expression shows that if the gas is expanding, that is, if $P_2 < P_1$, *then* $T_2 < T_1$, and the temperature of the gas will drop. This equation, together with the known decrease of pressure with height, allows calculation of the adiabatic lapse rate of dry air. This rate is a decrease of 1°C for each 100 m, or 5.5°F for each 1000 ft of altitude. The rising air cools faster than the temperature lapse rate of the environment, which involves a decrease of only 0.65°C for each 100 m or 3.5°F for each 1000 ft of elevation. Rising dry air will eventually reach a height where temperature and, therefore, the density matches that of the surrounding air. If the rising air contains a lot of moisture that condenses as the air rises and becomes cooler, then heat is given off during this condensation and the air cools more slowly with height, i.e., its temperature lapse rate is less than that of dry air. This moist air will therefore rise higher than dry air and will probably form clouds as it rises.

These conclusions are illustrated in Fig. 4-12. Fig. 4-12 shows what happens to heated dry air when the environment (the troposphere) is normal. The solid line shows the temperature of the lower atmosphere when the temperature of the surface of the earth is 24°C and when there are no temperature inversions in the atmosphere. If the sun is shining, the surface of the earth absorbs heat and becomes warmer; this will then heat the air just above it by conduction. The two dashed lines in Fig. 4-12 refer to two different "parcels" of air heated by the surface of the earth, to 25°C and 26°C, respectively. Both parcels of air will begin to rise as discussed above; they rise with the temperature lapse rate of rising dry air. Since the temperatures of these two parcels of air are decreasing faster, 1°C/100 m, than the temperature of the air through which they are rising, 0.65°C/100 m, each parcel will eventually reach a height at which its temperature equals that of the air already present. Its pressure and density are then also equal to that of the surrounding atmosphere, and the parcel will stop rising. Fig. 4-12 shows that the parcel that was heated to 25°C will rise to a height just below 300 m, while the parcel that was heated to 26°C by the surface of the earth will rise to a height just below 600 m. Hotter air rises to a higher altitude.

If a temperature inversion exists in the atmosphere at an altitude below that to which the heated air would normally rise, then the heated air may be trapped below the inversion along with any pollutants that were present in the heated air. Fig. 4-13 shows how heated air may be trapped when there is a temperature inversion in the troposphere. In this figure, the solid line that shows the variation of atmospheric temperature with altitude is not linear; it has a region in which temperature increases with increasing altitude between approximately 400 and 600 m altitude. This region of increasing temperature is known as a temperature inversion; above and below this region the air exhibits its normal lapse rate of 0.65°C/100 m elevation. If the sun shines and the earth's surface is heated, the resulting ground-level heated air rises as shown by the dashed lines in Fig. 4-13. Three parcels of air are shown, one heated to 25°C, the second to 26°C, and the third to 27°C. The parcel of air heated to 25°C rises to the same altitude, just below 300 m, as the similar parcel in Fig. 4-12; this happens because the temperature inversion occurs above this altitude and can affect only air that is attempting to rise through the inversion. Both the 26°C and the 27°C parcels of air are trapped by the inversion; each reaches an altitude at which its temperature equals that of the air already present within the temperature inversion zone. The 26°C parcel of air rises to less than 500 m, considerably lower than the 600 m it would ordinarily reach (Fig. 4-12).

These temperature inversions that trap rising air are naturally occurring phenomena that may have two different causes, i.e., they are of two different types, radiation inversions and subsidence inversions explained in the following paragraphs. London tends to have radiation inversions in the winter; one of these trapped a highly polluted fog in 1952, producing the notorious London killer fog, which was considered responsible for over 3000 deaths. Los Angeles tends to have subsidence inversions that trap pollutants in the summer and fall.

A radiation inversion generally occurs at night, when air near the ground is cooled by conduction from the cooled surface, especially on a clear night when radiation cooling occurs. Higher up in the troposphere, the air is not cooled appreciably, so, in the absence of winds, there is colder air underneath warmer air, an inversion by definition.

A subsidence inversion occurs in a region where high altitude air is pushed down nearer the ground because of topographical features in the path of prevailing winds or weather systems. This air is being pushed to regions of higher pressure and therefore is being compressed adiabatically. The same equation that held for rising air also holds for this descending air, but now $P_2 > P_1$ and $T_2 > T_1$, i.e., the air warms up as it nears the ground. Figure 4-14 shows how this subsiding air creates an inversion; again horizontal winds must be absent for the inversion to be reasonably stable. The two mechanisms for forming an inversion have exactly the same results, trapping of heated air and any pollutants that may be present.

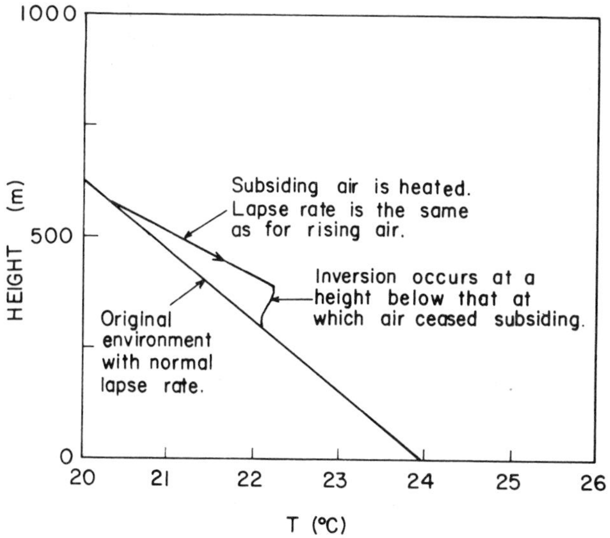

FIGURE 4-14. Mechanism by which a subsidence inversion occurs.

BIBLIOGRAPHY

General

"Inadvertant Climate Modification. Report of the Study of Man's Impact on Climate," (SMIC). The MIT Press, Cambridge, Massachusetts, 1971.

W. H. Matthews, W. W. Kellogg, and G. D. Robinson (Eds.), "Man's Impact on the Climate." The MIT Press, Cambridge, Massachusetts, 1971.

W. J. Humphreys, "Physics of the Air" (3rd ed.). McGraw-Hill, New York, 1940. Reprinted by Dover, New York, 1964.

T. L. Brown, "Energy and the Environment." Charles E. Merill, Columbus, Ohio, 1971.

A. Turk, J. Turk, and J. T. Wittes, "Ecology, Pollution, Environment." W. B. Saunders, Philadelphia, Pennsylvania, 1972.

I. J. Winn (Ed.), "Basic Issues in Environment. Studies in Quiet Desperation." Charles E. Merill, Columbus, Ohio, 1972.

"Chemistry in the Environment, Readings from Scientific American." W. H. Freeman, San Francisco, California, 1973

Climate and Climate History

J. E. Oliver, "Climate and Man's Environment. An Introduction to Applied Climatology." Wiley, New York, 1973.

H. Shapley, (Ed.), "Climatic Change, Evidence, Causes, and Effects." Harvard University Press, Cambridge, Massachusetts, 1953.

E. LeRoy Ladurie, "Times of Feast, Times of Famine: A History of Climate Since the Year 1000." Doubleday, Garden City, New York, 1971.

R. A. Bryson, A perspective on climate change, *Science 184,* 753, (1974).

R. A. Schroeder and J. L. Boda, Glacial-postglacial temperature difference deduced from aspartic acid racemization in fossil bones, *Science 182,* 479 (1973).

G. J. Kukla and R. K. Matthews, When will the present interglacial end?, *Science 178,* 190 (1972).

H. H. Lamb, "The Changing Climate." Methuen (Harper and Row), New York, 1972.

N. Calder, "The Weather Machine." Viking, New York, 1974.

A. L. Hammond, Modeling the climate: A new sense of urgency, *Science 185,* 1145 (1974).

W. W. Kellogg and S. H. Schneider, Climate stabilization: For better or for worse?, *Science 186,* 1163 (1974).

P. E. Damon and S. M. Kuner, Gobal cooling?, *Science 193,* 447 (1976).

S. H. Schneider and C. Mass, Volcanic dust, sunspots, and temperature trends, *Science 190,* 741 (1975).

J. A. Eddy, The maunder minimum, *Science 192,* 1189 (1976).

W. C. Wang, Y. L. Yunp, A. A. Lacis, T. Mo, and J. E. Hamsen, Greenhouse effects due to man-made perturbations of trace gases, *Science 194,* 685 (1976).

R. A. Reck, Stratospheric ozone effects on temperature, *Science 192,* 557 (1976).

S. D. Silverstein, Effect of infraraed transparency on the heat transfer through windows: A clarification of the greenhouse effect, *Science 193,* 229 (1976).

J. Otterman, Surface albedo and desertification, *Science 189,* 1012 (1975).

W. S. Broecker, Climatic change: Are we on the brink of a pronounced global warming?, *Science 189,* 460 (1975).

V. Ramanathan, Greenhouse effect due to chlorofluorocarbons: Climatic implications, *Science 190,* 50 (1975).

W. Lepkowski, Carbon dioxide: A problem of producing usable data, *Chem. Eng. News,* October 17, 1977, p. 26.

R. A. Kerr, Carbon dioxide and climate: Carbon budget still unbalanced, *Science 197,* 1352 (1977).

NAS panel is concerned over atmospheric CO_2 buildup, *Physics Today,* October, 1977, p. 17.

5 PETROLEUM HYDROCARBONS AND COAL

5.1 INTRODUCTION

Organic compounds are abundant in the environment, both as the products of natural, mainly biological, activity and as waste materials and pollutants from industrial processes. Organic compounds generally may be considered as substituted hydrocarbons which may be made up of straight- or branched-chains, rings or combinations of these structures. The hydrocarbon skeletons are modified by the presence of multiple bonds and oxygen, nitrogen, halogen, etc., substituents that markedly influence the reactivity of the carbon compounds. Unsubstituted hydrocarbons are relatively inert, particularly toward the polar compounds in the environment. All hydrocarbons are thermodynamically unstable with respect to oxidation to CO_2 and H_2O. The rate of this process is very slow except at elevated temperatures as we shall see later. However, the energy released by reaction with oxygen is very important in our industrial society and is about 12 kcal/gm (4500 J/gm) for saturated hydrocarbons.

Organic compounds are the basis for life on earth. The synthesis of organic compounds in living systems is catalyzed by other organic substances called enzymes. The degradation of organic compounds often follows a pathway that is the reverse of the synthesis and the reactions are catalyzed by the same or similar enzymes. Many microorganisms utilize simple organic compounds for their growth and their enzymes are responsible for these degradation reactions. As a consequence, any compound that is synthesized by a living system is readily degraded biologically and is said to be "biodegradable." However, many of the carbon compounds that have been formed by geochemical reactions (crude oil, coal), or industrial chemical processes (polyethylene, DDT) are not readily degraded. These compounds will accumulate in the environment unless their degradation just happens to be catalyzed by a microbial enzyme or they are rendered more susceptible to enzymatic degradation by some other environmental process.

Such processes include photodegradation by sunlight, oxidation by oxygen, hydrolysis by water, or other reactions catalyzed by acids, bases, clay minerals, or metal ions. These environmental processes are effective only in-so-far as they render the pollutant more susceptible to microbial degradation such as by introducing appropriate functional groups that can serve as points of microbial attack.

Since hydrocarbons are considered to be the parent compounds for organic materials, we shall consider them in this chapter. Coal, a potential source of hydrocarbons, is also discussed. Other organic compounds will be considered in later chapters. No attempt will be made to deal with the general chemistry of the various classes of compounds and functional groups except as these reactions are important environmentally. It is assumed that the reader has some familiarity with the "laboratory" properties of these materials.

5.2 THE NATURE OF PETROLEUM

Crude oil is the most predominant source of the hydrocarbon compounds used for combustion and for industrial processes, although as the cost of crude oil continues to rise, coal, oil shale, and tar sands may become more important sources of liquid and gaseous hydrocarbons. Crude oil is also a major source of environmental pollution. It has been estimated that over 6 million metric tons of crude oil and hydrocarbons enter the environment each year.

Crude petroleum is a complex mixture of hydrocarbons that is separated into fractions of differing boiling ranges by the refining process shown schematically in Fig. 5-1. The various fractions either are used directly as energy sources ("straight-run gasoline") or they are modified chemically to yield more efficient energy sources such as high octane gasoline. In addition, they may be converted to petrochemicals, compounds such as polymers and detergents that are derived from petroleum.

Sulfur, oxygen, nitrogen, and metal derivatives are also present in petroleum. Sulfur is the most important minor constituent because it interfers with the antiknock action of tetraethyl lead and it is emitted into the atmosphere as sulfur oxides when petroleum is burned. The sulfur content of crude oil ranges from 0-1% for low sulfur to greater than 3-7% for high sulfur petroleum. The sulfur is present mainly in the form of thiophene or thiophene derivatives such as those shown below.

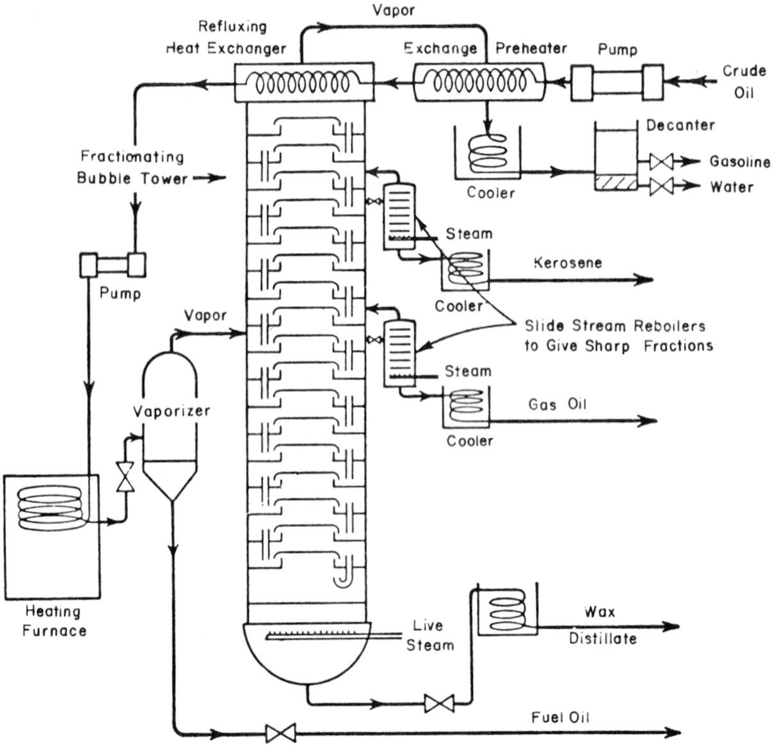

FIGURE 5-1. Diagram of a modern crude oil distillation unit. (It will be noted that the crude oil enters at the upper right-hand corner of the diagram and the products are taken off below on the right.) From J. Conant and A. H. Blatt, "The chemistry of organic compounds," (4th ed.), p. 21. Macmillan, New York, 1952.

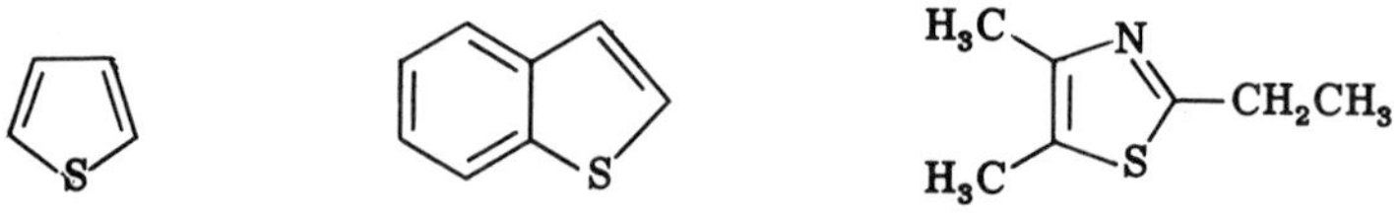

thiophene **2, 3-benzthiophene** **2-ethyl-4, 5-dimethylthiazole**

Nitrogen compounds constitute a much smaller percentage (0-0.8%) of petroleum than do sulfur compounds. The nitrogen is also present in the form of derivatives of the heterocyclic ring systems such as thiazole or quinoline.

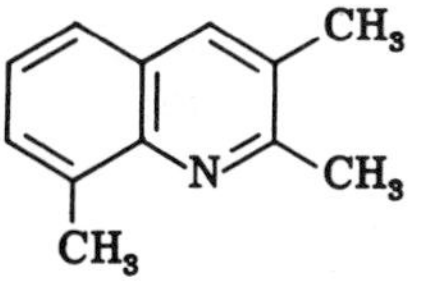

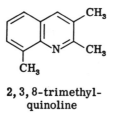

2,3,8-trimethyl-quinoline

Metals are present in petroleum either as salts of carboxylic acids or as porphyrin chelates (see Chapter 12). Aluminum, calcium, copper, iron, chromium, sodium, silicon, and vanadium are found in almost all samples of petroleum in 0.1-100 ppm range. Lead, manganese, barium, boron, cobalt, and molybdenum are observed occasionally. Chloride and fluoride also are encountered commonly. Some of these metallic components are probably introduced into the petroleum when it is removed from the ground. The use of emulsified brines and drilling muds (mixtures of iron containing minerals and aluminum silicate clays) may account for the presence of some of these trace elements.

5.3 USE OF PETROLEUM IN THE INTERNAL COMBUSTION ENGINE

One of the major uses of petroleum as an energy source is in the automobile internal combustion engine. Most automobiles are driven by a four-stroke-cycle engine (Fig. 5-2). The gasoline is mixed with air in the carburetor and drawn into

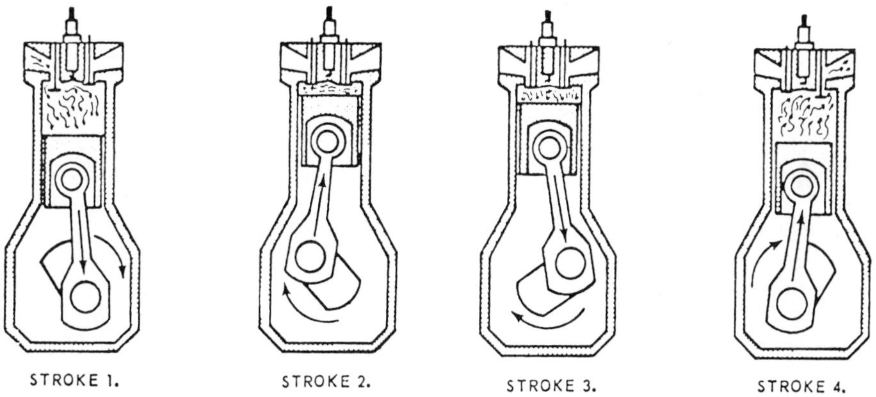

FIGURE 2. Schematic of a four-stroke cycle engine.

the cylinder on the first stroke. The oxygen-gasoline vapor mixture is compressed on the second stroke and ignited by the spark plug. The expanding gases formed by combustion then drive the piston down on the third stroke, the stroke that provides the power to drive the car, and the combustion products are forced out of the cylinder on the fourth stroke.

The first automobiles had low compression engines that could use straight-run gasoline, but as the automobile evolved, more powerful engines were developed. These high compression engines required a more sophisticated fuel than straight-run gasoline. In high compression engines, the hydrocarbons of simple petroleum distillates preignite before the top of the stroke of the compression cycle. Preignition decreases engine power because it results in an increase in the pressure on the piston before it has completed the upward stroke of the compression cycle. The preignition usually occurs explosively (knocking) which jars the engine and hastens its eventual demise.

The octane rating of gasoline is a measure of its tendency for preignition and detonation. The higher the research octane number (RON) the less likely that knocking will occur. The octane rating is an empirical number that is measured on a standard test engine. It relates to the performance of the automobile engine when it is operating at low-to-medium speeds. The RON values of some representative compounds are listed in Table 5-1. The values of 0 for *n*-heptane and 100 for isooctane were established as arbitrary standards.

TABLE 5-1

Research Octane Numbers (Octane Ratings) for Some Representative Hydrocarbons

Compound	*RON*
n-*octane*	*-19*
n-*heptane*	*0*[a]
n-*hexane*	*24.8*
n-*pentane*	*61.7*
2,4-dimethylhexane	*65.2*
cyclohexane	*83*
n-*butane*	*93.8*
2,2,4-trimethylpentane (isooctane)	*100*[a]
2,2,4-trimethylpentene-1	*102.5*
benzene	*105.8*
toluene	*120*

[a] *Standards for RON measurement.*

A variation of the octane scale designated motor octane number (MON) is also used to rate gasolines. The MON scale was established when it was observed that there was not good agreement between the RON determined in a test engine and that of the motor run under road conditions. The higher speeds and heavier loadings in contemporary automobiles resulted in lower octane ratings in road tests than were observed in the test engine. The same test engine is used for both the MON and RON but the operating conditions for MON determination give octane ratings that correlate more closely with those observed with an engine running at medium-to-high speed. Currently "regular" gasoline has an octane of about 94 RON and 87 MON and premium is 100 RON and 92 MON. The unleaded grade designated for catalyst-equipped vehicles has an octane value of about 92 RON, 84 MON. The octane rating currently listed on gasoline pumps is an average of the RON and MON. It should be emphasized that octane ratings do not measure the energy released on combustion - all hydrocarbons release about the same amount of energy per gram when burned completely - but instead are a measure of the tendency for preignition and explosive combustion.

A number of ways have been developed to increase the RON rating (55-72) of straight-run gasoline. It was discovered in 1922 that the RON of straight-run gasoline could be increased to 79-88 by the addition of 3 gm of "lead" per gallon. The "lead" or "ethyl fluid" that is added to gasoline is a mixture of about 60% tetraethyl lead [$Pb(CH_2CH_3)_4$], and/or tetramethyl lead [$Pb(CH_3)_4$], about 35-40% $BrCH_2CH_2Br$ and $ClCH_2CH_2Cl$ and 2% dye, solvent, and stabilizer. The tetraethyl lead prevents preignition and uneven burning by its conversion to lead oxide, a compound that quenches the free radicals, reactive intermediates with unpaired electrons, formed in the combustion process. This quenching decreases the combustion rate and prevents detonation and preignition. The ethylene dibromide and ethylene dichloride facilitate the removal of the lead from the cylinder by its conversion to volatile lead halides that are vented from the engine into the atmosphere.

5.4 CHEMICAL MODIFICATION OF PETROLEUM DISTILLATES

The yield of hydrocarbons suitable for use in modern automobiles and the octane rating of these hydrocarbons can be increased by chemical transformations. The yield of the gasoline fraction can be increased from 25% to 60% by converting the high boiling, higher molecular weight hydrocarbons to smaller ones (cracking) and by condensing the low boiling small molecules (C_3, C_4) together (alkylation) to form C_5-C_9 hydrocarbons. The octane rating is increased by isomerization of straight chain molecules into branched chain or aromatic compounds. Cracking is not as prevalent in the fall and winter because the higher boiling materials (fuel oil) are used for home heating.

The simplest cracking process involves heating the hydrocarbons to 600°C under pressure. The reactions that take place are illustrated with hexane [reactions (5-1) to (5-4)] but in practice the process is carried out on C_{16} and C_{18} hydrocarbons that yield a gasoline fraction and smaller olefins that are either alkylated or used in chemical synthesis.

$$CH_3CH_2CH_2CH_2CH_2CH_3 \xrightarrow{\Delta} H_2 + CH_2{=}CHCH_2CH_2CH_2CH_3 \quad (5\text{-}1)$$
$$\xrightarrow{\Delta} CH_4 + CH_2{=}CHCH_2CH_2CH_3 \quad (5\text{-}2)$$
$$\xrightarrow{\Delta} CH_3CH_3 + CH_2{=}CHCH_2CH_3 \quad (5\text{-}3)$$
$$\xrightarrow{\Delta} CH_3CH_2CH_3 + CH_2{=}CHCH_3 \quad (5\text{-}4)$$

These reactions take place through a free radical mechanism as illustrated in reactions (5-5) to (5-8). The high reaction temperature results in homolytic fission [reactions (5-5) and (5-6)] of the hydrocarbons to yield the radicals. The final products are probably a result of the disproportionation (5-7) of or hydrogen atom abstraction (5-8) by radicals.

$$RCH_2CH_3 \longrightarrow R\dot{C}HCH_3 + H\cdot \quad (5\text{-}5)$$
$$RCH_2CH_3 \longrightarrow R\cdot + \cdot CH_2CH_3 \quad (5\text{-}6)$$
$$2R\dot{C}HCH_3 \longrightarrow RCH{=}CH_2 + RCH_2CH_3 \quad (5\text{-}7)$$
$$CH_3CH_2\cdot + RCH_2CH_3 \longrightarrow CH_3CH_3 + R\dot{C}HCH_3 \quad (5\text{-}8)$$

In catalytic cracking, a catalyst such as aluminum silicate allows for much lower reaction temperatures than simple thermal cracking. The reaction product is a mixture of C_4-C_9 paraffins and olefins that has a RON of 92-94.

More recently the process of hydrocracking, which is a combination of catalytic cracking and catalytic hydrogenation of the olefins formed in the cracking process, has been developed. This procedure is advantageous because catalytic hydrogenation converts the sulfur and nitrogen compounds to hydrogen sulfide and ammonia, respectively. These are readily scrubbed from the hydrocarbon mixture with water. Hydrocracking gives a product with a RON of 86. The RON is lower than that of the product of catalytic cracking because of the absence of olefins, but addition of 3 gm Pb/gal[1] gives a gasoline with a RON of 98.

The formation of gasoline by alkylation gives a high octane gasoline (∿93) because it yields hydrocarbons with extensive chain branching. Alkylation is illustrated with isobutylene:

$$(CH_3)_3CH + CH_2{=}C(CH_3)_2 \xrightarrow{H^+} CH_3\underset{\substack{|\\ CH_3}}{\overset{\substack{CH_3\\ |}}{C}}CH_2\overset{\substack{CH_3\\ |}}{C}HCH_3 \qquad (5\text{-}9)$$

Reaction (5-9)proceeds in the following steps:

$$CH_2{=}C(CH_3)_2 + H^+ \rightarrow (CH_3)_3C^+ \qquad (5\text{-}10)$$

$$(CH_3)_3C^+ + CH_2{=}C(CH_3)_2 \rightarrow (CH_3)_3CCH_2\underset{\substack{|\\ CH_3}}{\overset{\substack{CH_3\\ |}}{C^+}} \qquad (5\text{-}11)$$

$$(CH_3)_3CCH_2\underset{\substack{|\\ CH_3}}{\overset{\substack{CH_3\\ |}}{C^+}} + (CH_3)_3CH \rightarrow (CH_3)_3CCH_2\underset{H}{\overset{\substack{CH_3\\ |}}{C}}CH_3 + (CH_3)_3C^+ \qquad (5\text{-}12)$$

The acid catalysis results in the generation of the more stable tertiary carbonium ion (4-10). In principle (and in fact) the condensation could continue to yield polymers. As will be noted later (see discussion of detergents in Chapter 6) a similar reaction process is used to give higher molecular weight hydrocarbons. The ratio of acid catalyst to olefin is one of the factors that governs the molecular weight of the alkylated product.

Higher octane gasoline is obtained by the isomerization of straight chain hydrocarbons to their branched chain isomers. The rearrangement shown by reaction (5-13) occurs because, as noted previously, the more highly substituted tertiary carbonium ion is more stable than a primary or secondary carbonium ion. The process is illustrated in (5-13) in the simple case of butane. In practice it is applied to the C_7-C_9 hydrocarbons to give a product with a RON of about 82.

$$CH_3CH_2CH_2CH_3 \xrightarrow{AlCl_3} CH_3\underset{\displaystyle CH_3}{\underset{|}{C}}HCH_3 \qquad (5\text{-}13)$$

Reaction (5-13) is initiated by a trace of olefin and acid:

$$CH_2 = CHR + HCl + AlCl_3 \rightleftarrows CH_3\overset{+}{C}HR + AlCl_4^- \qquad (5\text{-}14)$$

Rearrangement then proceeds in several steps.[1]

$$CH_3\overset{+}{C}HR + CH_3\text{-}CH_2CH_2CH_3 \longrightarrow RCH_2CH_3 + CH_3\overset{+}{C}HCH_2CH_3 \qquad (5\text{-}15)$$

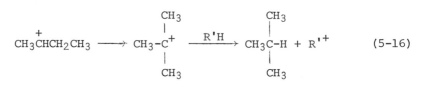

$$CH_3\overset{+}{C}HCH_2CH_3 \longrightarrow CH_3\text{-}\overset{\displaystyle CH_3}{\underset{\displaystyle CH_3}{C^+}} \xrightarrow{R'H} CH_3\overset{\displaystyle CH_3}{\underset{\displaystyle CH_3}{C}}\text{-}H + R'^+ \qquad (5\text{-}16)$$

Catalytic reforming, the conversion of aliphatic compounds to aromatic compounds (5-17), is also an isomerization process. The reaction follows a different course because it occurs on the surface of a platinum or platinum-rhenium catalyst. A product with a RON of 90-95 is obtained. The cyclic compounds proposed as intermediates in the catalytic reforming process are illustrated with *n*-heptane in reaction (5-17):

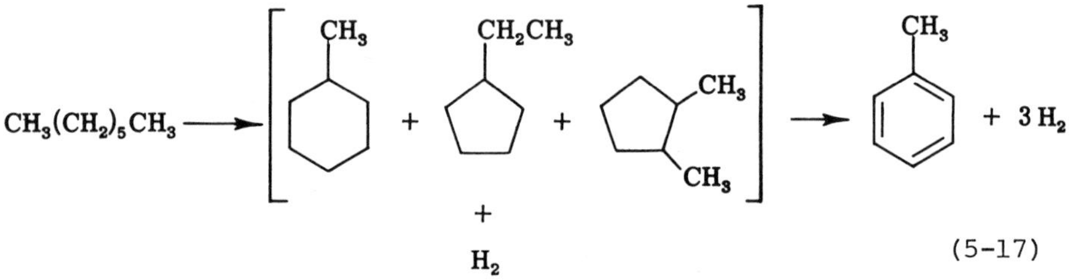

(5-17)

Catalytic reforming is also one of the major pathways for the preparation of the aromatic compounds that are used by the chemical industry.

[1] *G. J. Karabatsos* et al., <u>J. Amer. Chem. Soc</u>. *85, 729, 733 (1963)*.

5.5 ENVIRONMENTAL REACTIONS OF PETROLEUM AND SOME PETROLEUM DERIVATIVES

5.5.1 Introduction

As stated earlier, hydrocarbons are relatively inert under typical environmental conditions. However, they do react in a variety of ways. Microorganisms are the most effective agents for the total degradation of such molecules. The mechanism of this process will be discussed in detail in Chapter 6. Alkanes and olefins react with oxygen and ozone, but these reactions are not effective in removing hydrocarbons from the environment; they are significant in other ways, however. Ozone reactions are important at ambient temperatures in smog formation (Chapter 10), while the formation of oxygen-containing compounds (other than CO or CO_2) is important in combustion. Five to ten percent of the total organic emissions from automobiles contain oxygen substituents. These compounds are formed by the incomplete oxidation of hydrocarbons in the internal combustion engine. Aldehydes make up the major portion of these oxygenated compounds, with ketones, alcohols, ethers, esters, and nitroalkanes also being formed [reactions (5-18) to (5-33)]. The aldehydes are especially important in the formation of photochemical smog (see Chapter 10).

Reactions of hydrocarbons with oxygen may involve the latter in the ground state or in excited electronic states (singlet oxygen, also discussed in detail in Chapter 10). These forms may react differently. Some of the types of reactions with ground state and excited oxygen molecules and with ozone are discussed in the next section.

5.5.2. Reactions with Ground State Molecular Oxygen

Some of the initial reaction steps involved in the gas phase oxidation of hydrocarbons with oxygen molecules in the ground state are listed below. The free radical nature of oxidation is expected since one of the reactants (oxygen) is a ground state diradical; that is, it contains two unpaired electrons. These reactions proceed at a very slow rate at 25-100°C. Reactions (5-21)-(5-23) are much faster than (5-24)-(5-28) so that appreciable concentrations of peroxides develop slowly below 100°C. Reaction (5-18) will take place at an appreciable rate only if the radicals initially formed are stable. At temperatures above 100°C, reactions (5-18) and (5-24) proceed at significant rates so that the chain process involving reactions (5-19) and (5-24) predominates. As the reaction temperature reaches 250°C, the oxidation becomes autocatalytic. This is probably because of the increased importance of reaction (5-28) in which the accumulated peroxides decompose to form two radicals. Temperatures in the 345-500°C range favor hydrogen atom abstractions [reactions (5-20)] and radical decomposition to olefins [reactions (5-29), (5-30)].

$$O_2 + RCH_2CH_2R \rightarrow R\dot{C}HCH_2R + HO_2\cdot \qquad (5\text{-}18)$$

$$R\dot{C}HCH_2R + O_2 \rightarrow R\overset{O_2\cdot}{\overset{|}{C}}HCH_2R \qquad (5\text{-}19)$$

$$R\dot{C}HCH_2R + O_2 \rightarrow RCH{=}CHR + HO_2\cdot \qquad (5\text{-}20)$$

$$2R\overset{O_2\cdot}{\overset{|}{C}}HCH_2R \rightarrow 2R\overset{O\cdot}{\overset{|}{C}}HCH_2R + O_2 \qquad (5\text{-}21)$$

$$R\overset{O_2\cdot}{\overset{|}{C}}HCH_2R + HO_2\cdot \rightarrow R\overset{O_2H}{\overset{|}{C}}HCH_2R + O_2 \qquad (5\text{-}22)$$

$$R\overset{O_2\cdot}{\overset{|}{C}}HCH_2R + RCH_2\overset{O\cdot}{\overset{|}{C}}HR \rightarrow R\overset{O_2H}{\overset{|}{C}}HCH_2R + RCH_2\overset{O}{\overset{\|}{C}}R \qquad (5\text{-}23)$$

$$R\overset{O_2\cdot}{\overset{|}{C}}HCH_2R + RCH_2CH_2R \rightarrow R\overset{O_2H}{\overset{|}{C}}HCH_2R + RCH_2\dot{C}HR \qquad (5\text{-}24)$$

$$R\overset{O_2\cdot}{\overset{|}{C}}HCH_2R \rightarrow RCH{=}CHR + HO_2\cdot \qquad (5\text{-}25)$$

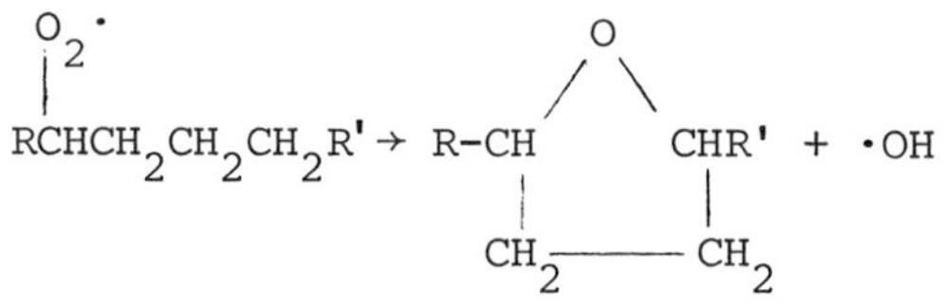

(5-26)

$$R\overset{O_2\cdot}{\overset{|}{C}}HCH_2R' \rightarrow R\cdot + R'H \quad \text{(shorter chain olefin or alkane)} \qquad (5\text{-}27)$$

$$R\overset{O_2H}{\overset{|}{C}}HCH_2R \rightarrow R\overset{\dot{O}}{\overset{|}{C}}HCH_2R + \cdot OH \qquad (5\text{-}28)$$

$$R\dot{C}HCH_2R \rightarrow RCH{=}CH_2 + R\cdot \qquad (5\text{-}29)$$

$$R\dot{C}HCH_2R \rightarrow RCH{=}CHR + H\cdot \qquad (5\text{-}30)$$

The actual products observed when neopentane (2,2-dimethylpropane) and *n*-pentane are oxidized with an equimolar amount of oxygen in the gas phase are shown in reactions (5-31)-(5-33). More extensive oxidative degradation takes place in the internal combustion engine where a much larger proportion of oxygen is used.

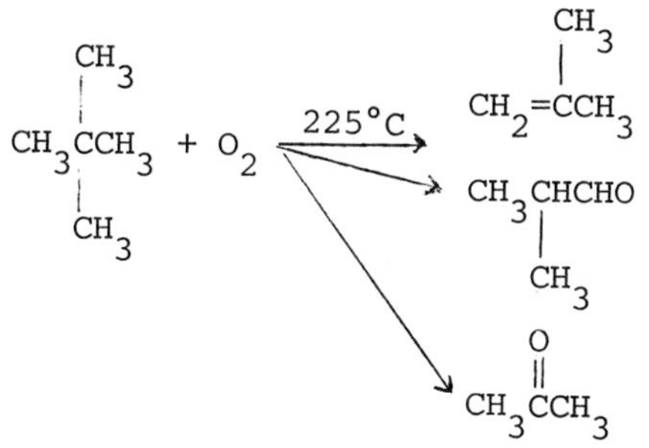

(5-31)

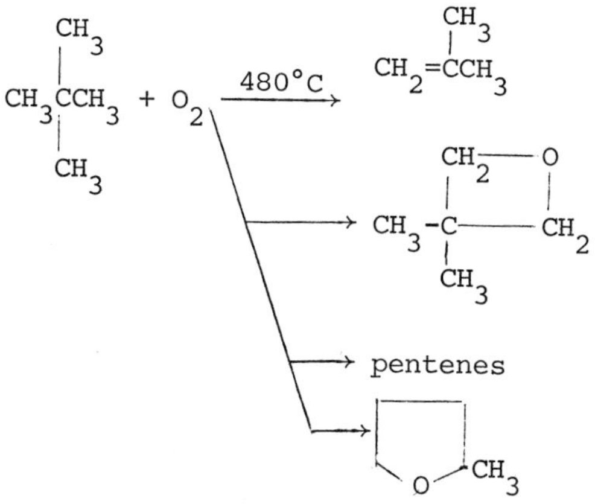

(5-32)

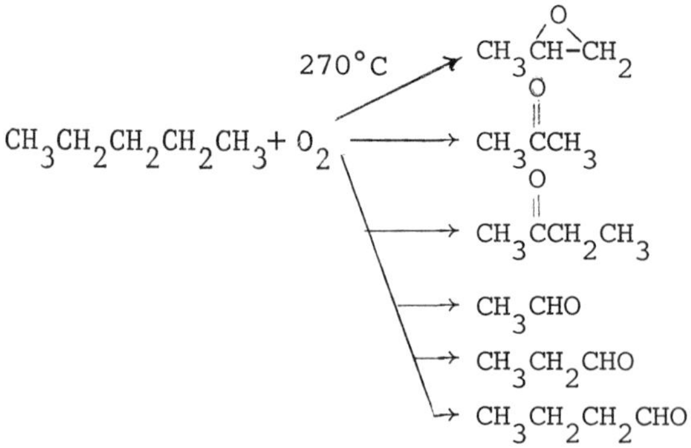

(5-33)

Similar products are obtained from the liquid phase oxidation of hydrocarbons. A peroxide is formed when a compound that forms a stable radical intermediate is oxidized. Reactions (5-34)-(5-36) occur at low temperatures and are of preparative value. Similar reactions occur with straight chain hydrocarbons.

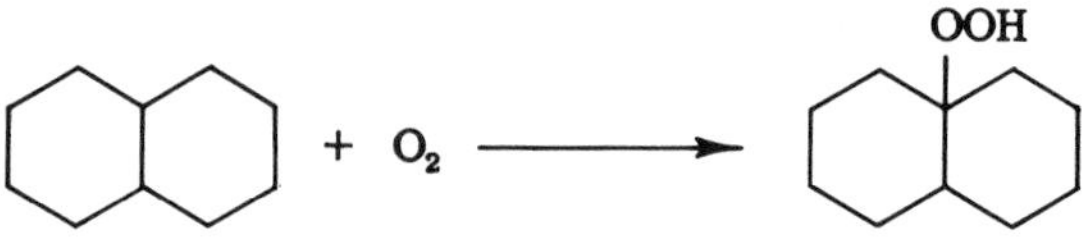

(5-34)

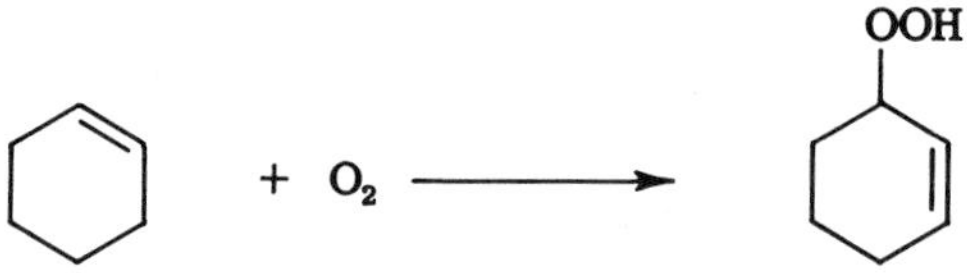

(5-35)

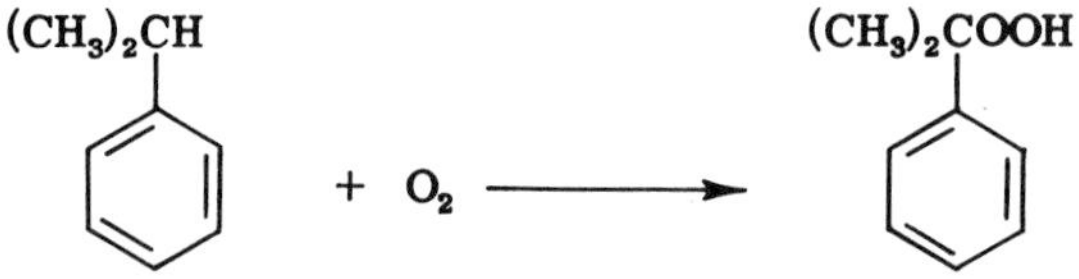

(5-36)

The following reaction mechanism is consistent with these reaction products:

$$RCH{=}CHCH_2R' + O_2 \rightarrow [RCH{=}CH\dot{C}HR' \leftrightarrow R\dot{C}HCH{=}CHR'] + HO_2^{\bullet} \tag{5-37}$$

$$RCH{=}CH\dot{C}HR' + O_2 \rightarrow RCH{=}CH\underset{}{\overset{O_2^{\bullet}}{\overset{|}{C}}}HR' \tag{5-38}$$

$$\underset{\text{RCH=CHCHR'}}{\overset{O_2^{\bullet}}{|}} + \text{RCH=CHCH}_2\text{R'} \rightarrow \underset{\text{RCH=CHCHR'}}{\overset{O_2H}{|}} + \text{RCH=CH}\dot{\text{C}}\text{HR'} \tag{5-39}$$

Peroxide decomposition is observed at higher temperatures with the resultant formation of alcohols, aldehydes, and ketones.

5.5.3 Reactions with Ozone

Ozone is present in the atmosphere in trace amounts (0.02 ppm) at sea level. However, it can build up to much larger concentrations under special conditions and can play an important role in atmospheric chemistry. For example, the blue haze that forms over some densely vegetated areas is attributed to aerosols formed by the ozone oxidation of the hydrocarbons (mainly terpenes) released by the plants. A similar but less picturesque effect observed over Los Angeles is due in part to the reaction of ozone with the hydrocarbons emitted from automobiles (see Chapter 3). It is not uncommon for the ozone concentration to reach 0.2 ppm in the Los Angeles basin and other metropolitan areas. Oxidant levels (including ozone) greater than 0.15 ppm cause eye irritation to a significant portion of the population.

Ozone is a very reactive substance that can oxidize most organic compounds. It is especially dangerous because it can form free radical intermediates that can react with the biomolecules of the body. For example, ozone concentrations above those normally present in air can produce lipid oxidation and it has been suggested that ozone and oxidants are responsible for the pulmonary damage caused by smog. Ozone does not contribute significantly to the environmental degradation of pollutants, but the chemical reactions of ozone merit discussion because of their significance in photochemial smog.

In laboratory reactions, the reaction of ozone with olefins yields an ozonide [reaction (5-40)]. The latter compound can be decomposed under reducing conditions to yield the corresponding aldehydes or ketones while oxidative workup results in the corresponding acids [equation (5-40)].

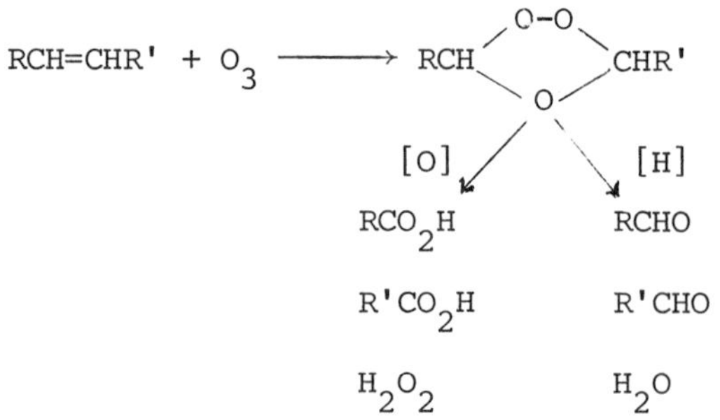

(5-40)

The structure of ozone may be represented by the two dipolar resonance forms shown in (5-41). In solution the mechanism of ozonide formation involves the initial 1,3-dipolar addition of ozone to the olefin to form the molozonide (5-42) followed by the dissociative rearrangement of this initially formed adduct (5-43).

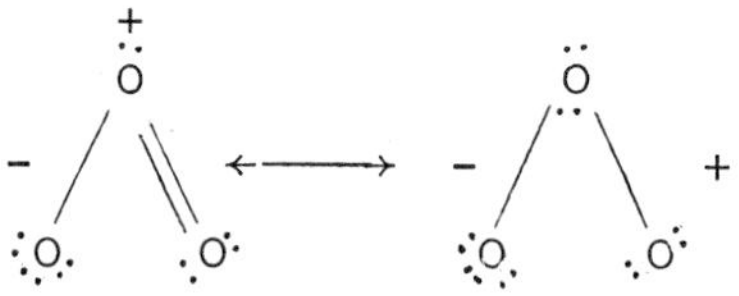

(5-41)

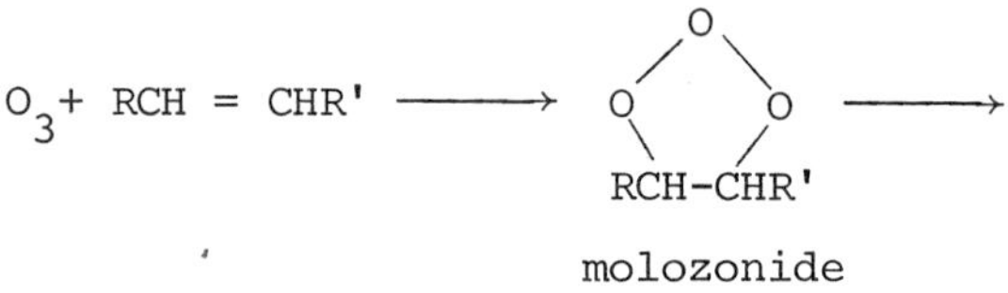

(5-42)

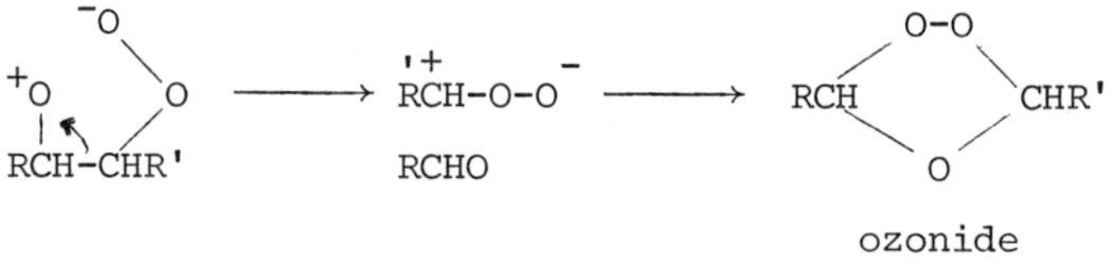

(5-43)

The ionic mechanism above is not observed in the gas phase because the zwitterion proposed in (5-43) would be a very high energy species in the absence of solvent stabilization. In addition, the ozonide is not formed because there is less chance of two fragments remaining in close enough proximity for recombination in the absence of surrounding solvent molecules (cage effect). Instead it has been suggested that the molozonide dissociates to a diradical (5-44) which in turn decomposes to an aldehyde and/or ketone and another diradical (5-45). This diradical reacts with oxygen to give radicals that are similar to those postulated previously in the reaction of oxygen with hydrocarbons [(5-46) to 5-51)].

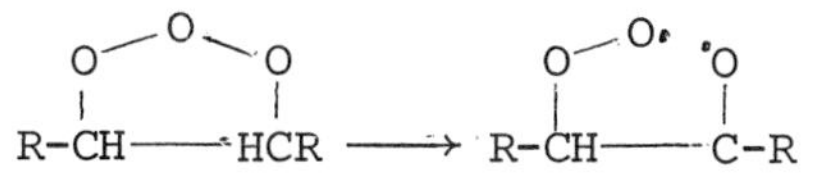
(5-44)

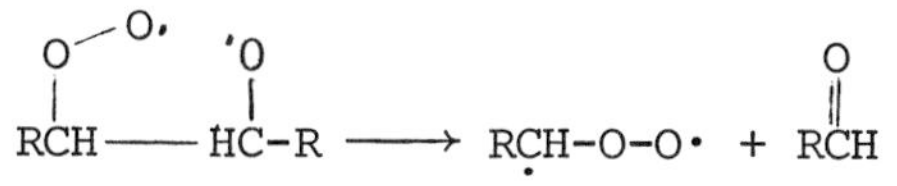
(5-45)

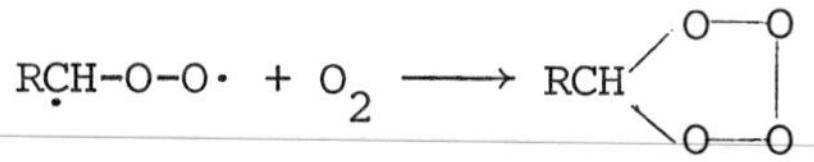
(5-46)

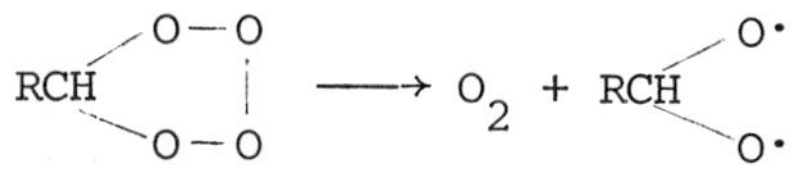
(5-47)

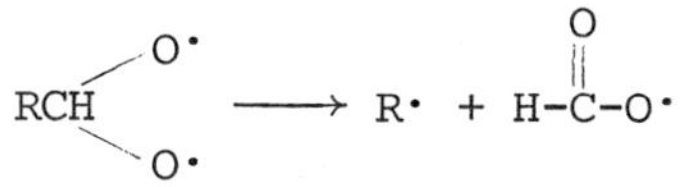
(5-48)

$$R^{\cdot} + O_2 \longrightarrow RO_2^{\cdot} \qquad (5\text{-}49)$$

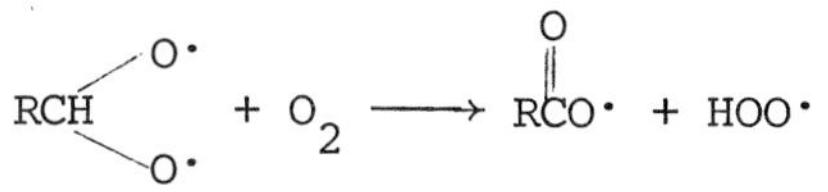
(5-50)

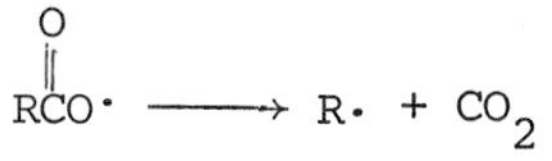
(5-51)

Ozone reacts with hydrocarbons at 25-50°C to yield products that are similar to those obtained by the reaction of hydrocarbons and molecular oxygen at 125-250°C. For example, reactions (5-52) to (5-57) were carried out at 25-50°C.

$$CH_4 + O_3 \rightarrow HCO_2H + CO_2 + CH_3OH \qquad (5\text{-}52)$$

$$CH_3CH_2CH_3 + O_3 \rightarrow HCO_2H + CO_2 + CH_3OH + CH_3COCH_3 \qquad (5\text{-}53)$$

$$CH_3CH(CH_3)CH_3 + O_3 \rightarrow HCO_2H + CO_2 + CH_3OH + CH_3COCH_3 + (CH_3)_3COH \qquad (5\text{-}54)$$

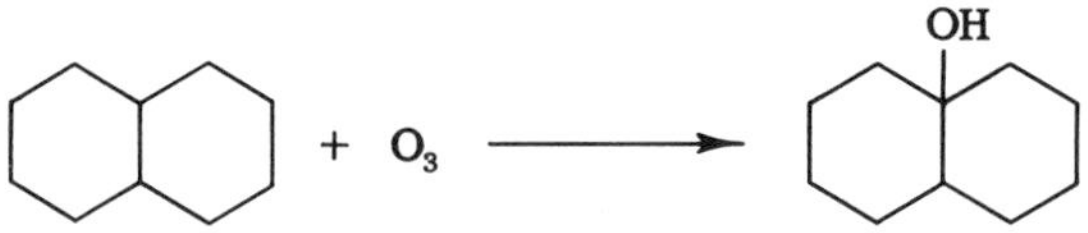

(5-55)

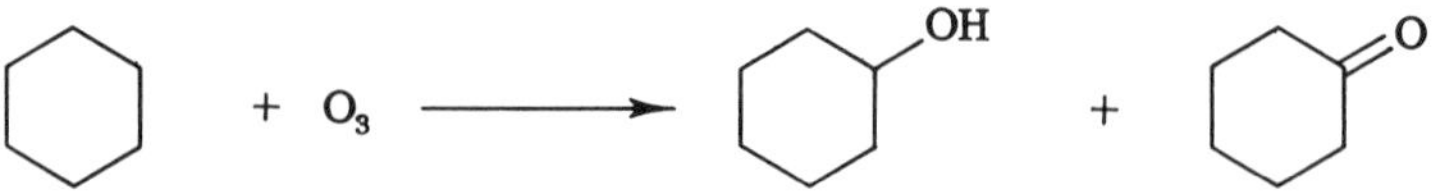

(5-56)

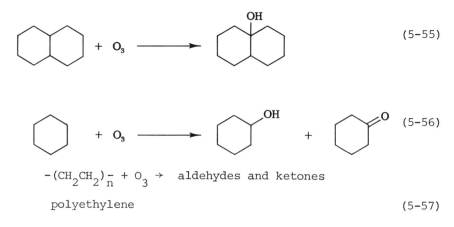

$$-(CH_2CH_2)_n- + O_3 \rightarrow \text{aldehydes and ketones}$$

polyethylene (5-57)

Ozone is reacting as if it were a diradical in these reactions. Several free radical mechanisms have been postulated to explain the reaction products and one of them is given in reactions (5-58) to (5-62).

$$RH + O_3 \rightarrow RO\cdot + HO_2\cdot \quad (5\text{-}58)$$

$$RO\cdot + RH \rightarrow ROH + R\cdot \quad (5\text{-}59)$$

$$RO\cdot \rightarrow R'(=O) \text{ (aldehyde or ketone)} + R'' \text{ (olefin)} \quad (5\text{-}60)$$

$$R\cdot + O_2 \rightarrow RO_2\cdot \quad (5\text{-}61)$$

$$RO_2\cdot + RH \rightarrow RO_2H + R\cdot \quad (5\text{-}62)$$

5.5.4 Reactions with Singlet Oxygen

Singlet oxygen [$O_2(^1\Delta g)$], an excited state of molecular oxygen, may be formed either chemically or by the photosensitized reaction of oxygen. The photochemical formation is probably the most significant from the environmental viewpoint and is discussed in Chapter 10. Singlet oxygen adds readily to dienes and olefins [(5-63)-(5-65)]. These reactions are reminiscent of the Diels-Alder and "ene" additions of electrophylic olefins with unsaturated compounds. The bulk of the experimental data are consistent with a concerted addition reaction; that is, no reaction intermediates are formed. For example, the absence of the trans-isomer in the reaction of singlet oxygen with the octohydrobiphenyl derivative in equation (5-66) is consistent with a concerted mechanism.

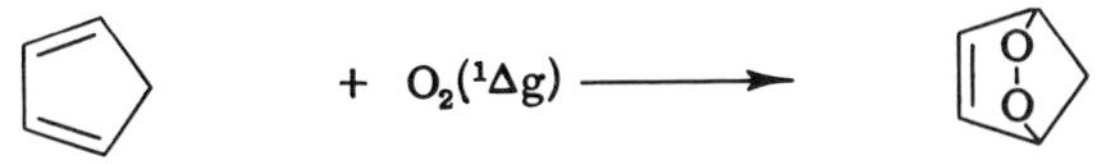

(5-63)

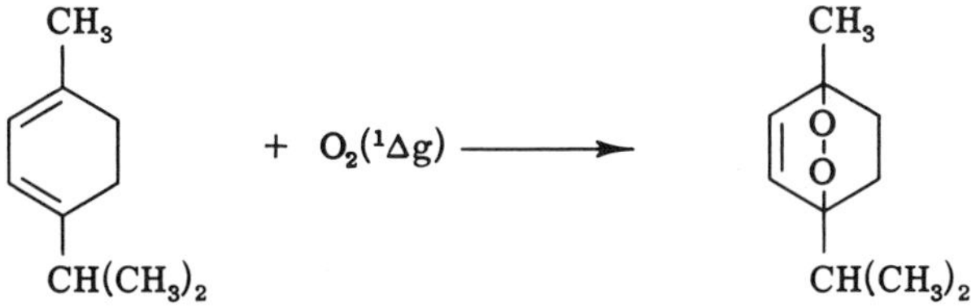

(5-64)

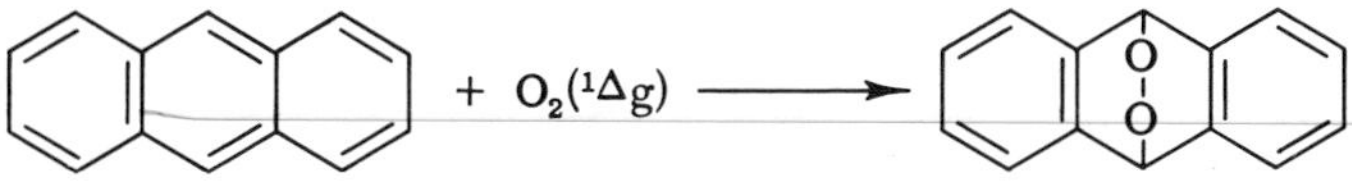

(5-65)

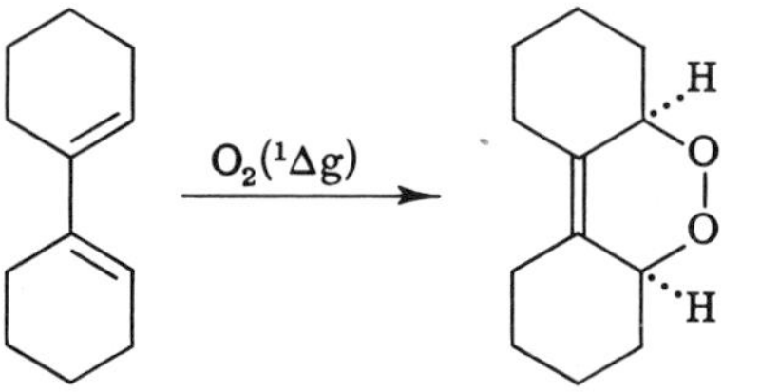

(5-66)

The reaction of singlet oxygen with simple olefins may be of greater significance than the reactions with conjugated olefins from an environmental viewpoint [reactions (5-67) and (5-68)]. However, the present evidence suggests that singlet oxygen is not important in photochemical smog formation (see Chapter 10).

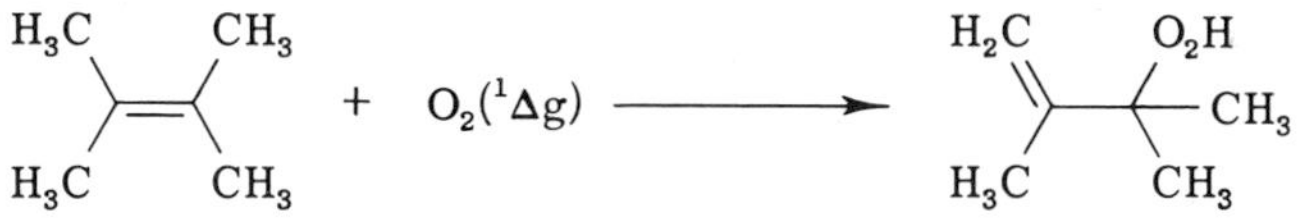

(5-67)

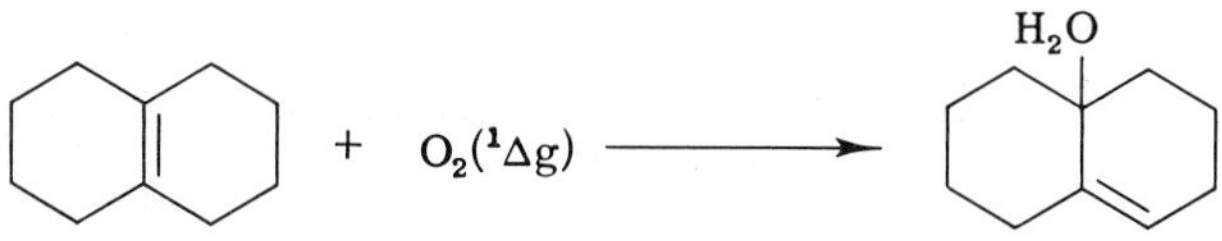

(5-68)

Two mechanisms have been proposed for the reaction of singlet oxygen with olefins. The "ene" mechanism (5-69) is the simplest conceptually and was proposed first.

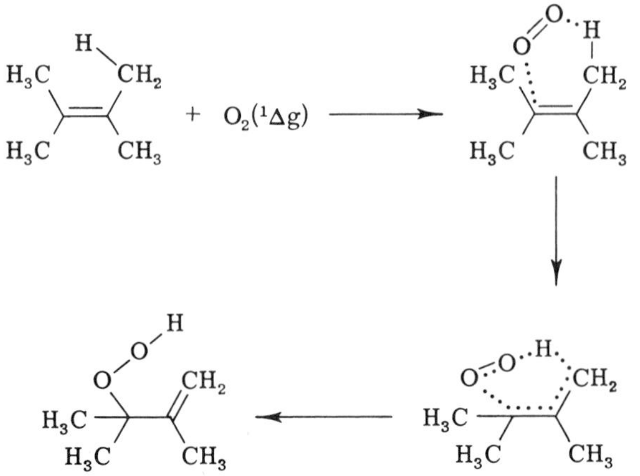

(5-69)

This concerted mechanism is supported by the observation that it is not possible to inhibit the reaction with compounds that react with free radicals. However, the course of the oxidation is changed (in solution) by the addition of azide [reaction (5-70)]. This experiment suggests that the reaction is not concerted and an intermediate is formed that is trapped by azide ion. Consequently, a nonconcerted mechanism given in reaction (5-71) was postulated to account for this finding. In the absence of azide, the hydroperoxide is formed by the rearrangement of the proposed three-membered ring intermediate. Further experimental data are needed to define precisely the reaction mechanism.

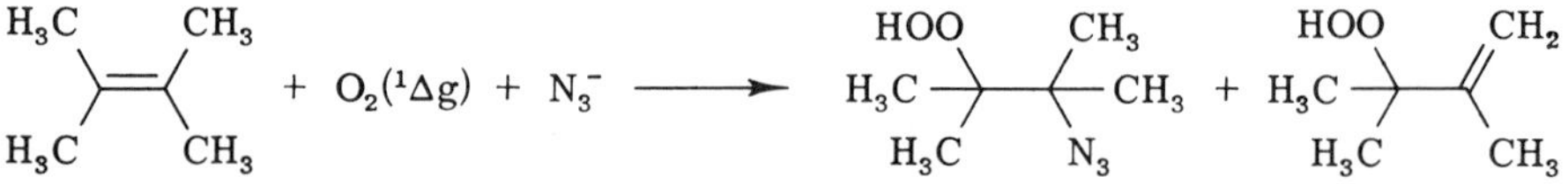

(5-70)

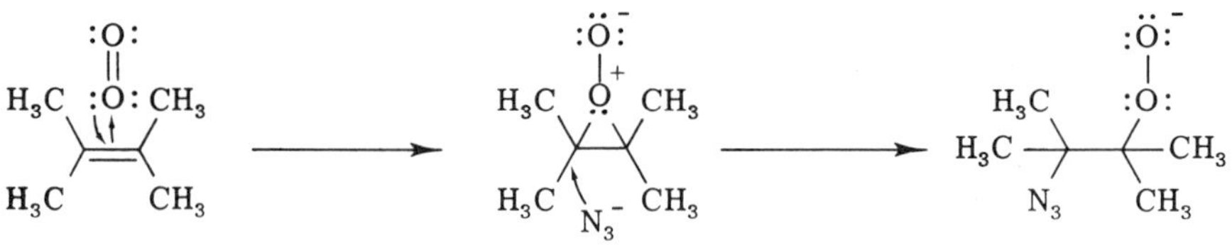

(5-71)

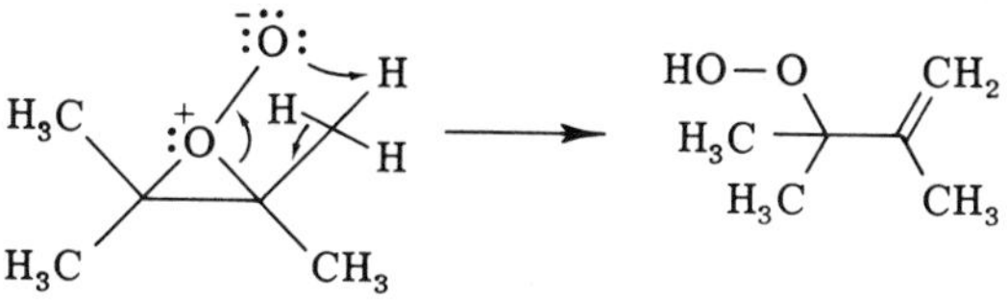

(5-72)

The reaction of singlet oxygen with electron-rich olefins follows a different reaction pathway to yield a carbonyl product [equations (5-73) and (5-74)]. Stereospecific cis-addition occurs as demonstrated by the fact that the stereochemistry of the 1,2-diethoxyethylene is maintained in the initially formed four-membered ring (dioxetane) product (5-74).

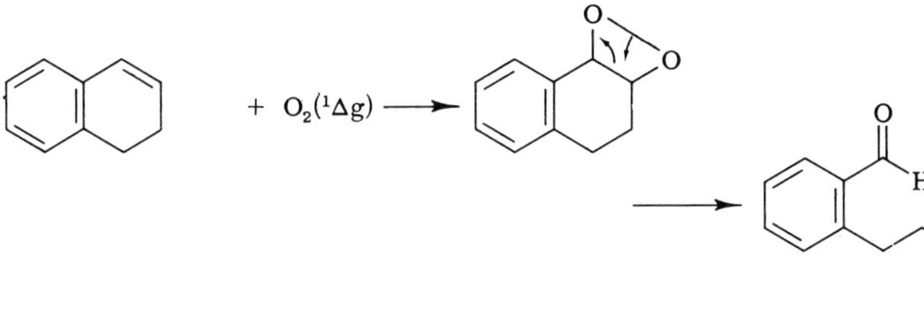

(5-73)

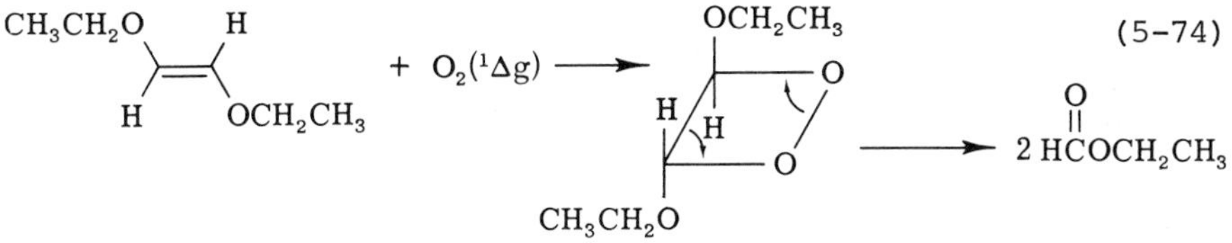

(5-74)

5.6 SOURCES OF CRUDE OIL AND HYDROCARBONS IN THE SEA

5.6.1 Petroleum in the Environment

The major environmental effects of petroleum are due either to crude oil itself or to the automobile hydrocarbon emissions from the gasoline fraction of the crude oil. It is estimated that 6 million metric tons of petroleum hydrocarbons enter the environment each year from both sources. Oil spills generally are associated with the release of hydrocarbons to the oceans or other water bodies, where much of the oil floats and spreads over large areas of the water surface. Spills on land tend to be more easily contained to small areas. The primary environmental problems are associated with the effects of the slowly degraded petroleum residues on aquatic life. Automobile emissions, on the other hand, cause atmospheric problems. Much spilled oil also enters the atmosphere by evaporation, but the problems associated with oil spills and automobile emissions are sufficiently different to warrant separate discussion. The additional environmental problems of the detergents and polymers synthesized from petroleum will be discussed in subsequent chapters.

5.6.2. Natural Oil Seepage

Natural oil seeps led to the discovery of most of the major oil fields in the world. The first oil well in the U.S. was drilled in Titusville, Pennsylvania because oil was observed seeping into the Oil River at that point. These oil seeps may be responsible for local environmental problems but often the rate of seepage is so slow that the oil is effectively dispersed by environmental forces. The most abundant oil seep, 3600 gal/day, is at Coal Oil Point on the coast of Southern California. This seep produces an extensive oil slick on the ocean and often results in tar accumulation on the shore. Most seeps produce less than 40 gal/day and the environmental effect appears to be quite limited. It has been estimated, on the basis of very incomplete data, that $50\text{-}1600 \times 10^6$ gallons ($160\text{-}62{,}000 \times 10^6$ liters) of oil per year are released into the environment from natural oil seeps with 160×10^6 gal/year suggested as the best estimate. Areas of high oil seepage that might cause environmental problems account for about 45% of this total amount.

5.6.3 Oil Spills

Oil spills can have dramatic local environmental effects and because of this they are well documented in the national press and television. This problem first arose in the early stages of World War II when German submarines sank over 60 oil tankers off the eastern U.S. seaboard, releasing 200×10^6 gallons of oil over a period of three years. Since these tankers were traveling close to the shore in an attempt to avoid the submarines, much of this oil was eventually washed up on the beaches. In 1972 the accidental spillage of oil is estimated to have been 16.5×10^6 gallons of oil from 8000 spills, 8% of the total resulting from the German submarines over a three year period. In 1972 there were 26 tanker spills of over 10,000 gallons; 11 of them in U.S. waters. These large crude oil spills reflect the fact that 60% of the world crude oil production is transported by sea. Oil spills on land are more readily contained and are usually less of an environmental problem. One notable exception was the flooding caused by Hurricane Agnes in 1972 which resulted in the release of 6-8 × 10^6 gallons of sludge and oil from storage tanks in Pennsylvania.

Although the local effects of oil spills can be disasterous, these spills are not the major source of environmental crude oil. The hundreds of oil tankers that deliver crude oil are designed to ride low in the water when carrying a full load. After they unload their cargo they must take on sea water as ballast for their return trip to the oil fields. This ballast, along with the residual crude oil in their tanks, is pumped into the ocean before the next load of crude oil is loaded. It is estimated that 800 × 10^6 gal of oil/year are introduced into the sea from this source and another 130 × 10^6 gal are added from the bilges of all other ocean vessels. The oil washed from tankers is an especially bad environmental problem because it is released close to the shore where it may be carried onto the beaches.

5.6.4 Off-Shore Wells

Leakage from off-shore oil wells is a third source of oil in the sea. This problem was dramatized in the U.S. by the well that "blew" off Santa Barbara, California, in January 1969. The well "blew" because the oil was pressurized by methane gas in the ground and this pressure forced out the oil when the well was drilled. The oil flow could not be stopped by capping the well because of the high pressure. It was necessary to drill additional wells to dissipate the gas pressure so that the oil flow could be controlled. A more recent example is the North Sea blowout in 1977. With the exception of the instances where the oil is pressurized, and these can be identified by geologists, the danger of major leaks from off-shore oil wells appears to be minimal. Currently there are 6000 wells off the coasts of Texas and Louisiana with no reports of major environmental disasters. One study has concluded that drilling off-shore wells along the Atlantic coast of the U.S. offers less environmental risk than bringing the oil in large tankers from the Gulf coast or the Middle East because of the problems with tankers noted in Section 5.6.3.

5.6.5 Lubricating and Crankcase Oil

Used lubricating oil from automobiles and industrial applications is another source of environmental oil pollution. About 2400 × 10^6 gal of lubricating oil are produced each year. About half of this is consumed in use. Of the remainder, about 600 × 10^6 gal are recovered and burned as fuel in incinerators and boilers and 110 × 10^6 gal are purified for reuse. The remaining 500 × 10^6 gal are a major source of environmental pollution. The bulk of it is unaccounted for and is probably washed into the rivers and oceans through storm sewers, although 200 × 10^6 gal of this oil is used as binder on dirt roads and parking lots. However, this relatively light oil does not stay in place and causes soil and water pollution. Perhaps most waste oil will eventually be rerefined. Direct combustion of it is not desirable because such combustion results in the introduction of heavy metals, mainly lead from leaded gasoline, into the environment. It has been estimated that 1-1.5% by weight of waste crankcase oil is lead so that its recovery could be an important source of this metal.

5.7 MARINE EFFECTS OF CRUDE OIL

Since crude oil is composed of chemically unreactive hydrocarbons, it may persist in the ocean for many years. When oil is spilled in lakes or oceans, the lower boiling fractions evaporate rapidly so that 25-50% of a crude oil spill may be lost in a week due to evaporation. The residue which remains is slowly degraded by microorganisms. This is a slow process because the other nutrients needed for growth, especially nitrogen, are quite limited in the ocean. The availability of these nutrients governs the rate of petroleum degradation. Furthermore, the very high molecular weight petroleum fractions (asphalt) are too insoluble in water for assimilation and degradation by microorganisms. It is this asphalt material, which still contains some low molecular weight hydrocarbons, that aggregates into the "tar balls" that are often seen floating in the ocean and washed up on beaches.

Until fairly recently, it was felt that although oil spills caused local environmental problems for a short period of time these effects were not permanent. For example, there was very little tar visible on the beaches of England and France a few months after the Torrey Canyon oil tanker ran aground and broke up off England in 1967. In addition, although many sea birds are killed by oil spills, those birds remaining are usually quite prolific and rapidly repopulate the effected area. Long-term effects of crude oil on marine life were discovered by the careful investigations of a spill in the town of Falmouth on Cape Cod, Massachusetts. It was observed that the aromatic compounds present in petroleum are more toxic to marine life than the saturated compounds. These aromatics are especially toxic to juvenile forms of marine life. The survivors of an oil spill often suffer long-term effects. For example, birds that have received sublethal doses of petroleum are more susceptible to infection. Mussels that survive develop no eggs or sperm and the shellfish remain contaminated with petroleum because the oil is not washed out by the sea water.

The indirect effects of oil spills are not well known, but these are potentially as important as the direct effects noted above. It has been discovered recently that fish and other forms of marine life are stimulated in their search for food, in their escape from predators, and in their mating by the organic compounds in the sea. The message received by the fish may be masked by the presence of petroleum or incorrect messages may be conveyed by the compounds present in the crude oil.

The procedures that were used before 1968 to clean up oil spills also proved to be damaging to sea life. The detergents used to solubilize some of the crude oil released from the Torrey Canyon proved to be especially toxic to the marine life along the seashore. The other approach that has been used is the addition of chalk to the crude oil so that its density increases and it sinks to the bottom. However, it has been found that the petroleum which sinks to the bottom covers up marine life and prevents its growth for many years. This oil decomposes very slowly because of the diminished oxygen supply on the ocean bottom. Furthermore, this crude oil may react with the oxygen that is present and thereby convert an aerobic environment to an anaerobic one and in this way prevent the growth of microorganisms. Precipitation should only be used as a last resort in dealing with oil spills, and only when there is little danger of destroying marine life on the sea bottom.

5.8 SOURCES OF HYDROCARBONS AND OTHER COMPOUNDS IN THE ATMOSPHERE

5.8.1 Introduction

Some hydrocarbons are natural components of the atmosphere. For example, most of the atmospheric methane is produced by anaerobic bacteria in swamps and bogs; a much smaller percentage is released by natural gas wells. It has been proposed that many of the hydrocarbons found in the Gulf of Mexico may be due to the marsh plants indigenous to the Gulf Coast. Many plants, especially conifers and citrus, release terpenes into the atmosphere. It has been suggested that the six-day air pollution alert in the Washington, D.C. area in August 1974 was due mainly to the presence of terpenes originating from plants in the Appalachian Mountains. In the absence of any man-made pollution, there would be parts-per-million concentrations of methane (see Table 3-1) and parts-per-billion amounts of ethane, ethylene, acetylene, and propane in the atmosphere.

The automobile is the main nonnatural source of hydrocarbons in the air. It was estimated that in 1968 the automobile was responsible for over 50% (160 million tons) of the hydrocarbon pollutants with industry contributing 14%, solid waste disposal 5%, gasoline marketing 4%, organic solvent evaporation 10%, and fires 11%. In addition to the hydrocarbons, it was estimated that the automobile added 60 million tons of CO and 7 million tons of nitrogen oxides (NO_x) to the atmosphere in the United States.

Olefinic hydrocarbons probably make an important contribution to the formation of photochemical smog (see Chapter 10) and this class of compound constitutes 45% of the hydrocarbon emissions from cars. Aromatics and paraffins are the next most important in smog formation and these constitute 20% and 24% of the automobile hydrocarbon emissions, respectively. However, the aromatic emissions are directly dependent on the percent aromatics in the gasoline. These emissions may increase as the lead in gasoline is replaced by aromatic compounds.

5.8.2 Hydrocarbons and Other Emissions from Automobiles

In 1960, before any federal regulations were imposed on emissions from automobiles, 20% of the automobile emissions were from the crankcase ventilator ("blowby"), 60% were from the exhaust pipe, and 20% were from gasoline evaporation from the gas tank and carburetor. The emissions from these sources in 1960 and the changes in these amounts for new automobiles as a result of federal regulation is shown in Table 5-2.

"Blowby" is the general term applied to those hydrocarbons that get past the piston rings of the internal combustion engine and into the crankcase. Prior to 1961 these were vented to the atmosphere. Automobile emissions were reduced by about 20% when these hydrocarbons were recycled back through carburetor and into the engine (positive crankcase ventilation).

Emissions resulting from evaporation can be controlled by the proper design of the carburetor and gas tank to minimize the escape of hydrocarbons. The addition of these design

TABLE 5-2

Maximum Emissions from New Automobiles (gm/mi)[a]

	Hydrocarbons: Crankcase	Hydrocarbons: Exhaust	Evaporation	CO	NO_x
1960[b]	3.3	11	3.0	79	3.9
1963	0	11	3.0	79	3.9
1968	0	3.3	3.0	34	5.8
1970	0	2.2	3.0	23	7.0
1970[c]	0	4.1	3.0	36	4.0
1974	0	3.3	0.1	39	3.0
1975	0	1.5	0	15	3.1
1977[d]	0	1.5	0	15	2.0
1978	0	0.4	0	3.4	0.4
1982	0	0.4	0	3.4	0.4

[a] *Adapted in part from D. J. Paterson and N. A. Herrein, "Emissions from combustion engines and their control." Ann Arbor Science Publications, Ann Arbor, Michigan, 1972.*

[b] *Average values before legislation.*

[c] *The standards from 1970 to present are based on a new procedure. The two sets of data for 1970 illustrate the differences in the two methods.*

[d] *California requires 0.4, 9.0, and 1.5 gm/mi for hydrocarbons, CO and NO_x, respectively.*

changes to the automobiles produced since 1971 should eventually reduce hydrocarbon emissions by another 20%.

The most difficult source of emissions to control is that from the exhaust pipe. The reason for this is that the internal combustion engine, as it is presently designed, will always emit some noncombusted hydrocarbons. The main source of these hydrocarbons is due to "wall quench." The burning of the gasoline close to the walls of the cylinder is quenched because the zone next to the wall is much cooler than the combustion zone in the center of the cylinder. This relatively cold area slows the rate of combustion so that the hydrocarbons in this region are swept out by the piston before combustion is complete. The crevice above the piston rings between the cylinder wall and the piston is also a zone where wall quench occurs (Fig. 5-3).

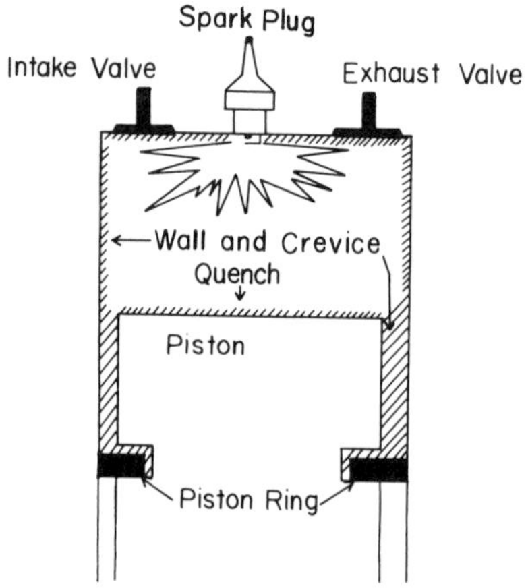

FIGURE 5-3 Areas of wall and crevice quench in the cylinder of a four-stroke cycle internal combustion engine.

There would still be hydrocarbon emissions from the automobile even if it were possible to eliminate wall quench. This is because there is insufficient time for the complete combustion of gasoline to carbon dioxide and water. Addition of more air (oxygen) to the engine does effect more complete combustion, but this results in an increase in the NO_x emissions (Fig. 5-4) and a decrease in the power (Fig. 5-5). The increase in NO_x is due to the higher temperature of the combustion zone in the engine in the presence of more oxygen.

Since the present designs of internal combustion engines cannot be built with satisfactorily low pollution levels, some type(s) of emission control will have to be added to the automobile if both NO_x and hydrocarbon emissions are to be controlled. In 1973, recirculation of the exhaust gases was introduced to help control NO_x emissions. In addition, in some cars air is injected into the hot exhaust gases to facilitate continued oxidation of hydrocarbons. In 1975, 85% of the new cars sold in the U.S. were equipped with a platinum-palladium catalytic reactor in the exhaust stream to catalyze the oxidation of CO and hydrocarbons and, if the 1978 NO_x standards are enforced, a reduction catalyst installed prior to the oxidation catalyst will probably be required. The present catalytic reactor is effective in reducing hydrocarbon emissions, but early reports that 10-30% of the sulfur in the petroleum was oxidized to SO_3 and emitted as sulfates and sulfuric acid mists raised concerns about the localized potentially hazardous concentrations of these aerosols that

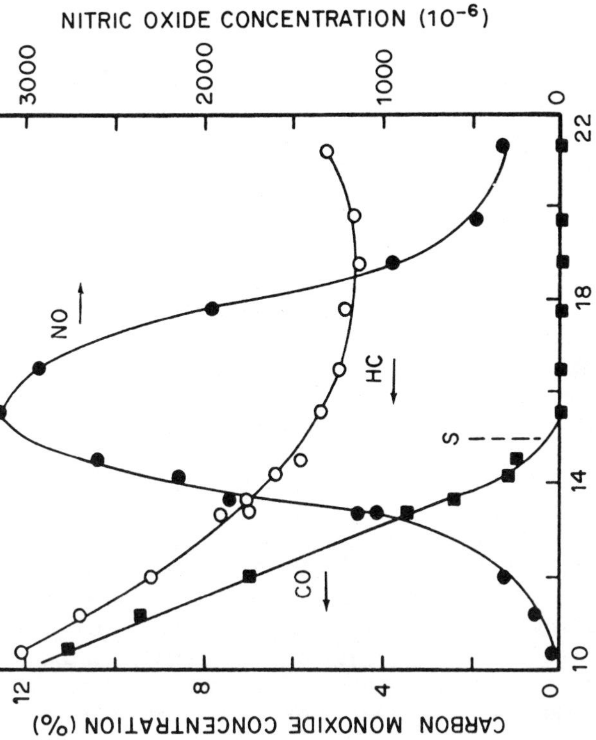

FIGURE 5-4. Effect of air-fuel ratio on hydrocarbon, carbon monoxide, and nitric oxide exhaust emissions. From W. Agnew, Proc. Roy. Soc. Ser. A. 307, *153 (1968).*

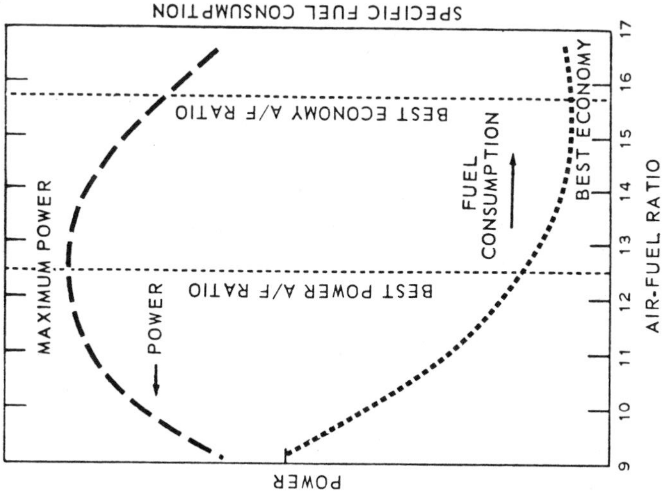

FIGURE 5-5. Effect of air-fuel ratio on power and economy.

might be generated from this source. In the absence of a catalytic converter, most of the sulfur is emitted as SO_2 and this is only slowly converted to SO_3 in the atmosphere (see Chapter 10). More recent reports suggest that sulfuric acid emissions from automobiles are not an environmental problem, however. A second problem with the catalytic converter is the uncertainty of its useful lifetime. Some mechanism must be devised that will indicate when the catalyst has been poisoned and is no longer effective. The possibility that the lead from high octane gasoline will poison the catalyst has been overcome by the requirement that all 1975 model cars equipped with a catalytic reactor use lead-free gasoline. To insure that no leaded gasoline is used, cars equipped with a catalytic converter have a special inlet on the gas tank so that it can only be filled at a gasoline pump which dispenses lead-free gasoline. Some progress has been made in the design of a catalytic converter that is not poisoned by lead. Results reported in 1977 demonstrate a 25% reduction in CO, particulate matter, and lead along a Los Angeles freeway since the introduction of the catalytic converter in 1974. The oxides of nitrogen have increased 50% over the same time period.

5.8.3 Gasoline and Lead

A gradual decrease in the lead content of gasoline was to begin in 1975. The Environmental Protection Agency instigated this decrease not only because of the adverse effect of lead on the catalytic converter but also because of the possibility that the lead emissions from automobiles may be hazardous. Ninety-eight percent of the 185,000 tons (166,500,000 kgm) of atmospheric lead resulted from leaded gasoline in 1968. Although it is well known that lead compounds are toxic (see Chapter 14), it has not been directly established that there is sufficient lead in the atmosphere to constitute a health hazard. It seems very likely that this large amount of lead would have some undesirable effects. The average lead content of all grades of gasoline coming from each refinery was scheduled to be decreased from 1.7 gm/gal or less in 1975 to 1.4 in 1976, 1.0 in 1977, 0.8 in 1978, and 0.5 in 1979. The standards for 1976 and 1977 were deleted by the EPA in 1976, and the 1978 standard will be suspended if the refiner can show that he will meet the value of 0.5 gm/gal in 1979.

The deletion of lead from gasoline will require the use of aromatics to maintain the present octane rating. This will result in increased exhaust emissions of aromatics, but it is assumed that the catalytic reactor will complete the oxidation of these compounds. Addition of more unsaturated compounds will increase the cost of gasoline. It was estimated in 1970 that the removal of lead while maintaining the same octane rating would increase the cost of gasoline by \$0.02/gal. However, it should be noted that the octane rating of the "no-lead" gasoline is equal to or lower than "regular" so there really should not be a large cost differential.

5.8.4 Low Emission Automobile Engines

A major unexpected problem resulting from the total emission control program is the lower operating efficiency of the automobile. Tuning the engine to produce fewer emissions has resulted in a greater consumption of gasoline - a serious problem in view of the diminishing oil reserves. However, the use of a catalytic reactor on 1975 cars made it possible to retune the engine to give better fuel economy since the control of emissions did not have to be done with the engine.

At the present time, all emission control technology has centered on the use of petroleum distillates in a piston-driven internal combustion engine. Alternative engine designs such as the rotary (Wankel), Stirling, turbine, and Rankine (steam) are being tested. (The references give further discussion of these engines.) Other sources of fuel such as diesel fuel (kerosene or light oil), methane, methanol, hydrogen, and in one instance chicken manure (really methane formed by the action of bacteria on the manure) are being investigated. A battery-powered automobile may be the ultimate low pollution vehicle. The extensive use of these vehicles awaits the development of a high energy density fuel cell or battery. Electricity from a power plant would be required to recharge batteries. However, the pollution from a few power plants is controlled more readily than the emissions from thousands of automobiles. Currently, automobiles with Wankel engines are manufactured in Japan and buses powered by Stirling engines are used in Sweden. The widespread use of alternatives to the four-stroke cycle internal combustion engine probably will not occur before 1985.

5.9 COAL

5.9.1 Coal Formation and Structure

Coal is probably the first of the fossil fuels to be used for energy. It was recognized as an energy source by the Chinese around 1100 B.C. while the ancient Greeks were probably the first of the western cultures to be aware of coal. The Romans reported that the "flammable earth" was being mined in Gaul when they captured that section of Europe. The first known coal mines in North America were operated by the Hopi Indians of Arizona some 200 years before Columbus.

Coal occurs mainly in the north temperate regions of the earth. Small deposits are located at the poles but there is almost none in the tropics. The bulk of the deposits are located in North America (50%) while significant amounts are found in the USSR (20%) and Asia (15%).

Coal was formed from the debris of giant tree ferns and other vegetation that was growing in swamps and bogs 300 million years ago. When the plants died they were covered with sediment that prevented their oxidation by atmospheric oxygen. Further build up of sediment and other geologic processes subjected these materials to high pressures that resulted in their conversion to coal.

There are several classes of coal that reflect different stages in the metamorphosis from the carbohydrate and lignin of plants to the carbon of anthracite coal. Peat, a substance of a low heat value, is formed in the first stage of the process as a result of anaerobic microbial transformations of plant material. The lignin and microbial remains in the peat gradually lose H_2O and CO_2 as a result of high temperatures and pressures and are converted successively to lignite, bituminous coal, and anthracite. Anthracite, or hard coal, is about 85% carbon with the remainder being mainly inorganic material and water.

The carbonization of plant carbohydrate results in the formation of a highly condensed aromatic ring structure similar to graphite. Aliphatic rings and chains as well as hydroxyl, carbonyl, carboxyl, and ether groupings are attached to these aromatic structures. Nitrogen is present as the cyclic amine function and sulfur as thiol (RSH) and thioether (RSR) groupings. Some of these structural elements are given in Fig. 5-6.

Coal contains a variety of inorganic ions that ultimately end up in the ash residue after the coal is burned. Some 36 elements were detected in the ash resulting from West Virginia

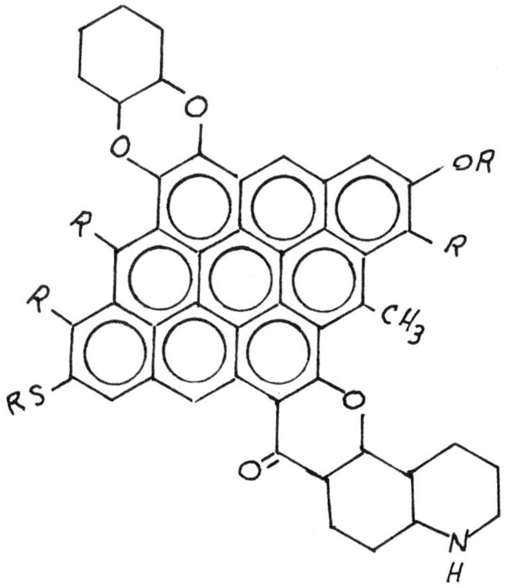

FIGURE 5-6. A structural unit present in coal.

bituminous coal with the major constituents being Na, K, Ca, Al, Si, Fe, and Ti.

The environmental problems associated with the use of coal are discussed in Chapter 3.

5.9.2 Coal Gasification

The gasification of coal was carried out extensively in the 1920s in the U.S. but the discovery of natural gas wells resulted in the shift to the cleaner, cheaper natural gas as a fuel. As a consequence only small advances have been made since 1920 in the gasification of coal.

Gasification of coal using oxygen and steam at temperatures of 900°C results in the formation of a low Btu gas ($\sim$100 Btu/ft^3; 1 Btu = 1054 J). This low Btu gas can be converted to a high Btu gas ($\sim$1000 Btu/ft^3) which is mainly methane. A variety of different reactors and catalysts have been designed for the gasifications of coal. The chemistry is similar in each and occurs in the following stages. First, the coal is heated to 600-800°C to form coke and a mixture of volatiles which contain mainly methane (5-75). In the second step, oxygen is added and combustion is initiated to raise the temperature to 900°C. Steam is then added to the mixture and CO and H_2 are formed (5-76). These gases together with the methane and other volatiles formed in the first step are the energy-containing constituents of low Btu gas. If high Btu gas is desired, the low Btu gas is passed into another reactor where a portion of the CO reacts with water to give hydrogen (5-77). Impurities such as H_2S are removed at this stage and then the CO and H_2 are transferred to a final reactor where the mixture is catalytically converted to methane (5-78).

The H_2S is removed from the gas mixture by reaction with half-calcined dolomite (5-80), and recovered by reversing reaction (5-80) by the addition of CO_2. Partial oxidation results in the conversion of H_2S to sulfur [reactions (5-81) and (5-82)].

$$\text{coal} \xrightarrow{600\text{-}800^\circ C} \text{coke} + CH_4 \text{ and other volatiles} \quad (5\text{-}75)$$

$$\text{coke} + H_2O \xrightarrow{900^\circ C} CO + H_2 \quad (5\text{-}76)$$

$$CO + H_2O \rightleftarrows CO_2 + H_2 \quad (5\text{-}77)$$

$$CO + 3H_2 \xrightarrow[\text{catalyst}]{\text{Ni}} CH_4 + H_2O \quad (5\text{-}78)$$

$$CaCO_3 \cdot MgCO_3 \overset{\Delta}{\rightleftarrows} \underset{\text{half-calcined dolomite}}{CaCO_3 \cdot MgO} + CO_2 \quad (5\text{-}79)$$

$$CaCO_3 \cdot MgO + H_2S \rightleftarrows CaS \cdot MgO + H_2O + CO_2 \quad (5\text{-}80)$$

$$2H_2S + 3O_2 \rightarrow 2SO_2 + 2H_2O \quad (5\text{-}81)$$

$$2H_2S + SO_2 \rightarrow 3S + 2H_2O \quad (5\text{-}82)$$

5.9.3 Coal Liquefaction

The technology of coal liquefaction is at a more primitive stage of development than gasification. At present the only coal liquefaction plant in the world is located in South Africa. In this plant the coal is first converted to low Btu gas [(5-76), (5-77)] and this mixture is hydrogentated at 200°C in the presence of an iron catalyst to a mixture of hydrocarbons. Methanol can be produced instead of the hydrocarbons if a copper catalyst is used.

Since the liquefaction procedure used in South Africa is not efficient, other processes that involve the direct conversion of coal to hydrocarbons are being tested in the U.S. In one process the pulverized coal is mixed with a solvent, hydrogen is added, and the mixture is heated until the coal dissolves. A high boiling oil is obtained that can be used as a boiler fuel. Variations on this process are being tested that involve the use of catalysts in the hydrogenation step. There are many problems to be resolved before a commercial plant can be constructed. One of these involves developing an inexpensive process for the production of hydrogen from a separate portion of coal or from the unreacted solids remaining from the hydrogenation process. An experimental plant designed to treat 600 tons per day of high sulfur eastern U.S. coal is scheduled for completion in Castleburg, Kentucky in 1978. If successful, it can be enlarged to a commercial scale of 10,000 tons per day.

BIBLIOGRAPHY

(Also see the references given at the end of Chapter 2.)

Oil Spills

T. J. Maugh, Rerefined oil: An option that saves oil, minimizes pollution, *Science 193,* 1108 (1976).

R. D. Wilson, P. H. Monaghan, A. Osanik, L. C. Price, and M. A. Rogers, Natural marine oil seepage, *Science 184,* 857 (1974).

M. Blumer, Scientific aspects of the oil spill problem, *Environ. Affairs 1,* 54 (1971).

Oil spill technology takes a big step forward, *Chem. Eng. News,* April 9, 1973, pp. 12-14.

Cleaning oil spills isn't simple, *Environ. Sci. Tech 7,* 398 (1973).

C. E. Steinhart and J. S. Steinhart, "Blowout. A Case Study of the Santa Barbara Oil Spill." Duxbury, Belmont, California, 1972.

J. E. Smith, "'Torry Canyon' Pollution and Marine Life." Cambridge University Press, London, 1968.

R. A. Brown and H. L. Huffman, Jr., Hydrocarbons in open ocean waters, *Science 191,* 847 (1976).

M. Blumer, Polycyclic aromatic compounds in nature, *Sci. Amer. 234* (3), 35 (1976).

W. B. Travers and P. R. Luney, Drilling, tankers and oil spills on the Atlantic outer continental shelf, *Science 194*, 791 (1976).

J. E. Moss, in "Impingement of Man on the Ocean," D. Hood (Ed.), p. 381. Wiley, New York, 1971.

E. J. Ledet and J. L. Laseter, Alkanes at the air-sea interface from offshore Louisiana and Florida, *Science 186*, 261 (1974).

C. W. Bird and J. M. Lynch, Formation of hydrocarbons by microorganisms, Chem. Soc. Rev. 3, 309 (1974).

National Academy of Sciences, "Petroleum in the Marine Environment," Workshop on Inputs, Fate, and the Effects of Petroleum in the Marine Environment, National Academy of Sciences, Washington, D.C., 1975.

Gasoline - Automobile Emissions

B. F. Greek, Gasoline. Antipollution forces bring marked changes to petroleum refining industry, *Chem. Eng. News*, Nov. 9, 1970. pp. 52-60.

D. J. Patterson and N. A. Henein, "Emissions from Combustion Engines and their Control," Ann Arbor Sc. Pub., Ann Arbor, Michigan, 1972.

W. G. Agnew, Effects of air fuel ratios on hydrocarbon, carbon monoxide, and nitric oxide exhaust emissions, *Proc. Roy. Soc. Ser. A 307*, 153 (1968)

"Control techniques for carbon monoxide, nitrogen oxide and hydrocarbon emissions from mobile sources," *National Air Pollution Control Publication A*, 66, (1970).

G. Walker, The Stirling engine, *Sci. Amer. 229* (2), 80 (1973).

D. E. Cole, The Wankel engine, *Sci. Amer. 227* (2), 14, (1972).

W. Worthy, New auto engines still decade away, *Chem. Eng. News,* May 26, 1975, pp. 23-25.

R. A. Hites and K. Biemann, Water pollution: Organic compounds in the Charles River, Boston, *Science 178,* 158 (1972).

R. S. Morse, "The automobile and air pollution: A program for progress." U.S. Dept. of Commerce, 1967.

R. H. Ebel, "Catalytic removal of potential air pollutants from auto exhaust," in "Advances in Environmental Sciences," Vol. 1 (J. N. Pitts and R. L. Metcalf, eds.), p. 237, Wiley, New York, 1969.

C. Holden, Auto emissions: EPA decision due on another clean-up delay, *Science 187,* 818 (1975).

L. Gibney, Court challenges EPA data for unleaded gas, *Chem. Eng. News,* Feb. 10, 1975, p. 13 (see also p. 15).

"Catalysts for the control of automotive pollutants." Advances in Chemistry Series No. 143, American Chemical Society, Washington, D.C., 1975.

E. E. Wigg, R. J. Campion, and W. L. Peterson, "The effect of fuel hydrocarbon composition on exhaust hydrocarbon and oxygenate emissions." Society of Automotive Engineers, Automotive Engineering Congress, 1972, Paper 720251.

K. L. Dermerjian, J. A. Kerr, and J. G. Calvert, "The mechanism of photochemical smog formation," in "Advances in Environmental Science and Technology," Vol. 4 (J. N. Pitts, Jr., R. L. Metcalf, and A. C. Lloyd, eds.), p. 1. Wiley-Interscience, 1974.

J. B. Heywood and J. C. Keck, Formation of hydrocarbons and oxides of nitrogen in automobile engines, *Environ. Sci. Tech 7,* 216 (1973).

W. A. Gruse, "Motor fuels, performance and testing," p. 280 Reinhold, New York, 1967.

T. H. Maugh II, Air pollution: Where do hydrocarbons come from?, *Science 189,* 277 (1975).

T. H. Maugh II, Sulfuric acid from cars: A problem that never materialized, *Science 198,* 280 (1977).

Oxidation of Hydrocarbons

H. A. Leedy, "Ozone chemistry and technology." Advances in Chemistry Series #21, American Chemical Society, Washington, D.C., 1959.

B. Weinstock, "Photochemical smog and ozone reactions." Advances in Chemistry Series #113, American Chemical Society, Washington, D.C., 1972.

F. R. Mayo, "Oxidation of organic compounds I, II, III." Advances in Chemistry Series #75, 76, 77, American Chemical Society, Washington, D.C., 1968.

P. S. Bailey, The reactions of ozone with organic compounds, *Chem. Rev. 58,* 925 (1958).

R. W. Murray, The mechanism of ozonolysis, *Acct. Chem. Res. 1,* 313 (1968).

F. R. Mayo, Free-radical autoxidations of hydrocarbons, *Acct. Chem. Res. 1,* 193 (1968).

D. R. Kearns, Physical and chemical properties of singlet molecular oxygen, *Chem. Rev. 71,* 395 (1971).

E. E. Royals, "Advanced organic chemistry," p. 80. Prentice-Hall, Englewood Cliffs, New Jersey, 1954.

W. A. Pryor, Free radical pathology, *Chem. Eng. News,* June 7, 1971, pp. 34-51.

W. A. Pryor, "Free radicals in biological systems," in "Chemistry in the Environment, Readings from Scientific American, p. 311. W. H. Freeman, San Francisco, California, 1973. (From *Sci. Amer.,* August, 1970.)

H. E. O'Neal and C. Blumstein, A new mechanism for gas phase olefin - ozone reactions, *Inter. J. Chem. Kinetics 5,* 397 (1973).

P. D. Bartlett, Four-membered rings and reaction mechanisms, *Chem. Soc. Revs. 5,* 149 (1976).

Coal

D. W. Van Krevelen, "Coal-typology-chemistry-physics-constitution," p. 514. Elsevier, New York, 1961.

S. M. Monskaya and T. V. Drozdova, "Geochemistry of organic substances," p. 345. Translated by L. Shapiro and I. A. Breger, Pergamon Press, New York, 1968.

"Coal gasification," in "Modern Energy Technology," Vol. II (H. Fogiel, ed.), Chapter 26. Research and Education Association, New York, 1975.

"Coal gasification techniques," in "Modern Energy Technology," Vol. II (H. Fogiel, ed.), Chapter 36. Research and Education Association, New York, 1975. (See also Chapters 37-43.)

"Conversion of coal to oil," in "Modern Energy Technology," Vol II (H. Fogiel, ed.), Chapter 44. Research and Education Association, New York, 1975.

"Chemical desulfurization of coal," in "Modern Energy Technology," Vol. II (H. Fogiel, ed.), Chapter 48. Research and Education Association, New York, 1975. (See also Chapter 49.)

6
SOAP AND DETERGENTS

6.1 INTRODUCTION

Most detergents are petrochemicals, compounds formed by the chemical transformation of petroleum. Consequently, the environmental degradation of many of these compounds is similar to that of petroleum. Soaps, the sodium, potassium, or ammonium salts of fatty acids, are prepared by the alkaline hydrolysis of plant and animal fats (saponification). Soaps and detergents merit separate consideration because of their widespread household use and because the mechanisms of their environmental degradation are well understood. As far as the organic constituents are concerned, the problem regarding their environmental degradation has been solved. However, the pollution problem due to the polyphosphate ($Na_5P_3O_{10}$) added to the detergents remains to be resolved (see Chapter 13).

The environmental problem caused by nonbiodegradable detergents was probably the first to be solved as a result of the pressure of public opinion. The foam covering many otherwise scenic rivers and streams was a source of much public concern for many years. Some indication of the nature of the problem was apparent as early as 1947 when the aeration tanks at the sewage disposal plant at Mt. Penn, Pennsylvania were inundated with suds. This spectacular effect was the result of a promotional campaign where housewives were given free samples of a new detergent at local stores. Everyone used the new product at about the same time but unfortunately the microorganisms residing at the disposal plant could not metabolize this detergent very efficiently. Despite this early warning, nonbiodegradable detergents were manufactured up to 1965 before the detergent industry finally responded to public pressure and voluntarily switched to the manufacture of biodegradable products.

6.2 PROPERTIES OF SOAPS AND DETERGENTS

Soaps and detergents have similar structures and their mechanisms of cleansing action are similar. Both contain a hydrophobic portion (usually a long hydrocarbon chain) to which a hydrophilic (polar) group is attached. Such materials are surface active, that is they concentrate at the surface of an aqueous solution. This lowers the surface tension of the water so that it penetrates the surface and interstices of the object being cleaned. If the binding of the detergent to the substance being cleaned is greater than the binding of the soiling agent, the latter will then be displaced into the water phase. Oils and greases will also adsorb the hydrophobic end of the surface-active agent. The soiling agent can be washed away because the affinity of the polar group of adsorbed soap or detergent for water keeps the dirt suspended in the water.

The main disadvantage of the use of salts of carboxylic acids as cleansing agents is the insoluble precipitates that form with Ca^{2+} and Mg^{2+}. This curdy grey precipitate deposits on clothes and leaves a "ring" in the bathtub. The deposits are not only undesirable but also greater amounts of soap must be used to make up for that lost by precipitation. Another disadvantage of soaps is that they are weak acids that protonate in mildly acidic or neutral solutions. The polarity of the "head" of the detergent decreases when protonated and as a consequence it does not solubilize the oils and greases as rapidly.

The main advantage of detergents is they are not precipitated by hard water. However their cleansing power is decreased markedly by calcium and magnesium ions so that it is necessary to add "builders" (commonly, polyphosphates, see Chapter 13) to chelate these ions and to prevent "water spotting" on dishes caused by insoluble calcium and magnesium salts, when automatic dishwaters are used. Sodium sulfate, sodium silicate, and sodium borate are other common components in detergent formulations. These inorganic compounds often exceed the amount of organic surface active agent present. Smaller quantities of brighteners, soil redeposition preventatives, foam control agents, and other materials also may be present.

6.3 SYNTHESIS OF SOAPS

Soaps are prepared by the saponification (base hydrolysis) of animal or vegetable fats. Hydrolysis converts the triglycerides in the fats to the sodium salts of carboxylic acids (soaps). The carboxylate group is the polar, water-soluble end of the soap.

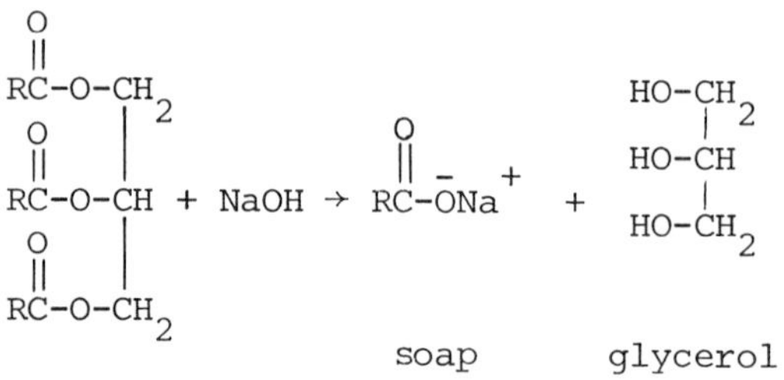

(6-1)

Stearic acid, $CH_3(CH_2)_{16}CO_2H$, is the major carboxylic acid released on hydrolysis of animal fat while oleic acid, $CH_3(CH_2)_7CH{=}CH(CH_2)_7CO_2H$, is the main acid released on hydrolysis of olive oil. Hydrolysis of palm oil yields approximately equal amounts of oleic acid and palmitic acid, $CH_3(CH_2)_{14}CO_2H$.

Since long chain ("fatty") acids occur naturally, it is not surprising to find some foam and suds at waterfalls and other places of turbulence in streams and rivers due to the presence of natural soaps in the water.

6.4 SYNTHESIS OF DETERGENTS

6.4.1 The General Nature of Detergents

It has been possible to synthesize a wide variety of detergents by varying the structure of both the polar and nonpolar ends of the molecule. The polar end most commonly used is a sulfonic acid group RSO_3^- and consequently such detergents are anionic in nature. Some alkyl sulfates $R\text{-}OSO_3^-$ also are in use. Cationic detergents, based on the hydrophilic quarternary ammonium group ($RN(CH_3)_3^+$) and nonionic detergents with hydroxyl groups as the hydrophilic sites, also are encountered. Cationic detergents have poor cleaning properties but are good germicides and frequently are encountered in specialty products where this property is important. In 1962, 70% of the detergents sold were alkylbenzene sulfonates (ABS), which were prepared from propene obtained by the cracking of hydrocarbons and from benzene made by the catalytic reforming of petroleum. Linear alkyl sulfonates (LAS) are presently in greatest use. It should be noted that a synthetic detergent is not a pure chemical entity but is a mixture of very similar compounds. This is due to the use of a mixture of olefins in the feed stock and to variation in the extent of oligomerization of these olefins.

6.4.2 Synthesis of ABS Detergents

The synthesis of the ABS detergents is outlined in reactions (6-2) to (6-4). The starting material propene is oligomerized:

$$CH_3CH{=}CH_2 \xrightarrow[300\text{-}500°C]{H_3PO_4} (CH_3)_2CH(CH_2\overset{\overset{CH_3}{|}}{C}H)_2CH{=}CH\text{-}CH_3 \qquad (6\text{-}2)$$

This propene oligomer is abbreviated as $RCH{=}CHCH_3$ in reaction (6-3), the reaction with benzene.

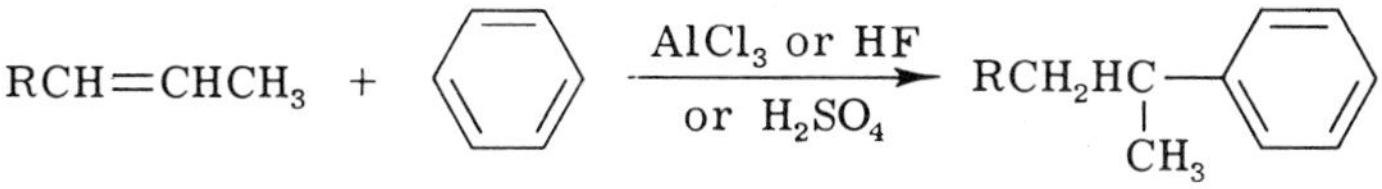

(6-3)

The sulfonate group is then introduced into the aromatic ring:

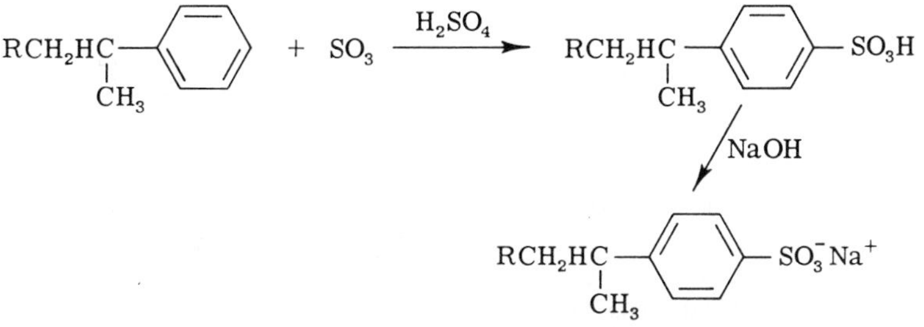

(6-4)

ABS detergent

The propene oligomer produced in reaction (6-2) is probably formed in the sequence outlined in (6-5).

$$CH_3CH{=}CH_2 + H^+ \rightarrow CH_3C^+HCH_3 \xrightarrow{CH_3CH=CH_2} (CH_3)_2CHCH_2\overset{\overset{CH_3}{|}}{C}H^+$$

$$\xrightarrow{2CH_3CH=CH_2} (CH_3)_2CH(CH_2\overset{\overset{CH_3}{|}}{C}H)_2CH_2C^+HCH_3 \xrightarrow{-H^+}$$

$$(CH_3)_2CH(CH_2\overset{\overset{CH_3}{|}}{C}H)_2CHCH{=}CH_2 \qquad (6\text{-}5)$$

The alkylation of benzene [reaction (6-3)] involves the attack of a carbonium ion intermediate on the electron-rich benzene ring [reactions (6-7) and (6-8)]. The sulfonation of benzene [reaction (6-9)] follows a pathway which is completely analogous to the alkylation of benzene [reaction (6-3)].

$$HCl + AlCl_3 \rightarrow H^+AlCl_4^- \qquad (6\text{-}6)$$

$$RCH{=}CH_2 + H^+AlCl_4^- \rightarrow R\overset{+}{C}HCH_3 + AlCl_4^- \qquad (6\text{-}7)$$

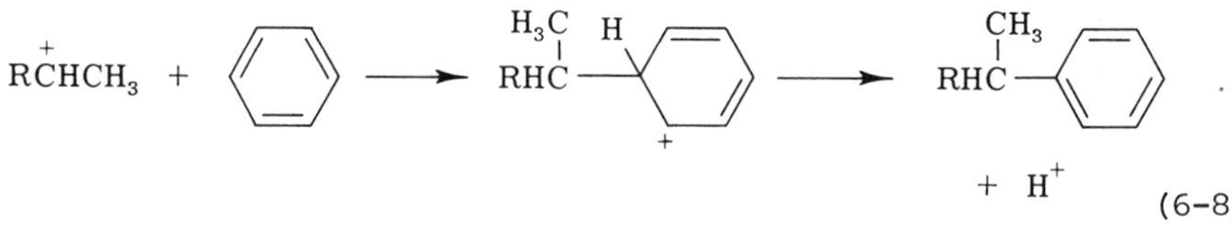

(6-8)

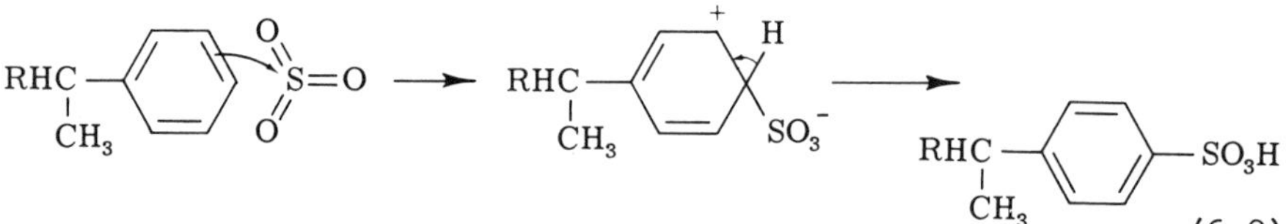

(6-9)

Reactions (6-3) and (6-8) are oversimplified because the carbonium ion formed initially rearranges by hydride shifts to yield a mixture of carbonium ions. Hence a mixture of alkylated products is obtained in (6-3) and (6-8).

The ABS detergent is not biodegradable because microorganisms are not able to metabolize a hydrocarbon chain containing numerous side chains. Linear side chains are biodegradable and since 1965, the emphasis has been directed toward the synthesis of linear alkyl sulfonate (LAS) detergents. The linear alkyl sulfonates account for the bulk of the detergents used today.

6.4.3. Synthesis of Linear Detergents

It is not possible to synthesize linear hydrocarbon molecules by the acid catalyzed obligomerization of olefins because the more substituted secondary carbonium ion is formed, which results in the formation of a branched chain hydrocarbon as shown in reaction (6-2). However, linear hydrocarbons can be separated from petroleum or prepared by other synthetic methods.

Urea crystallizes in a unique fashion from hydrocarbon solutions and this is the basis of one method of separating linear and branched hydrocarbons. The urea crystallizes as helices and the linear hydrocarbons are encased in the central channel of the tubes formed by these helices. Branched chain hydrocarbons do not fit into the 5- Å diameter tube so the linear hydrocarbons can be separated by merely filtering the urea crystals. "Lightly branched" hydrocarbons with more than ten carbon atoms will also form urea inclusion compounds and these are also present in the mixture of "linear" hydrocarbons separated by this procedure.

The use of molecular sieves has proved to be a cheaper method for the separation of linear and branched hydrocarbons. Molecular sieves are semipermeable inorganic zeolites (a clay mineral) that have a channel structure that is similar to that of urea when crystallized from hydrocarbons. The molecular sieves with a cross section of 5 Å are used to screen linear hydrocarbons from petroleum selectively.

Linear hydrocarbons can be synthesized from olefins by use of catalysts such as the complex of triethylaluminum and titanium tetrachloride. The product can be an alkane or a derivative. In one process, the "alfol process," linear alcohols are formed by the air oxidation and subsequent hydrolysis of the oligomers formed initially [reactions (6-10) and (6-11)]. Olefins can be formed by the thermolysis of the adduct of the organometallic derivatives (6-12).

$$-\overset{|}{M}-CH_2CH_3 + nCH_2{=}CH_2 \longrightarrow -\overset{|}{M}-(CH_2CH_2)_nCH_2CH_3 \qquad (6\text{-}10)$$

$$-\overset{|}{M}-(CH_2CH_2)_nCH_2CH_3 \xrightarrow{O_2} -\overset{|}{M}-O(CH_2CH_2)_nCH_2CH_3$$

$$\xrightarrow{H^+} CH_3CH_2(CH_2CH_2)_nOH \qquad (6\text{-}11)$$

$$-\overset{|}{M}-(CH_2CH_2)_nCH_2CH_3 \xrightarrow[Ni]{\Delta} CH_3CH_2(CH_2CH_2)_{n-1}CH{=}CH_2 \qquad (6\text{-}12)$$

The syntheses of the biodegradable linear detergents are completed by the addition of a "polar head" to the linear hydrocarbon $R'CH_2CH_3$. The alkane can be chlorinated as in (6-13) and converted to a LAS [reactions (6-14)-(6-16)] by procedures similar to those used in the synthesis of the ABS detergents [reactions (6-3) and (6-4)]. The chain branching next to the benzene ring does not impede the microbial degradation of these compounds (see Section 6.5).

$$n\text{-}R'CH_2CH_3 + Cl_2 \longrightarrow R'CH(CH_3)Cl \qquad (6\text{-}13)$$

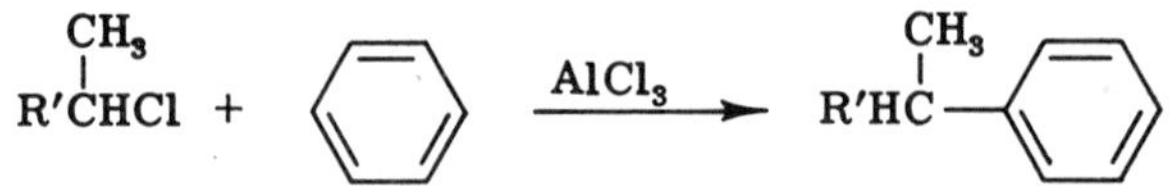

(6-14)

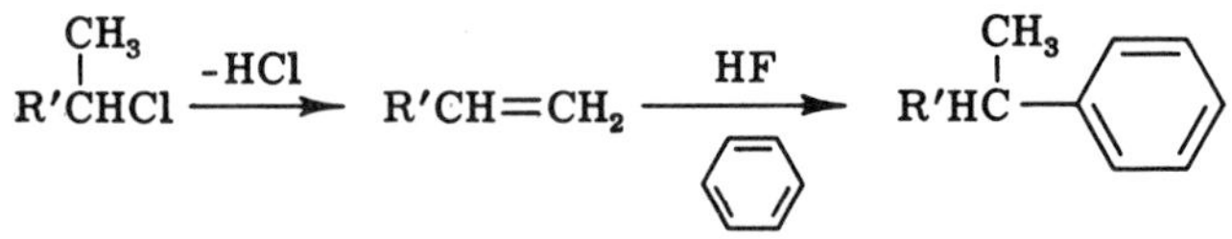

(6-15)

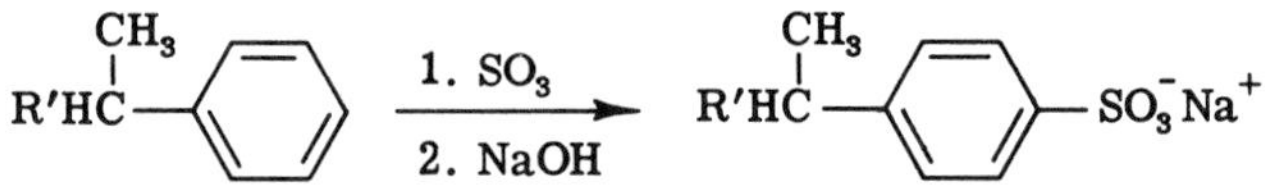

(6-16)

LAS

(The difference between the ABS and LAS structures is that the R group in the former is a branched chain while the R' group in the latter is not branched.) Alternatively the alkanes can be oxidized to a mixture of alcohols as in (6-17), and these compounds can be converted to sulfate esters (6-18) or to nonionic detergents by reaction with ethylene oxide (6-19).

$$n\text{-}R'CH_2CH_3 \xrightarrow[\Delta]{O_2} R'CH(CH_3)OH \text{ (a mixture of alcohols)} \qquad (6\text{-}17)$$

$$\mathrm{R'CH(CH_3)OH + H_2SO_4 \rightarrow R'CH(CH_3)OSO_3H \xrightarrow{NaOH} R'CH(CH_3)OSO_3^-Na^+} \qquad (6\text{-}18)$$

sulfate ester

$$\mathrm{R'CH(CH_3)OH + 8\ \overset{O}{CH_2CH_2} \xrightarrow{NaOH} R'CH(CH_3)O\text{-}(CH_2CH_2O)_7CH_2CH_2OH} \qquad (6\text{-}19)$$

a nonionic detergent

The mechanism of the formation of the nonionic detergent involves the anionic oligomerization of ethylene oxide by the initial attack of the alcohol anion.

$$\mathrm{ROH + OH^- \rightleftharpoons RO^- + HOH} \qquad (6\text{-}20)$$

$$\mathrm{RO^- + \overset{O}{CH_2CH_2} \longrightarrow ROCH_2CH_2O^-} \qquad (6\text{-}21)$$

$$\mathrm{ROCH_2CH_2O^- + n\overset{O}{CH_2CH_2} \longrightarrow RO(CH_2CH_2O)_nCH_2CH_2O^-} \qquad (6\text{-}22)$$

$$\mathrm{RO(CH_2CH_2O)_nCH_2CH_2O^- + HOH \rightleftharpoons RO(CH_2CH_2O)_nCH_2CH_2OH + OH^-} \qquad (6\text{-}23)$$

6.5 THE METABOLISM OF HYDROCARBONS, SOAPS, AND DETERGENTS

Microorganisms degrade hydrocarbons, soaps, and detergents because they can use the energy and organic fragments released for the synthesis of biomolecules essential for their growth. The metabolism of these compounds is affected by enzymes, high molecular weight proteins which catalyze a broad range of chemical transformations in living systems. Enzymes are characterized by their extreme catalytic efficiency (the catalyzed reactions typically proceed about 10^9 times as fast as the uncatalyzed reaction) and specificity (there is usually a specific enzyme for each chemical transformation that takes place in a cell). Many enzymes require a small, nonprotein prosthetic group, called a coenzyme, for their catalytic activity. These prosthetic groups are usually bound to the enzyme during the course of the reaction and sometimes they combine with one or more of the reacting chemical species.

Fatty acids (protonated soaps) occur widely in biological systems and are therefore readily degraded by microorganisms in enzyme-catalyzed reactions. Linear hydrocarbons and linear detergents are similar in structure to soaps and these are degraded by the same pathway.

The steps in the oxidative metabolism of fatty acids are outlined in reactions (6-24)-(6-28). Each of these reactions is catalyzed by an enzyme even though the enzyme is not specified in the reaction. Coenzyme A, abbreviated as CoASH, is a complex structure that contains a thiol group (-SH) as the reactive center. The acetyl coenzyme A ($CH_3\overset{O}{\overset{\|}{C}}SCoA$) produced in reaction (6-28) may be used by the microorganism for the synthesis of the specific fatty acids needed for growth by a pathway that is essentially the reverse of the one shown here. FAD and NAD^+ are the oxidized forms of hydrogen transfer coenzymes while $FADH_2$ and NADH are the respective reduced forms.

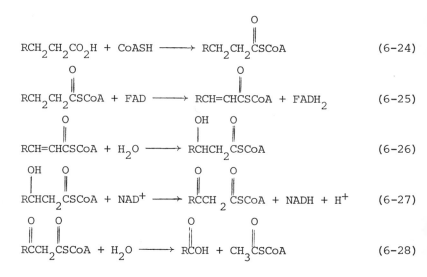

$$RCH_2CH_2CO_2H + CoASH \longrightarrow RCH_2CH_2\overset{O}{\overset{\|}{C}}SCoA \qquad (6\text{-}24)$$

$$RCH_2CH_2\overset{O}{\overset{\|}{C}}SCoA + FAD \longrightarrow RCH{=}CH\overset{O}{\overset{\|}{C}}SCoA + FADH_2 \qquad (6\text{-}25)$$

$$RCH{=}CH\overset{O}{\overset{\|}{C}}SCoA + H_2O \longrightarrow R\overset{OH}{\overset{|}{C}}HCH_2\overset{O}{\overset{\|}{C}}SCoA \qquad (6\text{-}26)$$

$$R\overset{OH}{\overset{|}{C}}HCH_2\overset{O}{\overset{\|}{C}}SCoA + NAD^+ \longrightarrow R\overset{O}{\overset{\|}{C}}CH_2\overset{O}{\overset{\|}{C}}SCoA + NADH + H^+ \qquad (6\text{-}27)$$

$$R\overset{O}{\overset{\|}{C}}CH_2\overset{O}{\overset{\|}{C}}SCoA + H_2O \longrightarrow R\overset{O}{\overset{\|}{C}}OH + CH_3\overset{O}{\overset{\|}{C}}SCoA \qquad (6\text{-}28)$$

This scheme shows why the highly branched ABS detergents are degraded very slowly by microorganisms. If a branched chain acid is oxidized, a tertiary alcohol will be produced as in reaction (6-29), a reaction equivalent to (6-26).

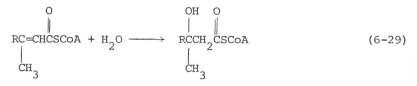

$$R\underset{CH_3}{\underset{|}{C}}{=}CH\overset{O}{\overset{\|}{C}}SCoA + H_2O \longrightarrow R\underset{CH_3}{\underset{|}{\overset{OH}{\overset{|}{C}}}}CH_2\overset{O}{\overset{\|}{C}}SCoA \qquad (6\text{-}29)$$

The tertiary alcohol cannot be oxidized to a ketone; thus further reaction is inhibited by the presence of the side chain.

The microbial breakdown of detergents and petroleum requires the terminal oxidation of the hydrocarbon chain to a carboxylic acid group to initiate the β-oxidation procedure outlined above. This terminal oxidation is well-documented process, and in some instances it has been established that the corresponding alcohol and aldehyde are reaction intermediates. Molecular oxygen is the oxidizing agent and the oxygen is activated by binding to the iron-containing enzyme cytochrome P-450. The reaction follows the sequence shown in (6-30):

$$RCH_2CH_2CH_3 \longrightarrow RCH_2CH_2CH_2OH \longrightarrow RCH_2CH_2CO_2H \longrightarrow RCO_2H + CH_3CO_2H \qquad (6\text{-}30)$$

As discussed in Chapter 5, these reactions are carried out in microorganisms, but are slow, especially with high molecular weight materials.

The microbial oxidation of the cyclic hydrocarbons and aromatic compounds present in detergents and petroleum is also catalyzed by the enzyme cytochrome P-450. It has been established in some cases that arene oxide intermediates are formed as in the reaction sequence given in (5-31) for the oxidation of benzene.

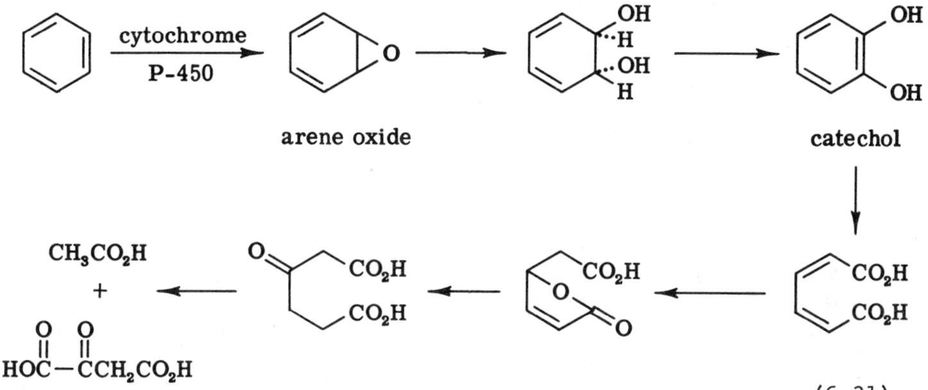

(6-31)

The microbial oxidation of the more highly condensed aromatics such as naphthalene or of the alkyl-substituted benzines proceeds via salicyclic acid which is then oxidized to catechol [reaction (6-32)] and finally to oxaloacetate.

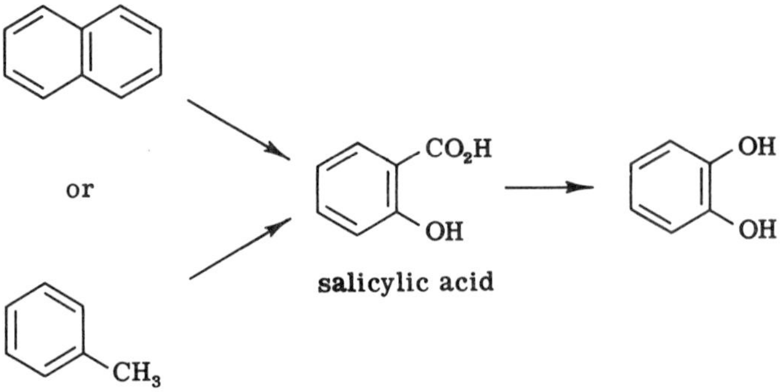

(6-32)

The sulfonic acid residues present in the sulfonic acid detergents are subject to hydrolysis by a variety of aquatic bacteria. Therefore, these sulfonic acid groups do not slow the environmental degradation of these detergents.

6.6 ENVIRONMENTAL EFFECTS OF BIODEGRADABLE ORGANIC COMPOUNDS

The presence of biodegradable organic compounds in lakes and rivers can result in an environmental problem. Oxygen is required for the microbial degradation of these compounds and relatively small quantities of organic material can deplete the supply of dissolved oxygen. The amount of oxygen consumed in the microbial degradation of organic compounds can be estimated if the crude assumption is made that the organic compounds are pure carbon and that they are all converted to carbon dioxide.

$$C + O_2 \longrightarrow CO_2 \qquad (6\text{-}33)$$

From this reaction and the ratio of the molecular weight of oxygen to the atomic weight of carbon, it can be seen that 32/12 or 8 parts of oxygen are required to oxidize 3 parts of carbon. Since water can dissolve only about 10 ppm of oxygen, this means, in the absence of further dissolution of atmospheric oxygen, that only 4 ppm of carbon compounds can deplete all the dissolved oxygen in a body of water. This calculation is an oversimplification since the microbial oxidation does not proceed instantly and the oxygen is always being replenished. However, the oxygen near the bottom of deep lakes can be depleted by dissolved organics because of the slow rate of oxygen transport to these low levels (Chapter 13). Fish have difficulty breathing when the oxygen level drops to 5 ppm (one-half the maximum level).

The biological oxygen demand (BOD), a measure of the biodegradable organic material in water, is determined by the amount of oxygen required for the microbial oxidation of the dissolved organic content of a water sample. The test measures the amount of dissolved oxygen lost from a water sample kept in a sealed container at 20°C over a five-day period. Optimally about 75% of the organic material is oxidized under these conditions. Chemical oxygen demand (COD) is another measure of dissolved organic compounds. It is determined by measuring the equivalents of acidic permanganate or dichromate necessary for the oxidation of organic constituents. Neither BOD nor COD measures the total oxidizable carbon content. Total oxidizable carbon is measured by the amount of CO_2 formed on combustion after the water and the carbonate in the sample are removed. Obviously even biodegradable compounds such as detergents reduce the oxygen level in lakes and rivers. However, the major sources of dissolved organics in lakes and rivers are the effluent from sewage disposal plants, manure from animal feed lots, industrial wastes, and decomposing plants and algae.

BIBLIOGRAPHY

T. E. Brenner, "Biodegradable detergents and water pollution," in "Advances in Environmental Science," Vol. 1 (J. N. Pitts and R. L. Metcalf, eds.), p. 147. Wiley-Interscience, New York, 1969.

A. L. Hammond, Problems with the washday miracle, *Science 172*, 361 (1971).

R. D. Swisher, The chemistry of surfactant biodegradation, *J. Amer. Oil Chemists Soc. 40*, 650 (1963).

P. R. Dugan, "Biochemical Ecology of Water Pollution." Plenum Press, New York, 1972.

D. W. Schindler, Eutrophication and recovery in experimental lakes; implications for lake management, *Science 184*, 897 (1974).

O. Hayaishi and M. Nozaki, Nature and mechanisms of oxygenases, *Science 164,* 389 (1969).

A. H. Conney and J. J. Burns, Metabolic interactions among environmental chemicals and drugs, *Science 178,* 576 (1972).

P. K. Stumpf, "Metabolism of fatty acids," in "Annual reviews of biochemistry," Vol. 38 (E. E. Snell, ed.), p. 159. Ann. Reviews, Inc., Palo Alto, California, 1969.

E. J. McKenna, "Microbial metabolism of normal and branched chain alkanes," in "Degradation of synthetic organic molecules in the biosphere." National Academy of Sciences, Washington, D.C., 1972.

D. T. Gibson, "Initial reactions in the degradation of aromatic hydrocarbons," in "Degradation of synthetic organic molecules in the biosphere." National Academy of Sciences, Washington, D.C., 1972.

W. W. Niven, "Industrial Detergency." Reinhold, New York, 1955.

C. E. Meloan, Detergents, soaps, and syndets, *Chemistry 49* (7), 6 (1976).

7 PESTICIDES, POLYCHLORINATED BIPHENYLS, AND OTHER CHLORO-ORGANIC COMPOUNDS

7.1 INTRODUCTION

The scope of the environmental problems associated with the extensive use of slowly degrading pesticides has been appreciated for many years. Large quantities of these compounds have been manufactured (Table 7-1) and deliberately distributed in the environment to control plant and insect pests. The environmental stability of many of these compounds is appreciable (Fig. 7-1). However, the toxicity and environmental problems associated with the commercial use of other haloorganics that are accidentally released into the environment has only recently been appreciated. For example, polychlorinated biphenyls (PCBs) (Table 7-1) are extremely stable industrial compounds with an environmental half-life of about 15 years. Another example is vinyl chloride (chloroethene), which is probably almost as stable as the PCBs (see Section

TABLE 7-1

U.S. Production of Organochlorine Compounds in 1969

Compound	*Class Compound*	*Production (million pounds)*
DDT	*Organochlorine insecticide*	*87*
Methylparathion	*Organophosphorous insecticide*	*48*
Malathion	*Organophosphorous insecticide*	*10*
2,4 -D	*Organochlorine herbicide*	*44*
PCB's	*Organochlorine dielectric and plasticizer*	*76*[a]

[a] *Figure is for the U.S. production by the Monsanto Co. The 1972 production was about half this amount due to a voluntary cutback by Monsanto Co. PCB's are also manufactured in Japan, Europe, and the USSR.*

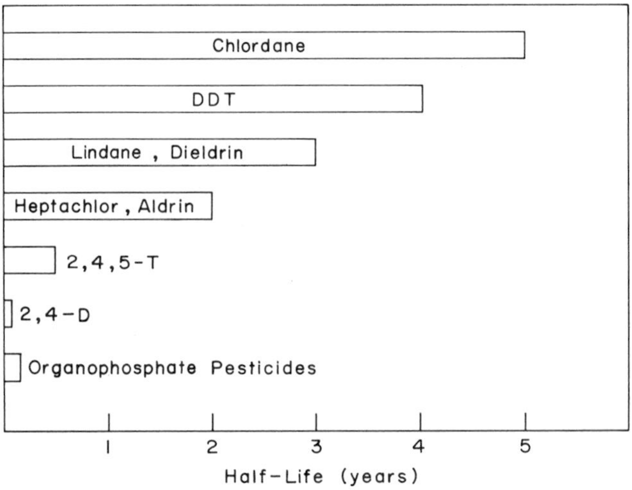

FIGURE 7-1. Half-lives in the environment of some pesticides.

7.2.2), and it has been estimated that 200 million pounds are lost to the environment each year in the process of manufacturing 5 billion pounds of polyvinyl chloride plastic. Since vinyl chloride is a gas at room temperature, this compound probably occurs mainly as an atmospheric contaminent. Vinyl chloride was used as an aerosol propellant but this use was banned by the Environmental Protection Agency in 1974. Recently it was discovered that the vinyl chloride monomer was responsible for a rare type of liver cancer and may be responsible for birth defects. There is suspicion that other vinyl chloride derivatives, 1,1-dichloroethylene and trichloroethylene, are also carcinogens. 1,1-dichloroethylene is used in the manufacture of the plastic Saran Wrap while trichloroethylene is used extensively for the degreasing of metal parts in metal fabricating plants. No estimates have been made on the release of trichloroethylene in the environment but 434 million pounds were manufactured in the U.S. in 1974.

Yet another example of a hazardous halocarbon is 1,2-dibromo-3-chloropropane. The production of this nematocidal agent (Section 7.2.2) was discontinued in 1977 because it caused sterility in the males exposed to this chemical in the course of its manufacture.

A potential environmental problem of a different kind may result from the extensive use of fluorochlorocarbons (Freons[1]) as propellants in aerosol cans and as the refrigerants in refrigeration systems. Some examples of Freons include trichlorofluoromethane (Freon 11[2]) and dichlorodifluoromethane (Freon 12). These compounds are colorless, odorless, and essentially stable in the biosphere. Because of the volatility of these fluorocarbons, they are present almost exclusively in the atmosphere. An illustration of their stability is demonstrated by the observation that pure dichlorodifluoromethane decomposes at the rate of 1% a year at 500°C. Carbon tetrachloride, an important industrial solvent, is almost as stable as the Freons. Freons and carbon tetrachloride are not susceptible to environmental degradation until they reach the stratosphere where they are photodissociated by ultraviolet light. This process may lead to the partial destruction of the ozone layer as described in Chapter 10. Even a partial loss of the ozone layer could result in a dangerous increase in the amount of ultraviolet light reaching the surface of the earth. Because of this potential danger, the U.S. Food and Drug Administration decreed that aerosol products containing fluorocarbons will no longer be sold in interstate commerce.

A different source of environmental chloroorganic compounds is the chlorination of drinking water. Some of the organic compounds present in water can be chlorinated by this procedure. In particular it has been established that chloroform, bromodichloromethane, dibromochloromethane, and bromoform are formed in the process of chlorination if appropriate materials are present. In addition both carbon tetrachloride and 1,2-dichloroethane have been detected in the drinking water of several cities in the U.S. Although it is believed that both chloroform and carbon tetrachloride are carcinogens, the amounts present in the drinking water (up to 311 ppb for chloroform) are probably not sufficient to induce tumor formation.

While chlorinated organic materials have been among the most widely used pesticides, their overall environmental

[1]*DuPont registered trademark.*

[2]*In the numbering system used for Freon nomenclature, the last digit is the number of F atoms; the next-to-last digit is the number of H atoms plus one; and the third-from-last digit (if present) is the number of C atoms minus one. All other atoms present are Cl.*

impact coupled with the development of insect resistance to their effects, has led to increasing use of other classes of compounds for pesticidal purposes. Some aspects of the chemistry of the most important and typical of these other classes will be discussed in this chapter.

7.2 CHEMISTRY OF CHLORINATED ORGANIC COMPOUNDS

7.2.1 Synthesis

Organochlorine derivatives are prepared industrially by the direct chlorination of hydrocarbons.

$$RH + Cl_2 \longrightarrow RCl + HCl \qquad (7\text{-}1)$$

The mechanisms of this reaction is discussed in elementary chemistry texts so it will only be briefly outlined here. It is a free radical process initiated by atomic chlorine.

$$Cl_2 \xrightarrow[\Delta]{h\nu \text{ or}} 2Cl \qquad (7\text{-}2)$$

$$RH + Cl \longrightarrow R\cdot + HCl \qquad (7\text{-}3)$$

$$R\cdot + Cl_2 \longrightarrow RCl + Cl \qquad (7\text{-}4)$$

$$R\cdot + Cl \longrightarrow RCl \qquad (7\text{-}5)$$

Chlorination is useful in the preparation of organochlorine compounds such as 1,3-dichloro-1-propene, which are effective in killing nematodes (soil worms). The overall reaction is

$$CH_2{=}CHCH_3 + 2Cl_2 \longrightarrow ClCH{=}CHCH_2Cl + 2HCl \qquad (7\text{-}6)$$

The first step in this process is the formation of the relatively stable allylic-free radical by means of a hydrogen abstraction reaction with a chlorine atom. This is followed by further reactions with Cl_2.

$$CH_2{=}CHCH_2^{\cdot} + HCl \quad \text{from} \quad CH_2{=}CHCH_3 + Cl \longrightarrow CH_2{=}CHCH_2^{\cdot} + HCl \qquad (7\text{-}7)$$

$$CH_2{=}CHCH_2^{\cdot} + Cl_2 \longrightarrow CH_2{=}CHCH_2Cl + Cl \qquad (7\text{-}8)$$

$$CH_2{=}CHCH_2Cl + Cl_2 \longrightarrow ClCH_2\underset{\substack{|\\ Cl}}{CH}CH_2Cl \qquad (7\text{-}9)$$

The trichloro compound is then dehydrohalogenated:

$$ClCH_2CH(Cl)CH_2Cl \longrightarrow ClCH{=}CHCH_2Cl + HCl \qquad (7\text{-}10)$$

The degree and position of halogenation of hydrocarbons will depend on the ease of hydrogen abstraction (or stability of the corresponding radical) at each carbon atom and the reaction conditions, for example, the ratio of chlorine to hydrocarbon.

Aromatic compounds generally are more difficult to chlorinate than aliphatic compounds. The aromatic chlorination reaction proceeds by an ionic pathway. Polychlorinated biphenyls are prepared from biphenyl by direct chlorination using ferric chloride as a catalyst. The mechanism involves the initial formation of a chloronium ion [reaction (7-12)] that undergoes an electrophilic addition to the aromatic ring system as illustrated in (7-13) with benzene.

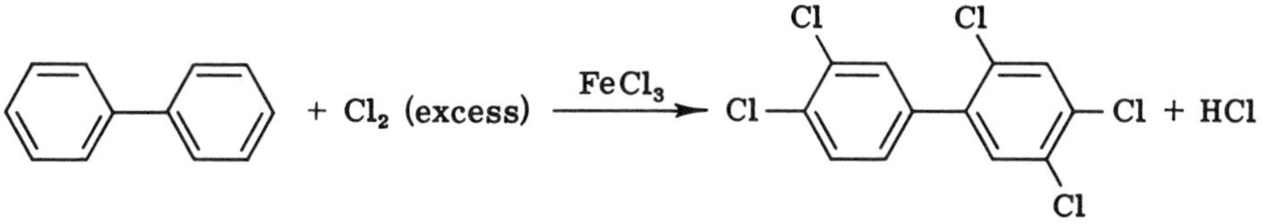

(one of many products)

(7-11)

$$Cl_2 + FeCl_3 \rightleftharpoons FeCl_4^- + Cl^+ \qquad (7\text{-}12)$$

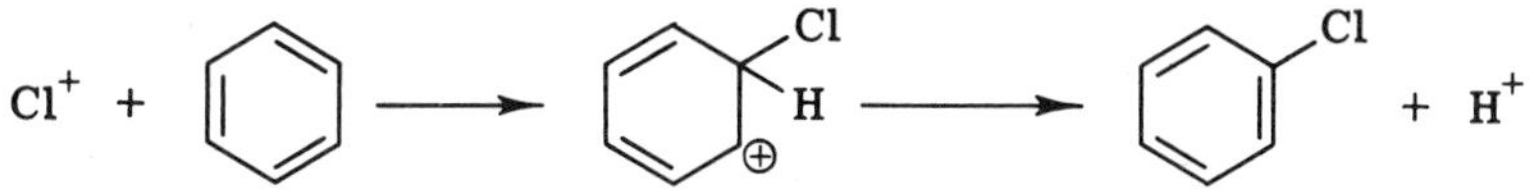

(7-13)

Polychlorinated biphenyls were formed in one instance by the chlorination of wastes in sewage disposal plants that had an appreciable input of biphenyl from a textile mill. This result suggests that the use of chlorine in waste treatment (Chapter 15) may be dangerous because environmentally stable organochlorine compounds may be formed.

7.2.2 Reactions

The aliphatic organochlorine compounds are generally more reactive than the corresponding aromatic derivatives. Two limiting reaction pathways (displacement and elimination) are observed for aliphatic chloro derivatives in the presence of nucleophilic reagents. Displacement may proceed by a unimolecular (S_N1) [reactions (7-14) and (7-15)] or biomolecular (S_N2)[reaction (7-16)] reaction pathway. The rate limiting step of the S_N1 reaction is ionization of the alkyl halide to the planar carbonium ion (7-14) while the rate of the S_N2 reaction is proportional to the concentrations of both alkyl halide and nucleophile (7-16).

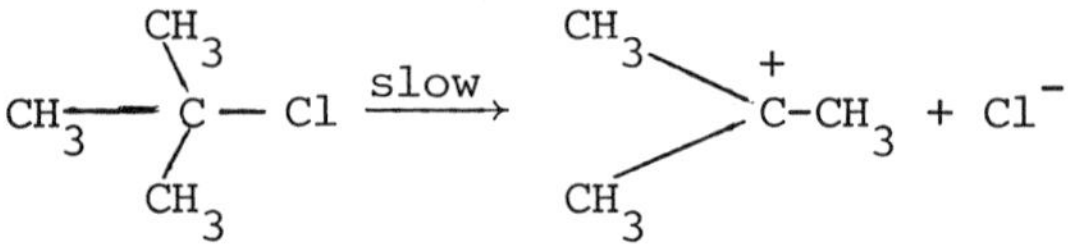
(7-14)

(7-15)

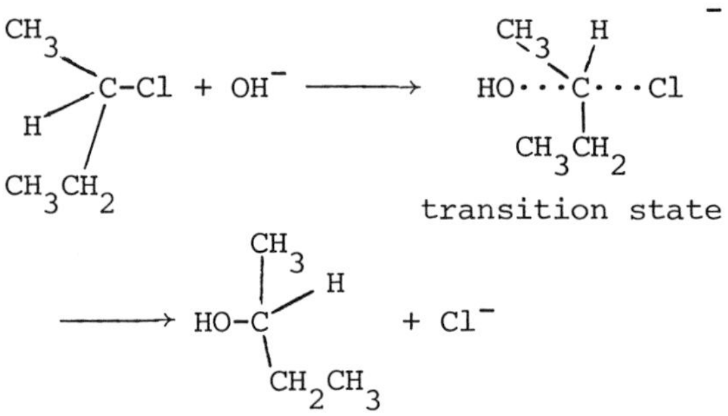
(7-16)

The mechanism of the displacement reaction is determined by the solvent, displacing agent, and the structure of the organic chlorine derivative. The S_N1 (carbonium ion) mechanism is favored by polar solvents that stabilize the ionic intermediates or when poor nucleophiles such as water are the displacing agent. Carbon compounds with sterically hindered halogens resistant to direct (S_N2) displacement or which contain tertiary, allylic, and benzylic halogens (which ionize to stabilized carbonium ions) usually react by an S_N1 pathway. The formation of an intermediate carbonium ion is the most likely pathway under environmental conditions. Usually the environmental reactions occur in a polar solvent (water) near pH 7 with a poor nucleophile (water) or with low concentrations of nucleophile displacing agent, i.e., S_N1 conditions.

It has been observed that the toxicity of nematocidal agents is proportional to their S_N2 reactivity determined by measuring the rate of reaction of the nematocidal agent with idodide ion in acetone (Table 7-2). A good correlation with the amount of allyl halide required to immobilize half of the nematodes (ED_{50}) and the reaction of the allyl halide with iodide was observed, suggesting that toxicity is associated with an S_N2 reaction in the nematode. It might be expected that the allyl halides in Table 7-2 would ionize readily to form carbonium ions, but the S_N1 reaction is not favored under these conditions because acetone is a weakly polar solvent and iodide is a very good nucleophile.

TABLE 7-2

S_N2 Reactivity and Nematocidal Action[a]

Compound	*ED_{50}*[b]	*S_N2 Reactivity (I^- in acetone)*
$CH_2{=}CHCH_2Cl$	1.5×10^{-3}	1.0
$CH_2{=}C(Cl)CH_2Cl$	2.6×10^{-4}	0.72
$CH_2{=}C(CH_3)CH_2Cl$	2.8×10^{-4}	1.58
$HC{\equiv}CCH_2Cl$	2×10^{-4}	1.78
<u>trans</u>-$ClCH{=}CHCH_2Cl$	7.7×10^{-5}	2.9
<u>cis</u>-$ClCH{=}CHCH_2Cl$	3.3×10^{-5}	8.6
$CH_2{=}CHCH_2Br$	7.5×10^{-5}	506
$HC{\equiv}CCH_2Br$	4.5×10^{-6}	909

[a]Adapted from W. Moje, Adv. Pest Control Research 3, 181 (1966).

[b]Molar concentration required to produce 50% inhibition of the mobility of the citrus nematode T. semipenetrans.

Two limiting reaction mechanisms are available for the elimination reaction; a unimolecular (E1) [reactions (7-14) and (7-17)] and biomolecular (E2) [reaction (7-18)] pathway.

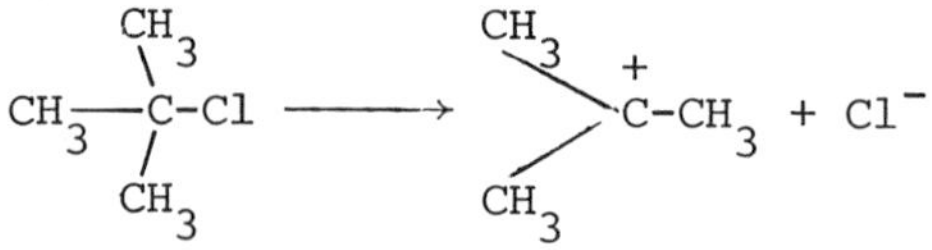

(7-14)

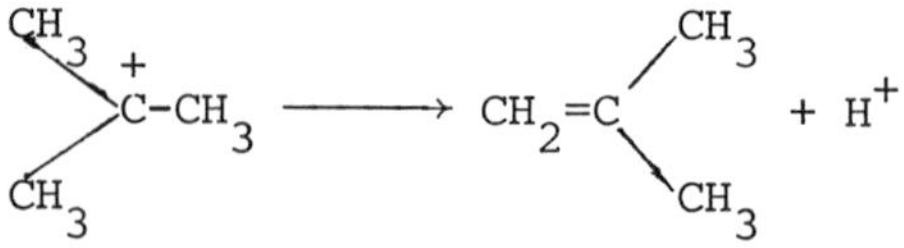

(7-17)

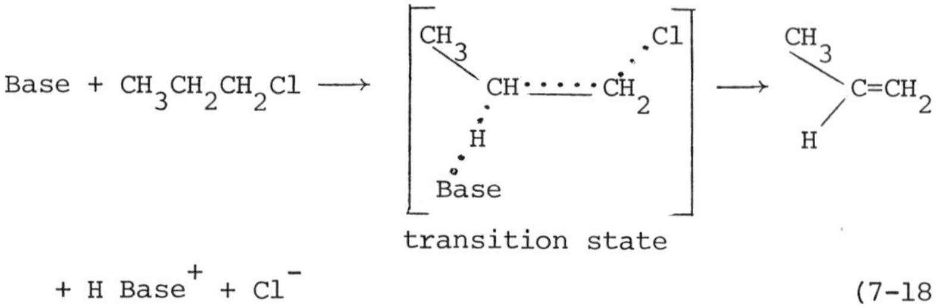

(7-18)

Those factors that favor the S_N1 reaction also favor E1 elimination because the same carbonium ion intermediate is involved (7-14). The relative proportion of elimination to displacement is determined by the relative nucleophilicity and basicity of the group attacking the carbonium ion. The E2 elimination is observed with strong bases in nonpolar solvents and is not likely to take place under environmental conditions. For example, the dehydrochlorination of DDT and DDT analogs with sodium hydroxide proceeds by an E2 mechanism in 92.6% ethanol (7-19). The second-order rate constant decreases by 50% in going from 92.6% ethanol to 76% ethanol. Therefore the elimination probably proceeds by an E1 mechanism at neutral pH in aqueous solution.

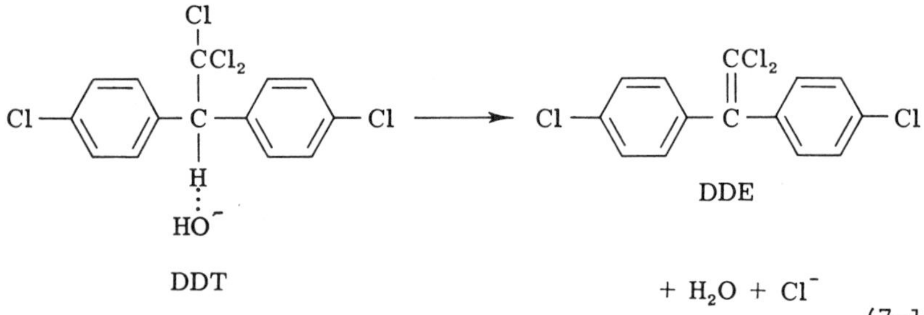

(7-19)

Nucleophilic displacement of aryl and vinyl halides is a very slow process that does not occur readily under environmental conditions. The S_N1 or E1 reactions are not observed because the sp^2 hybrid carbon atom of the aryl or vinyl halide is more electronegative than the sp^3 hybrid and cannot readily take on the positive charge necessary if it were to become a carbonium ion. Furthermore, the lone pairs of electrons on the chlorine participate in the bonding and strengthen the bond. Thus the ground states of the vinyl and aryl halides

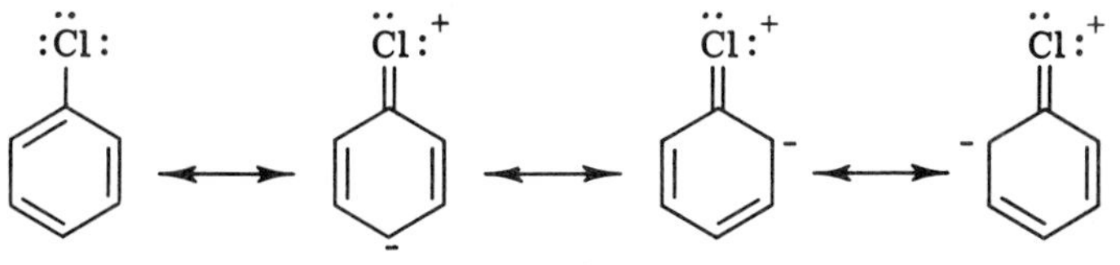

are more stable than the corresponding ground states of aliphatic halides. This results in a greater energy of activation (E_A) for carbonium ion formation and a slower rate of reaction. This is shown qualitatively in Fig. 7-2.

The reactivity of aryl halides is enhanced by electron-withdrawing groups, for example, two nitro substituents greatly accelerate the rate of displacement of aryl bound halogens (7-20). This is due to the formation of an

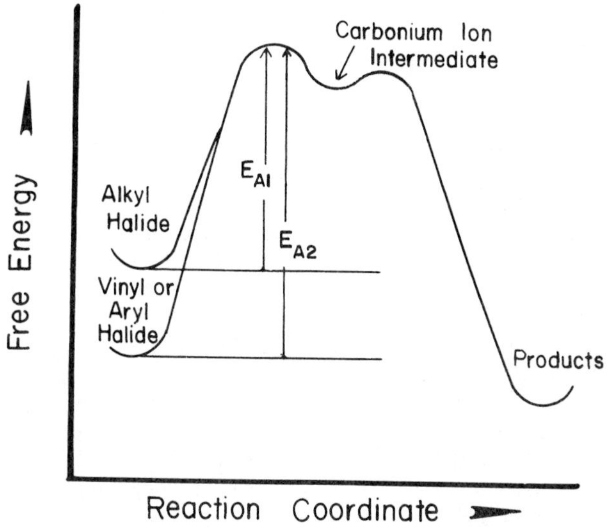

intermediate that is stabilized by the electron-withdrawing substituents. Halide ion displacement from an aromatic ring containing electron-withdrawing groups might occur under environmental conditions.

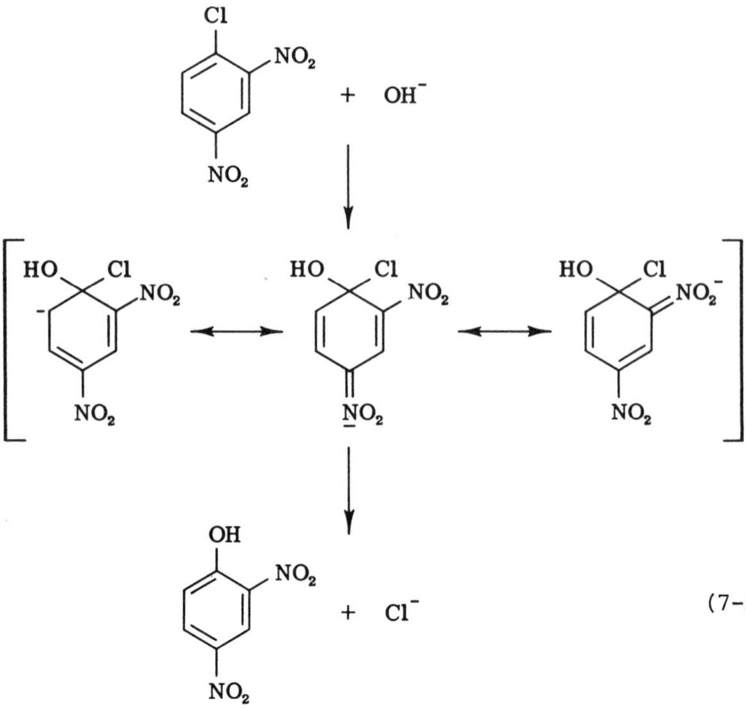

(7-20)

The stability of the Freons and other fluorocarbon compounds is due mainly to the special characteristics of the C-F bond. This is very strong with a bond energy of approximately 115 kcal (4.81×10^5 J) as compared to 80 kcal (3.34×10^5 J) for the C-Cl bond. Also the C-F bond distance (1.3-1.4 Å) is much shorter than the C-Cl bond distance (∿1.8 Å). Since bond energies provide a rough estimate of the rate of S_N1 reactions [compare reaction (7-14)], we would predict that fluorocarbons would not undergo rapid reaction by this pathway. The S_N2 or E2 reactivity of fluorocarbons is low because the fluorine atom is not as readily polarized as the other halogens, thus there cannot be much charge separation in the transition state [reaction (7-16) and (7-18)]. In addition, there is greater repulsion of the incoming nucleophile by fluorine than by other halogens because of the short C-F bond. Finally, the steric effects for displacement by an S_N2 pathway are very large for the polyhalogenated Freons.

7.3 POLYCHLORINATED BIPHENYLS

7.3.1 Synthesis and Properties

Polychlorinated biphenyls have been available commercially for over 40 years. The synthesis from chlorine and biphenyl yields a mixture of derivatives with the extent of chlorination being dependent on the chlorine-biphenyl ratio. These compounds were sold in the U.S. by Monsanto Co. under the trade name of Aroclor. Aroclor products containing various amounts of chlorine are differentiated by a four-digit number; the last two digits indicate the percentage of chlorine.

PCBs are resistant to acid, base, heat, and oxygen. This extreme stability makes them especially useful as a dielectric material in capacitors and transformers. They have also been used as plasticizers and solvents in plastics and printing inks. However, since 1970, Monsanto has refused to sell PCBs to customers in plasticizer operations where disposal of the end product is not controlled. As a consequence they are now only being used in "captive" applications such as transformers where they can be recovered.[3] Monsanto decided to stop the manufacture of PCBs in 1977. This decision by Monsanto should result in an appreciable decrease in the accumulation of PCBs in the environment, especially if manufacturers in other countries adopt similar policies of control. Closely related compounds, polybrominated biphenyls, are used as flame retardants.

7.3.2 Sources of PCBs in the Environment

Although PCBs may be formed in some instances in sewage treatment plants and by the photolyses of DDT derivatives, undoubtedly industrially produced PCBs are the main environmental source. It is estimated that 4000 tons per year enter the environment from the dumping and leaking of lubricants, heat transfer fluids, and hydraulic fluids into rivers and streams. Another 1000-2000 tons per year were discharged into the atmosphere by the combustion of plastics containing PCBs as plasticizers. Since PCBs are resistant to combustion, they were merely volatilized during the incineration. Presumably this source of environmental PCBs has almost been eliminated in the U.S. now that Monsanto Co. no longer sells these compounds for use as plasticizers.

[3]*PCBs are not used as insecticides.*

7.3.3. Environmental Degradation

The environmental breakdown of PCBs is exceedingly slow with a half-life estimated to be 10-15 years. The mechanism for this breakdown is not known. Recently it has been reported that lightly chlorinated PCBs are hydroxylated by rats and pigeons but not by brook trout, as indicated in reactions (7-21) and (7-22). The hydroxylation reaction is reminiscent of the hydroxylations catalyzed by the aryl oxidase enzyme cytochrome P-450 (Chapter 5). Since similar oxidations are effected by microorganisms, it is conceivable that lightly chlorinated PCBs are also degraded in this way environmentally, although the highly chlorinated ones are certainly much more resistant.

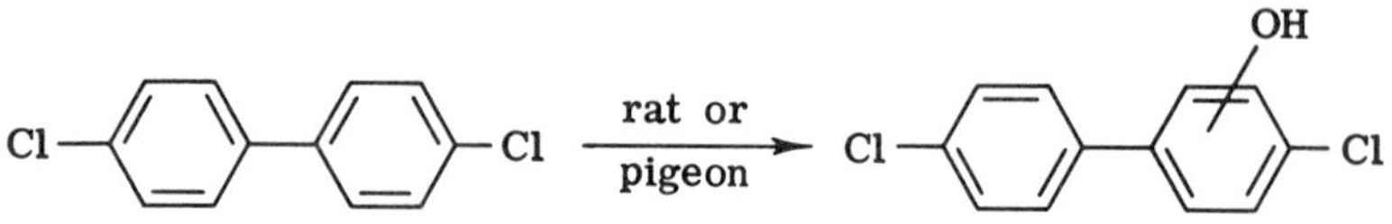

(7-21)

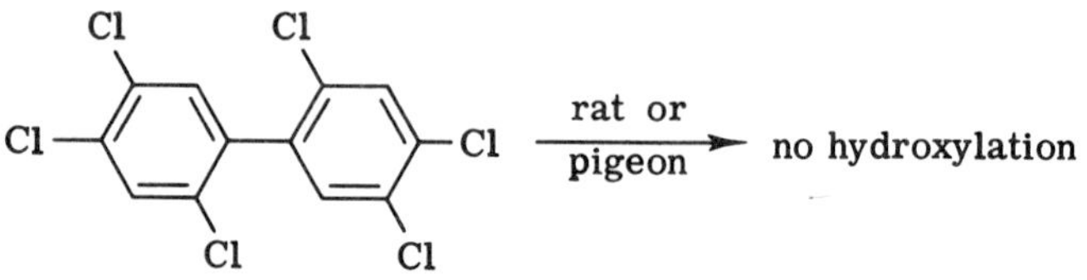

(7-22)

7.3.4 Toxicity

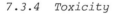

PCBs do not exhibit significant immediate (acute) toxicity and have been considered to be nontoxic. However, toxic effects were amply demonstrated in 1968 when more then 1000 persons in Japan ate rice oil contaminated with PCBs that had leaked into the oil from a heat exchanger. Those persons who ate 0.5 gm or more (average consumption was 2 gm) developed darkened skin, eye damage, and severe acne. It is not sure if the subsequent deaths of some of the patients were due to the PCB poisoning. Recovery was slow with symptoms still present even after three years. Several infants were born with the same symptoms, demonstrating the PCBs can readily cross the placental barrier. PCBs have been found to be toxic to mink and some species of shellfish, shrimp, and fish. The chronic (long term) toxicity of PCBs and other chloroorganics will be discussed in Section 7.7.3.

7.4 DDT

7.4.1 Synthesis

DDT [1,1,1-trichloro-2,2-bis(p-chlorophenyl)ethane, also called dichlorodiphenyltrichloromethane, hence the name DDT], is prepared by the reaction of chloral with chlorobenzene (7-23). The mechanism of the reaction involves the electrophylic addition of a carbonium ion species to the aromatic ring--a Friedel-Crafts type reaction. The mechanism is shown in reactions (7-24) to (7-27).

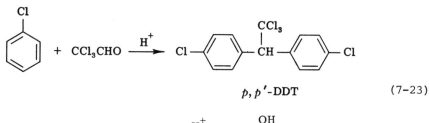

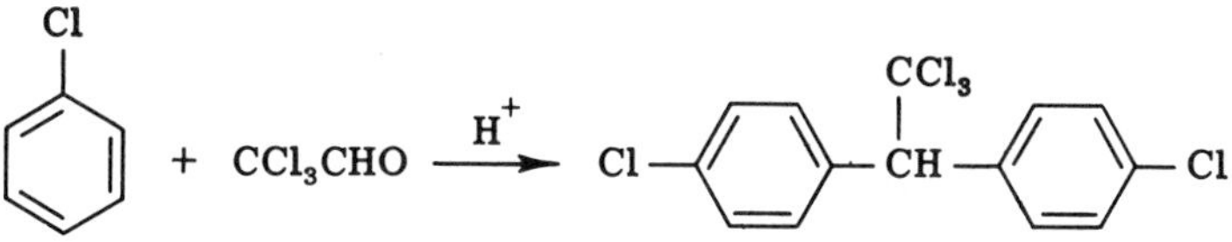

p,p'-DDT (7-23)

$$CCl_3CHO \xrightleftharpoons{H^+} CCl_3\overset{OH}{\underset{+}{C}}H \qquad (7\text{-}24)$$

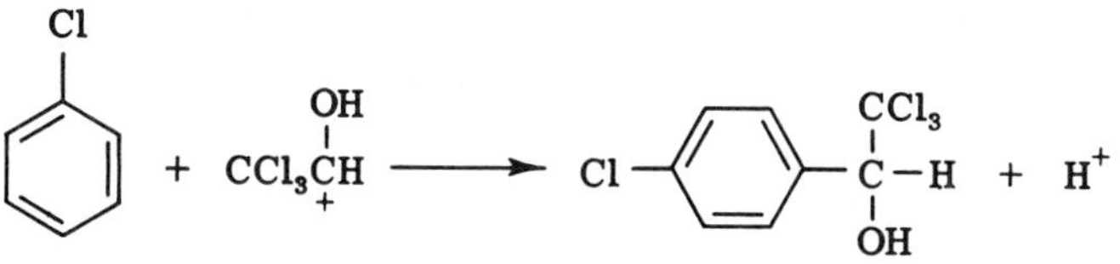

(7-25)

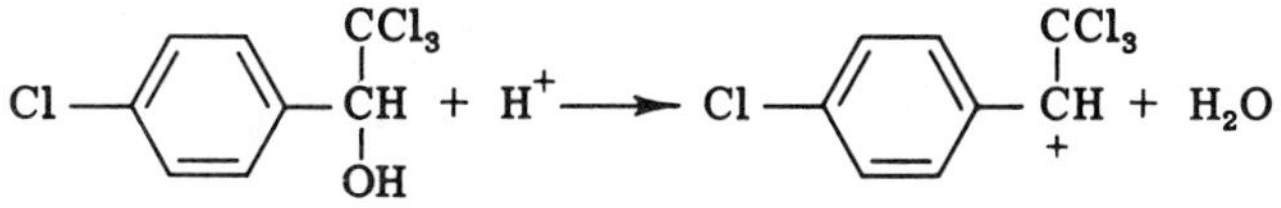

(7-26)

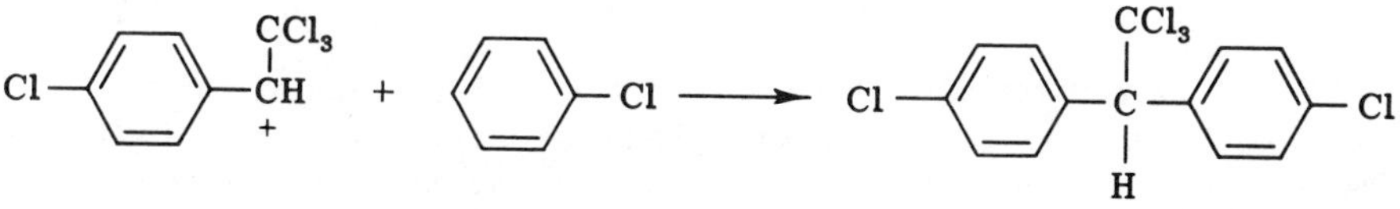

(7-27)

$+ H^+$

The *p,p'*-isomer shown is not the only reaction product. Appreciable amounts of *o,p'*- and *o,o'*-DDT are also obtained. This synthetic approach has been applied for the preparation of a great variety of DDT analogs. One important example, methoxychlor, is currently being used in place of DDT in many applications because it is biodegradable. However, it is less effective as an insecticide and about three times as costly to prepare because anisole (methoxybenzene) is a more expensive starting material than chlorobenzene.

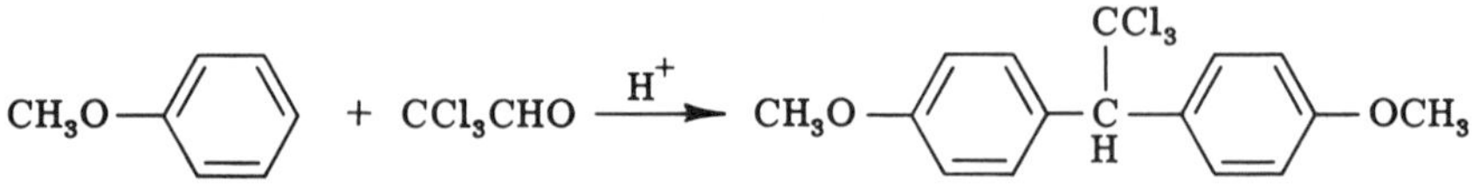

methoxychlor (7-28)

7.4.2 Environmental Degradation

DDT undergoes a relatively rapid elimination of HCl to yield DDE [1,1-dichloro-2,2-bis(*p*-chlorophenyl)ethylene] when heated in water. However, the subsequent hydrolysis of DDE is extremely slow [reaction (7-29)] because there are only unreactive vinyl and aryl chlorides in DDE. As a consequence, DDE is the principal DDT degradation product found in the environment.

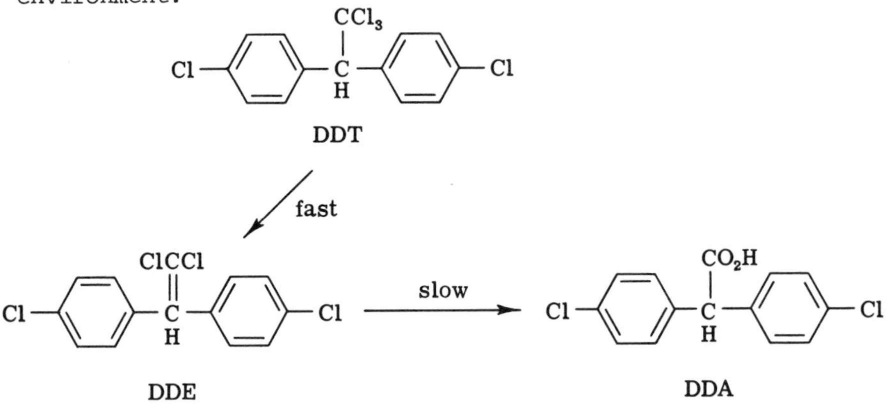

(7-29)

Methoxychlor and other DDT analogs that do not contain the *p,p'*-chloro group are much less persistent in the environment. This is due to their metabolism by soil microorganisms to phenols that are readily degraded further to acetate (see Chapter 6).

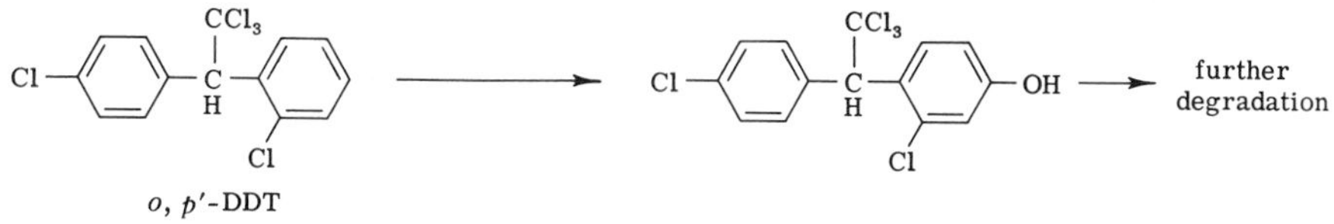

(7-30)

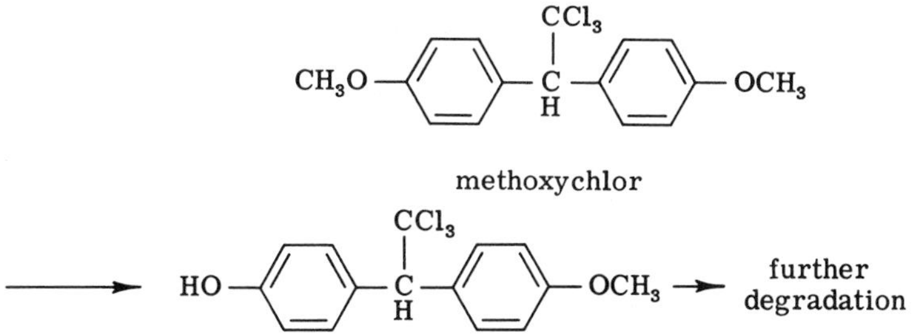

(7-31)

The enzymes involved in these oxidations are probably of the cytochrome P-450 type mentioned previously (Section 6.5). Similar enzymes are present in insects. The lower toxicity of methoxychlor as compared to DDT may reflect the effective oxidative detoxification of methoxychlor by some insects.

7.4.3 Mechanism of Action of DDT and DDT Analogs

Insects sprayed with DDT exhibit hyperactivity and convulsions consistent with the interaction of the DDT with the nervous system. Many theories have been suggested for the toxic effect of DDT and the exact mechanism is not known. A current theory that appears plausible and has some experimental basis is that the DDT molecules are of the correct size to be trapped in the pores of the nerve membranes. This distorts the membrane and sodium ions leak through and depolarize the nerve cell so that it can no longer transmit impulses. This theory states that the toxicity of DDT is not due to its chemical reactivity but rather to its size and geometry that allows for the blockage of the pores of the nerve membranes. This theory is supported by the observation that a variety of quite different compounds that are stereochemically similar to DDT exhibit DDT-like activity. The following are examples.

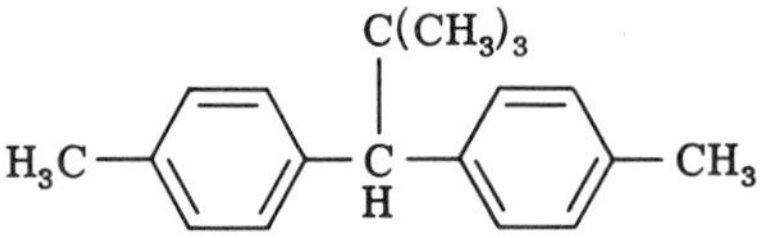

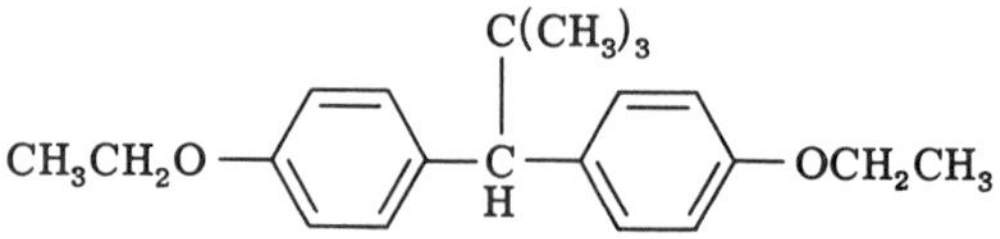

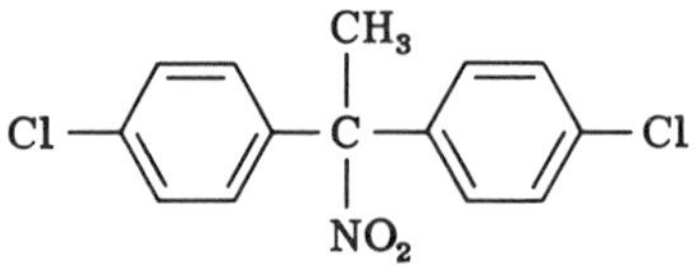

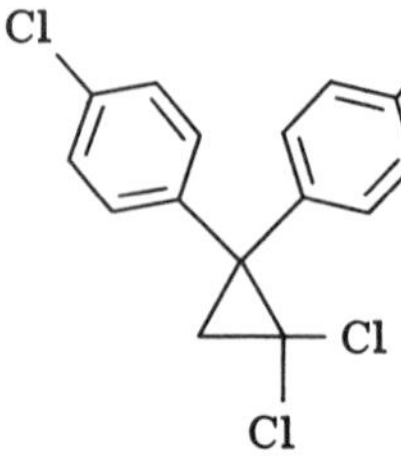

The special fit hypothesis is further supported by the observation that the biological activity of methoxychlor analogs decreases rapidly when R in the formula below contains five carbon atoms or more. Presumably the longer-chain methoxychlor analogs are not able to fit in the nerve pores.

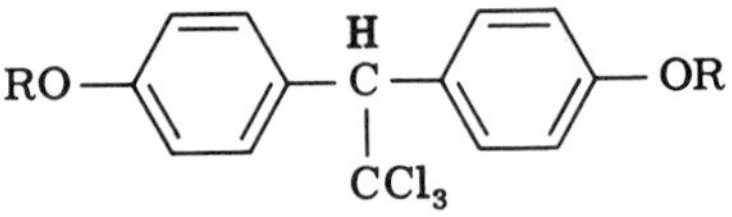

Insects are especially susceptible to DDT because it is readily absorbed through the insect cuticle. DDT is not appreciably absorbed through the protein skin of mammals. If the polarity of the DDT analog is increased by introducing $-NO_2$, $-CO_2H$, $-CO_2CH_3$, and $-OH$ substituents into the aromatic ring, the toxicity is lost. This loss is probably due in part to decreased absorption through the cuticle and in part to decreased adsorption on the nonpolar nerve membrane.

7.4.4 Insect Resistance to DDT

When DDT came into widespread use after World War II, it appeared to be the answer for the control of all insects. However, as early as 1948 there were reports that it was no longer as effective as it had previously been. It was subsequently found that DDT-resistant flies and mosquitoes were still susceptible to dieldrin (Section 7.6) and lindane (Section 7.5), although in a few years, resistance developed for these pesticides as well. The spectacular success obtained initially with DDT made it difficult to believe that we would not be able to use pesticides to control all insect pests. As a consequence, more and more pesticides were applied in the hope that greater amounts would be effective.

Unfortunately it was necessary to use 100 times as much DDT as was necessary when DDT was first used, to control resistant strains. This large usage resulted in a very rapid build up of DDT and DDE in the environment.

It was observed that DDT-resistant flies were "knocked down" by normal doses of DDT, but after some buzzing around on the floor they eventually recovered. This observation suggested that the resistant strains have a mechanism for rapidly detoxifying the DDT. Subsequent investigations revealed that DDT-resistant strains contained an enzyme (DDT dehydrochlorinase) that was not present in those flies that were susceptible to DDT, and that it catalyzed the conversion of DDT to DDE.

An interesting result that is consistent with the proposed action of DDT dehydrochlorinase is the observation that a deuterium-substituted DDT derivative shown below is more effective than DDT against resistant strains, yet it exhibits the same level of toxic effect against nonresistant strains as DDT. The enzymic dehydrochlorination reaction of DDT probably proceeds by an E2 mechanism with the C-H (or C-D) and C-Cl bonds broken in the rate-limiting step [reaction (7-19)]. However, removal of the deuterium by base is slower than proton removal because of the greater strength of the C-D bond (1.2 kcal/mol greater) as compared to the C-H bond (a deuterium isotope effect). Since the deuterated analog is converted to DDE more slowly by DDT dehydrochlorinase, it has a greater time to exert its toxic effect on the resistant strain.

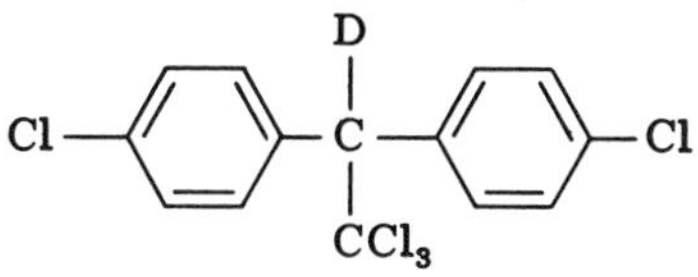

Insect resistance to insecticides is a commonly observed phenomenon. Resistance develops where 100% insect mortality does not occur and where just sublethal doses of the insecticide remain in the vicinity for long periods of time. The surviving insects breed under these conditions and only those progeny that are resistant to the residual insecticide survive. In this way a resistant strain develops that is able to survive in the presence of the insecticide. Resistance is not observed as frequently with those insecticides that are destroyed rapidly in the environment.

A nonresistant population of insects may build up again when the insecticide is no longer used. This occurs if the nonresistant strain (wild type) adapts more successfully to the insecticide-free environment than the resistant strain. However a population of resistant insects forms very rapidly if the insecticide is reintroduced.

Knowledge of the mechanism of toxic action, the pathways for environmental degradation, and the cause of insect resistance should allow for the syntheses of environmentally degradable DDT analogs that are active against resistant strains. More sophisticated pesticides can be designed once the complexities of the molecular reactions and interactions are understood. However, resistance will always be a problem and it is highly unlikely that insects can be controlled exclusively with synthetic insecticides.

7.5 LINDANE

7.5.1 Synthesis

Lindane is prepared by the free radical addition of chlorine atoms to benzene. Of the eight possible isomers, only four are formed in appreciable amounts [reaction (7-32)]. These isomers differ in the relative orientations of the chloro substituents. Axial (a) groups are perpendicular to the cyclohexane ring and are labeled in the α-isomer in (7-32). Equatorial groups are in the equatorial plane of the ring as shown in the β-isomer. Only one of these isomers, the γ-isomer, exhibits significant insecticidal action. Initially, the total reaction mixture was used as the

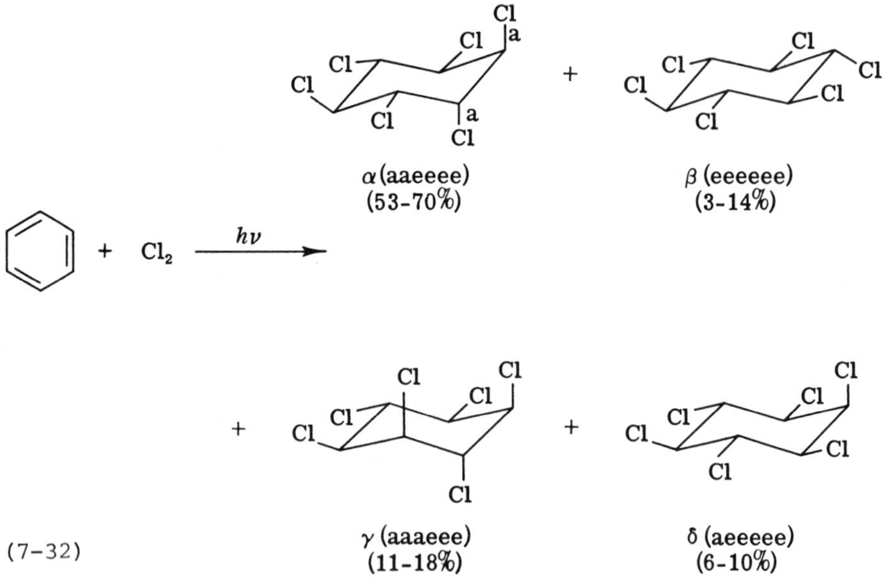

(7-32)

insecticide. However, that practice has been discontinued because this results in the distribution of large amounts of inactive chlorocarbon compounds in the environment. The use of lindane is quite restricted in the U.S. It tends to accumulate in the fatty tissues of mammals, thus it is no longer used in animal husbandry. It is used to control limited insect infestations such as borers, beetles, and hornets and it has been used in flea collars.

7.5.2 *Environmental Degradation*

Lindane is a very inert chloroorganic which undergoes a slow elimination of HCl in aqueous solution [reaction (7-33)]. Presumably this is an El reaction [see (7-14) and (7-17)].

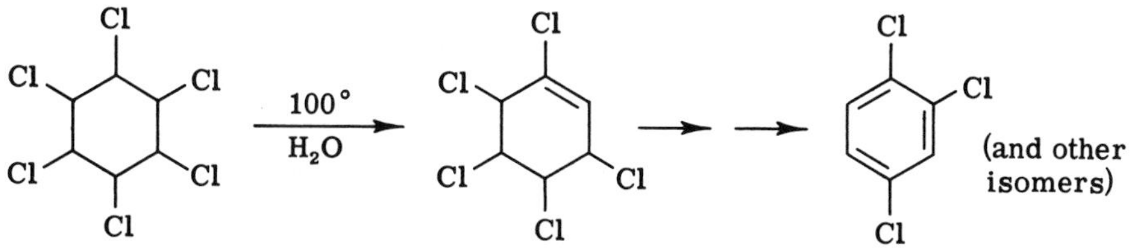

(7-33)

A series of what are probably E2 elimination reactions (7-18) occur in aqueous base to give the same trichloroaromatic compounds that are observed in neutral solution. From these limited data, it is clear why lindane and its aromatic degradation products persist for a long time in the environment.

7.6 POLYCHLORINATED CYCLOPENTADIENE DERIVATIVES

7.6.1 *Synthesis*

A series of bicyclic insecticides, chlordane, heptachlor, aldrin, and dieldrin, can be synthesized by the Diels-Alder addition of hexachlorocyclopentadiene to olefins. The preparation of these compounds is shown in reactions (7-34)-(7-36). Obviously it is possible to prepare a whole series of potential insecticides by this route by simply varying the diene and olefin.

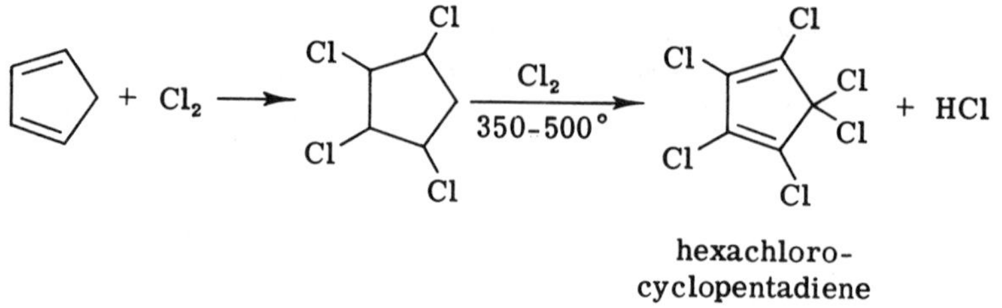

(7-34)

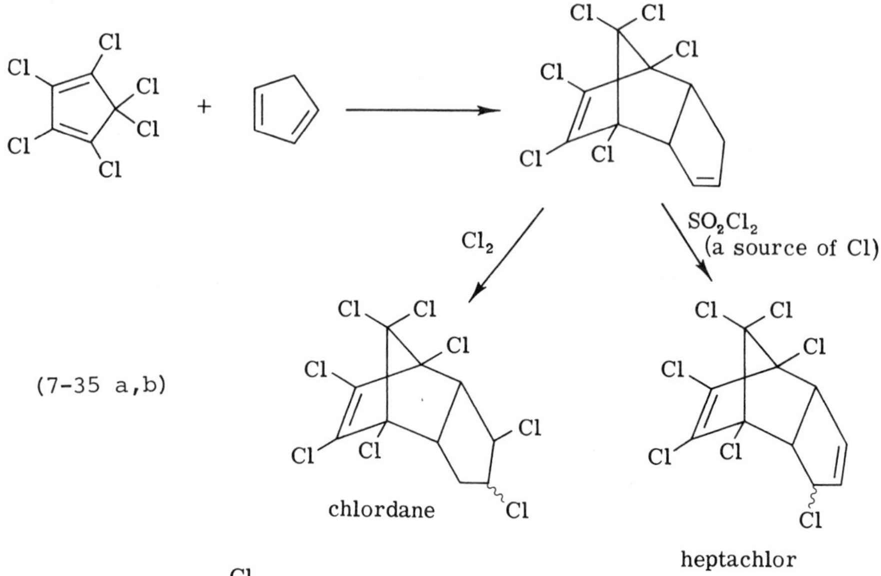

(7-35 a,b)

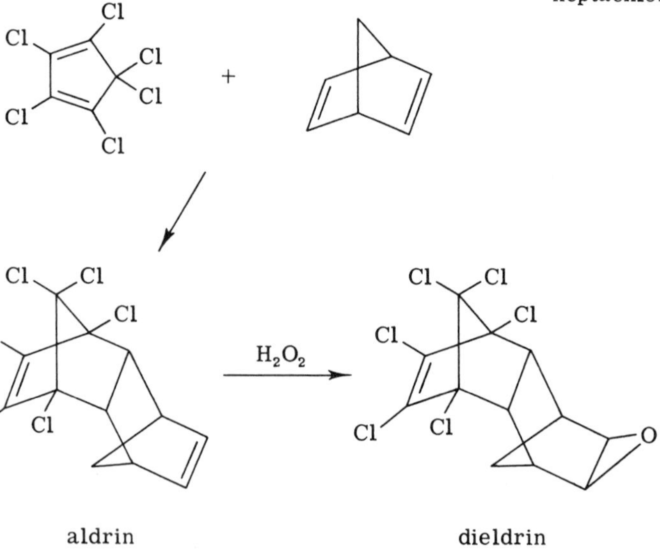

(7-36)

The endo-adduct (see structure below) is the major product resulting from the Diels-Alder addition of a diene and an olefin. No intermediate has been detected, but the reaction product can be rationalized as originating from a transition state in which there is maximum interaction between the *pi* bonds of the olefin and diene. This is illustrated in reactions (7-37) and (7-38), in which the upper cyclopentadiene may be regarded as the olefin, and the lower cyclopentadiene, the diene.

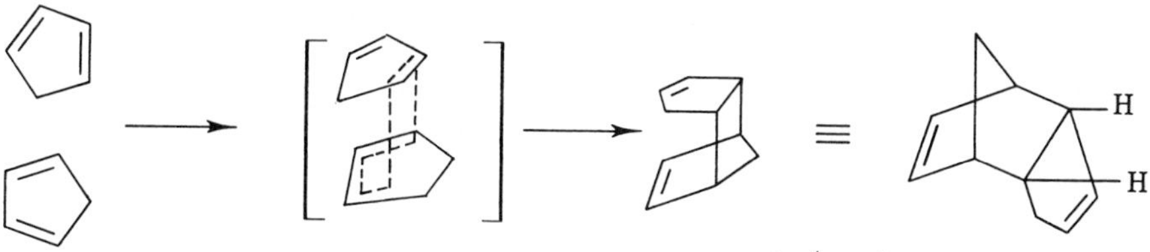

endo (predominant product) (7-37)

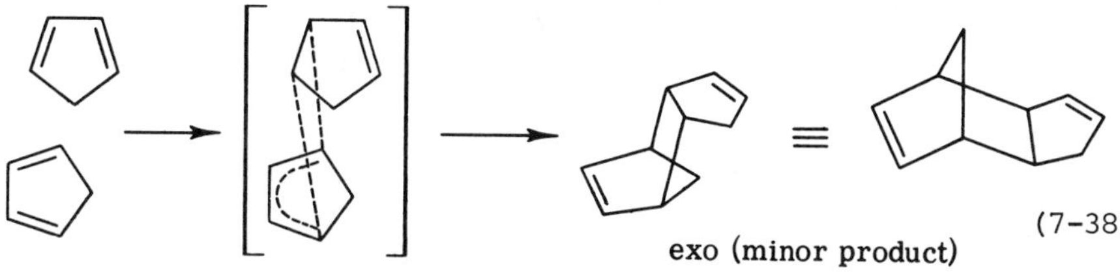

exo (minor product) (7-38)

7.6.2 Environmental Degradation

The cyclopentadiene insecticides are very resistant to environmental forces. One might expect that the tertiary, allylic chloro groups would be especially susceptible to S_N1 hydrolysis or E1 elimination (7-14). However, the rigid geometry of the bicyclic ring system prevents carbonium ion formation, as shown in reaction (7-39). A carbonium ion is more stable if the positive central carbon atom and the three substituents attached to it are in the same plane. It is not possible to obtain a planar carbonium ion intermediate at the bridgehead carbon atom of these compounds without introducing extraordinary strain in the bicyclic system.

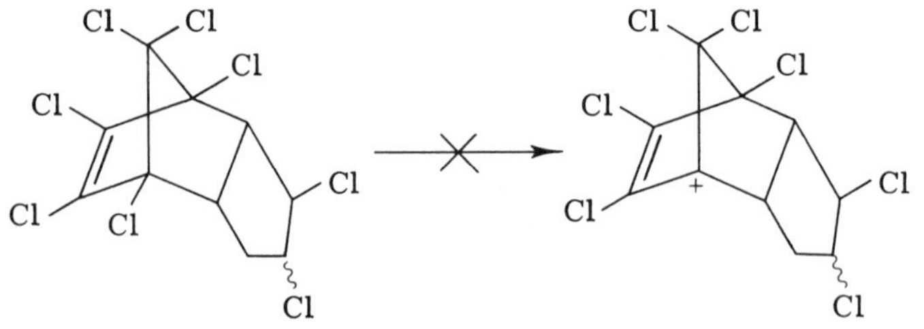

(7-39)

These compounds are susceptible to oxidative attack by soil microorganisms, plants, and animals. However, the resulting products, such as the one illustrated in reaction (7-40), are often more toxic than the starting pesticide. Photodieldrin (7-41) is formed by the action of ultraviolet light as well as by the action of oxidase enzymes on dieldrin.

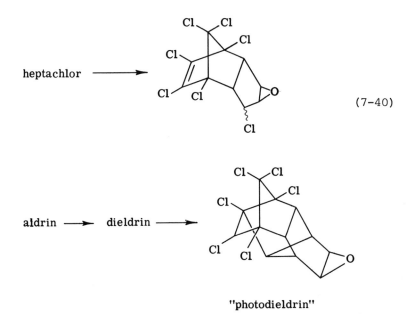

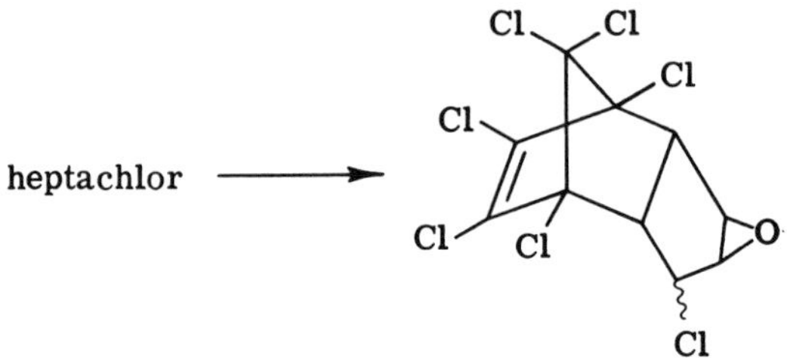

(7-40)

"photodieldrin"

(7-41)

The mechanism of action of these bicyclic organochlorine compounds is similar to that of DDT. However, the effects of these compounds suggest that the site of action is the ganglia of the central nervous system rather than the peripheral nerves that are affected by DDT. The manufacture of aldrin and dieldrin was banned in 1974 by the Environmental Protection Agency because of their potential danger as carcinogens. Preliminary evidence suggests that heptachlor and chlordane are also carcinogens--a not unexpected result in view of their structural similarity to aldrin and dieldrin.

7.7 2,4-DICHLOROPHENOXYACETIC ACID (2,4-D) AND 2,4,5-TRICHLOROPHENOXYACETIC ACID (2,4,5-T)

7.7.1 Introduction

2,4,-D and 2,4,5-T are herbicides that are especially effective in the control of broad-leafed plants yet they have little or no effect on grasses. Hence they are used to control weeds and other unwanted vegetation. These compounds mimic the plant growth hormone (auxin) indoleacetic acid, but the large amounts used (compared to the natural auxin) cause abnormal growth of the plant. In particular, little or no root growth, abnormal stem growth, and leaves with little or no chlorophyll are observed.

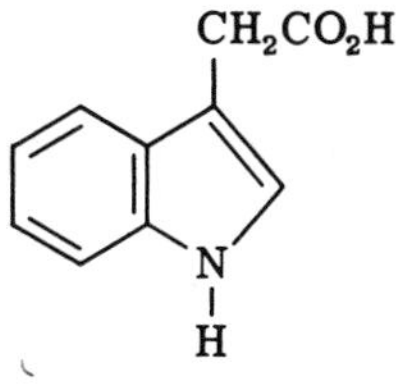

Indoleacetic acid

7.7.2 Synthesis

2,4-D and 2,4,5-T are manufactured by the S_N2 displacement of the chloro group of chloroacetate with a phenoxide anion or by the chlorination of phenoxyacetic acid as shown below.

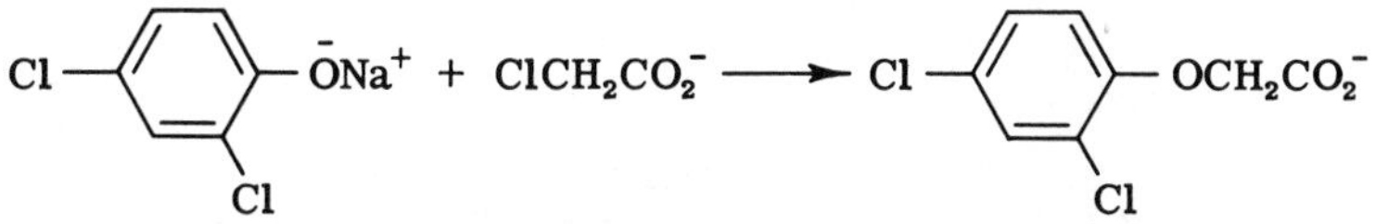

(7-42)

2,4-D

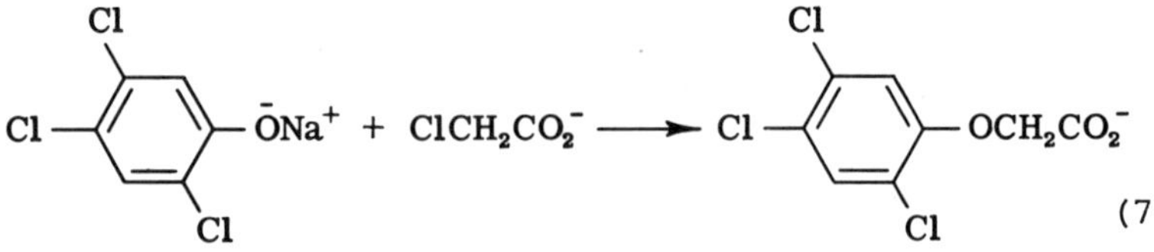

(7-43)

2,4,5-T

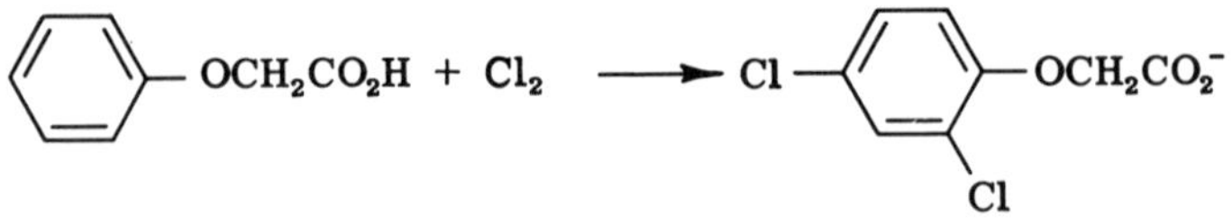

2,4-D

(7-44)

A variety of closely related 2,4-D analogs also shows herbicidal activity. The carboxyl can be replaced by the amide, nitrile, or ester groups. In addition, a longer side chain can be used in place of the acetic acid moiety. However, this side chain must contain an even number of carbon atoms to be active. This result is consistent with β-oxidation of the side chain in the plant (Section 6.5) to form the biologically active 2,4,-D.

(7-45)

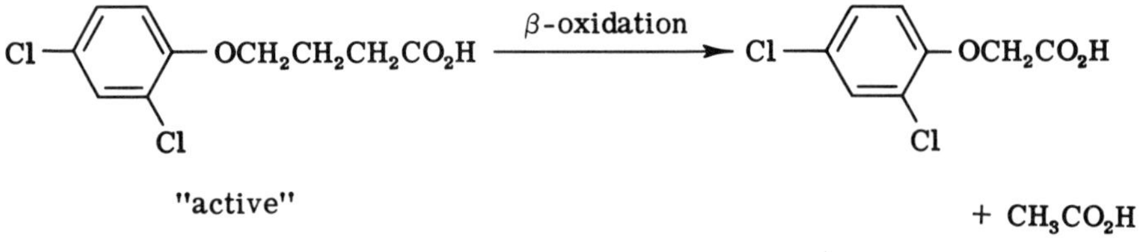

(7-46)

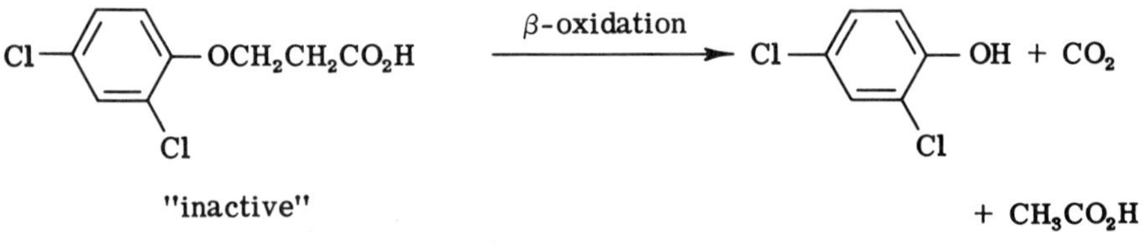

7.7.3 *Environmental Degradation*

Pure 2,4-D and 2,4,5-T are subject to rapid environmental degradation. These compounds are cleaved by microorganisms to the corresponding phenol, which is readily degraded further. Presumably, oxalic acid is the other reaction product. In addition, the photochemical cleavage of the acetic acid side chain and the photochemical hydrolysis of the aryl chloro groups has been demonstrated. These data suggest that the long-term environmental impact of these compounds is small.

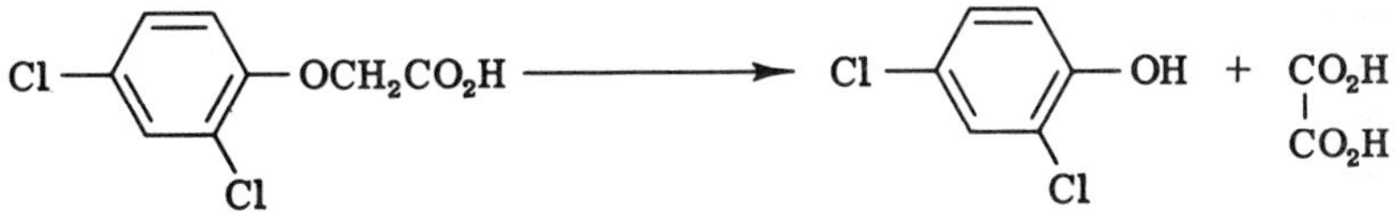

(7-47)

The environmental problem which has been associated with these compounds is due to dioxin impurities produced during their manufacture (7-49). Dioxins are formed in small amounts during the synthesis of 2,4-dichlorophenol and 2,4,5-trichlorophenol (7-48) and are carried along with the final product. The dioxins are formed because of the vigorous conditions that are required to hydrolyze the aryl chloro groups. Related toxic five membered ring derivatives

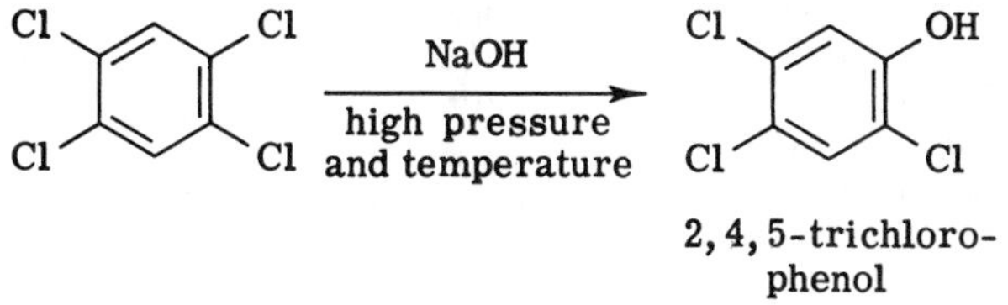

(7-48)

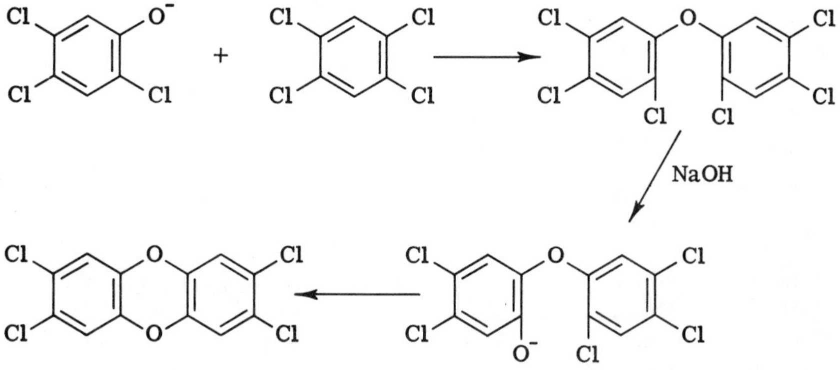

2,3,7,8-tetrachlorodioxin (7-49)

(dibenzofurans) formed in the manufacture of PCBs may be responsible for the toxicity of these compounds. Polybromobiphenyls (PBB) exhibit similar toxic effects to PCBs. This toxicity may be due in part to trace amounts of brominated dibenzodioxins and dibenzofurans in the PBBs, but this possibility requires verification.

The dioxins produced during the manufacture of 2,4,5-T have been shown to be toxic, teratogenic (cause birth defects) and acnegenic in laboratory animals. The toxicity to humans and a wide array of mammals was demonstrated when dioxins were accidentally applied to the soil in a horse arena. Contact with only 32 micrograms of dioxins per gram of soil was sufficient to kill the birds, cats, dogs, and horses that used the arena. Humans were less sensitive but one child developed hemorrhagic cystitis and several other skin lesions. The acnegenic effect has been observed in workers involved in the manufacture of 2,4,5-T. Similar toxic effects were noted in 1976 when 2-10 lb of tetrachlorodibenzodioxin was released over a 123-acre area in northern Italy from a plant which manufactures 2,4,5 trichlorophenol. High levels of dioxins have been found in fish in Vietnam where large amounts of 2,4,5-T were used as defoliants. These preliminary results suggest that dioxins may accumulate in the food chain and constitute a health hazard. This possibility is suggested by the observation that there appears to be a greater number of stillbirths, placental tumors, and malformations where the 2,4,5-T defoliants were used in Vietnam.

The biochemical cause of the toxicity of dioxins is not known, but two different acnegenic effects have been correlated with the induction of two different enzymes. Higher levels of cytochrome P-450 (Section 6.5) are induced by dioxins and structurally related tetrachloroazobenzene derivatives. A direct correlation between the ability to produce chloracne and the affinity for the binding sites specific for P-450 induction was observed. Acne associated with enhanced levels of porphyrins is also caused by dioxins (see Chapter 11 for the porphyrin structure) because of increased levels of δ-aminolevulinic acid synthetase, an enzyme responsible for one of the steps in porphyrin biosynthesis.

7.8 LONG-TERM EFFECTS OF ORGANOCHLORINE COMPOUNDS

The chronic (long-term) toxicity of chlorinated hydrocarbons is the main problem arising from their use. This toxic effect was first discovered when it was observed that the populations of eagles, hawks, falcons, pelicans, and other birds at the end of a food chain were decreasing. Closer examination revealed that the birds were mating later in the season and were producing eggs with much thinner shells than normal. Analyses of the birds and their eggs revealed that they contained large amounts of organochlorine compounds. Laboratory studies confirmed that when birds were fed DDT and PCBs, they produced eggs with thin shells.

The cause of the delayed mating has been traced to the enhanced levels of the cytochrome P-450 oxidase enzymes in the livers of those birds containing the organochlorine residues. Enhanced levels of oxidase enzymes are known to be induced by barbiturates and polycyclic aromatic compounds. It was found that DDT, DDE, dieldrin, and PCBs also cause such an enhancement. The enhancement of the oxidase enzymes affects mating behavior because these enzymes catalyze the oxidation of steroid hormones. Specifically it was found that an increase in the oxidase enzymes resulted in a decrease in the levels of estradiol in the birds. It is known that the birds will mate only if a certain level of estradiol is present. If breeding is delayed, then the offspring hatch at a time when there is less food available for their growth. This is an especially important factor with large birds that require a full season for the young to develop.

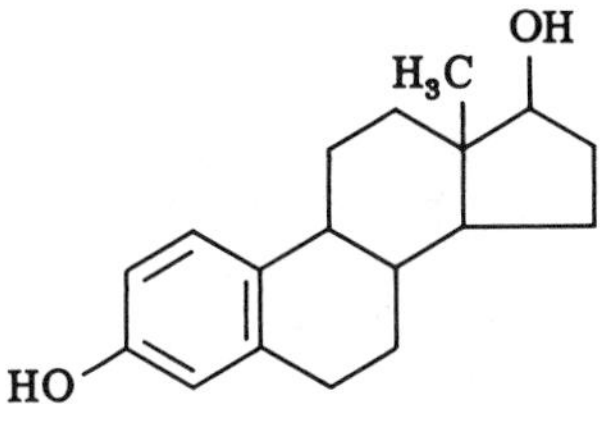

estradiol

The production of egg shells is also influenced by estradiol. The shell of an egg is only formed in the last day before laying. About 60% of the calcium needed to form the egg is obtained from the bird's food intake and the remainder comes from calcium stored in the bone marrow. Since the level of calcium deposited in the bone is regulated by levels of estradiol in the blood, a low level of estradiol results in a low level of stored calcium. The absence of stored calcium could be a significant factor in egg survival since a 20% decrease in shell thickness results in extensive egg breakage. A 20% decrease occurs with as little as 25 ppm of chloroorganics in the egg.

Finally, it has been observed that some pesticides inhibit the formation of eggshells in the presence of an adequate supply of calcium. Presumably these pesticides interfere with the enzyme carbonic anhydrase that is responsible for the conversion of carbon dioxide to carbonate. The latter is required to combine with calcium to form the calcium carbonate of the shell. So far it has been established that DDE but not dieldrin interfere with the formation of the shell in this way.

The chronic toxicity of chloroorganics is not limited to birds. Mink are highly susceptible to PCBs with levels of 5 parts per billion (ppb) sufficient to halt their reproduction. This is an economic problem because mink ranchers have been using Great Lakes coho salmon for mink feed. However this can no longer be used since the salmon contains 5 ppb PCBs.

At present there are no reports of chronic toxicity of PCBs and chlorinated insecticides in man, but there are some indications that it may exist. The first is the observation that pesticides induce high levels of liver enzymes in a wide array of vertebrates including the monkey. It seems very likely that such induction will be observed in man as well. Secondly, it has been observed that a dose of 46 mg/kg of DDT can produce a fourfold increase in tumors in animals. These data suggest a correlation between the ingestion of DDT and cancer. In addition, there is a growing list of chlorinated organic compounds that are known to produce tumors (Section 7.1). Obviously man is in as vulnerable a position as the predatory birds in the food chain so that it may be essential for his survival that he reduce the level of chlorinated hydrocarbons in the environment.

7.9 GLOBAL DISTRIBUTION OF CHLORINATED HYDROCARBONS

Chlorinated hydrocarbons are ubiquitous in the environment. Residues have been found in locations and in species where chloroorganics have never been used; for example, they have been found in the polar ice pack, in animals in arctic regions, and in the air over the center of the ocean. The chlorinated hydrocarbons must be carried through the atmosphere to achieve such wide distribution. The detection of chlorinated hydrocarbons in rainwater is consistent with this observation. Furthermore, DDT is sufficiently volatile that all that has been manufactured to date could be volatilized and still not saturate the atmosphere.

The global circulation of chlorinated hydrocarbons demonstrates that all nations have a stake in the reduction in the manufacture of these compounds. Banning the use of DDT, aldrin, and dieldrin in the U.S. will be of little help if we continue to manufacture them for use in other countries. Obviously some means must be devised for the international control of chloroorganics.

7.10 ORGANOPHOSPHORUS INSECTICIDES

7.10.1 Synthesis

The chemistry and mechanism of action of the organophosphorous insecticides are quite different from those of the organochlorine materials. Organophosphorus compounds merit consideration because they are being used more frequently in place of the organochlorine compounds. Some common phosphorus containing insecticides are listed in Table 7.3. A related compounds, tris(2,3 dibromopropyl) phosphate, has been widely used as a flame retardant in man-made fibers. As illustrated by the examples in the table, the majority of the organophosphorus insecticides are derivatives of phosphoric acid or the sulfur analogs of phosphoric acid. The phosphoric acid esters used in the synthesis of these insecticides are prepared industrially by the reaction of phosphoryl chloride with alcohols [reaction (7-50)]. Under suitable conditions diester chlorides (dialkyl phosphorochloridates) may be prepared. The same reaction pathway may be used for the preparation of corresponding sulfur analogs (dialkyl phosphorothiochloridates) [reaction (7-51)]. Both the phosphoric acid and thiophosphoric acid derivatives may also be prepared by the action of chlorine on the corresponding derivative of phosphorous acid [reaction (7-52)].

$$Cl-P(=O)(Cl)-Cl + 2\ ROH + 2R_3N \longrightarrow RO-P(=O)(OR)-Cl + 2\ R_3N \cdot HCl \qquad (7\text{-}50)$$

$$Cl-P(=S)(Cl)-Cl + 2\ ROH + 2\ R_3N \longrightarrow RO-P(=S)(OR)-Cl + 2\ R_3N \cdot HCl \qquad (7\text{-}51)$$

TABLE 7-3

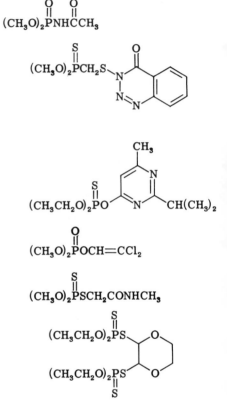

Acephate	$(CH_3O)_2\overset{O}{\overset{\|}{P}}NH\overset{O}{\overset{\|}{C}}CH_3$
Azinphosmethyl	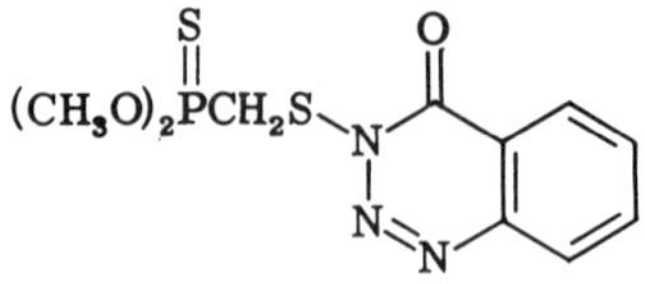
Diazinon	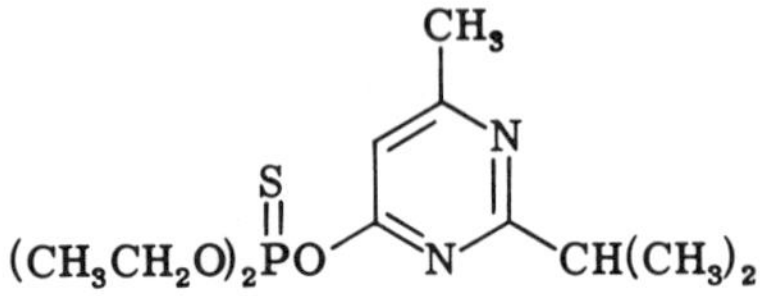
Dichlorvos	$(CH_3O)_2\overset{O}{\overset{\|}{P}}OCH{=}CCl_2$
Dimethoate	$(CH_3O)_2\overset{S}{\overset{\|}{P}}SCH_2CONHCH_3$
Dioxathion	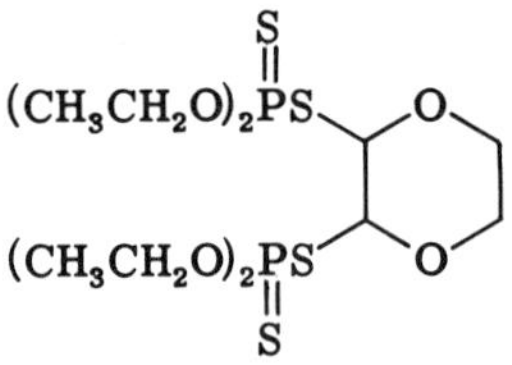
Disulfoton	$(CH_3CH_2O)_2\overset{S}{\overset{\|}{P}}SCH_2CH_2SCH_2CH_3$
Ethion	$(CH_3CH_2O)_2\overset{S}{\overset{\|}{P}}SCH_2S\overset{S}{\overset{\|}{P}}(OCH_2CH_3)_2$
Fenthion	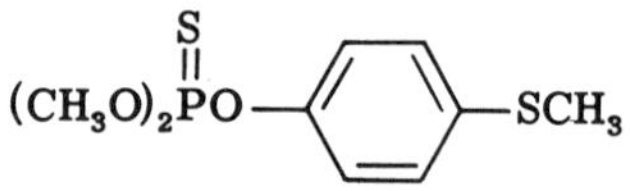

TABLE 7-3 (Continued)

Fonofos	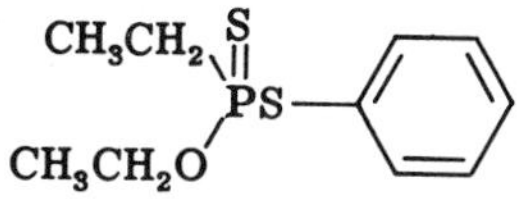
Malathion	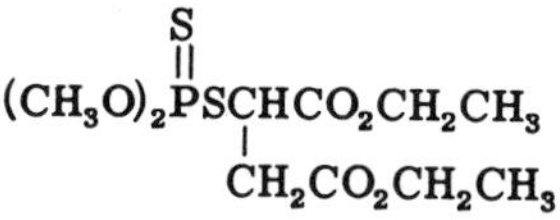
Parathion	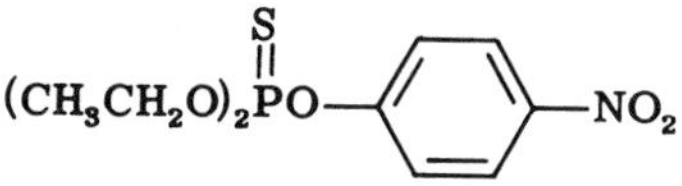
Phorate	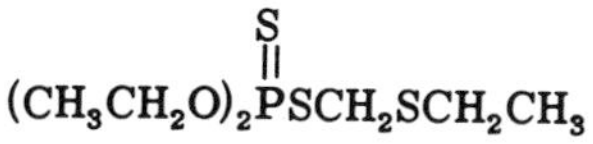
Phosdrin	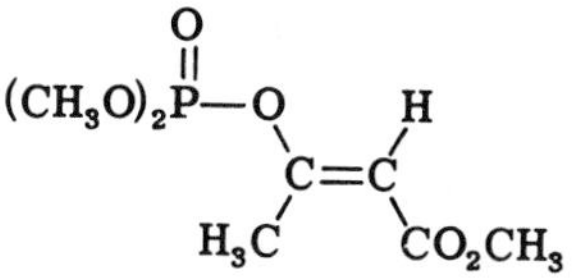
Phosphamidon	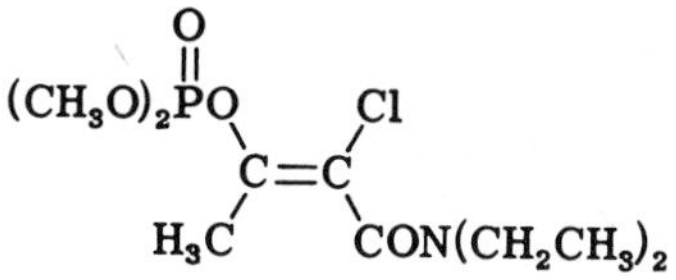
Tetrachorvinphos	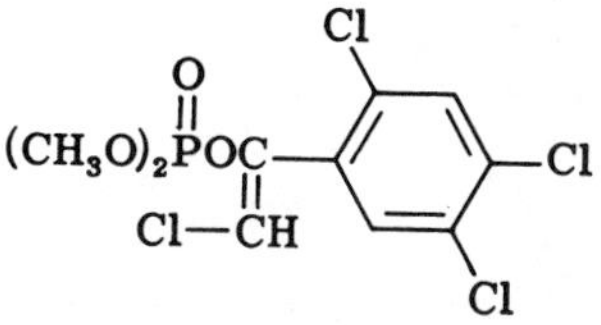
Trichlorfon	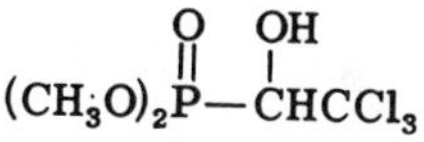

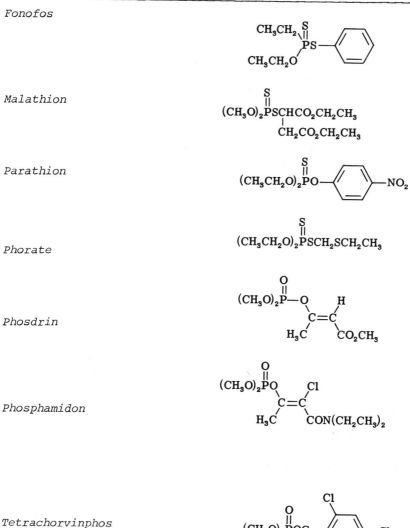

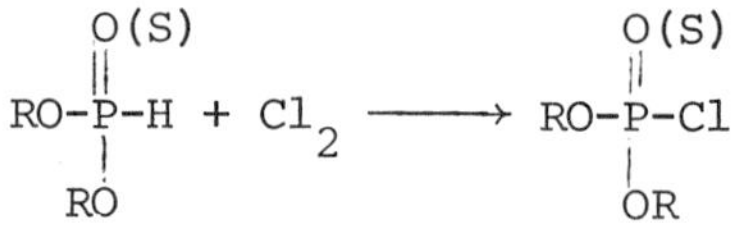

$$
\begin{array}{ccc}
& \overset{\displaystyle O(S)}{\underset{\displaystyle RO}{\underset{|}{\overset{||}{RO-P-H}}}} + Cl_2 & \longrightarrow & \overset{\displaystyle O(S)}{\underset{\displaystyle OR}{\overset{||}{RO-P-Cl}}}
\end{array}
\qquad (7\text{-}52)
$$

Most nucleophiles attack phosphorochloridates at the phosphorous atom and displace the chloro group [reaction (7-53)]. This reaction is analogous to the S_N2 displacement on carbon (Section 7.2.2). The transition state shown in (7-53) may actually be a stable intermediate although this has not been established with certainty.

$$
HS^- + \overset{\displaystyle O}{\underset{\displaystyle OR}{\overset{||}{RO-P-Cl}}} \longrightarrow HS\cdots\overset{\delta^-}{\overset{\displaystyle O}{\underset{RO\ \ OR}{\overset{||}{P}}}}\cdots Cl \longrightarrow \overset{\displaystyle O}{\underset{\displaystyle OR}{\overset{||}{HS-P-OR}}} + Cl^-
$$

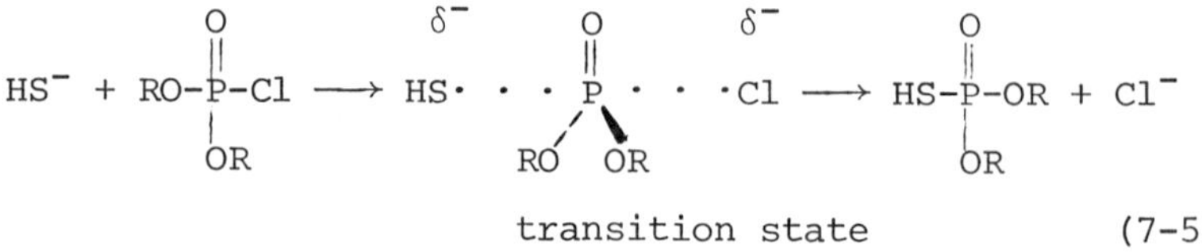

transition state   (7-53)

Most of the organophosphorus insecticides are triesters of phosphoric or thiophosphoric acid that are prepared by the attack of a nucleophile on the phosphochloridates or phosphorothiochloridates as shown in the examples which follow. Since the ester group is more difficult to displace than the chloro group, the triesters exhibit greater stability than the phosphorochloridates. As a consequence of the slow displacement of the ester, a reaction pathway different from that illustrated in (7-53) is sometimes observed. For example, in neutral and acid solutions water can attack the α-carbon atom of the phosphate ester grouping with resultant cleavage of the carbon-oxygen bond [reaction (7-54)]. It has been shown by isotope tracer studies using ^{18}O that the P-O bond is not broken. This pathway is more likely under environmental conditions than the direct displacement on phosphorus [(reaction (7-53)].

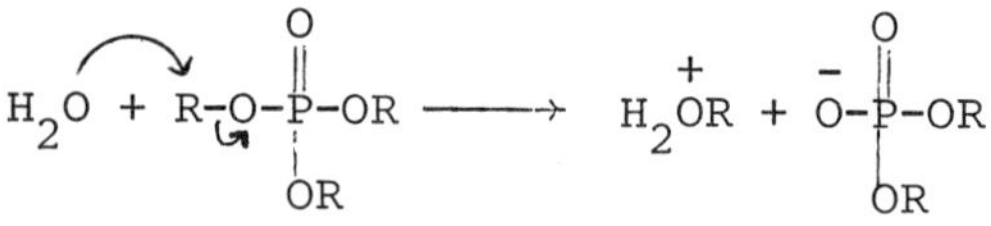

$$
H_2O + \overset{\displaystyle O}{\underset{\displaystyle OR}{\overset{||}{R-O-P-OR}}} \longrightarrow H_2\overset{+}{O}R + \overset{\displaystyle O}{\underset{\displaystyle OR}{\overset{||}{\overset{-}{O}-P-OR}}}
\qquad (7\text{-}54)
$$

The synthesis of parathion or methyl parathion involves an S_N2-like displacement reaction (7-55).

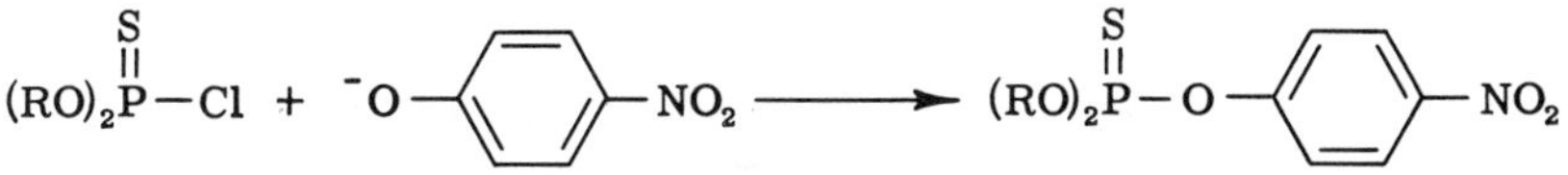

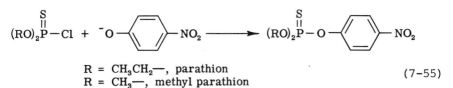

R = CH_3CH_2—, parathion
R = CH_3—, methyl parathion (7-55)

A similar type of reaction is used in the synthesis of demeton. However a rearranged product is also formed.

$$(CH_3CH_2O)_2\overset{S}{\overset{\|}{P}}Cl + HOCH_2CH_2SCH_2CH_3 \longrightarrow$$

$$(CH_3CH_2O)_2\overset{S}{\overset{\|}{P}}OCH_2CH_2SCH_2CH_3 + (CH_3CH_2O)_2\overset{O}{\overset{\|}{P}}\text{-}SCH_2CH_2SCH_2CH_3$$

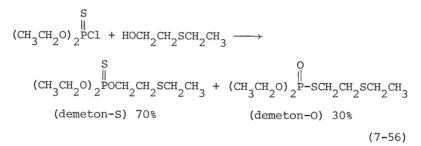

(demeton-S) 70% (demeton-O) 30%

(7-56)

The formation of the rearranged product probably involves the participation of the sulfur of the thioether since the oxygen analog does not rearrange. The following mechanism has been suggested.

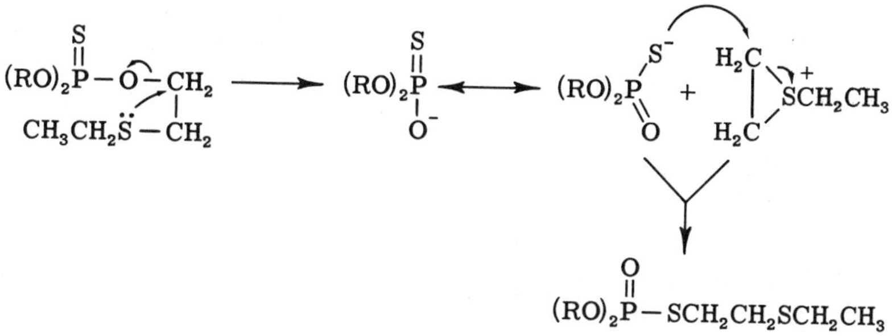

(7-57)

The synthesis of malathion involves a nucleophilic addition of a thioacid to a double bond that is conjugated to a carbonyl group (Michael addition).

$$(CH_3O)_2\overset{S}{\overset{\|}{P}}\text{-}SH + \begin{matrix} CHCO_2CH_2CH_3 \\ \| \\ CHCO_2CH_2CH_3 \end{matrix} \longrightarrow (CH_3O)_2\overset{S}{\overset{\|}{P}}\text{-}S\begin{matrix} CH_2CO_2CH_2CH_3 \\ | \\ CHCO_2CH_2CH_3 \end{matrix}$$

malathion (7-58)

7.10.2 Application as Pesticides

These organophosphorus compounds exhibit high insecticidal action against a wide variety of species. This is not really an asset because they kill beneficial as well as harmful insects. These compounds are highly toxic to vertebrates as well as to insects and, as a consequence, are much more dangerous to the workers using them than are chloroorganics. Demeton is unique in that it shows systemic action. That is, it is absorbed by the plant and its toxic effect is retained in the plant for 1-1.5 months. Obviously this insecticide is mainly useful for ornamental plants and shrubs that are not consumed as animal or human food. Malathion can be used safely by home gardeners because it is readily detoxified by vertebrates. The ester groups are readily hydrolyzed by the esterase enzymes present in mammals but absent in insects. The lethal dose for killing 50% of the mammals ingesting it (LD_{50}) is 500-1500 mg/kg of body weight for malathion as compared to an LD_{50} of 6-12 mg/kg for parathion and 25-50 mg/kg for methyl parathion.

The toxic effect of the organophosphorus insecticides is due to their interference in the transfer of nerve impulses from one nerve cell to the next. When a nerve impulse reaches the end of a nerve cell it triggers the release of a minute amount of the compound acetylcholine. The acetylcholine activates a receptor on an adjacent nerve cell causing it to carry the impulse to the next nerve cell. The acetylcholine is then hydrolyzed to choline and acetic acid by the enzyme cholinesterase as in reaction (7-59). Other enzymes then reconstitute the choline and acetic acid to acetylcholine on the nerve endings.

$$\underset{\text{acetylcholine}}{(CH_3)_3\overset{+}{N}CH_2CH_2O\overset{\overset{O}{\|}}{C}CH_3} \xrightarrow[\text{cholinesterase}]{H_2O} \underset{\text{choline}}{(CH_3)_3\overset{+}{N}CH_3CH_2OH} + \underset{\text{acetic acid}}{CH_3CO_2H} \qquad (7\text{-}59)$$

The organophosphorous pesticides bind chemically to the cholinesterase so that it can no longer catalyze the hydrolysis of acetylcholine. The resulting excess of acetylcholine hyperstimulates nerves and produces convulsions, irregular heartbeat, and choking in vertebrates.

7.10.3 Environmental Degradation

The organophosphorus insecticides undergo a very rapid hydrolysis in the environment. This is a mixed blessing. There is no buildup of residues; however, frequent application of the pesticide is required. Some typical hydrolytic reactions include reactions (7-60)-(7-62). The nitro group of parathion is reduced by soil microorganisms to give the nontoxic aminoparathion, which is stable and remains bound to the soil. It is not known if this "inactive" form of parathion constitutes an environmental problem.

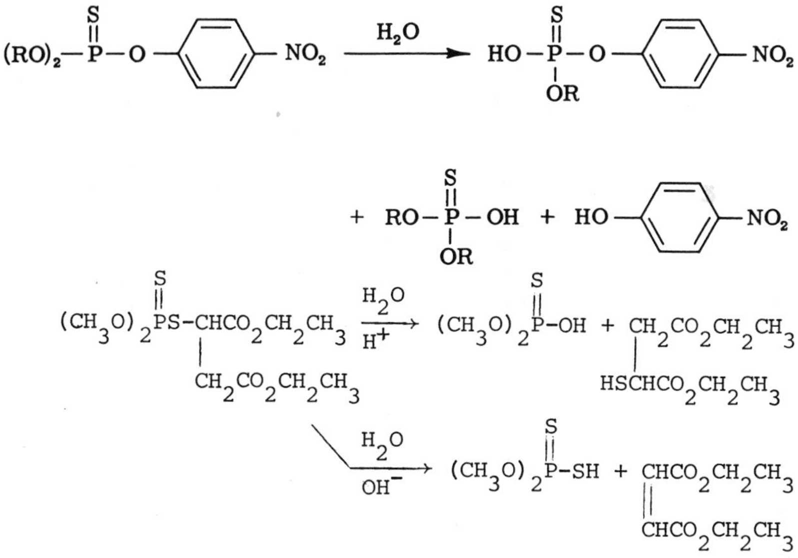

(7-60)

(7-61)

(7-62)

A problem associated with these organophosphorus insecticides is their environmental conversion to more toxic substances. For example, demeton-O, the by-product formed in the synthesis of demeton-S [reaction (7-56)] has an LD_{50} of 2 mg/kg for vertebrates as compared to an LD_{50} of 100 mg/kg for demeton-S. The demeton rearrangement [reaction (7-57)] takes place readily in plant tissues and may explain the long-term systemic action of demeton. If the rearranged product is really the toxic agent, then trace amounts would still be effective. Parathion is readily oxidized by atmospheric oxygen or enzymatically to a derivative that is four times as toxic as parathion itself. Since parathion exhibits no anticholinesterase activity itself, its insecticidal action may be due to its enzymatic conversion to the oxygen analog in the insect.

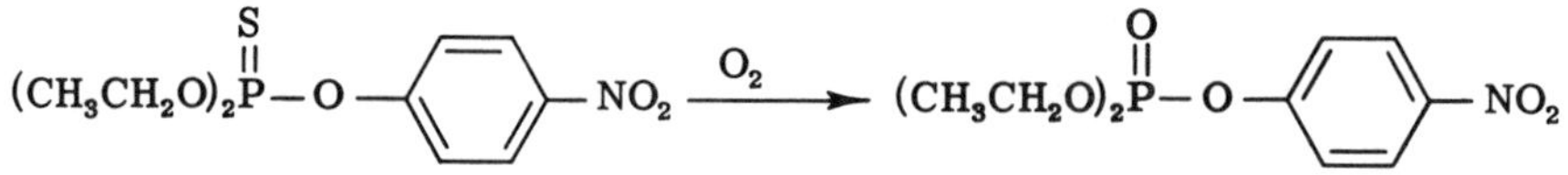

(7-63)

7.11 OTHER PESTICIDES IN COMMON USE

7.11.1 Carbaryl

The carbamate insecticide carbaryl has been used extensively as a replacement for DDT to control the gypsy moth as well as other insects. Its toxicity to insects is due to its anticholinesterase action. It is not very toxic to mammals (LD_{50} for rats is 0.5 gm/kg) presumably because it is readily oxidized and excreted. Carbaryl is not persistent in the environment because it is rapidly destroyed by a number of pathways and as a consequence it must be applied frequently for insect control. Some of the key reactions of this material are given in reactions (7-64)-(7-66).

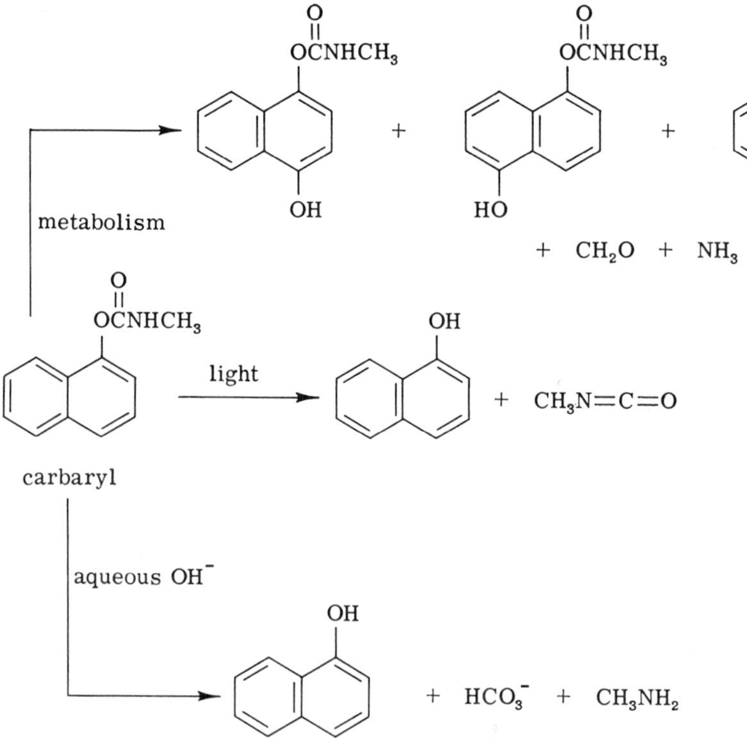

(7-64)

(7-65)

(7-66)

7.11.2 *Captan*

Captan is used as a broad-spectrum protective fungicide for the control of diseases of agricultural crops. It is essentially nontoxic with an LD_{50} for rats of 9 gm/kg. Captan undergoes ready hydrolysis by water [reaction (7-67)] so it does not persist in the environment. The fungicidal action may be due to alkylation of the various sulfhydryl groups in the fungus cell by the reactive trichloromethylthio group.

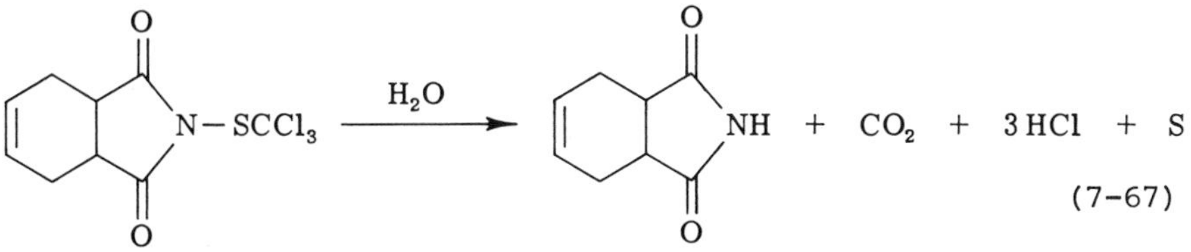

(7-67)

7.11.3 *Picrolam*

Picrolam is a herbicide which is used to control perennial and annual weeds that are difficult to control with other compounds. It is fairly long-lived with a half-life in soil of 18 months. It has been shown to be destroyed photochemically in laboratory studies. The decarboxylation product of picrolam has been isolated (7-68) along with two monodehalogenated products of unknown structure. Picrolam has a very low toxicity with an LD_{50} of 2-8 gm/kg in a variety of experimental animals.

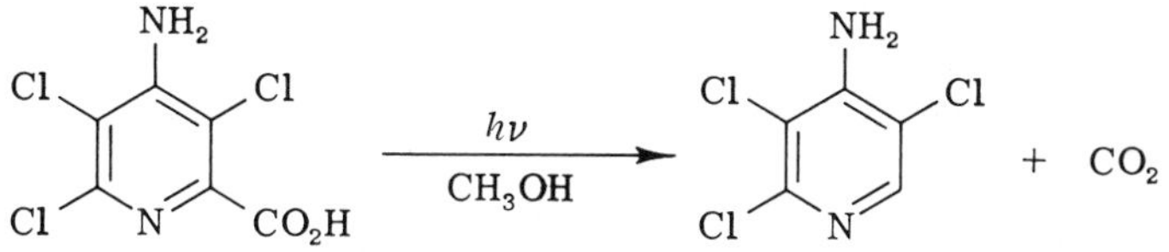

(7-68)

7.11.4 *Atrazine and Simazine*

Atrazine and simazine are derivatives of symmetrical triazine that are used for the control of weeds in corn. Symmetrical triazines are planar six-membered heterocyclic compounds with three angular nitrogen atoms. Atrazine and simazine are relatively nontoxic compounds with an LD_{50} for rats of 3 gm/kg or greater. Microorganisms are responsible for the environmental degradation of these compounds, but this is a slow process since their half-life in soil samples is 10-12 months. The first step is cleavage of the N-alkyl grouping [reaction (7-69)]. The hydrolytic cleavage of the C-Cl bond to form the C-OH group is another decomposition pathway. Recent studies have shown that atrazine may enhance or suppress the toxicity of DDT and parathion depending on the soil type, concentration of pesticides, and other environmental factors. The reason for this effect is not clear. It is often observed that the biological effect of a mixture of pesticides is not the sum of the effects of the individual components of the mixture.

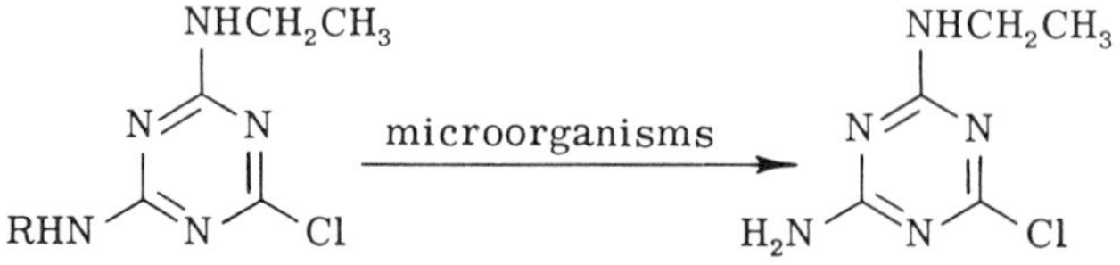

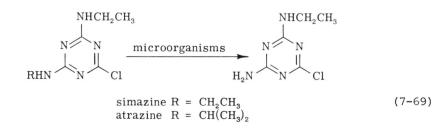

simazine R = CH_2CH_3 (7-69)
atrazine R = $CH(CH_3)_2$

7.11.5 Diquat and Paraquat

The bipyridylium herbicides dequat and paraquat are used for the nonselective eradication of plants prior to the planting of a crop and for the control of aquatic plants. They kill the green leaves on which they are applied but cannot penetrate mature bark and so they may be used to control weeds in orchards. They work most rapidly in the presence of strong sunlight. The compounds are relatively toxic to mammals with an LD_{50} of 30-300 mg/kg for oral doses. However, the quaternary salts are not absorbed through the skin so they are not hazardous to man.

These compounds serve as herbicides because they are capable of being reduced in plants to free radicals. The energy for this reduction comes from photosynthesis (see Chapter 11). The free radicals are reoxidized readily back to the original quaternary ions generating hydrogen peroxide or other radicals that destroy the plant cell.

Diquat and paraquat are destroyed or inactivated rapidly in the environment. They are inactivated by soils and are metabolized by some microorganisms. Diquat is readily destroyed by sunlight but the losses of paraquat under the same conditons are much less. This is because the bulk of the ultraviolet absorption of diquat occurs above 300 nm, wavelengths at which sunlight in penetrating the earth's atmosphere (see Chapter 4), while the bulk of the ultraviolet absorption of paraquat is below 300 nm.

The pathways observed for the photolysis of diquat and paraquat in laboratory studies are given below.

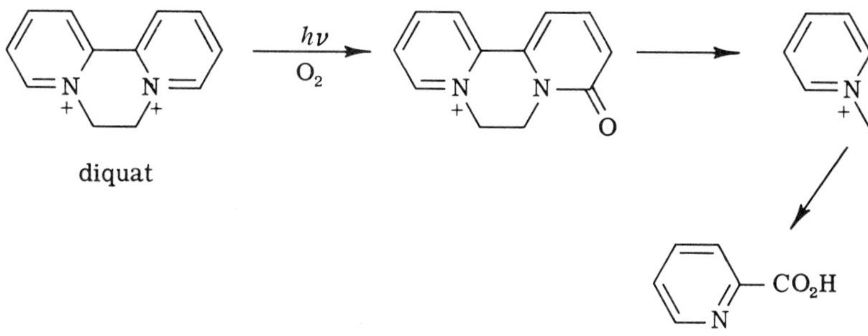

(7-70)

(7-71)

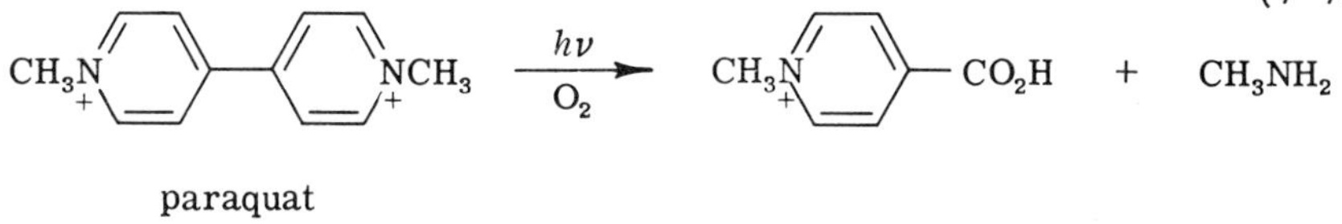

BIBLIOGRAPHY

General References

N. N. Melnikov, "Chemistry of Pesticides." Springer-Verlag, New York, 1971.

J. Robinson, Organochlorine compounds in man and his environment, *Chem. Britain 7*, 472 (1972).

W. Moje, The chemistry and nematocidal activity of organic halides, *Adv. Pest Control Res. 3*, 181 (1966).

R. White-Stevens, "Pesticides in the Environment," Vol. 1. Marcel Dekker, New York, 1971.

P.C. Kearney, E. A. Woolson, J. R. Plimmer, and A. R. Isensee, Decontamination of pesticides in soils, *Residue Reviews 29*, 137, (1969).

R. L. Metcalf, G. K. Sangha, and I. P. Kapoor, Model ecosystem for the evaluation of pesticide biodegradability and ecological magnification, *Environ. Sci. Technol. 5,* 709 (1971).

J. M. Barnes, Control of health hazards associated with the use of Pesticides, *Adv. Pest Control Res. 1,* 1 (1957).

G. P. Georghiou, Genetic studies on insect resistance, *Adv. Pest Control Res. 6,* 171 (1965).

R. Carson, "Silent Spring." Houghton Mifflin Co., Boston, Massachusetts, 1962.

W. R. Benson, The chemistry of pesticides, *Ann. New York Acad. Sci. 160,* 7 (1969).

R. D. O'Brien, "Insecticides Action and Metabolism." Academic Press, New York, 1967.

C. A. Edwards, Insecticide residues in soils, *Residue Reviews 13,* 83 (1966).

E. H. Marth, Residues and some effects of chlorinated hydrocarbon insecticides in biological material, *Residue Reviews* 9, 1 (1965).

C. A. Edwards, "Persistent Pesticides in the Environment." CRC Press, Cleveland, Ohio, 1970.

K. C. Walker, The role of pesticides in pollution management, *Residue Reviews 34,* 163 (1970).

D. G. Crosby, The nonmetabolic decomposition of pesticides, *Ann. New York Acad. Sci. 160,* 82 (1969).

"Degradation of synthetic organic molecules in the biosphere. Natural, pesticidal and various other man-made compounds." National Academy of Sciences, Washington, D.C., 1972.

L-Y. Gibney, EPA seeks substitutes for banned pesticides, *Chem. Eng. News,* June 9, 1975, pp. 15-16.

T. T. Liang and E. P. Lichtenstein, Synergism of insecticides by herbicides: Effect of environmental factors, *Science 186,* 1128 (1974).

L. J. Carter, Cancer and the environment (II): Groping for new remedies, *Science 186,* 242 (1974).

Water contaminated throughout U.S., *Chem. Eng. News,* April 28, 1975, pp. 18-19.

T. Page, R. H. Harris, and S. S. Epstein, Drinking water and cancer mortality in Louisiana, *Science 193,* 55 (1976).

G. T. Brooks, "Chlorinated Insecticides," Vols. I and II. CRC Press, Cleveland, Ohio, 1974.

EPA lists pesticides that may be too dangerous to use, *Chem. Eng. News,* June 14, 1976, p. 18.

C. Murray, Vinylidene chloride: No trace of cancer at Dow, *Chem Eng. News,* March 14, 1977, p. 21.

Phaseout set for fluorocarbon aerosols, *Chem. Eng. News,* May 16, 1977, p. 4.

N. Wade, Drinking water: Health hazards still not resolved, *Science 196,* 1421 (1977).

DDT

R. L. Metcalf, I. P. Kapoor, and A. S. Hirwe, Development of biodegradable analogs of DDT, *Chem. Technol.* 105 (1972).

I. P. Kapoor, R. L. Metcalf, A. S. Hirwe, J. R. Coats, and M. S. Khalsa, Structure activity correlations of biodegradable DDT analogs, *J. Agr. Food Chem. 21,* 310 (1973).

G. Holan, New halocyclopropane insecticides and the mode of action of DDT, *Nature 221,* 1025 (1969).

T. F. West and G. A. Campbell, "DDT and Newer Persistent Pesticides." Chemical Publishing Co., New York, 1952.

R. L. Metcalf, I. P. Kapoor, and A. S. Hirwe, Biodegradable analogs of DDT, *Bull. World Health Org. 44,* 363 (1971).

R. L. Metcalf and T. R. Fukuto, The comparative toxicity of DDT and analogs to susceptible and resistant houseflies and mosquitoes, *Bull. World Health Org. 38,* 633 (1968).

R. L. Metcalf, Insects versus insecticides, in Chemistry in the Environment, Readings from Scientific American," p. 108. W. H. Freeman, San Francisco, California, 1973. (From Sci. Amer., Oct 1952.)

W. Hom, R. W. Risebrough, A Soutar, and D. R. Young, Deposition of DDE and polychlorinated biphenyl in dated sediments of the Santa Barbara basin, *Science 184,*1107 (1974).

H. Lipke and C. W. Kearns, DDT-dehydrochlorinase, *Adv. Pest Control Res. 3,* 253 (1960).

R. Riemschneider, Chemical structure and activity of DDT analogues with special consideration of their spatial structures, *Adv. Pest Control Res. 3,* 307 (1958).

S. J. Cristol, A kinetic study of the dehydrochlorination of substituted 2,2-diphenylchloroethanes related to DDT, *J. Amer. Chem. Soc. 67,* 1494 (1945).

Polychlorinated and Polybrominated Biphenyls

G. G. Gustafson, PCBs--prevalent and persistent, *Environ. Sci. Technol. 4,* 814 (1970).

A. L. Hammond, Chemical pollution: Polychlorinated biphenyls, *Science 175,* 155 (1972).

T. H. Maugh II, Polychlorinated biphenyls: Still prevalent, but less of a problem, *Science 178,* 388 (1972).

T. H. Maugh II, Chemical pollutants: Polychlorinated biphenyls still a threat, *Science 190,* 1189 (1975).

T. H. Maugh II, DDT: An unrecognized source of polychlorinated biphenyls, *Science 180,* 578 (1973). See also G. R. Harvey, *Science 180,* 1122 (1973).

O. Hutzinger, D. M. Nash, S. Sage, A. S. W. DeFreitas, R. J. Norstrom, D. J. Wildish, and V. Zitko, Polychlorinated biphenyls: Metabolic behavior of pure isomers in pigeons, rats, and brook trout, *Science 178,* 312 (1972).

G. R. Harvey, W. G. Steinhauer, and J. M. Teal, Polychlorobiphenyls in North America ocean water, *Science 180,* 643 (1973).

M. Friend and D. O. Trainer, Polychlorinated biphenyl: Interaction with duck hepatitis virus, *Science 170,* 1314 (1970).

PCBs: Leaks of toxic substances raise issue of effects, regulation, *Science 173,* 899 (1971).

J. L. Mosser, N. S. Fisher, T.-C. Teng, and C. F. Wurster, Polychlorinated biphenyls: Toxicity to certain phytoplanktons, *Science 175,* 191 (1972).

N. S. Fisher, Chlorinated hydrocarbon pollutants and photosynthesis of marine phytoplankton: A reassessment, *Science 189,* 463 (1975).

G. W. Bowes, M. J. Mulvihell, B.R.I. Simoneit, and P. W. Risebrough, Identification of chlorinated dibenzofurans in American polychlorinated biphenyls, *Nature 256,* 305 (1975).

Monsanto releases PCB data, *Chem. Eng. News,* Dec. 6, 1971, p. 15.

Monsanto to quit PCB business next year, *Chem. Eng. News,* Oct. 11, 1976, p. 8.

L. J. Carter, Michigan's PBB incident: Chemical mix-up leads to disaster, *Science 192,* 240 (1976).

Polychlorinated Cyclopentadiene Insecticides

S. B. Soloway, Correlating between biological activity and molecular structure of the cyclodiene insecticides, *Adv. Pest Control Res. 6,* 85 (1965).

L. J. Carter, Cancer and the environment (I): A creaky system grinds on, *Science 186,* 186 (1974).

EPA suspends use of heptachlor, chlordane, *Chem. Eng. News,* August 4, 1975, p. 7.

D. A. Carlson, K. D. Kongha, W. B. Wheeler, G. P. Marshall, and R. G. Zaylskie, Mirex in the environment: Its degradation to kepone and related compounds, *Science 194,* 939 (1976).

2,4-D and 2,4,5-T

D. Shapely, Herbicides: AAAS study finds dioxin in Vietnamese fish, *Science 180,* 285 (1973).

Health hazards of dioxins still uncertain, *Chem. Eng. News,* April 16, 1973, p. 12.

A. Poland and E. Glover, 2,3,7,8-Tetrachlorodibenzo-*p*-dioxin: A potent inducer of α-aminolevulinic acid synthetase, *Science 179,* 476 (1973).

R. L. Wain, The relation of chemical structure to activity for the 2,4-D-type herbicide and plant growth regulator, *Adv. Pest Control Res. 2,* 263 (1958).

R. F. Gould, Ed., "Chlorodioxins-Origin and Fate," Advances in Chemistry Series No. 120, American Chemical Society, Washington, D.C., 1973.

R. H. Stehl and L. L. Lamparski, Combustion of several 2,4,5-trichlorophenoxy compounds: formation of 2,3,7,7-tetrachlorodibenzo-*p*-dioxin, *Science, 197,* 1008 (1977).

C. D. Carter, R. D. Kimbrough, J. A. Liddle, R. E. Cline, M. M. Zack, Jr., W. F. Barthel, R. E. Koehler, and Patrick E. Phillips, Tetrachlorodibenzodioxin, an accidental poisoning episode in horse arenas, *Science 188,* 738 (1975).

D. G. Crosby and A. S. Wong, Environmental degradation of 2,3,7,8-tetrachlorodibenzo-*p*-dioxin, *Science 195,* 1337 (1977).

J. Walsh, Seveso: The questions persist where dioxin created a wasteland, *Science 197,* 1064 (1977).

Other Chloro- and Bromoorganics

B. Dowty, D. Carlisle, J. L. Laseter, and J. Storer, Halogenated hydrocarbons in New Orleans drinking water and blood plasma, *Science 197,* 75 (1975).

J. L. Marx, Drinking water: Another source of carcinogens?, *Science 186,* 809 (1974).

R. J. Seltzer, Reactions grow to trichloroethylene alert, *Chem. Eng. News,* May 19, 1975, pp. 41-43. See also *Chem. Eng. News,* August 4, 1975, p. 7.

I. J. Selikoff and E. C. Hammond, Toxicity of vinyl chloride-polyvinyl chloride, *Ann. New York Acad. Sci. 246,* 1 (1975).

P. H. Howard and A. Hanchett, Chlorofluorocarbon sources of environmental contamination, *Science 189,* 217 (1975).

H. B. Singh, D. P. Fowler, and T. O. Peyton, Atmospheric carbon tetrachloride: Another man-made pollutant, *Science 192,* 1231 (1976). See also *Science 193,* 573 (1976).

A. Blum and B. N. Ames, Flame-retardant additives as possible cancer hazards, *Science 195,* 17 (1977)

Biochemical Effects of Chloroorganics

D. B. Peakall, Pesticides and the reproduction of birds, in "Chemistry of the Environment, Readings from Scientific American," p. 113. W. H. Freeman, San Francisco, California, 1973. (From Sci. Amer., April 1970.)

J. C. Street, Organochlorine insecticides and the stimulation of liver microsome enzymes, *Ann. New York Acad. Sci. 160,* 274 (1969).

R. L. Delong, W. G. Gilmartin, and J. G. Simpson, Premature births in California sea lions: Association with high organochlorine pollutant residue levels, *Science 181,* 1168 (1973).

A. H. Cooney and J. J. Burns, Metabolic interactions among environmental chemicals and drugs, *Science 178,* 576 (1972).

A. W. A. Brown, Pest resistance to pesticides in the environment, in "Pesticides in the Environment" (R. White-Stevens, ed.), p. 457. Marcel Dekker, New York, 1971.

J. L. Sell and K. L. Davidson, Changes in the activities of hepatic microsomal enzymes caused by DDT and dieldrin, *Federation Proceedings 32,* 2003 (1973).

J. E. Casida, Insecticide biochemistry, *Ann. Review Biochem. 42,* 259 (1973).

Effect of trace metals on sea lions probed (DDT levels are also discussed) *Chem. Eng. News,* Nov. 17, 1975, p. 34.

A. Poland, E. Glover, A. S. Kende, M. DeComp, and C. M. Giandomenico, 3,4,3',4'-Tetrachloro azoxybenzene and azobenzene: Potent inducers of aryl hydrocarbon hydroxylase, *Science 194,* 627 (1976).

D. B. Peakall, DDE: Its presence in peregrine eggs in 1948, *Science 183,* 673 (1974).

J. Bitman, H. C. Cecil, and G. R. Fries, DDT-induced inhibition of avian shell gland carbonic anhydrase: A mechanism for thin eggshells, *Science 168,* 554 (1970).

D. Peakall, p,p'-DDT: Effect on calcium metabolism and concentration of estradiol in the blood, *Science 168,* 529 (1970).

D. Kufer, Effects of some pesticides and related compounds on steroid function and metabolism, *Residue Reviews 19,* 11 (1967).

Global Distribution of Chlorinated Hydrocarbons

G. M. Woodall, P. P. Craig, and H. A. Johnson, DDT in the Biosphere: Where does it go?, *Science 174,* 1101 (1971); see also *Science 177,* 724 (1972).

J. Cramer, Model of the circulation of DDT on earth, *Atmos. Environ. 7,* 241 (1973).

M. M. Cliath and W. F. Spencer, Dissipation of pesticides from soil by volatilization of degradation products, *Envir. Sci. Technol. 6,* 910 (1972).

T. F. Bidleman and C. E. Olney, Chlorinated hydrocarbons in the Sargasso sea atmosphere and surface water, *Science 183,* 516 (1974).

R. W. Risebrough, P. Rieche, D. B. Peakall, S. G. Herman, and M. N. Kirven, Polychlorinated biphenyls in the global ecosystem, *Nature 220,* 1098 (1968).

Z. Jegier, Pesticide residues in the atmosphere, *Ann. New York Acad. Sci. 160,* 143 (1969).

K. L. E. Kaiser, Mirex: An unrecognized contaminant of fishes from Lake Ontario, *Science 185,* 523 (1974).

Organophosphorus Insecticides and Related Compounds

T. Fukuto, The chemistry and action of organic phosphorus insecticides, *Adv. Pest Control Res. 1,* 147 (1957).

E. Adams, Poisons in "Chemistry of the Environment, Readings from Scientific American," p. 295. W. H. Freeman, San Francisco, California, 1973. (From Sci. Amer. Nov. 1959.)

M. Eto, "Organophosphorus Pesticides: Organic and Biological Chemistry." CRC Press, Cleveland, Ohio, 1974.

C. Fest and K. J. Schmidt, "The Chemistry of Organophosphorus Pesticides." Springer-Verlag, Berlin, 1973.

J. Katan, T. W. Fuhremann, and E. P. Lichtenstein, Binding of [^{14}C] parathion in soil: A reassessment of pesticide persistence, *Science 193,* 891 (1976).

M. J. Prival, E. C. McCoy, B. Gutter, and H. S. Rosenkranz, Tris(2,3,dibromopropyl) phosphates. Mutagenicity of a widely used flame retardant, *Science 195,* 76 (1977).

Other Pesticides

C. I. Harris, D. D. Kaufman, T. J. Sheets, R. G. Nash, and P. C. Kearney, Behavior and fate of *s*-triazines in soils, *Adv. Pest Control Res. 8*,1 (1968).

A. Calderbank, The bipyridylium herbicides, *Adv. Pest Control Res. 8,* 127 (1968).

8
THIRD GENERATION INSECT CONTROL

8.1 INTRODUCTION

The third generation insecticides were defined by Williams[1] as the insect hormones that were developed from basic studies of insect physiology. He defined the first generation insecticides as those toxic inorganic compounds, such as lead arsenate, that were used for insect control prior to World War II. DDT and the other synthetic organic insecticides represent the second phase of insecticide development and were defined by Williams as the second generation insecticides. We shall extend the definition to third generation insect control systems to include attractants, pheromones, microbial predators such as viruses and bacteria, and the technique of insect sterilization. Presumably the fly swatter is an example of a zero generation method of insect control! The use of third generation insecticides involves completely biodegradable materials that will not accumulate in the environment. In general these insecticides, like the zero generation methods, are specific to the target species so that beneficial insects are not destroyed.

8.2 PHEROMONES AND ATTRACTANTS

8.2.1 Introduction

Pheromones are chemicals secreted by one species to affect the behavior of another of its own species. Insect pheromones are used mainly to indicate the location of food

[1]*C. M. Williams, Third-generation pesticides in "Chemistry in the Environment, Readings from Scientific American," p. 120. W. H. Freeman, San Francisco, California, 1973. (From Sci. Amer., July 1967.)*

or to attract a mate (sex pheromones). Sex pheromones are usually secreted by the female. The insect sex pheromones have been studied extensively with the goal of controlling the insect by disrupting mating behavior or by attracting the insects to traps where they are destroyed by a toxic agent. These pheromones are usually species-specific so that one can selectively control a specific insect pest. The pheromones of some insects are present in their food sources. Mating and feeding may be signaled by the same chemicals.

8.2.2 Isolation of Sex Pheromones

Pheromones are present only in trace amounts in insects so that special techniques coupled with considerable skill, ingenuity, and intuition are required to decipher the structures of these compounds. For example, only 900 μg (9×10^{-4} gm) of pure male sex attractant was isolated from 135,000 virgin female fall armyworm moths. Obviously the usual techniques of organic chemistry cannot be used for the structure analysis of these pheromones. A sensitive bioassay is required, which is usually based on the observation of the response of the insect to the pheromone. A common technique is to allow the effluent from a gas chromatograph to pass through a cage containing the insects. Compounds that are present in too small an amount to be detected by the detectors in the gas chromatograph can be detected by the insects as evidenced by their agitated behavior. Once a sensitive assay has been developed, large numbers of the insects can be bred or collected to provide a source of the pheromones. The whole insect or the glands that secrete the pheromones are extracted and the pure pheromone is separated from the hundreds of other compounds in the extract. The structure assignment of the trace amount of purified pheromone is usually based mainly on spectroscopic measurements. As a consequence, the postulated structure must then be confirmed by an unambiguous synthesis. The final objective of the project is to devise a synthesis for the preparation of large amounts of the pheromone to use for insect control.

This description is an oversimplification that assumes that the pheromone is one compound. However, the sex pheromone of the boll weevil, for example, is a mixture of four compounds. No attractive power is observed if one of the components is missing. Occasionally, masking agents are present that inhibit the effectiveness of the pheromone. A major problem is that the insect produces other compounds that are structurally similar and difficult to separate from the biologically active compound. Sometimes only after the completion of the total synthesis of a biologically inactive compound is it discovered that the pheromone was present as a lesser "contaminant" in the compound whose structure was determined. The pheromone literature is replete with controversy because of these chemical difficulties and in many instances incorrect structures have been assigned to the pheromones.

8.2.3 Examples of Sex Pheromones

The sex pheromone of the boll weevil has been studied extensively. This insect is important because of the extensive cotton crop losses that it causes each year. The four compounds that make up this pheromone were isolated from fecal material in the proportions indicated:

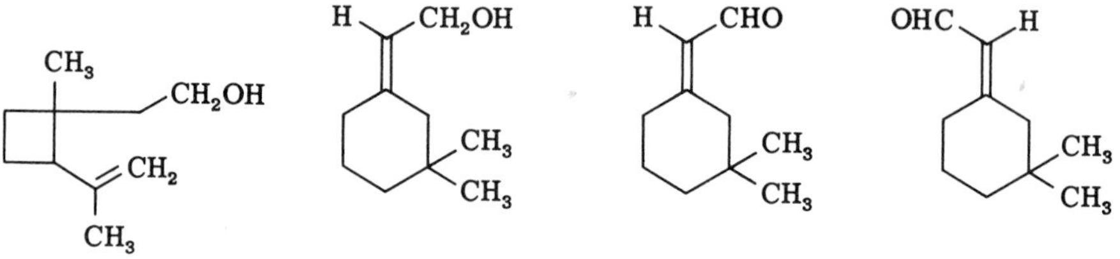

13 10 1 1

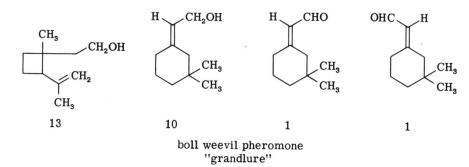

boll weevil pheromone
"grandlure"

The sex pheromone of the pink bollworm moth, another predator of cotton, has been identified as a mixture of *cis, cis* and *cis, trans* isomers of 7, 11-hexadecadienyl acetate (see following structures). *Cis*-7-hexadecyl acetate also attracts the male and was originally believed to be the pheromone, and it is currently being used as an attractant under the commercial name hexalure. Recent attempts to detect hexalure in the extract of 1.2 million pink bollworm moths proved unsuccessful.

$$CH_3\overset{O}{\overset{\|}{C}}OCH_2(CH_2)_5\underset{cis}{CH{=}CH}(CH_2)_2\underset{cis\ or\ trans}{CH{=}CH}(CH_2)_3CH_3$$

pink bollworm moth pheromone

$$CH_3\overset{O}{\overset{\|}{C}}OCH_2(CH_2)_5\underset{cis}{CH{=}CH}(CH_2)_7CH_3$$

hexalure

The structures of a number of other pheromones are listed in Table 8-1. It is remarkable that their structures are so similar yet the attractants are species specific.

8.2.4 Attractants which are not Pheromones

Not all insect attractants are pheromones. For example, it has been established that hexalure is not a pheromone, but it exerts a strong attraction to the pink bollworm moth. Some other attractants that have been used in insect control are listed in Table 8-2. Siglure and medlure were especially useful in the program for the eradication of the Mediterranean fruit fly in Florida.

8.2.5 Use of Pheromones and Attractants in Insect Control

Pheromones have been used in three different approaches to insect control.

(1) To monitor the population of a specific insect--the pheromone is placed in traps and the number of insects trapped is recorded. The population density of the insect can be monitored in this way to measure the effectiveness of the insecticides being used in the control program or to monitor an insect's migration into new areas.

(2) To trap males--a large number of traps containing the pheromone are used to trap all the males in an area so that they are not available for mating.

(3) For male confusion--large amounts of the pheromone are distributed so the air is permeated with the female sex pheromone. The male is surrounded by the attractant and is unable to locate the female.

These techniques are only effective if there is a low population of the insects to be controlled. The male confusion technique does not work where the insect population is sufficiently high for the male to see the female because it has been shown that males are attracted to any females in sight. The likelihood of spotting a female is obviously higher with a high population density of a particular insect.

All three approaches are currently being used or tested for the control of the gypsy moth. Traps are being used to monitor the spread of the gypsy moth from New England to

TABLE 8-1.

Examples of Insect Sex Pheromones

Insect	*Pheromone*
Gypsy moth	$CH_3\overset{CH_3}{\overset{\vert}{C}}H(CH_2)_4\overset{\quad O \quad}{CH—CH}(CH_2)_9CH_3$ *cis* (disparlure)
Fall armyworm moth	$CH_3\overset{O}{\overset{\Vert}{C}}OCH_2(CH_2)_7CH{=}CH(CH_2)_3CH_3$ *cis*
Cabbage looper moth	$CH_3\overset{O}{\overset{\Vert}{C}}OCH_2(CH_2)_5CH{=}CH(CH_2)_3CH_3$ *cis*
European corn borer moth[a] *Red-banded leaf roller moth*[b] *Smartweed borer*[c]	$CH_3\overset{O}{\overset{\Vert}{C}}OCH_2(CH_2)_9CH{=}CHCH_2CH_3$ *cis*
Oriental fruit moth[d]	$CH_3\overset{O}{\overset{\Vert}{C}}OCH_2(CH_2)_6CH{=}CH(CH_2)CH_3$ *cis*

TABLE 8-1. (Continued)

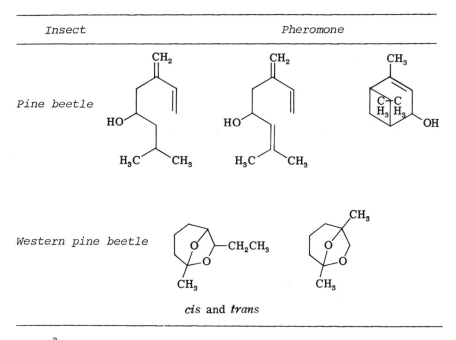

Insect	Pheromone
Pine beetle	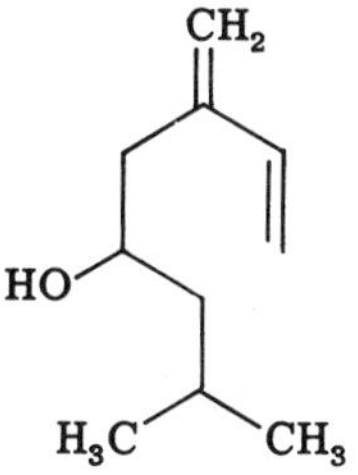 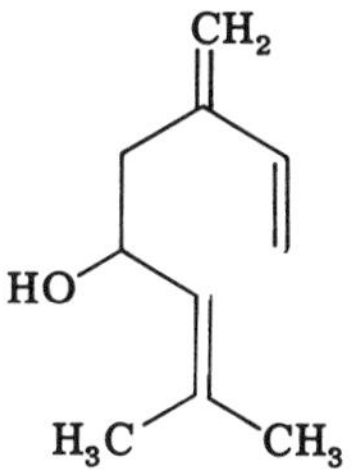 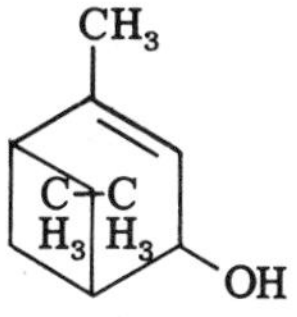
Western pine beetle	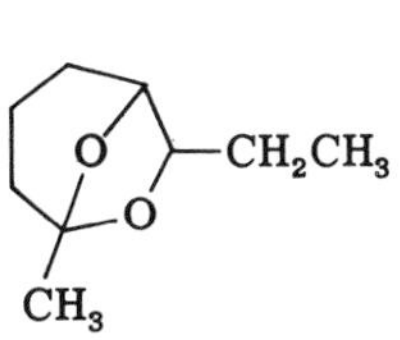 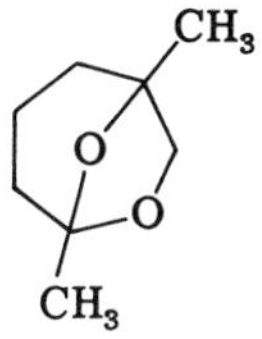 *cis* and *trans*

[a]The Iowa variety requires 4% trans-isomer while the New York variety requires 97% trans-isomer for maximal activity.

[b]6-7% trans-isomer is required for maximal activity.

[c]50% trans-isomer is required for activity.

[d]8% trans-isomer is required for maximal activity.

the South and West (Figs. 8-1 and 8-2). The traps are also being used to control the moth in areas of low infestation. The trap contains 20 μg of the attractant (disparlure) in a control-release agent (see below) that allows its gradual release over one season. The trap also contains a sticky material to hold the moth once it flys into the trap. The male confusion technique has been tested on a small scale with encouraging results. Significantly fewer females laid fertile eggs in the area where disparlure was sprayed. It should be emphasized that other methods of control will probably be required in the heavy areas of infestation in the northeastern U.S.

TABLE 8-2

Insect Attractants

Insect attracted	*Attractant*
Mediterranean fruit fly	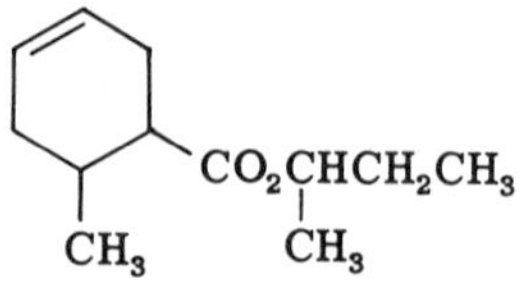 (Siglure)
Mediterranean fruit fly	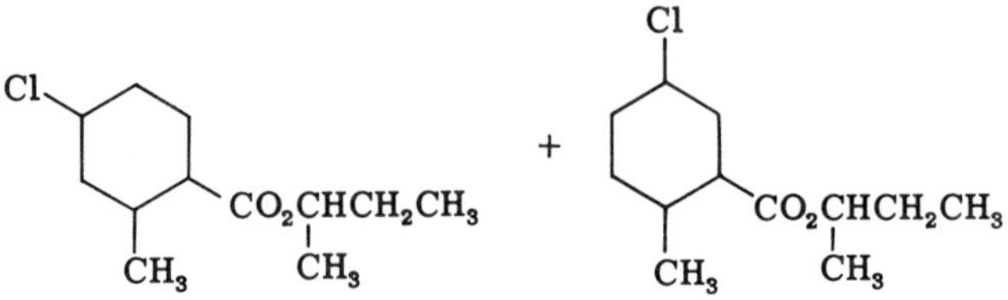 (Medlure)
Melon fly	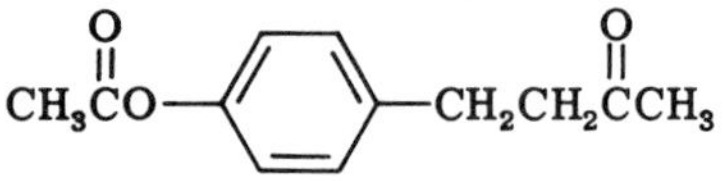
Ants	$CH_3\overset{O}{\overset{\|}{C}}OCH_2CH_2CH_2CH_2CH_3$
Sugar beet wireworm	$CH_3CH_2CH_2CH_2CO_2H$
June beetle	$CH_3CH_2CH_2\overset{CH_3}{\overset{\|}{C}}HNH_2$

FIGURE 8-1. A gypsy moth. Photograph courtesy of Dr. Morton Beroza, U.S. Dept. of Agriculture, Beltsville, Maryland.

FIGURE 8-2. A trap for the gypsy moth. Photograph courtesy of Dr. Morton Beroza, U.S. Dept. of Agriculture, Beltsville, Maryland.

Pheromones and other pesticides have been utilized more effectively when encapsulated or absorbed by compounding in rubber or plastic. The pesticide is then released slowly over an extended time period in this control-release formulation. This allows for the continued presence of an effective amount of the pesticide without requiring continued application. This procedure could eliminate the buildup of undesirable pesticide residues because much less pesticide will be required. This technique has already been shown to be very effective for the application of the male confusion technique for insect control. Using the control-release formulation, it is possible to maintain a constant effective concentration of the pheromone in the air during the entire mating period. The rapid development of resistant strains is the one danger of extended exposure of low levels of an insecticide (see Section 7.4.4). Whether resistance will be a problem with pheromones remains to be established.

Pheromones have the advantage of being very specific for one insect and in most cases only very small amounts are required. Some have expressed the opinion that the insects will not develop resistant strains because the pheromone is produced naturally by the insect. However, others note that insects have rapidly developed resistance to the synthetic pesticides and see no reason why they should not develop varieties that do not respond to these pheromones. However, even the pessimists agree that it will probably take longer for the insects to develop resistance to pheromones than it took for them to develop resistance to the second generation insecticides.

8.3 JUVENILE AND MOLTING HORMONES

8.3.1 Introduction

Insect hormones regulate growth and maturation of the animal from the larva to the adult. The juvenile hormone controls the development of the immature larva through successive growth stages. However, it must be absent for the metamorphosis of the larva to the adult to take place or a deformed larva or adult is formed that dies. The molting hormone is required for the differentiation to the adult. If the immature larva is injected with the molting hormone, it passes through its life cycle at a rapid rate and dies prematurely. Insect growth and development could be controlled by the use of these hormones.

8.3.2 Structure of Natural and Synthetic Hormones

The general approach for the isolation of the insect hormones is similar to that of the pheromones. A large supply of insects is required for the isolation of a very small amount of the purified hormone. The juvenile hormone was first isolated from the silkworm because silkworms are available in large quantities from breeders who supply them to the silk industry. $\underline{\alpha}$-Ecdysone has been isolated as the molting hormone from the tobacco hornworm. However, three other ecdysone derivatives are known to be produced at different stages of embryonic development. The structural similarity between the molting hormone and the mammalian hormones (e.g., estradiol, Section 7.8) is striking.

(CH_3 may be substituted for either or both CH_2CH_3 groups)

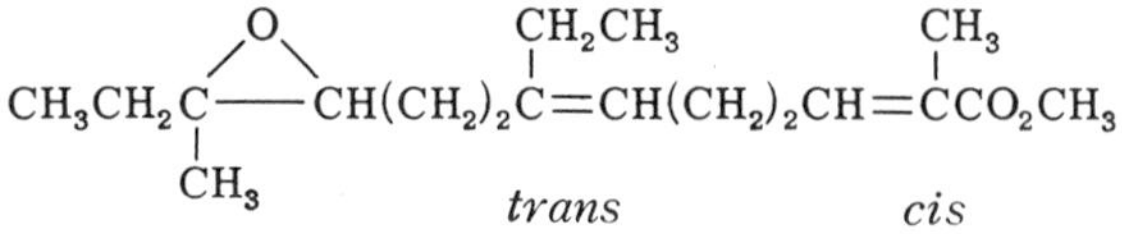

juvenile hormone

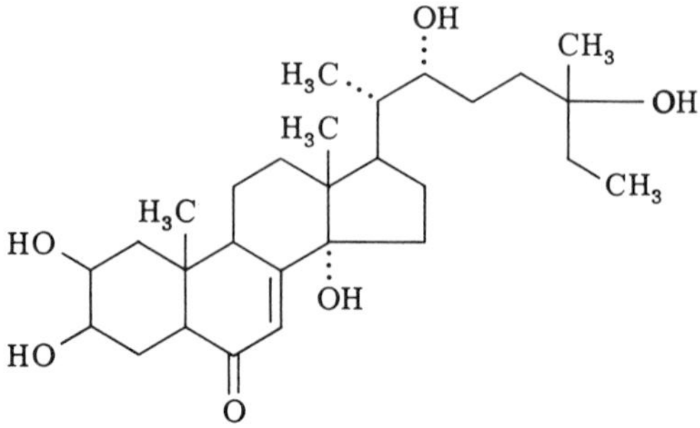

α-Ecdysone
(molting hormone)

Although it has been possible to synthesize the juvenile hormone, it is too unstable (at the epoxide ring) and too difficult to prepare in large amounts to be used commercially. However, a number of analogs have been prepared that show commercial promise (Table 8-3). Surprisingly, some are considerably more active than the natural hormone.

A juvenile hormone mimic is even produced by plants. This was discovered accidentally when it was observed that the European linden bug did not mature when grown in the presence of paper prepared from North American trees; however it did grow in the presence of paper manufactured in Europe. The compound juvabione, responsible for the growth retardation, was traced to the balsam fir and its structure has been determined.

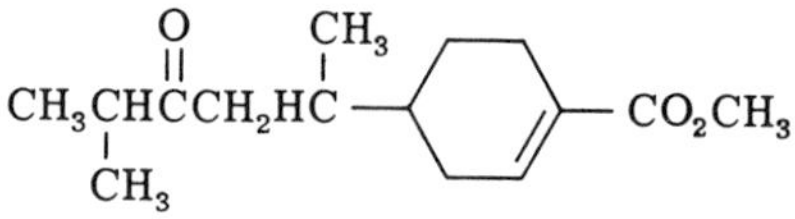

juvabione

It is surprising that the balsam fir contains a juvenile hormone, which affects the growth of a family of insects that are not natural predators. It has been suggested that juvabione is the juvenile hormone of a former predator from that insect family that is either extinct or has learned to

TABLE 8-3 Synthetic Juvenile Hormones

	Potency[a]
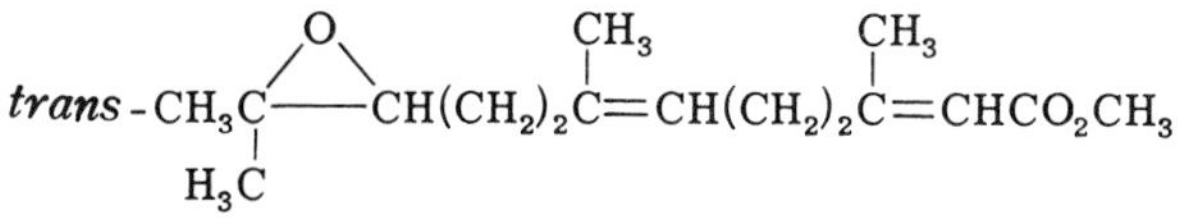	1
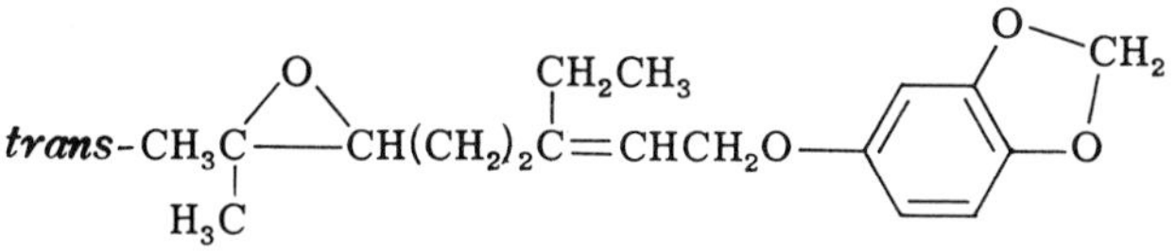	70
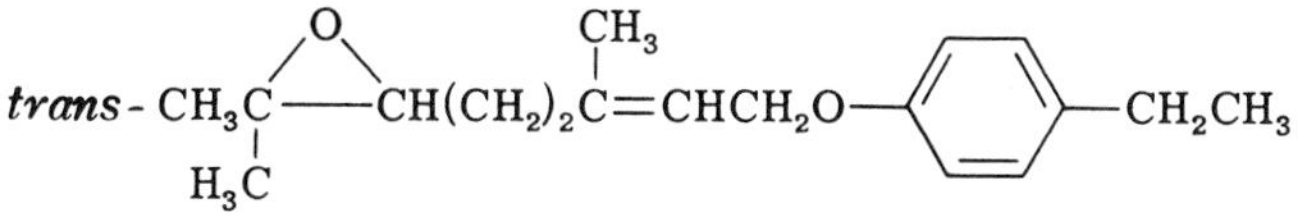	450
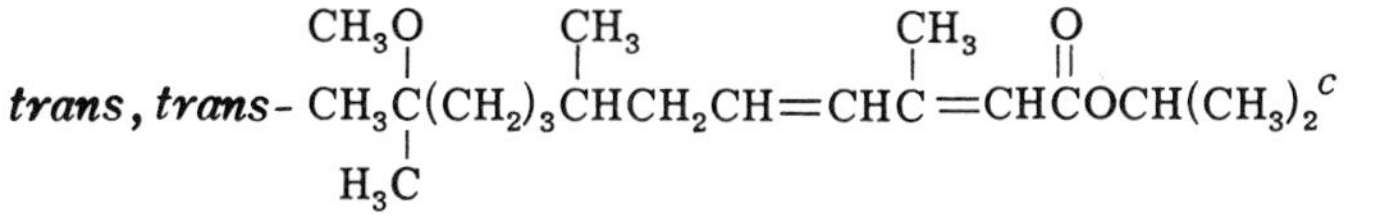	1900
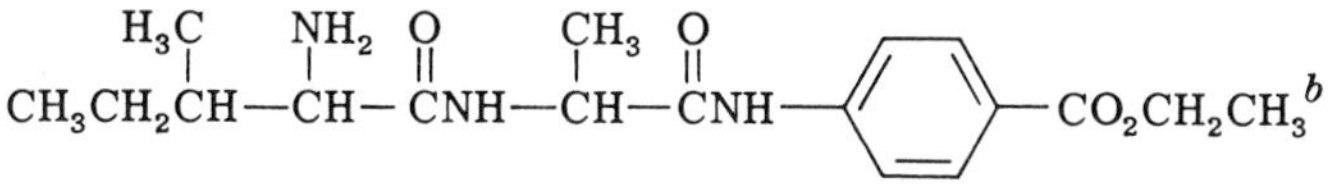	
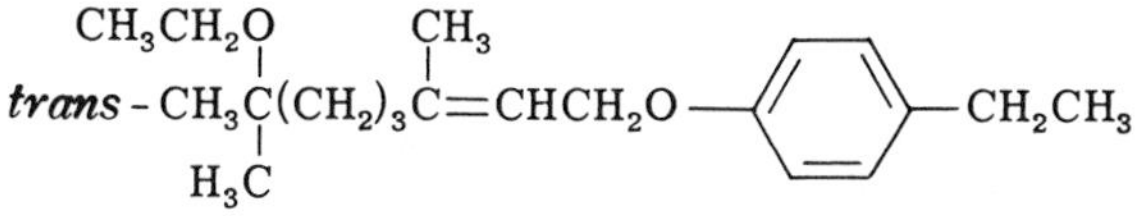	
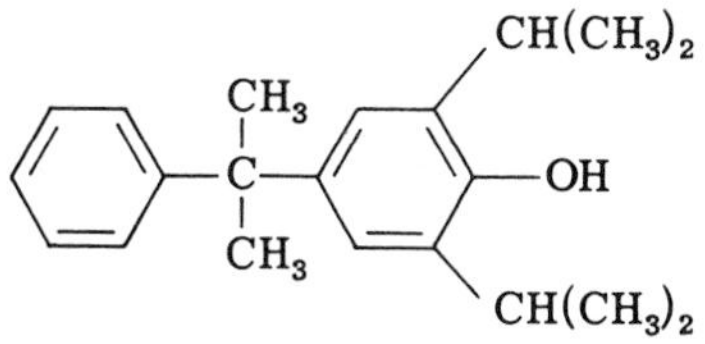	

[a]*The natural hormone has a potency of 1.4 on this scale.*
[b]*Has systemic action.*
[c]*Commercially known as Altosid.*

avoid the balsam fir. While the juvenile hormone affects the growth of all insects, juvabione is specific for the Pyrrhocoridae family. This suggests that it may be possible to develop more specific juvenile hormone analogs that will affect only harmful insects.

8.3.3 Use of Hormones in Insect Control

Insect hormones must be applied at a specific stage in the insect's life to influence its development. The effective time periods are often very brief so that the hormone must be present for a long enough time to insure that a significant proportion of the insect population is exposed during the sensitive period. One application of juvenile hormone analogs is for the control of floodwater mosquitoes. The eggs of this mosquito are dormant until they are in water. Since the farmer controls the time at which he irrigates the field, he can apply the juvenile hormone analog at the precise time when all the mosquitoes hatch and are in their susceptible growth stage. The hormone analog is also formulated in controlled-release polymer spheres that slowly release the compound over the life cycle of the mosquito (seven to ten days). Precise timing of the application of the pesticide is not required with this formulation.

The observation that some plants produce juvenile hormone analogs suggested that some of these compounds might be stable when absorbed in plant tissues (systemic). This has been observed with the juvenile hormone analog ethyl pivaloyl-l-alanyl-*p*-aminobenzoate. It is absorbed and retains its juvenile hormone activity in the sunflower plant for 1-2 weeks.

As in the case of pheromones, it is not known whether insects will develop resistance to these hormones. Those who argue that no resistance will develop note that insects need these hormones at specific times in their own development. If resistance developed, it would result in the death of the insect. Those who argue that resistance will develop note that as the insect develops to an adult, it must be able to inactivate these hormones. The insect could develop more efficient ways of doing this and in this way negate the effect of the added hormone.

There are several problems associated with the use of insect hormones. One is the lack of specificity. This will not be a problem if the hormones are applied only at certain times during the year when the target insect is going through metamorphosis. Another problem is the hormones are only effective when the insect is changing from the final larval stage to the adult. However, insects are usually most destructive during their larval stage so that the hormone would not curtail the most destructive period. This means that the hormones are best suited for the treatment of light infestations where the damage is tolerable or against flies and mosquitoes where the adult is the pest. A third problem associated with the use of hormones is their effect on the development of arthropods (members of the insect genus) in the sea. For example, the molting hormone also affects the molting of crabs and crayfish and a juvenile hormone analog affects the development of the acorn barnacle. Obviously these hormones will have to be used judiciously to avoid destroying beneficial arthropods.

8.4 ANTIJUVENILE HORMONES FROM PLANTS

Two compounds have been isolated from the common ageratum which block the action of juvenile hormones. Compounds with this general biological effect are called antiallatrotropins. The particular compounds isolated from ageratum were named precocene I and precocene II because they caused premature metamorphosis of a number of insects. The adult females that develop after treatment with the precocenes are sterile and only a few of the males are capable of successful mating with normal females. In addition, treatment of newly developed adult females with precocenes prevents the development of ovaries and in the case of the cockroach, prevents the secretion of sex pheromones. Treatment of the adult potato beetle with precocenes caused it to prematurely stop feeding and go into a dormanent stage. Precocene II also prevented the development of the eggs of the milkweed bug and the Mexican bean beetle.

The wide range of biological activity exhibited by the precocenes suggest that the antiallatrotropins should be very effective agents for insect control. They have the decided advantage over juvenile hormones in that their toxic properties are effective on both the adult and egg as well as on the larva. In addition, their effect on the larval stage is to accelerate its development to the adult and thereby shorten the larval stage which is generally the most destructive.

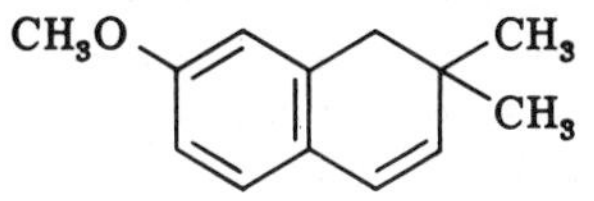

precocene I

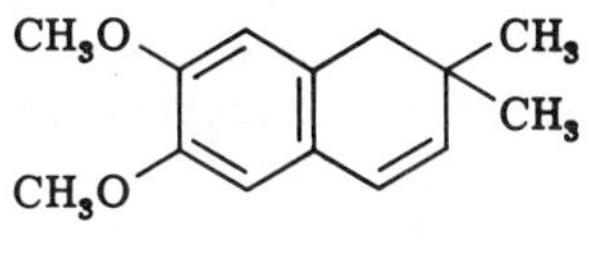

precocene II

8.5 INSECT PREDATORS: VIRUSES AND BACTERIA

8.5.1 Introduction

The natural enemies of insects are obvious agents for insect control. These are species-specific and it will probably be a long time before the insect develops defense mechanisms to ward off the predators. Viruses, bacteria, fungi, and protozoa have been suggested as possible natural control agents, although only viruses and bacteria are currently in use.

8.5.2 Viruses

The first virus was licensed as an insecticide in 1973 for use against the cotton bollworm. The virus was demonstrated to be almost as effective as methyl parathion in field tests, yet it has the advantage of not killing the other predators of the bollworm. Promising field tests have also been carried out on viruses for the control of the alfalfa caterpillar, wattle bagworm, cabbage looper, European cabbage worm, cotton leafworm, and the Great Basin tent caterpillar.

Viruses have only been used recently because of a concern for possible human infection. However, these viruses are already present in the environment and in the food we eat. It has been noted that when a cabbage looper succumbs to viruses, its body dissolves and releases large numbers of viruses on the cabbage. By October the cabbage is laden with viruses, so much so that the average bowl of coleslaw contains four billion live cabbage looper viruses. If these viruses were harmful to man, a "coleslaw syndrome" would have been observed by now. However, more research is necessary so that the molecular basis of the virus specificity for its host is understood. If this is known, then one can determine if there is any possibility of infection or chronic toxicity to humans.

Viruses and bacteria have the advantage over other methods of pest control in that they are highly infectious and spread rapidly. Often they become established in the environment so that they only need to be applied once. Therefore, in the ideal case it will not be necessary to mass produce the virus but rather rely on its reproductive ability in the presence of the insect host. This ideal was achieved in the case of the European spruce sawflies in Canada. This insect has been virtually eliminated by a virus predator.

Some viruses must be applied each year because the natural population does not build up rapidly enough to prevent significant crop damage. The production of viral pesticides is difficult because the viruses can only be grown in the tissues of their hosts. Consequently, the production of viruses also requires the growth of large numbers of host insects. This is a difficult and expensive project and as a consequence, only two U.S. companies are producing virus insecticides. Current research is centered on techniques for the growth of viruses in cell cultures. This would permit the use of sterile conditions and greatly decrease the possibility of contamination by unwanted viruses and bacteria.

8.5.3 Bacteria

The first successful application of a bacterium for the control of insects was the use of the spores of *Bacillus popilliae* against the Japanese beetle in 1939. When this bacterium invades the beetle grub, it multiplies rapidly and causes the grub to turn an abnormal white color, hence the name "milky disease." This bacterium, together with other predators, succeeded in reducing the Japanese beetle to a minor pest in the U.S. by 1945.

Another bacterium, *Bacillus thuringiensis*, is effective against over a hundred species of caterpillars including the cabbage looper and the tomato hornworm. The bacterium paralyzes the gut and inhibits feeding. The bacterium requires a pH of 9 or higher for effective growth and is therefore specific for high pH caterpillars.

As in the case of viruses, the use of bacteria can be a "one shot" process if the bacteria become permanently established in the ecosystem. *B. popilliae* is an example of this type. *B. thuringiensis* requires annual application because the bacterium in the soil does not establish itself rapidly enough in the insect to prevent heavy crop damage.

8.6 INSECT STERILIZATION

8.6.1 Introduction

The insect itself is the insecticide when insect sterilization is used for pest control. Large numbers of viable but sterile males are released into the infected areas in this approach, and the sterile males compete with the nonsterilized males for the females. If the number of sterile males is significantly greater, it is then possible to control or in some cases eradicate an insect population. This approach is obviously species-specific and it appears unlikely that resistant strains will develop.

For a sterilization program to be successful, one must be able to raise large numbers of healthy, vigorous insects that can be sterilized without affecting their vigor and mating competitiveness. The sterilization is usually accomplished with X-rays or gamma rays from a cobalt-60 source (see Chapter 17). The biological effects of both types of irradiation are similar. A dosage of 2500 roentgens (R) was effective for the sterilization of the male screwworm fly (Section 8.6.2) although the fly can tolerate 20,000 R without any other apparent side effects. Then these sterilized insects must be released in suitable numbers to be effective but not destructive.

It seems unlikely that this approach will be used to eliminate major insect infestations because of the large numbers of sterile insects that will have to be reared. However, it can be used to eliminate or manage insect populations that have been suppressed by other means. It could also be used to prevent the spread of insects into a new area. It has been especially effective in peninsulas and islands where there is a natural barrier to prevent the influx of nonsterile insects once control has been established.

8.6.2 The Screwworm Fly and the Mexican Fruit Fly

Sterilization has been used effectively to control or eradicate the screwworm fly in the U.S. The larvae of these insects hatch from eggs laid in animal wounds. The larvae feed on the tissues of the animals for five to seven days before dropping to the ground and burrowing into the soil to pupate. The life cycle is completed in about a week when the fly emerges from the pupa. This fly has been especially damaging to livestock in the South and Southwest.

Two environmental factors made it possible to use the sterile insect approach for the control of the screwworm. The first factor is that the screwworm population is limited by the number of animals with open wounds and the second is that the pupa stage does not survive in the ground in cold weather. The insect can only survive the winters in the extreme southern parts of the U.S. and in Mexico. However, the insect does migrate into the midwestern states in the summer. The winter range of the screwworm fly can be estimated from the severity of the winter so it is possible to determine where to release the sterilized insects and prevent its annual northern migration.

A combination of a severe winter, strict inspection of livestock, and sterile insects made it possible to eliminate the screwworm in Florida. The sterile insects were released in southern Florida in the winter while a severe cold wave was killing the pupae in northern Florida. In addition, all livestock shipped north were strictly inspected to be sure no infected animals were being shipped out of the infested area. In this way it was possible to eliminate the screwworm and prevent its annual summer migration into Georgia and Alabama. The same approach was used for its eradication on the island of Curacao. In this case the influx of virile insects was prevented by the water barrier.

The same approach has been used to prevent the annual northern migration of the screwworm fly from Mexico. Here the problem is much greater because the sterile insects must be released over an 1800-mile front along the Mexican border. Ten million sterile insects are released each year in the program. Plans are underway to attempt to eliminate the screwworm fly in North America south to the Isthmus of Panama. If this could be achieved, then the animal control program would be much less costly and potentially more effective because there would be a much shorter border to control.

The same technique is used to control the spread of the Mexican fruit fly into Southern California. The fruit fly is concentrated in the border towns when the farmers bring their produce to market. The largest numbers of the flies cross the border near these towns. A few million sterilized flies are released each week near these border towns to prevent the northern migration of this insect. This procedure, plus rigorous inspection and control of all fruit brought in from Mexico, has eliminated the fruit fly from California.

The spectacular successes obtained by the sterilization technique are the result of its application only after a careful study of the life cycle and range of the insect. In addition, the population of the target insect was limited either by natural forces or by insecticides so that the sterile insects could compete successfully with the virile insects. One problem has been the mass breeding of sterile flies that are as successful as the wild flies in competing for mates.

8.7 INTEGRATED PEST CONTROL

It has been emphasized in this chapter that the third generation insecticides are not usually amenable to the control of major infestations of pests which cover a large land area; e.g., the gypsy moth in the northeastern U.S. or the boll weevil in the southern U.S. It has already been demonstrated that these pests cannot be eradicated with first and second generation insecticides. The present strategy is to control the insect by use of a number of techniques in an integrated approach rather than use of massive quantities of insecticides in an attempt to eradicate the insect. Attempted control rather than eradication would also limit the chance for the development of resistant mutants. In addition, a variety of control techniques would make it more difficult for the insect to develop resistance to these quite different control methods. Finally, more emphasis must be placed on the use of natural forces for insect control. Crop rotation prevents the build up of one particular pest. Growing plants that the insect prefers next to the desired crop has been effective in diverting the interests of the insect to these plants. Predatory insects such as the ladybug, praying mantis, and lacewing (aphid lion) can be used to control the pest insect. These techniques coupled with the use of third generation insecticides and the *judicious* use of chemical sprays could lead to the control of harmful insects without destroying the environment.

BIBLIOGRAPHY

General References

"Pest Control Strategies for the Future," National Academy of Science, Washington, D.C. 1972.

C. B. Huffaker, "Biological Control." Plenum Press, New York 1971.

W. Worthy, Integrated insect control may alter pesticide use pattern, *Chem. Eng. News*, April 23, 1973, pp. 13-19.

L. J. Carter, Eradicating the boll weevil: Would it be a no-win war?, *Science 183*, 494 (1974).

C. Djerassi, C. Shih-Coleman, and J. Dickman, Insect control of the future: Operational and policy aspects, *Science 186*, 596 (1974).

Controlled-release pesticides attract interest, *Chem. Eng. News*, Sept. 30, 1974, pp. 20-22.

H. J. Sanders, New weapons against insects, *Chem. Eng. News*, July 28, 1975, pp. 18-35.

J. L. Marx, Applied ecology: Showing the way to better insect control, *Science 195*, 860 (1977).

Pheromones

M. Jacobson, "Insect Sex Pheromones." Academic Press, New York, 1972.

D. L. Wood, R. M. Silverstein, and M. Nakajima, "Control of Insect Behavior by Natural Products." Academic Press, New York, 1970.

C. Berkoff, Insect control with hormones, *J. Chem. Ed 98*, 577 (1971).

H. H. Shorey, L. K. Gaston, and R. N. Jefferson, Insect sex pheromones, *Adv. Pest Control Res. 8*, 57 (1968).

E. O. Wilson, Pheromones, in "Chemistry in the Environment, Readings from Scientific American," p. 125. W. H. Freeman, San Francisco, California, 1973. (From Sci. Amer., May, 1963).

J. G. MacConnell and R. M. Silverstein, Recent advances in insect pheromone chemistry, *Agnew. Chem. Internat. Edit. 12*, 644 (1973).

M. Beroza and E. F. Knipling, Gypsy moth control with the sex attractant pheromone, *Science 177*, 19 (1972).

D. Shapley, "Insect control (I): Use of pheromones, *Science 181*, 736 (1973).

M. Beroza, B. A. Bierl, and H. R. Moffitt, Sex pheromones: (E,E)-8,10-Dodecadien-1-ol in the codling moth, *Science 183*, 89 (1974).

D. A. Carlson, M. S. Mayer, D. L. Silhacek, J. D. James, M. Beroza, and B. A. Bierl, Sex attractant pheromone of the house fly: Isolation, identification, and synthesis, *Science 171*, 76 (1971).

H. E. Hummel, L. K. Gaston, H. H. Shorey, R. S. Kaae, K. J. Byrne, and R. M. Silverstein, Clarification of the chemical status of the pink bollworm sex pheromone, *Science 181*, 873 (1973).

R. J. Seltzer, Role of sex lure isomers probed, *Chem. Eng. News*, August 20, 1973, pp. 19-20.

M. Beroza, "Chemicals Controlling Insect Behavior." Academic Press, New York, 1970.

N. Green, M. Beroza, and S. A. Hall, Recent developments in the chemical attractants for insects, *Adv. Pest Control Res. 3*, 129 (1957).

D. Schneider, The sex-attractant receptor of moths, *Scientific Amer. 231* (1), 28 (1974).

J. H. Tumbinson, C. E. Yonce, R. E. Doolittle, R. R. Heath, C. R. Gentry, and E. R. Mitchell, Sex pheromones and reproductive isolation of the lessor peachtree borer and the peachtree borer, 614 (1973).

M. Beroza, Insect sex attractant pheromones, a tool for reducing insecticide contamination in the environment, *Toxicol. Environ. Chem. Reviews 1*, 190 (1972).

M. Beroza, Insect sex attractants, *Amer. Scientist 59*, 320 (1971).

J. H. Tumlinson, M. G. Klein, R. E. Doolittle, T. L. Ladd, and A. T. Proveaux, Identfication of the female japanese beetle sex pheromone: Inhibition of male response by an enantiomer, *Science 197*, 789 (1977).

L. K. Gaston, R. S. Kaae, H. H. Shorey and D. Sellers, Controlling the pink bollworm by disrupting sex pheromone communication between adult moths, *Science 196*, 904 (1977).

Juvenile and Molting Hormones

J. J. Menn and M. Beroza, "Insect Juvenile Hormones: Chemistry and Action." Academic Press, New York, 1972.

C. M. Williams, "Third-generation pesticides," in "Chemistry in the Environment, Readings from Scientific American," p. 121. W. H. Freeman, San Francisco, California, 1973. (From Sci. Amer. July 1967.)

J. L. Marx, Insect control II: Hormones and viruses, *Science 181,* 833 (1973).

Insect hormone mimics near market, *Chem. Eng. News,* Nov. 29, 1971, pp. 33-34.

R. Sarmiento, T. P. McGovern, M. Beroza, G. D. Mills, Jr., and R. E. Redfern, Insect juvenile hormones: Highly potent synthetic mimics, *Science* 179, 1342 (1973).

M. Zaoral and K. Sláma, Peptides with juvenile hormone activity, *Science 170,* 92 (1970).

T. H. Baba and K. Sláma, Systemic activity of a juvenile hormone analog, *Science 175,*78 (1972).

J. N. Kaplanis, W. E. Robbins, M. J. Thompson, and S. R. Dutky, 26-Hydroxyecdysone: New insect molting hormone from the egg of the tobacco hornworm, *Science 180,* 307 (1973).

E. D. Gomez, D. J. Faulkner, W. A. Newman, and C. Ireland, Juvenile hormone mimics: Effect on cirriped crustacean metamorphosis, *Science 179,* 813 (1973).

Antijuvenile Hormones

W. S. Bowers, T. Ohta, J. S. Cleere, and P. A. Marsella, Discovery of insect antijuvenile hormones in plants, *Science 193,* 542 (1976).

T. H. Maugh, Plant biochemistry: Two new ways to fight pests, *Science 192,* 874 (1976).

W. S. Bowers and R. Martinez-Pardo, Antiallatotropins: Inhibition of corpus allatun development, *Science 197,* 1369 (1977).

Insect Viruses

N. Wade, Insect viruses: A new class of pesticides, *Science 181,* 935 (1973).

J. L. Marx, Insect control (II): Hormones and viruses, *Science 181,* 833 (1973).

L. A. Bulla, Jr., Regulation of insect populations by microorganisms, *Ann. New York Acad. Sci. 27,* 1 (1973).

Sterilized Insects

R. C. Bushland, Male sterilization for the control of insects, *Adv. Pest Control Res. 3,* 1 (1960).

R. H. Smith and R. C. von Borstel, Genetic control of insect populations, *Science 178,* 1164 (1972).

G. L. Bush, R. W. Neck, and G. B. Kitto, Screwworm eradication: Inadvertent selection for noncompetitive ecotypes during mass rearing, *Science 193,* 491 (1976).

9
PRINCIPLES OF PHOTOCHEMISTRY

9.1 INTRODUCTION

Visible and ultraviolet solar radiation is essential for all life on earth as we know it, and photochemical reactions are some of the most important processes taking place in man's environment. Several examples of photochemical influences on our environment have already been given in earlier chapters. Direct utilization of solar energy as a possible alternative to fossil fuel combustion is discussed in Chapter 2, and its efficient usage is certain to be an increasingly important goal of research and development over the next few decades as we seek solutions for the world-wide energy shortage. In Chapter 3 it is pointed out that atmospheric oxygen comes from photodissociation of water vapor in the upper atmosphere and from photosynthesis in the biosphere, a very important photochemical reaction that also leads to food production and storage of solar energy. Photochemical decomposition of pesticides has been discussed in Chapter 7.

Photochemical processes occurring at high altitudes in the mesosphere and stratosphere are essential for the maintenance of the thermal and radiation balance at the surface of the earth (see Chapter 3) and they also provide phenomena such as night glow. The reactions following light absorption in the lower atmosphere that generate photochemical smog from atmospheric pollutants are less desirable! In association with living matter, light-induced reactions lead to such phenomena as vision, cyclic activities of plants and animals, and both cellular damage *and* regeneration (or reactivation). In a more historic sense, it has been suggested that the origin of life may have occurred via the photochemical synthesis of purines.

As we shall show later, absorption of light by a molecule results in excitation of the molecule to a higher electronic energy level followed in many cases by reactions such as dissociation or interaction with other molecules, leading to the overall photochemical process. We shall see that the specific electronic state reached by the absorption of light often determines the types of following reactions, so that the study of the absorption process is important to our understanding of photochemistry. In this chapter, we shall develop the basic principles of light absorption, electronic excitation, and subsequent photochemical and photophysical processes. As these are being developed, they will be applied to specific examples of photochemical processes occurring in our environment.

9.2 PRINCIPLES OF PHOTOCHEMISTRY AND THE ABSORPTION OF LIGHT

9.2.1. Reaction Energies

Photochemistry is the study of the interaction of a "photon" or "light quantum" of electromagnetic energy with an atom or molecule, and of the resulting chemical and related physical changes that occur. The energy necessary for the reaction (or at least for it to be initiated) is thus gained from the photon. This is in contrast to thermal reactions, in which the energy needed for the reaction is distributed among the molecules and among the internal vibrational and rotational motions of the molecule according to the Boltzmann distribution law (Fig. 9-1).

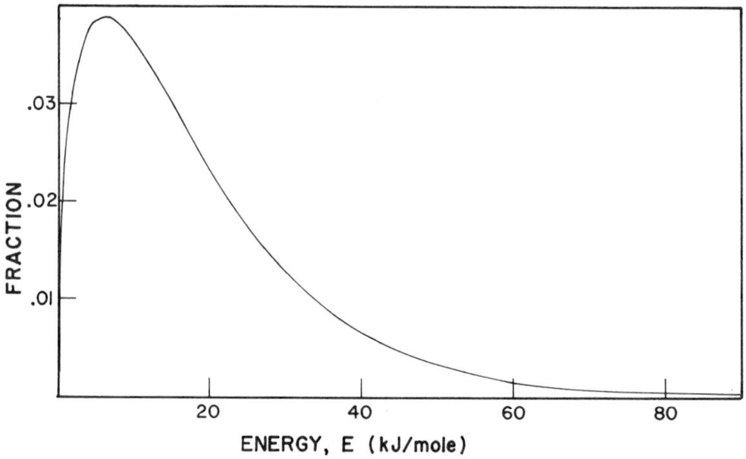

FIGURE 9-1. Distribution of kinetic energy for O_2 at 1500K, expressed as the fraction of O_2 molecules having energies between E *and* (E+1) *kJ/mole.*

As an example, consider the simple dissociation reaction of molecular oxygen,

$$O_2 \longrightarrow 2\ O \tag{9.1}$$

a reaction we might assume could be taking place in the atmosphere. The fraction of molecules with thermal kinetic energy equal to or greater than an energy E_c is approximately equal to the Boltzmann factor

$$\exp(-E_c/RT) \tag{9.2}$$

The energy required to break an oxygen bond is 5.1 eV, which is equivalent to 118 kcal/mole (492 kJ/mole). The fraction of O_2 molecules with energies E_c equal to or greater than 5.1 eV at 1500 K is approximately 7×10^{-18} from Eq. (9.2), which is too small to lead to even the most efficient thermal reactions involving oxygen atoms. Many reactions such as this one that are not feasible thermally can, however, be initiated by light, as will be shown below.

9.2.2 *Properties of Light*

Monochromatic light is an electromagnetic wave, that is, it is made up of mutually perpendicular in-phase electric and magnetic fields, of wavelength λ traveling at speed c ($= 3 \times 10^{10}$ cm/sec in a vacuum), the frequency ν thus being equal to c/λ. One of the important characteristics or properties of an electromagnetic wave for photochemical purposes is that it can transport energy, as it does from the sun to the earth, and from a light bulb to an object. In addition, an electromagnetic wave can transport momentum and hence exert a pressure (radiation pressure). Although so small that it is not felt by us under ordinary conditions, this radiation pressure has been measured under very carefully controlled conditions and is in complete agreement with classical theories of electromagnetic radiation.

Although most of the experimental techniques used in photochemistry are based on the wave properties of light (such as light polarization, and prisms and gratings to produce monochromatic light), it turns out that many aspects of radiation, particularly its absorption and emission by matter and the effects of such absorption or emission, are not adequately explained in this manner. Examples are the photoelectric effect and blackbody radiation, which has been described in Chapter 3. The development of the Planck quantum hypothesis (1901) was necessary to resolve these failures of classical wave theory, and has led to the concept of the dual nature of matter and radiation and the theory of quantum mechanics.

In essence, quantum theory says that energy interacts with molecules only in discrete amounts rather than continuously, so that a light beam may be considered to be made up of a stream of discrete units (photons or quanta). Each photon has zero rest mass and possesses, and thus transports, a "packet," or quantum, of energy ε given by the Einstein relationship

$$\varepsilon = h\nu = hc/\lambda \tag{9.3}$$

and momentum p

$$p = h/\lambda \tag{9.4}$$

where λ is the wavelength of the corresponding electromagnetic wave and h is Planck's constant, a universal natural constant equal to 6.63×10^{-27} erg sec/ molecule (6.63×10^{-34} J sec/molecule). Referring back now to the energetics of the oxygen dissociation reaction [Eq. (9.2)], it is easily shown that the bond dissociation energy of 118 kcal/mole corresponds to a wavelength of 243 nm:

$$\lambda = hc/\varepsilon$$

$$= \frac{6.63 \times 10^{-27}\ \text{erg sec/molecule})(3 \times 10^{10}\ \text{cm/sec})}{(118\ \text{kcal/mole})}$$

$$\times \frac{6.02 \times 10^{23}\ \text{molecule/mole}}{(4.184 \times 10^{10}\ \text{ergs/kcal})}$$

$$= 2.43 \times 10^{-5}\ \text{cm} = 243\ \text{nm}$$

Thus, absorption of one photon of light of wavelength 243 nm by an oxygen molecule will give it sufficient energy to break apart or dissociate.

9.2.3. Absorption of Light

The above rather simple example can be used to point out one of the basic laws of photochemistry first formulated early in the nineteenth century: In order for light to be effective in producing photochemical transformations, not only must the photon possess sufficient energy to initiate the reaction, it must also be absorbed. Thus, we know for example that molecular oxygen does not absorb (to any measurable extent) radiation of 243 nm wavelength, so that even though the photon possesses enough energy to break the oxygen bond, no dissociation will occur because the photon is not absorbed.

A second important law of photochemistry, which follows directly from quantum theory, is that absorption of radiation is a one-photon process: Absorption of one photon excites one atom or molecule in the primary (or initiating) step, and all subsequent physical and chemical reactions follow from this excited species (although, as shown below, the overall effect from secondary reactions may be much greater than simply one molecule reacting per photon absorbed). This law is not strictly obeyed with the extremely high intensities possible from high-energy artificial light sources such as flash lamps or pulsed lasers. Using such high-intensity radiation, the simultaneous absorption of two photons is possible. However, single-photon absorption probably occurs in all environmental situations involving solar radiation.

In principle, electromagnetic radiation can extend essentially over an infinite range of wavelengths, from the very long wavelength (thousands of miles) and low energy [the relationship between the two given by Eq. (9.3)] to the very short wavelengths (10^{-12} m, which is on the order of magnitude of nuclear dimensions). It is convenient to divide this electromagnetic spectrum into the following regions:

very low (<3 cal/mole) energies ($\lambda < 1$ cm): radio waves
3 cal/mole - 3 kcal/mole (1 cm - 10^{-5} m): microwave or far infrared
3 kcal/mole - 40 kcal/mole (10^{-5} m - 700 nm): infrared
40 kcal - 150 kcal/mole (700 nm - 200 nm): visible and ultraviolet
150 kcal - 300 kcal/mole (200 nm - 100 nm): far, or vacuum ultraviolet
300 kcal - 1 MeV (100 nm - 10^{-3} nm): x rays and γ rays

Photochemistry is generally limited to absorption in the visible, ultraviolet, and vacuum ultraviolet spectral regions, corresponding primarily to electronic excitation and (at sufficiently high energies) atomic or molecular ionization.

Absorption in the low-energy radio, microwave, or infrared spectral regions results in no direct photochemistry (unless very high intensity laser radiation is used), the excitation for the most part only increasing the rotational and vibrational energy of the molecule and eventually being dissipated as heat. Some deleterious physiological effects do result from overexposure to microwave and infrared radiation (as for example at radar installations or with improperly controlled electronic ovens), but these presumably are due to excessive localized heating. The chemical effects following illumination with very high energy radiation (x rays and γ rays) are considered in Chapter 17.

The absorption of monochromatic electromagnetic radiation is given by the Beer-Lambert law, which says that the probability of light being absorbed by a single absorbing species is directly proportional to the number of molecules in the light path, which in turn is proportional to the concentration c (Beer's law) and the thickness of the absorbing sample dx (Lambert's law):

dI/I = probability of light of intensity I being absorbed in incremental thickness dx

= fraction that the intensity I is reduced in thickness dx

$$= -\alpha\ c\ dx \tag{9.5}$$

The negative sign is included to account for the fact that the intensity is reduced, and thus dI is negative. Assuming that c and α are constant, integration of (9.5) from the incident light intensity I_0 at $x = 0$ to I at x leads to

$$\int_{I_0}^{I} dI/I = \int_{I_0}^{I} d(\ln I) = -\int_{0}^{X} \alpha\ c\ dx \tag{9.6a}$$

$$\ln\ (I/I_0) = 2.303\ \log\ (I/I_0) = -\alpha\ c\ x \tag{9.6b}$$

Or,

$$\text{Absorbance} = A = \log\ (I_0/I) = \epsilon\ c\ x \tag{9.7}$$

where ϵ (= $\alpha/2.303$) is the extinction coefficient.[1] The fraction of light absorbed I_a/I_0 is given by

$$I_a/I_0 = 1 - I/I_0 = 1 - 10^{-\epsilon cx} \tag{9.8}$$

which holds for all concentrations. For small fractions of light absorbed, for example, at low concentrations of absorbers in the atmosphere, the fraction is directly proportional to the concentration of the absorbing species since $e^{-y} \cong 1-y$ when y is much less than unity:

$$I_a/I_0 = 1-10^{-\epsilon cx} = 1-\exp(-2.303\ \epsilon cx \cong 2.303\ \epsilon cx \tag{9.9}$$

This relationship cannot hold at high concentrations as is shown by going to the extreme in the opposite direction, i.e., where essentially all of the light is absorbed,

$$I_a/I_0 \cong 1 \tag{9.10}$$

indicating that at high concentrations all of the light is absorbed independent of absorber concentration.

The light intensity I is normally expressed as the rate at which energy is transmitted through the cell or column of material, so that I_0 is then the light energy incident at the cell face per unit time, i.e., it is the rate at which photons pass the cell face. For reasons that will be apparent later, it is often convenient in photochemical systems to express the light absorbed I_a as the "concentration" of photons absorbed per unit time (photons volume^{-1} time^{-1}) or alternately of a mole of photons (6.02×10^{23} photons, which is an einstein) absorbed per unit time (einsteins volume^{-1} time^{-1}); in these cases, I_0 is then the concentration of photons (or einsteins) passing through the cell face per unit time.

The units of the extinction coefficient ϵ will be (concentration^{-1} length^{-1}), but obviously a variety of units of concentration may be used leading to different values for ϵ. If c is given in molarity M (moles/liter) and x in centimeters, ϵ has the units M^{-1} cm^{-1}, and is called the molar extinction coefficient. On the other hand, if the concentration is

[1]*The constant α is also sometimes called the extinction coefficient. In usage here, however, we shall always refer to the extinction coefficient as the constant ϵ in the Beer-Lambert law in the decadic form.*

expressed in molecules per cubic centimeter, then the extinction coefficient has the units square centimeter per molecule and is the molecular absorption cross section σ. Frequently, however, in gaseous systems such as the atmosphere, it is convenient to express the concentration in pressure units; these are generally related to concentration units at low pressures by the ideal gas equation $P = (n/V)RT = cRT$, and therefore the temperature must be specified.

Strictly speaking, the Beer-Lambert law applies only to monochromatic radiation since ϵ is a function of wavelength. The extent to which use of nonmonochromatic light leads to significant error in the determination of concentration depends on the spectral characteristics of the absorbing and illuminating system. If the extinction coefficient and incident intensity are known as functions of wavelength, however, the total amount of light absorbed may be obtained by integrating over all wavelengths. Also, if there are i absorbing species or components present, the Beer-Lambert becomes

$$I/I_0 = 10^{- \Sigma_i \epsilon_i c_i x)} \quad (9.11)$$

9.3 KINETICS OF THERMAL PROCESSES

Most chemical reactions are kinetically complex, that is, they take place by a series of two or more consecutive steps (the *mechanism* of the reaction) rather than by a single encounter of the reacting species. Nevertheless, the overall reaction between, say, two reactants A and B can be represented by the generalized equation

$$a\mathrm{A} + b\mathrm{B} \longrightarrow c\mathrm{C} + d\mathrm{D} \quad (9.12)$$

Where the stoichiometry of the reaction is given by the coefficients a, b, etc.

The extent of a chemical reaction ξ is defined such that change in ξ from ξ to $(\xi + d\xi)$ means that $a\, d\xi$ moles of A have reacted with $b\, d\xi$ moles of B to produce $c\, d\xi$ moles of C and $d\, d\varepsilon$ moles of D. The change in the number of moles of A, dn_A, is then $dn_A = -a\, d\xi$; similarly, $dn_B = -b\, d\xi$, $dn_C = +c\, d\xi$, and $dn_D = +d\, d\xi$.

The rate of reaction v is given by the differential equation

$$v = d\xi/dt \quad (9.13)$$

so that for the generalized reaction (9.12) the rate becomes

$$v = -\frac{1}{a}\frac{dn_A}{dt} = -\frac{1}{b}\frac{dn_B}{dt} = +\frac{1}{c}\frac{dn_C}{dt} = +\frac{1}{d}\frac{dn_D}{dt} \qquad (9.14)$$

For reactions in which the volume of the system (V) is constant (independent of time), the molar concentration of A is

$$[A] = \frac{n_A}{V}; \quad [B] = \frac{n_B}{V}; \text{ etc.} \qquad (9.15)$$

and

$$dn_A = V\,d[A]; \; dn_B = V\,d[B]; \text{ etc.} \qquad (9.16)$$

it follows that the rate per unit volume $r = v/V$ (also usually referred to simply as the rate of the reaction) is

$$r = -\frac{1}{a}\frac{d[A]}{dt} = -\frac{1}{b}\frac{d[B]}{dt} = \frac{1}{c}\frac{d[C]}{dt} = \frac{1}{d}\frac{d[D]}{dt} \qquad (9.17)$$

In certain cases, it is convenient to describe the rate with respect to the rate of change in concentration of A only:

$$r_A = \text{rate of change of A} = -\frac{d[A]}{dt} \qquad (9.18)$$

However, it is important to remember that in general this will not be equal to the rate of change of B, r_B, although the two rates are related by the proportionality factor b/a.

Although the expression for the experimentally determined rate of reaction in terms of the composition of the system can be quite involved, in many cases at constant temperature it is of the form

$$r = k\,[A]^{\alpha}\,[B]^{\beta}\,[C]^{\gamma}\,[D]^{\delta} \cdots [X]^{\chi} \cdots \qquad (9.19)$$

where k is the temperature-dependent rate constant, X is neither a reactant nor a product (i.e., a catalyst), the sum of the exponents $\alpha + \beta + \gamma + \delta$ is the overall order of the reaction, and the order with respect to A is α, etc. A special situation for Eq. (9.19) is the kinetically simple reaction which involves only the single step of reactants coming together; in this case the rate of the reaction is proportional to the concentration of the reacting species. The overall order of the reaction is then the number of reactant molecules that must come together in a single encounter in order for the reaction to occur; such reactions are called unimolecular, bimolecular, or trimolecular, depending on whether the number of reactant molecules involved is one, two, or three. (The probability that more than three molecules may come together in a single collision is so small that such events are not considered in kinetic treatments.) For kinetically complex reactions, which take place by two or more kinetically simple steps, there is in general no relationship between the exponents in Eq.(9.19) and the coefficients in the stoichiometric equation (9.12).

The dependence of the rate on temperature is embodied in the rate constant k. The most useful relationship expressing this dependence is the empirical Arrhenius equation

$$k = A \exp(-E_a/RT); \ \ln k = \ln A - E_a/RT \tag{9.20}$$

In this treatment the preexponential factor A and the activation energy E_a are assumed to be temperature-independent, so that a plot of $\ln k$ versus $1/T$ should be linear with a slope equal to $-E_a/R$. This behavior is found to be the case for many reactions, and this method involving the above plot is the most common one for determining E_a. However, there is no reason to expect that Arrhenius behavior should necessarily be followed for kinetically complex reactions, and indeed reactions with very marked deviation from that predicted by Eq.(9.20) are encountered.

The simple collision theory of chemical kinetics, applicable to kinetically simple bimolecular gas-phase reactions, considers the rate of reaction to be proportional to the frequency of bimolecular collisions Z_{12}. The proportionality constant includes the Boltzmann factor [Eq.(9.2)], which as we saw before, is the fraction of molecules in two dimensions (defined by the trajectories of motion of two colliding particles) with energies equal to or greater than a potential barrier E_c (see Fig. 7-2). Note that this is the same form of exponential term found in the Arrhenius equation (9.20); the difference between the two is that E_a is the experimentally determined empirical activation energy, whereas E_c may be associated with a specific potential energy barrier that must be overcome in order for a kinetically simple reaction to occur. The rate is also proportional to a temperature-independent steric factor p which allows for the possibility that not all colliding particles with sufficient energy to overcome the potential barrier do react. (For example, a particular orientation of the molecules might be necessary in order for the molecules to react.) Thus,

$$r = pZ_{12}\ \exp(-E_c/RT) \tag{9.21}$$

$$k = pZ'_{12}\ \exp(-E_c/RT) \tag{9.22}$$

where Z'_{12} is the specific collision frequency proportional to $T^{\frac{1}{2}}$. The preexponential term is therefore slightly temperature-dependent, and this is in fact observed for reactions with low or zero activation energies, although any small temperature effect in this term is completely masked by the exponential term for reactions with large activation energies.

9.4 KINETICS OF PHOTOCHEMICAL PROCESSES

The second law of photochemistry, taken together with quantum theory, requires that each quantum that is absorbed brings about a change in that one molecule. In essence, this may be considered to be a "bimolecular" process involving interaction of one photon and one molecule. However, the overall number of molecules reacting chemically as a consequence of this absorption of a single photon may be virtually any value ranging from zero (the excited molecule simply deactivates) to a very large number ($>10^6$). This latter is interpreted as arising from subsequent chain reactions involving reactive intermediates.

The efficiency of a photochemical reaction is the quantum yield Φ, which is the number of molecules undergoing a change per photon absorbed. Generally the quantum yield relates to a specific species; thus, for the generalized reaction (9.12):

$$\Phi_A = \frac{\text{number of A molecules reacted}}{\text{number of photons absorbed}} \tag{9.23}$$

$$= \frac{\text{number of moles of A reacted}}{\text{number of einsteins absorbed}} \tag{9.24}$$

and

$$\Phi_C = \frac{\text{number of C molecules produced}}{\text{number of photons absorbed}} \tag{9.25}$$

$$= \frac{\text{number of moles of C produced}}{\text{number of einsteins absorbed}} \tag{9.26}$$

Since r_A is the rate that A reacts, r_C the rate that C is produced, and I_a by definition the rate of light absorption, it follows that

$$\Phi_A = r_A/I_a; \quad \Phi_C = r_C/I_a; \text{ etc.} \tag{9.27}$$

The quantum yield Φ represents the overall efficiency of the reaction initiated by the absorption of light. It involves changes not only brought about in the initial, or primary, act of light absorption but also changes occurring in subsequent thermal reactions by molecules not directly associated with the light absorption process (for example, changes induced by energy transfer). In contrast to this, the primary quantum yield ϕ is the fraction of the absorbing molecules decomposed in the kinetically simple primary process in which the photon of light is absorbed:

$$\text{A} \xrightarrow{h\nu} \text{photoproduct(s)} \tag{9.28}$$

$$\phi = \frac{\text{number of A molecules reacted in (9.28)}}{\text{number of photons absorbed}} \tag{9.29}$$

$$= \frac{\text{number of moles of A reacted in (9.28)}}{\text{number of einsteins abosrbed}} \tag{9.30}$$

$$= \text{rate } (r) \text{ of reaction (8.28)}/I_a \tag{9.31}$$

so that the rate of a primary process is $\phi\ I_a$. The symbol $h\nu$ above the arrow in (9.28) indicates that this is the light-absorbing reaction; although this symbol should be used only for the kinetically simple absorbing step, it is sometimes used even in the overall stoichiometric equation to signify a light-initiated reaction. Since light absorption in the primary process is a one-for-one event, the maximum value for ϕ is unity; physical decay processes (to be described later) may lead to $\phi < 1$.

9.5 DESIGNATION OF SPECTROSCOPIC STATES

9.5.1 Introduction

We are concerned for the most part with electronic excitation as the primary process of absorption leading to photochemical transformations. Absorption in different spectral regions may lead to different electronic states of the absorbing molecule and electronically different intermediate species, as well as to different products of the photochemical reaction. It is therefore important to understand the designations of electronic states of atoms and molecules insofar as they relate to (or indicate) possible electronic transitions and transition probabilities. The following is a brief summary of the recipes used in arriving at atomic and molecular state descriptions; the reader is referred to standard textbooks of quantum mechanics and spectroscopy, some of which are given in the bibliography, for the mathematical formulations of these principles.

9.5.2. *Atoms*

The state of the electron in the hydrogen atom or hydrogenlike ion (for example, He^+ and Li^{2+}, each containing only one electron) is specified by the three quantum numbers n, ℓ, and m_ℓ. The principal quantum number n can have the values 1, 2, 3 ...; it defines the shell of the atom occupied by the electron, and therefore the total energy of the single bound electron (neglecting relativistic effects) is specified completely by n. The azimuthal or orbital angular momentum quantum number ℓ can have the values 0, 1, 2 ..., $(n-1)$; it determines the total angular momentum of the electron, which is quantized and can be represented by a vector having magnitude $\hbar\ [\ell(\ell + 1)]^{1/2}$ (where $\hbar = h/2\pi$). The states (or *atomic orbitals*) with ℓ = 0, 1, 2, 3 ..., are designated, respectively, s, p, d, f, ..., orbitals, so that the electron in the lowest electronic state of the hydrogen atom ($n = 1$, $\ell = 0$) is called a 1s electron. The quantum number m_ℓ is the magnetic quantum number (so named because it is related to the effects of a magnetic field on atomic spectra), and can have the values $-\ell$, $-(\ell-1)$,..., 0,..., $(\ell-1)$, ℓ; it determines the allowed components of the angular momentum in a definite z direction, which are also quantized and can only have values of $m_\ell\hbar$.

In addition to the three quantum numbers given, another factor, electron spin, has to be considered in order to explain the fine structure observed in atomic spectra. This fourth parameter is also a natural consequence of quantum mechanics if relativistic effects are included in the detailed wave mechanical treatment. In addition to orbital angular momentum described by the quantum number ℓ, an electron has an intrinsic magnetic moment as if (classically) it were spinning about its own axis in addition to orbiting the nucleus. Associated with this moment is a quantum number s, such that the total spin angular momentum vector has a magnitude $\hbar\ [s(s+1)]^{1/2}$; however, s can only have the single value of ½. Analogous to the orbital angular momentum, the z component of the spin angular momentum is also quantized with allowed values of magnitude $m_s\hbar$, where $m_s = \pm\frac{1}{2}$ (so that there are only two possible orientations of the spin angular momentum). The electron spin and orbital angular momentum magnetic moments interact so that ℓ and s are coupled to give a total electronic angular momentum vector of magnitude $\hbar\,[j(j+1)]^{\frac{1}{2}}$, where $j = \ell \pm s = \ell \pm \frac{1}{2}$.

With polyelectronic atoms, the complexities of the mathematics involved in describing the system as a result of electrostatic interactions among the electrons as well as between the electrons and the nucleus are such that only approximate solutions of the quantum mechanical treatment are possible.[2] The concept of hydrogenlike orbitals is retained with the principal quantum number n designating the electronic shells of the atom, but now the energy of the state is a function of the total angular momentum of the electrons as well as the quantum number n. The number of electrons in a given orbital is governed by the Pauli exclusion principle, which states that no two bound electrons can exist in the same quantum state, i.e., no two electrons in the atom can have the same values of n, ℓ, m_ℓ, and m_s. Thus, an atomic orbital characterized by a specific set of n, ℓ, and m_ℓ values can be occupied by two electrons only if their spins are opposed with $m_s = +\frac{1}{2}$ and $-\frac{1}{2}$. Subject to this constraint, the orbitals are filled in the order of increasing energy (the Aufbau principle). If two or more orbitals have the same energy (i.e., they are degenerate), then the electrons go into different orbitals as much as possible, thus reducing electron-electron repulsion by keeping them apart, with spins parallel and in the same direction rather than opposed (Hund's rule).

For light atoms of the type we shall be concerned with here (nuclear charge less than 40), the orbital angular momenta of all of the electrons strongly couple together to give a total, or resultant, orbital angular momentum. This is designated by the quantum number L, which is obtained by the combination of the moments of the individual electrons. Similarly, the spin angular momenta combine to give a resultant spin, designated by the quantum number S. It follows that the total orbital angular momentum in the z direction is characterized by a quantum number $M_L = \Sigma m_\ell$, and the total

[2] *For a more thorough treatment of many-electron atoms than is given here, the reader is referred to the excellent books by G. Herzberg given in the bibliography.*

spin angular momentum in the z direction has associated with it a quantum number $M_S = \Sigma m_S$. Allowed values of M_L are -1, $-(L - 1)$, ..., 0, ..., $(L - 1)$, L (i.e., $|M_L| \leqslant L$) and allowed values of M_S are $-S$, $-(S - 1)$, ..., 0, ..., $(S - 1)$, S (or $|M_S| \leqslant S$). Closed (completely filled) shells and subshells have to a good approximation zero net resultant orbital and spin angular momenta, so that the combination need be carried out over only the electrons in partially filled subshells. Examples of this process are given later. The L and S momenta then couple together (Russell-Saunders coupling) to form a total atomic angular momentum J. Possible values of J are

$$J = L + S,\ L + S - 1,\ L + S - 2\ \ldots,\ |L - S + 1|,\ |L - S| \tag{9.32}$$

where $|L - S|$ is equal to $L - S$ if $L \geqslant S$, and equals $S - L$ if $S \geqslant L$ (i.e., J can only have positive values). The complete term symbol, which designates the electronic state of the atom, is written

$$n^{2S+1}L_J \tag{9.33}$$

The quantity $(2S + 1)$ is the multiplicity of the atom (called singlet, doublet, triplet, etc., respectively, for values of 1, 2, 3, ...); it gives the total number of possible spin orientations. Often the values of n and J are omitted from the term symbol, particularly if the energy differences among the different J states are so small that they are not a factor in photochemical considerations. Analogous to hydrogen or hydrogenlike atoms, the atomic states of $L = 0, 1, 2, 3, \ldots$ are designated, respectively, S, P, D, F, ... states.

As a specific example of the above, consider the oxygen atom with eight electrons in an electronic configuration $1s^2 2s^2 2p^4$. The 1s shell and 2s subshell are completely filled, hence only the four 2p ($\ell = 1$) electrons need be considered. The state of lowest energy, and hence the most stable, is called the ground state. According to Hund's rules, this state is (1) the state of maximum multiplicity; (2) for a given multiplicity, the state of maximum L; and (3) for given L and S, the state of minimum J if the partially filled subshell is less or half occupied, or conversely the state of maximum J if the partially filled subshell is more than half occupied. As is shown below, the ground state of oxygen is thus given by the four electrons occupying the three 2p orbitals in the following manner.

The quantum numbers for the four electrons in this unfilled p subshell are

electron	n	ℓ	m_ℓ	m_s
1	2	1	+1	+1/2
2	2	1	+1	-1/2
3	2	1	0	+1/2
4	2	1	-1	+1/2

For this configuration M_S is $\Sigma m_S = 1/2 - 1/2 + 1/2 + 1/2 = 1$, which is the maximum possible for 4 electrons in 3 orbitals to be consistent with the Pauli exclusion principle, and therefore $S = 1$ (since $|M_S| \leqslant S$) and the multiplicity $(2S + 1) = 3$. Similarly, $M_L = \Sigma m_\ell = 1$, which is also the maximum possible value, and hence L must also $= 1$ since $|M_L| \leqslant L$. The quantities S and L are maximum allowed values, and therefore they give the state of lowest energy which is a 2^3P (a "triplet-P") state. Possible J values are 2, 1, and 0; since the subshell is more than half-filled, maximum J gives the most stable state, and therefore the complete term symbol for the oxygen atom in its lowest energy state is 2^3P_2.

Electronic excitation may result in change of J (generally unimportant in photochemistry, an exception being the rather large separations between the halogen $^2P_{3/2}$ and $^2P_{1/2}$ states), transition to another orbital within the subshell, or to a higher energy subshell. For example, two possible excited states of the oxygen atom within the same subshell (same n and ℓ) are

m_ℓ +1, +1   0, 0

m_S $+\frac{1}{2}, -\frac{1}{2}$ $+\frac{1}{2}, -\frac{1}{2}$

and

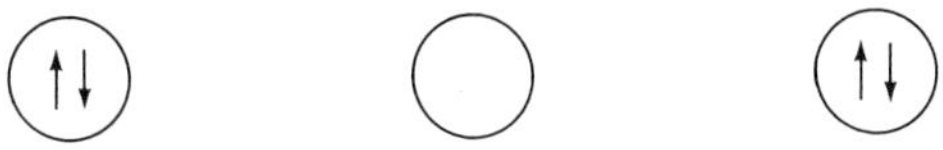

m_ℓ +1, +1   −1, −1

m_s +½, −½   +½, −½

with term symbols 2^1D_2 and 2^1S_0, respectively, the former being of lower energy because of maximum *L*.

Similarly, it can readily be verified using the same procedure that the total momenta for nitrogen with seven electrons ($1s^2 2s^2 2p^3$) in the ground-state configuration

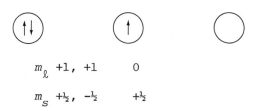

m_ℓ +1   0   −1

m_s +½,   +½   +½

is $S = 3/2$ (multiplicity = 4), $L = 0$, and $J = 3/2$, and hence the term symbol is $2^4S_{3/2}$. A possible excited state is

m_ℓ +1, +1   0

m_s +½, −½   +½

for which the term symbol is $2^2D_{3/2}$.

It should be pointed out that some excited electronic configurations are excluded by the Pauli exclusion principle for atoms with electrons with the same values of *n* and *ℓ* (called *equivalent* electrons), because simply interchanging the order in the designation does not lead to different states.[3]

[3]*G. Herzberg, "Atomic Spectra and Atomic Structure," 2nd ed., p. 130. Dover Publications, New York, 1974.*

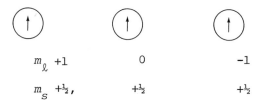

9.5.3 Diatomic Molecules

For diatomic (or linear polyatomic molecules), it is possible to designate the electronic states by a set of quantum numbers analogous to atoms. Thus, electrons with atomic orbital angular momentum quantum number ℓ can have quantized molecular values λ from 0 to ℓ, and these combine in a manner analogous to polyelectronic atoms to give a quantum number Λ representing the total orbital angular momentum along the internuclear axis. Λ can have values 0, 1, 2, ..., L (where L is the quantum number for the resultant orbital angular momentum for all of the electrons in the molecule), and are designated, respectively, Σ, π, Δ, ..., states, analogous to s, p, d, ..., for the hydrogen atom and to S, P, D, ..., for polyelectronic atoms. Similarly, the total molecular spin quantum number is S, and the component of S along the internuclear axis is Σ with possible values S, S-1, ..., 0, ..., -(S-1), -S. The sum of these two quantum numbers is a quantum number analogous to J, $\Omega = |\Lambda + \Sigma|$, and the total term symbol for a specific diatomic or linear polyatomic electronic state is

$$^{(2S\,+\,1)}\Lambda_{\Omega} \tag{9.34}$$

In addition to these quantum numbers Λ, S, and Ω, there are two other properties of homonuclear diatomic species dealing with the symmetry of its wavefunction, properties that will be shown later to be important in determining the probability of a molecule absorbing a photon of light, and hence of being excited to a specific electronic state:

(a) If inversion of all electrons through a center of symmetry of the molecule leads to no change in sign of the electronic wavefunction, then the state is <u>even</u>, or gerade (g); if a change of sign results, the state is <u>odd</u>, or ungerade (u). The symbol g or u is included in the molecular term symbol as a second subscript to Λ.

(b) If reflection in a plane of symmetry passing through the nuclei (containing the internuclear axis of symmetry) leads to no change in sign of the electronic wavefunction, then the state is <u>positive</u> (+); if it changes sign, the state is <u>negative</u> (-). The symbol + or - is written as a second superscript to Λ. (This element of symmetry applies to $\Lambda = 0$, or Σ states.)

Molecular oxygen provides a practical atmospheric photochemical example for illustrating these state designation factors. The electronic configuration for ground-state O_2 (16 electrons) is

$$(1s\sigma_g)^2\ (1s\sigma_u{*})^2\ (2s\sigma_g)^2\ (2s\sigma_u{*})^2\ (2p_x\sigma_g)^2$$

$$(2p_y\pi_u)^2\ (2p_z\pi_u)^2\ (2p_y\pi_g{*})^1\ (2p_z\pi_g{*})^1$$

The $(1s\sigma_g)$, etc., designate molecular orbitals, the asterisk (*) refers to an antibonding orbital (i.e., the density of electrons between the two oxygen atoms in this type of orbital is less than that for the two free atoms, and therefore there is a repulsive force between them), and the superscript gives the number of electrons in the molecular orbital. Since there are two unpaired electrons in the equivalent (degenerate) $(2p_y\pi_g{*})$ and $(2p_z\pi_g{*})$ orbitals, by Hund's rule of maximum multiplicity the total spin is 1 and the ground state is a triplet. The quantum number Λ can be 2 or 0; however, if $\Lambda = 2$, both electrons would have the same $\lambda = \pm 1$ value, which is forbidden by the Pauli exclusion principle for electrons in equivalent orbitals also with the same spins, and therefore only the $\Lambda = 0$, or Σ, state is allowed for $S = 1$. Fig. 9-2 shows the wavefunctions for the two electrons in the $2p_y\pi_g{*})$ and $(2p_z\pi_g{*})$ orbitals for the lowest energy configuration, from which it is seen that interchange through the

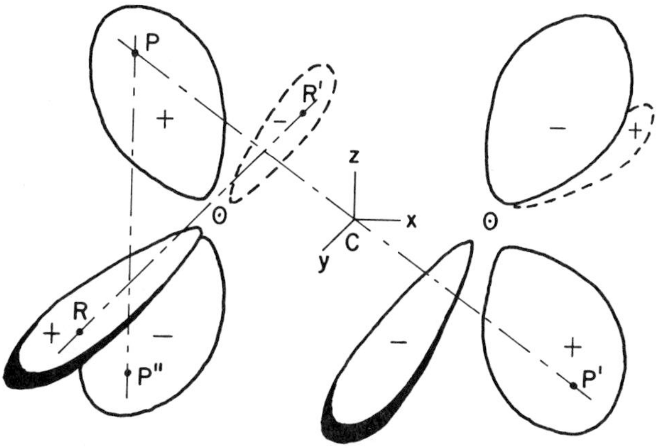

FIGURE 9-2. Schematic drawing of the $(2P_y\pi_g{*})$ and $(2p_z\pi_g{*})$ orbitals of O_2. The $(2p_y\pi_g{*})$ orbital is in the x-y plane, and the $(2p_z\pi_g{*})$ orbital is in the x-z plane. Inversion through the center of symmetry C from P to P' does not change sign, whereas inversion through a plane of symmetry passing through the two nuclei such as the x-y plane (from P to P") or the x-z plane (from R to R') does lead to a change in sign.

center of symmetry does not lead to a change in sign, whereas interchange through the *plane* of symmetry does. The ground state of oxygen is therefore gerade and negative, and the term symbol is $^3\Sigma_g^-$.

While the ground electronic state of O_2 is the $^3\Sigma_g^-$ state, there are two other possible states (of somewhat higher energies) for the electrons in these same two equivalent orbitals in which the total spin is zero (multiplicity = 1). In this case, both electrons can now have either the same λ values (each +1 or -1), leading to $\Lambda = 2$ and a $^1\Delta$ state, or different λ values (+1 and -1) giving a net angular momentum component $\Lambda = 0$ and hence a $^1\Sigma$ state. From the same symmetry considerations based on Fig. 9-2 that were used to obtain the term symbol for the ground state, it follows that the term symbols for these two electronically excited states are $^1\Delta_g$ and $^1\Sigma_g^+$. On the other hand, if an electron in a lower energy orbital [such as the $(2p_y\pi_u)$ state] is excited to the $(2p_y\pi_g*)$ orbital, then two nonequivalent orbitals [$(2p_z\pi_u)$ and $(2p_z\pi_g*)$] are singly occupied; the Pauli exclusion principle no longer applies since the two unpaired electrons occupy separate orbitals, and six states are possible: $^1\Sigma_u^+$, $^1\Sigma_u^-$, $^1\Delta_u$, $^3\Sigma_u^+$, $^3\Sigma_u^-$, and $^3\Delta_u$.

9.5.4 Polyatomic Molecules

Although the concepts developed for atoms and diatomic species can be extended to polyatomic molecules, the designation of electronic states for more complex molecules can become quite complex if the maximum spectral information is to be retained. However, many of the more subtle points that are especially important for fine-structure spectroscopic characterizations in most cases simply do not affect understanding of photochemical transformations involving complex species, and therefore for our purposes we can get by with relatively simple terms involving orbitals for only a single optical electron. The optical electron is the one electron promoted in the light absorption process.

Of importance for these orbitals are their symmetry and multiplicity characteristics, plus their involvement in bonding within the molecules. These may be of the following types involving both bonding and antibonding characteristics:

(a) σ and $\sigma*$ orbitals; these are associated with two atoms, and involve two tightly bound electrons;

(b) π and $\pi*$ orbitals; for example, the contribution of p electrons to the double bond in ethylene, or the conjugated electrons in benzene.

(c) *n* orbitals; these are nonbonding orbitals occupied by lone-pair electrons in heteroatomic molecules.

Excited states are then designated by the initial and final orbitals associated with the transition of the optical electron. Thus, the first excited state of ethylene is formed by promoting an electron from a bonding π orbital to an antibonding π^* orbital, leading to two unpaired electrons. Hund's rule of maximum multiplicity again suggests the spins of these electrons should be in the same direction, so that $S = 1$ and the multiplicity is 3. The designation of this state is $^3(\pi, \pi^*)$. For a molecule involving a carbonyl group,

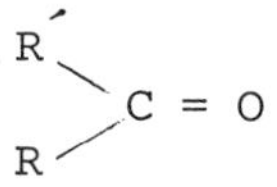

an additional low-lying state is the $^3(n, \pi^*)$ state formed by exciting an electron from the nonbonding n orbital to the antibonding π^* orbital. The state of lower energy, whether the $^3(n, \pi^*)$ state or the $^3(\pi, \pi^*)$ state, will depend somewhat on substituent (R and R′) groups and on the physical environment of the molecule.

9.5.5 *Selection Rules for Radiative Transitions*

Although many electronic states are possible for atoms, diatoms, and polyatomic molecules, the number of radiative transitions that can actually occur between states is limited by *selection rules*. The following rules showing changes for which electronic transitions are allowed are given here without proof or justification, these being found however in standard quantum mechanics and spectra reference books.[4]

(a) For atoms:

$\Delta S = 0$

$\Delta L = 0, \pm 1$

$\Delta J = 0, \pm 1$ (but $J = 0 \nleftrightarrow J = 0$)

[4] *See, for example, M. D. Harmony, "Introduction to Molecular Energies and Spectra," pp. 450-455. Holt, Reinhart, and Winston, New York, 1972.*

(b) For diatomic and polyatomic molecules (light nuclei):

$\Delta S = 0$

$\Delta\Lambda = 0, \pm 1$

$u \leftrightarrow g$

$+ \leftrightarrow +$

$- \leftrightarrow -$

Transitions that follow these rules are called "allowed" transitions. Strictly speaking these selection rules are valid only for small molecules made up of light atoms; "forbidden" transitions can occur in larger molecules, particularly those containing heavy atoms, but they are generally of very low intensity.

9.6 PHOTOPHYSICAL EFFECTS OF LIGHT ABSORPTION

We shall be most concerned in following chapters with the photochemical consequences of light absorption in our environment. However, there are several areas (such as energy transfer in photosynthesis) where photophysical effects will be important.

As pointed out before, primary absorption of a photon in the visible or ultraviolet spectral region by a molecule leads to excitation or promotion of an electron (the optical electron) to a higher electronic state of the species, or to ionization at sufficiently high energies such as encountered in vacuum ultraviolet absorption. Most ground states of molecules are singlet (unit multiplicity, or zero net electronic spin) states, so that the spin selection rule ($\Delta S = 0$) dictates that the excited state reached by a completely allowed radiative transition will be a singlet state. (An important exception is triplet ground state molecular oxygen, with two unpaired electrons.) In the case of molecules generally, electronic excitation also results in vibrational and rotational changes, but vibrational relaxation to the lowest vibrational level and rotational equilibration among the rotational states in the excited state are very fast processes (of the order of picoseconds, except for isolated gas-phase molecules) so that reactive properties are determined solely by the specific electronic state. The excited molecule can undergo the following processes in addition to direct photoionization:

(a) If the excited state is completely repulsive, that is, there are no attractive forces strong enough to lead to a potential energy minimum, then dissociation into two or more molecular fragments, free radicals, or ions will occur within approximately one vibration (10^{-12} - 10^{-13} sec). (There are, of course, no rotational or vibrational states associated with a repulsive state.)

(b) Even if the excited electronic state is "bound," i.e., it possesses a potential energy minimum, the absorbed photon may be of sufficient energy to raise the vibrational energy to a dissociative limit. Although this type of direct photodissociation may also be extremely rapid (as for example, with diatomic molecules), with the more complex species the large number of vibrational modes and energy "sinks" within the molecule itself may make its dissociative lifetime very long and competitive with deexcitation or chemical reactions.

(c) The molecule may return to the stable ground state by the radiative emission of light (fluorescence) from the lowest vibrational level, as illustrated in Fig. 9-3. In general, this emission will be of lower energy (longer wavelength) than the excitation energy for molecules because of the nonradiative dissipation of vibrational energy in the excited state before fluorescence. If the probability of absorption of a photon is high (large ϵ), then so also in general will be the probability of fluorescence emission, so that the radiative lifetime of the excited state will be very short.

(d) The excited molecule may return to the vibrationally excited ground state by a nonradiative (internal conversion IC) process, also illustrated in Fig. 9-3. In general, this type of process (without an accompanying chemical change) is quite rare, but does occur in some simple aromatic molecules and in cyanine dyes.

(e) The excited system may cross over in a nonradiative process to another excited state of different multiplicity (intersystem crossing ISC), as shown in Fig. 9-3. The most common example of this type of crossing is singlet → triplet. Since this is a nonradiative process, it is not limited by the $\Delta S = 0$ limitation; by the very nature of ISC transition, however, the resulting new excited state can no longer undergo a rapid radiative decay back to the ground state because of this spin selection rule. Long-lived radiation of this "forbidden" type (phosphorescence) is observed mainly in the solid state and at low temperatures where other decay modes are even slower.

(f) The excited (donor D) molecule may be deactivated by radiationless energy transfer to another (acceptor A) molecule.

Usually, excited singlet states are very short-lived, of the order of 10-100 nsec, so that electronic energy transfer by collision is unlikely for these systems. Long-range (many molecular diameters) singlet-singlet energy transfer can occur, however, through coupling of the dipoles of the

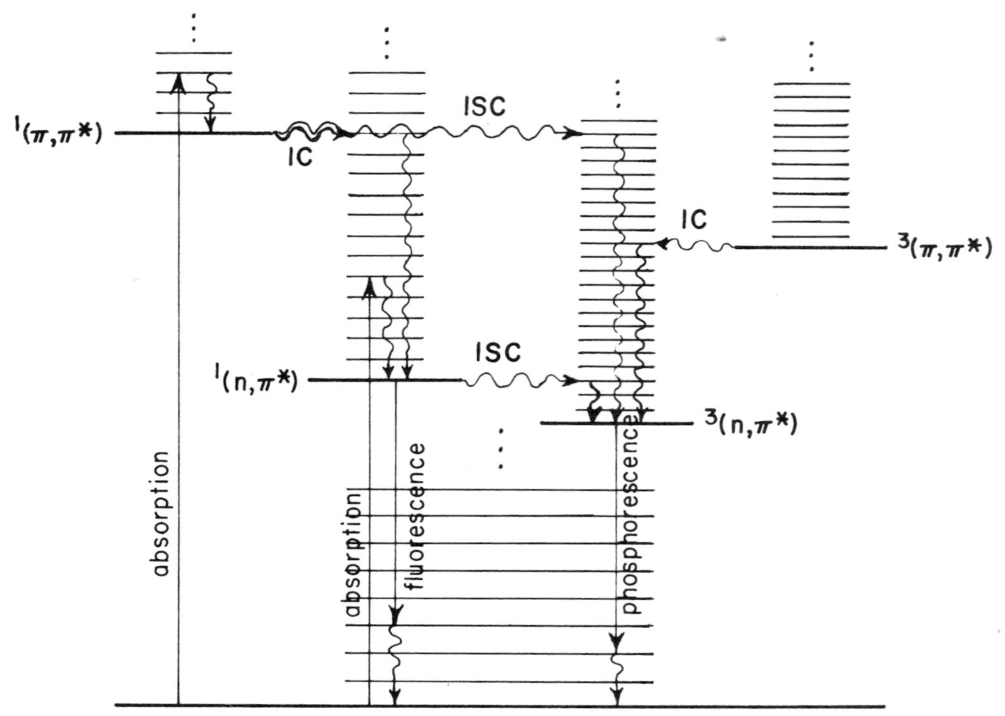

FIGURE 9-3. Schematic molecular energy level diagram showing radiative (⟶) and nonradiative (∿→) processes between vibrational and electronic energy levels.

molecules. Although this type of transfer, called *resonance energy transfer,* has an inverse sixth power dependence on separation, appreciable transfer can occur over 5 - 10 nm separation if the transitions in both the D and A molecules are fully allowed and if there is close matching of donor fluorescence and acceptor absorption spectra. Furthermore, since an actual physical encounter between D and A is not involved, this type of energy transfer occurs even in the solid state or in high-viscosity solvents.

Because of the spin-forbidden character of phosphorescence decay to the ground state, the first excited triplet state for most molecules is longer-lived (of the order of microseconds to milliseconds, or even longer) than the excited singlet state. It can therefore readily undergo collisional energy transfer at normal encounter distances by the spin-allowed reaction

$$\text{D(triplet)} + \text{A(singlet)} \rightarrow \text{D(singlet)} + \text{A(triplet)}$$

provided the transfer is exothermic ($\Delta E_{acceptor} < \Delta E_{donor}$) or at least less than approximately 10 kJ endothermic. Triplet states, and hence phosphorescence, are quenched in most fluid solutions by this type of energy transfer to dissolved oxygen and solvent impurities even in the absence of added quenchers.

These physical processes such as the radiative transitions fluorescence and phosphorescence, and radiationless internal conversion, collisional deactivation, etc., will all lead to a primary quantum yield ϕ (see Section 9.2) less than one. However, the sum of all primary quantum yields for the various processes (ϕ_f = quantum yield of fluorescence, ϕ_p = quantum yield of phosphorescence, etc.) is

$$\sum_i \phi_i = 1 \tag{9.35}$$

where the sum includes all i modes of disappearance of the photoexcited species including direct chemical reaction.

BIBLIOGRAPHY

General Photochemistry and Properties of Light

J. G. Calvert and J. N. Pitts, Jr., "Photochemistry." Wiley, New York, 1966.

J. P. Simons, "Photochemistry and Spectroscopy." Wiley, London, 1971.

R. K. Clayton, "Light and Living Matter": "The Physical Part," Vol. 1; "The Biological Part," Vol. 2. McGraw-Hill, New York, 1971.

R. B. Cundall and A. Gilbert, "Photochemistry." Appleton-Century-Crofts, New York, 1970.

D. R. Arnold, N. C. Baird, J. R. Bolton, J. C. D. Brand, P. W. M. Jacobs, P. de Mayo, and W. R. Ware, "Photochemistry, an Introduction." Academic Press, New York, 1974.

Light, *Sci. Amer. 219,* No. 3 (September, 1968).

Designation of Spectroscopic States

M. Karplus and R. N. Porter, "Atoms and Molecules." W. A. Benjamin, New York, 1970.

G. Herzberg, "Atomic Spectra and Atomic Structure," 2nd ed. Dover, New York, 1944.

G. Herzberg, "Spectra of Diatomic Molecules." Van Nostrand, Princeton, New Jersey, 1950.

G. Herzberg, "Electronic Spectra and Electronic Structure of Polyatomic Molecules." Van Nostrand, Princeton, New Jersey, 1966.

10
ATMOSPHERIC PHOTOCHEMISTRY

10.1 INTRODUCTION

We have already noted in earlier chapters some of the important photochemical reactions occurring in the atmosphere that are important to the nature of our environment. In Chapter 3 we saw that photodissociation of water vapor in the mesosphere is an important source of atmospheric oxygen. Molecular oxygen is photodissociated into oxygen atoms in the thermosphere, mesosphere, and upper stratosphere, and in so doing it absorbs most of the high-energy radiation below 180 nm. Ozone is produced in the stratosphere by the combination of an oxygen atom and an oxygen molecule. We have seen in Chapter 3 that O_3 extends the earth's shield against lethal ultraviolet radiation from 180 nm to about 310 nm. The troposphere connects the biosphere (that is, the earth's surface) to the stratosphere, and therefore is closely involved in the chemistry of both. For example, man-made emissions of oxides of nitrogen and hydrocarbons from internal combustion automobile engines and stationary furnaces have been discussed in Chapter 5; these substances are major contributors to photochemical reactions in the troposphere that drastically affect the quality of our environment. Strong vertical mixing occurs in the troposphere, so that species produced naturally (for example, methane and water) and man-made (oxides of nitrogen and chlorofluorocarbons) in the biosphere can contribute to photochemical processes occurring in the stratosphere.

10.2 REACTIONS IN THE UPPER ATMOSPHERE

10.2.1 Oxygen

By far the most abundant photochemical reaction in the upper atmosphere is the photolysis of molecular oxygen. Figure 10-1 shows the potential energy curves for the various electronic states of O_2 of importance to us; Fig. 10-2 is the

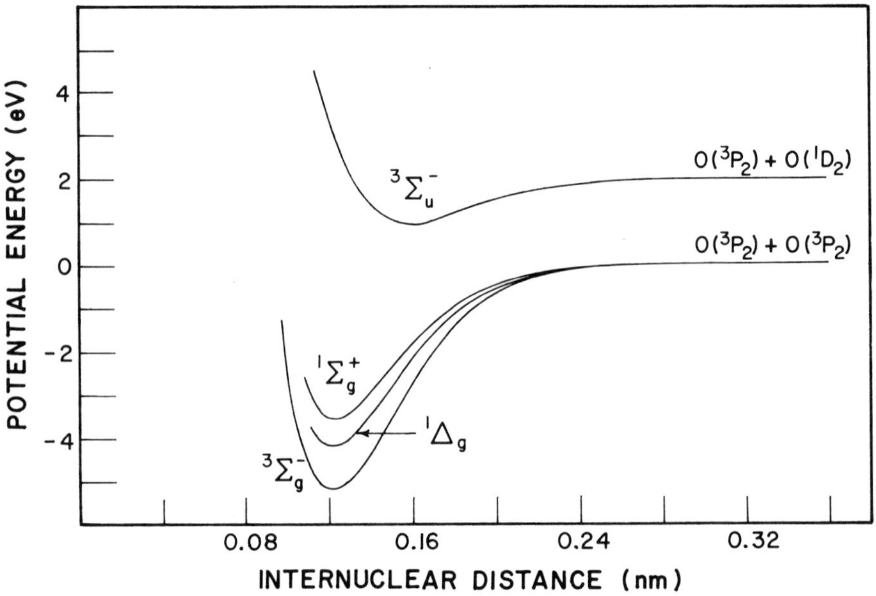

FIGURE 10-1. Potential energy curves for O_2. Curves drawn from data of F. R. Gilmore, J. Quant. Spectrosc. Radiat. Transfer 5, 369 (1965).

absorption in the far ultraviolet and vacuum ultraviolet regions. It has been shown in Section 8.2.1 that the bond dissociation energy of oxygen (5.1 eV) corresponds to a photon of wavelength 243 nm. Oxygen actually begins to absorb just below this wavelength in a spectral region known as the *Herzberg continuum*. Absorption is weak ($\epsilon \cong 10^{-3}$/atm cm at 200 nm), but the fact that the spectrum is a continuum in this region indicates that dissociation occurs ($\phi = 1$). Actually, excitation is to the weakly bound $^3\Sigma_u^+$ state which rapidly dissociates into two ground-state ($^3P_2$) atoms.

$$O_2(^3\Sigma_g^-) \xrightarrow{h\nu} O_2(^3\Sigma_u^+) \longrightarrow 2\ O(^3P_2) \qquad (10\text{-}1)$$

The very weak absorption shown in this case is a good example of the consequence of selection rule violation (the specific one being that the $- \longrightarrow +$ is a forbidden transition).

Below 200 nm, the absorption spectrum becomes much stronger and banded (the *Schumann-Runge absorption bands)*, as shown in Fig. 10-2. These bands get closer together as the wavelength decreases, and at 176 nm (the *convergence limit*) the bands converge together into a continuum known as the

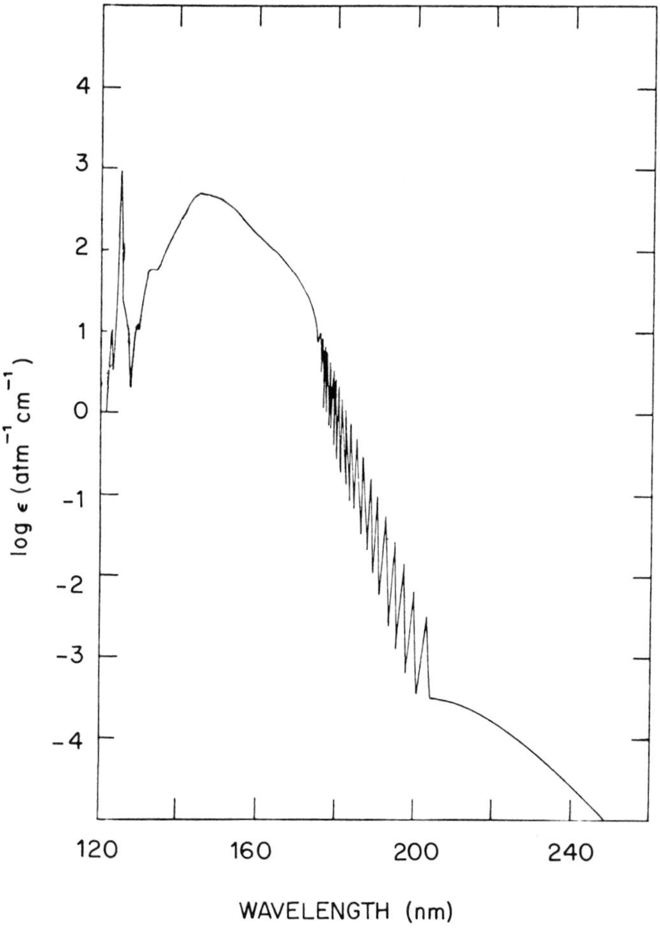

FIGURE 10-2. Absorption spectrum of O_2. (Note logarithmic scale for the extinction coefficient.) From J. R. McNesby and H. Okabe, in "Advances in Photochemistry" (W. A. Noyes, Jr., G. S. Hammond, and J. N. Pitts, Jr., eds.), Vol. 3, p. 174. Interscience, New York, 1964.

Schumann-Runge continuum which reaches maximum absorption at approximately 147 nm ($\epsilon_{max} \simeq 500$/atm cm). Excitation in this region is to the bound $^3\Sigma_u^-$ state (an allowed transition, hence large extinction coefficient), which dissociates below the 176 nm convergence limit into one ground-state ($^3P_2$) and one excited ($^1D_2$) oxygen atom:

$$O_2(^3\Sigma_g^-) \xrightarrow{h\nu} O_2(^3\Sigma_u^-) \qquad (10\text{-}2)$$

$$O_2(^3\Sigma_u^-) \xrightarrow{\lambda<176\ \text{nm}} O(^3P_2) + O(^1D_2) \qquad (10\text{-}3)$$

Below 130 nm there is sufficient energy to produce oxygen atoms in even higher electronic states, and some dissociation to the ($^1S_0$) state takes place

$$O_2(^3\Sigma_g^-) \xrightarrow[\lambda<130\ \text{nm}]{h\nu} O(^3P_2) + O(^1S_0) \qquad (10\text{-}4)$$

and below 92.3 nm, two excited atoms are produced

$$O_2(^3\Sigma_g^-) \xrightarrow[\lambda<92.3\ \text{nm}]{h\nu} 2\ O(^1S_0) \qquad (10\text{-}5)$$

However, it was pointed out in Chapter 9 that the ionization potential of O_2 (12.15 eV) corresponds to a photon wavelength λ = 102 nm, and therefore a more likely reaction below 102 nm is photoionization

$$O_2(^3\Sigma_g^-) \xrightarrow[\lambda<102\ \text{nm}]{h\nu} O_2^+\ (^2\Pi_g) + e^- \qquad (10\text{-}6)$$

followed by dissociative recombination. Reactions (10-7) and (10-8) are two possible examples of this type of process:

$$O_2^+ + e^- \rightarrow O(^1S_0) + O(^3P_2) + 64.3\ \text{kcal}\ (269\ \text{kJ}) \qquad (10\text{-}7)$$

$$O_2^+ + e^- \rightarrow O(^1S_0) + O(^1D_2) + 19.1\ \text{kcal}\ (80\ \text{kJ}) \qquad (10\text{-}8)$$

with (10-7) found to be the favored reaction.

Two other excited electronic states of O_2 are important in atmospheric photochemistry because of specific reactions to be discussed later in reference to their possible roles in the photochemical smog cycle, ozone photolysis, and photo-oxidation reactions with olefins. These are the $^1\Delta_g$ ("singlet oxygen") and $^1\Sigma_g^+$ states, with energies 22.5 kcal (94.1 kJ) and 37.5 kcal (157 kJ), respectively, above the ground state (Fig. 10.1). Direct excitation from the triplet ground state to either of these states is spin-forbidden, and oxygen absorbs only very weakly around 760 nm (37.5 kcal) in a completely banded structure (the *Fraunhofer bands*). The reverse radiative process is of course also spin-forbidden and these states, when produced, are therefore relatively stable in the atmosphere, a factor contributing to their importance; the radiative lifetime of the $^1\Sigma_g^+$ is 12 sec, while that of $^1\Delta_g$ is 60 min. Furthermore, collisional quenching is also inefficient, the rate constants of deactivation by ground-state O_2 being 5×10^{-16} and 2×10^{-18} cm^3/molecule sec for the $^1\Sigma_g^+$ and $^1\Delta_g$ states, respectively. In the absence of collisional deactivation, the emission from the normally "forbidden" radiative transitions

$$O_2(^1\Sigma_g^+) \longrightarrow O_2(^3\Sigma_g^-) + h\nu \text{ (762 nm)} \tag{10-9}$$

and

$$O_2(^1\Delta_g) \longrightarrow O_2(^3\Sigma_g^-) + h\nu \text{ (1270 nm)} \tag{10-10}$$

turn out to be two of the most intense bands in the atmospheric dayglow and in aural displays. (The $^1\Sigma_g^+$ state also relaxes to the $^1\Delta_g$ state by collisional deactivation.)

Ground state ($^3P_2$) oxygen atoms might be expected to combine directly to form O_2. However, simultaneous conservation of energy and momentum requires the presence of a third body in the nonradiative process

$$2\ O(^3P_2) + M \longrightarrow O_2(^3\Sigma_g^-) + M \tag{10-11}$$

where M can be another oxygen atom or any other atomic or molecular species such as O_2 or N_2. Reaction (10-11) is often considered to be kinetically simple and therefore trimolecular:

$$R = k[O]^2[M]; \quad k_{298} = 3 \times 10^{-33} \text{ cm}^6/\text{molecule}^2 \text{ sec} \tag{10-12}$$

The rate constant k is very nearly temperature independent; in fact it has a slightly inverse temperature behavior, the rate decreasing with increasing temperature implying a negative activation energy, which of course is impossible physically if activation energy is considered to be the height of a potential barrier.[1] For reaction (10-11), k is of the order

[1] *The reason for the negative activation energy is that oxygen atoms form an unstable intermediate complex with M*

$$O + M \rightleftarrows O{\cdot}M \tag{10-13}$$

$$O{\cdot}M + O \rightarrow O_2 + M \tag{10-14}$$

As the temperature is raised, O·M becomes more unstable so that its concentration is decreased, thereby decreasing the rate of reaction (10.14) and hence decreasing the overall rate of O_2 formation represented by reaction (10-11). Since reaction (10-11) is now kinetically complex, as pointed out in Section 9.3 the interpretation of the experimental activation energy as the height of a potential energy barrier is not valid. Assuming equilibrium between O, M, and O·M leads however to the same rate law as that for the trimolecular reaction (10-12).

of magnitude of the collisional frequency for triple collisions, indicating that recombination occurs essentially at every encounter. Even so, at the very low pressures encountered in the upper atmosphere (10^{-6} atm at 100 km), the rate of reaction (10-11) is small compared to the rate of O_2 photodissociation, and therefore above 100 km the primary oxygen species present is atomic oxygen. (It should be noted that radiative recombination processes are possible:

$$2\ O(^3P_2) \longrightarrow O_2(^1\Sigma_g^+) + h\nu \tag{10-15}$$

$$O(^3P_2) + O(^1D_2) \longrightarrow O_2(^3\Sigma_g^-) + h\nu \tag{10-16}$$

However, these reactions are also governed by spin conservation and atomic radiative selection rules, and they also are negligible compared to photodissociation.)

It has been seen that excited singlet ($^1D_2$ and $^1S_0$) atoms are formed from the vacuum ultraviolet photodissociation of molecular oxygen. These species also relax to triplet ground-state oxygen with emission of light:

$$O(^1D_2) \longrightarrow O(^3P_2) + h\nu\ (\lambda = 630\ \text{nm}) \tag{10-17}$$

$$O(^1S_0) \longrightarrow O(^3P_2) + h\nu\ (\lambda = 577.7\ \text{nm}) \tag{10-18}$$

Again, both are "forbidden" transitions and therefore show only weak emission contributing to the air glow and aurora. Below 140 km, however, collisional deactivation with any atomic or molecular species M may become the dominant reaction, particularly with $O(^1D_2)$:

$$O(^1D_2) + M \longrightarrow O(^3P_2) + M$$
$$k = 6 \times 10^{-11}\ \text{cm}^3/\text{molecule sec} \tag{10-19}$$

If $M = O_2\ (^3\Sigma_g^-)$, excitation is to the $^1\Sigma_g^+$ state which decays collisionally to the $^1\Delta_g$ state:

$$O(^1D_2) + O_2(^3\Sigma_g^-) \longrightarrow O(^3P_2) + O_2(^1\Sigma_g^+) \xrightarrow{M} O_2(^1\Delta_g) \tag{10-20}$$

An important aspect of this *energy transfer* or *sensitization* is that spin is conserved, that is, the oxygen atom undergoes a singlet-triplet transition while the reverse triplet-singlet reaction occurs with molecular oxygen, so that total reactant and product spins are the same. This is a very efficient process under exothermic conditions, probably occurring within an order of magnitude of collisional frequency, and therefore reaction (10-20) is one of the major sources of "singlet oxygen" $O_2(^1\Delta_g)$ in the atmosphere.

10.2.2 Ozone

Below 100 km, three-body recombination [reaction (10-11)] becomes potentially important. Because of the large proportion of O_2 in the atmosphere, however, another three-body process forming ozone also occurs, although the trimolecular recombination rate constant is less than that for the atom recombination:

$$O(^3P_2) + O_2\ (^3\Sigma_g^-) + M \longrightarrow O_3 + M$$

$$k = 5.6 \times 10^{-34}\ \text{cm}^6/\text{molecule}^2\ \text{sec} \qquad (10\text{-}21)$$

This reaction is the sole source of ozone in the stratosphere and mesophere as well as being a major source of destruction of O atoms. Ozone is one of the most important and remarkable minor constituents in our atmosphere. As pointed out in Chapter 3, not only does it provide a shield for the earth against damaging ultraviolet radiation, but also, through absorption of this radiation, it becomes the main energy reservoir in the upper atmosphere and hence a factor in climatic regulation. On the other hand, ozone is one of the most toxic inorganic chemicals known, and its presence at relatively high concentrations in lower-atmosphere polluted air makes it a potentially dangerous substance for man and other organisms (Chapter 5).

The bond dissociation energy of ozone is low (24.2 kcal, corresponding to a photon of wavelength λ = 1180 nm), and the primary quantum yield of dissociation is unity at all wavelengths below 1180 nm. However, ozone absorbs only weakly in the visible region (*Chappuis bands*), leading to ground-state species

$$O_3 \xrightarrow{h\nu} O_2(^3\Sigma_g^-) + O(^3P_2) \qquad (10\text{-}22)$$

The wavelength of maximum absorption in the visible is 600 nm, with $\epsilon \simeq 0.06$/atm cm.

The major absorption of ozone is in the ultraviolet (*Hartley continuum*) commencing at 334 nm with λ_{max} = 254 nm, ϵ_{max} = 140/atm cm, as shown in Fig. 10-3. This ultraviolet photolysis is quite complex, involving extensive secondary reactions. At $\lambda > 310$ nm, the predominant dissociative reaction is

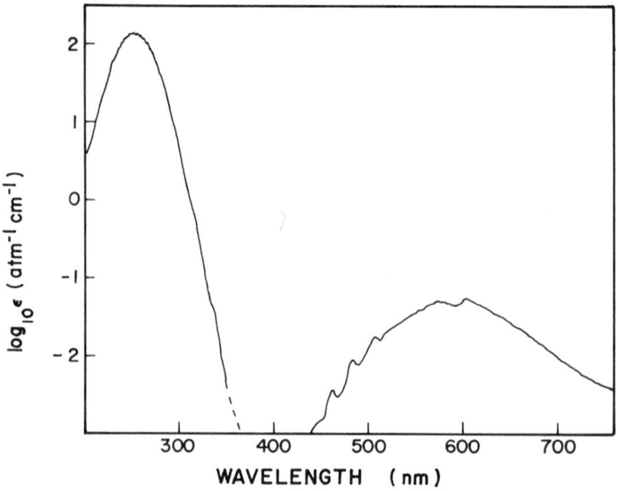

FIGURE 10-3. Absorption spectrum of O_3. (Note logarithmic scale for the extinction coefficient.) Drawn from data of E. C. Y. Inn and Y. Tanaka, in "Ozone chemistry and technology," <u>*Adv. Chem. Ser.*</u> *<u>21</u>, p. 263. American Chemical Society, Washington, D.C. 1959.*

$$O_3 \xrightarrow{h\nu} O(^3P_2) + O_2(^1\Delta_g \text{ or } ^1\Sigma_g^+) \tag{10-23}$$

but below 310 nm it is

$$O_3 \xrightarrow{h\nu} O(^1D_2) + O_2(^1\Delta_g) \tag{10-24}$$

and this is another primary source of singlet oxygen in the atmosphere. Further reactions of ozone will be considered later in connection with atmospheric reactions involving oxides of nitrogen and components of polluted air.

10.2.3 Water

Water absorbs radiation below 186 nm, the absorption spectrum consisting of a continuum region (λ_{max} = 165 nm) and a banded region between 125 and 143 nm superimposed on a second continuum. The presence of two continuums indicates that two different excited states are involved in the photochemistry of water.

Energetically, water could dissociate into hydrogen and ground-state oxygen atoms by absorbing below 245 nm:

$$H_2O \xrightarrow{h\nu} H_2 + O(^3P_2) \tag{10-25}$$

However, there is no experimental evidence that this primary process does actually occur. Dissociation into an excited O atom

$$H_2O \longrightarrow H_2 + O(^1D_2) \qquad (10\text{-}26)$$

is energetically possible below 178 nm, and may contribute in a minor way in both continuum regions. The major primary process, however, is dissociation into a hydrogen atom and hydroxyl radical, either by reaction (10-27) in the long-wavelength continuum region or by (10-28) in the lower-wavelength continuum:

$$H_2O \xrightarrow[\lambda<242\text{ nm}]{h\nu} H(^2S_{1/2}) + HO\cdot(^2\Pi) \qquad (10\text{-}27)$$

$$H_2O \xrightarrow[\lambda<136\text{ nm}]{h\nu} H(^2S_{1/2}) + HO\cdot(^2\Sigma^+) \qquad (10\text{-}28)$$

The precise mechanism is not clear for the formation of oxygen, which is a major product in water photolysis (see Chapter 3). The most likely mechanism appears to be

$$H + HO\cdot \longrightarrow H_2 + O(^3P_2) \qquad (10\text{-}29)$$

followed by reaction (10-11), although reaction of two hydroxyl radicals has been proposed:

$$HO\cdot + HO\cdot \longrightarrow H_2 + O_2 \qquad (10\text{-}30)$$

10.2.4 Nitrogen

So far we have been concerned with the photolysis of oxygen and its compounds, even though the more prevalent species in the atmosphere is molecular nitrogen. The reason that nitrogen is not so important photochemically is its large bond energy (7.373 eV, corresponding to a photon of $\lambda = 169$ nm), which limits its photochemistry to above the ozone layer. Molecular nitrogen absorbs only weakly between 169 and 200 nm, absorption in this region leading to a spin-forbidden excitation from ground-state $N_2(^1\Sigma_g^+)$ to the $(^3\Sigma_u^+)$ excited state

$$N_2(^1\Sigma_g^+) \xrightarrow{h\nu} N_2(^3\Sigma_u^+) \qquad (10\text{-}31)$$

which is a source of highly excited oxygen atoms by collisional energy transfer:

$$N_2(^3\Sigma_u^+) + O(^3P_2) \longrightarrow N_2(^1\Sigma_g^+) + O(^1S_0) \qquad (10\text{-}32)$$

Below 100 nm, photodissociation occurs, possibly through photoionization and dissociative recombination involving O_2 and producing nitric oxide:

$$N_2(^1\Sigma_g^+) \xrightarrow{h\nu} N_2^+ + e^- \tag{10-33}$$

$$N_2^+ + O_2(^3\Sigma_g^-) \longrightarrow NO^+ + NO \tag{10-34}$$

$$NO^+ + e^- \longrightarrow N(^2D_{3/2}) + O \tag{10-35}$$

Deactivation of energetically excited nitrogen atoms is a source of singlet oxygen:

$$N(^2D_{3/2}) + O_2(^3\Sigma_g^-) \longrightarrow N(^4S_{3/2}) + O_2(^1\Delta_g) \tag{10-36}$$

Nitric oxide may also form at every collision by the trimolecular recombination

$$N(^4S_{3/2}) + O(^3P_2) + M \longrightarrow NO + M \tag{10-37}$$

although the major source of NO at stratospheric heights is from N_2O, either by reaction with excited oxygen atoms

$$N_2O + O(^1D_2) \longrightarrow 2\ NO \tag{10-38}$$

or by direct photolysis below 250 nm:

$$N_2O \xrightarrow[\lambda<250\ nm]{h\nu} NO(^2\Pi) + N(^4S_{3/2}) \tag{10-39}$$

Besides being an important constituent in photochemical smog in the lower atmosphere, nitric oxide plays a significant role in upper atmosphere photochemistry. For example, NO reacts with ozone in a two-step mechanism

$$NO + O_3 \longrightarrow NO_2 + O_2(^1\Delta_g)$$

$$k_{298} = 1.8 \times 10^{-14}\ cm^3/molecule\ sec \tag{10-40}$$

(which is another atmospheric source of singlet oxygen) followed by

$$NO_2 + O \longrightarrow NO + O_2$$

$$k_{298} = 9.5 \times 10^{-12}\ cm^3/molecule\ sec \tag{10-41}$$

The net effect of these two reactions (10-40) and (10-41) is thus the NO-catalyzed decomposition of ozone:

$$O_3 + O \xrightarrow{NO} 2\ O_2 \tag{10-42}$$

It is well known that jet aircraft, since they are air breathing, produce copious amounts of oxides of nitrogen, and the possibility that projected fleets of supersonic transports flying within the ozone layer (at heights greater than 20 km) could release sufficient nitrogen oxides (as well as water vapor, hydrocarbons, carbon monoxide) to significantly alter the ozone layer is currently being seriously considered in several research groups.[2] Similar effects might be expected from atmospheric nuclear explosions because of the high temperature generated, thereby also producing nitrogen oxides. Reaction (10-42) is a gross oversimplification of the reaction, and more thorough treatments involve not only many more reactions of other oxides of nitrogen, but also both vertical and horizontal atmospheric transport of ozone accompanying photochemical production.[3] Although the results are inconclusive and it appears that the effects may be strongly dependent on the elevation of flight and time of year, it is concluded that a fleet of 500 transatlantic Concorde-type aircraft flying routinely at 28 km *could* reduce the ozone layer by the order of 5%, thereby increasing UV radiation by up to 10%. Such an increase may lead to increased incidences of skin cancer and premature aging of the skin, as well as a change in the stratospheric temperature and possible effects on the earth's climate and vegetation.

In the same sense, serious concern has recently been expressed over possible consequences of introduction of chlorofluorocarbon gases (Freons) into the ozone layer or shield.[4] Examples of such compounds are fluorocarbon-11 ($CFCl_3$) and fluorocarbon-12 (CF_2Cl_2), used primarily as refrigerants and aerosol sprays. In 1973, approximately 48×10^8 kg of these two chlorofluorocarbons were produced in the world. These are released for the most part into the lower atmosphere from which eventually they may diffuse into

[2] *H. S. Johnston, Science 173, 517 (1971).*

[3] *H. S. Johnston, Catalytic reduction of stratospheric ozone by nitrogen oxides, in "Advances in Environmental Science and Technology" (J. N. Pitts, Jr. and R. L. Metcalf, eds.), Vol. 4, p. 263. Wiley, New York, 1974.*

[4] *M. J. Molina and F. S. Rowland, Nature 249, 810 (1974).*

the ozone layer, although very little data are currently available for this atmospheric diffusion process. The chlorofluorocarbons are extremely inert, as seen in Chapter 7, even to ground-state $O(^3P_2)$ atoms. They are not soluble in water, and are not involved in any biological reactions. It is their extreme inertness, in fact, that has made them useful as tracers of atmospheric motions. However, they react at almost collisional frequency with $O(^1D_2)$ oxygen giving COF_2. Furthermore, direct photolysis with UV light (175 to 220 nm) leads to decomposition producing one chlorine atom, and the remaining radical fragment reacts further with O_2 to produce another chlorine entity (either Cl or ClO·). We have seen that radiation as low as 220 nm is absorbed by both O_2 and O_3, although there is a "window" between the Schumann-Runge continuum of O_2 and the Hartley continuum of O_3 that partially transmits in the chlorofluorocarbon absorption band, and therefore the chlorofluorocarbons must diffuse well into the stratosphere (>20 km) in order to undergo photodecomposition. The chlorine atoms then react with ozone by analogous reactions to those of NO:

$$Cl + O_3 \longrightarrow ClO\cdot + O_2$$

$$k_{298} = 1.1 \times 10^{-11}\ \text{cm}^3/\text{molecule sec} \qquad (10\text{-}43)$$

$$ClO\cdot + O(^3P_2) \longrightarrow Cl + O_2$$

$$k_{298} = 5.3 \times 10^{-11}\ \text{cm}^3/\text{molecule sec} \qquad (10\text{-}44)$$

The net reaction again is the catalyzed decomposition of O_3:

$$O_3 + O \xrightarrow{Cl} 2\ O_2 \qquad (10\text{-}45)$$

The reaction between ClO· and NO

$$ClO\cdot + NO \longrightarrow Cl + NO_2 \qquad (10\text{-}46)$$

couples the ClO and NO cycles, and is important in stratospheric chlorine chemistry.

Reaction (10-45) will have the same depleting and deleterious effect on the ozone layer as is suggested for NO, and may be even more serious because transport into the stratosphere will continue long after emission from sources at the surface of the earth have ceased. It is estimated, for example, that there may be a delay of from 10 to 25 years before maximum effect is experienced, even if current usage of the chlorofluorocarbons is strongly curtailed. It is also estimated that continued usage at the 1973 rate could reduce stratospheric ozone from 2 to 20%.[5]

While the calculations so far on these "ozone sink" systems must be considered speculative at this stage, primarily because actual concentrations of the reactants in the atmosphere are very difficult to obtain, they do serve to show that man's pollution of the atmosphere may have far-ranging consequences on the environment that cannot be adequately evaluated without reliable photochemical, kinetic, and thermodynamic equilibrium data. In 1976 the National Research Council Committee on Impacts of Stratospheric Change recommended drastic curtailment of nonessential uses of the chlorofluorocarbons, such as in aerosol cans, unless new data are obtained mitigating the danger.[5]

10.3 PHOTOCHEMISTRY IN THE LOWER ATMOSPHERE PHOTOCHEMICAL SMOG

10.3.1 The General Nature of Smog

As we progress to lower altitudes and higher molecular concentrations, other minor constituents in the air may contribute to the overall oxygen and ozone balance. For example, such minor constituents may provide sources for intermediate free radical species that can participate in catalytic destruction of oxygen atoms and ozone. In polluted lower atmospheres, particularly in so-called civilized urban environments, a number of complex photochemical reactions may occur producing a variety of eye and throat irritants, aerosols and reduced visibility, and other irritating or destructive species manifesting themselves in the general category of *photochemical smog*.

Unlike the notorious "killer smogs" (smoke plus fog) that were a part of London life, as well as that of many other highly industrial urban areas, for centuries before massive cleanup efforts were undertaken in the past few decades, photochemical smog is of fairly recent origin. It was first

[5]*"Halocarbons: Effects on Stratospheric Ozone," Report of the National Research Council (NRC) Panel on Atmospheric Chemistry (1976), and "Halocarbons: Environmental Effects of Chlorofluoromethane Release," Report of the NRC Committee on Impacts of Stratospheric Change (1976). Available from National Academy of Sciences, Printing and Publishing Office, 2101 Constitution Avenue, Washington, D.C. 20418.*

detected through plant damage in Los Angeles in 1944, but has now spread to so many metropolitan areas that it must be considered one of the most pressing actual or potential world-wide problems with global aspects transcending national boundaries, particularly in the Northern Hemisphere in a belt between 30° and 60° latitude. Also in contrast to the poisonous London-type smog, (for example, approximately 4000 deaths are attributed to the four-day smog in London in 1952), to date there do not appear to be major human disasters associated with photochemical smog; the principal deleterious effects so far appear to be its nuisance value and reduced human efficiency from eye and throat irritation, damages to the environment (trees and plants), and decreased visibility. Potentially, however, photochemical smog can have very serious long-range global effects on the earth's radiation balance through submicron aerosol formation and large general distribution that may cause unfavorable changes in weather patterns in addition to its influence on an affected local basis. It should also be emphasized that long-range effects of extended exposure to photochemical smog on the very young, the elderly, and those suffering from chronic respiratory illnesses are not fully characterized.

Photochemical smog requires somewhat special meteorological and geographical conditions to be produced in large quantities. For example, the photochemical reactions must occur in a stable air mass, which can result when there is a temperature inversion (i.e., cool air at low elevations is trapped by a layer of warmer air, see Chapter 4). The Los Angeles basin is particularly suited for this to happen, and in fact, temperature inversions occur there about 80% of the time in hot summer months.

There are other significant differences between "London" and photochemical smogs that suggest different remedial actions that might (or should) be taken in each case. London-type smogs occur mainly in the early morning hours in winter, where there is relatively high humidity, and virtually no solar photochemistry taking place; sulfur dioxide is the major contaminant, and this creates a chemically reducing atmosphere because of the ease of sulfur dioxide oxidation. On the other hand, photochemical smogs usually occur on warm sunny days (summer *and* winter) with initially clear skies and low humidity. Primary contaminants (those that are emitted directly into the atmosphere) are nitric oxide (NO), hydrocarbons, and carbon monoxide. Secondary contaminants, largely ozone and nitrogen dioxide, are produced by reactions of the primary contaminants in the atmosphere; they begin to build up during morning traffic hours and reach peak intensity in midday, although hourly variations depending on automobile traffic/time distributions and even time variations for different constituents do occur (Fig. 10-4). Photographs taken at the University of California (Riverside), Fig. 10-5, show the dramatic difference in visibility that can occur over only a few hours of smog generation. These pictures also show that, while peak smog effects may be generated around noontime in downtown Los Angeles, they may not show up until several hours later "downwind."

We have seen in Section 5.8 that internal combustion automobile engines and stationary power plants (petroleum and coal) are the major man-made sources of nitrogen oxides, hydrocarbons (alkanes, alkenes, and aromatics), and products of partial oxidation of hydrocarbons (such as aldehydes). It has also been pointed out in Sections 3.2 and 5.8 that there are also natural sources of these materials, and on a global basis these latter sources may lead to far greater emissions than from the man-made sources. For example, it is estimated that automobiles produce 5×10^{10} kg of nitrogen oxides per year (60 kg per uncontrolled car per year), while natural sources lead to more than 70×10^{10} kg per year; comparable numbers are also estimated for hydrocarbon emissions. The importance of the artificial sources is that they generate sufficiently high concentrations of the pollutants in an urban environment with the appropriate climatic, geographic, and solar radiation conditions to undergo the photochemical and secondary reactions leading to photochemical smog.

Much progress has been made over the past 25 years in the understanding of the role of these substances in the formation of the noxious products in photochemical smog,

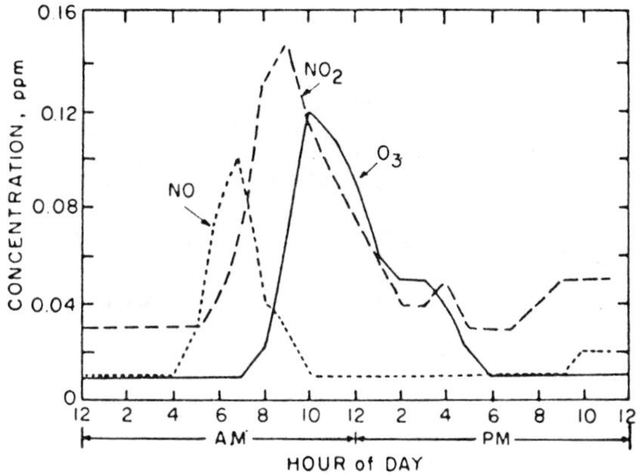

FIGURE 10-4. Photochemical smog buildup (Los Angeles, California, July 10, 1965). From K. J. Demerjian, J. A. Kerr, and J. G. Calvert, Adv. Environ. Sci. Technol. 4, 1 (1974).

FIGURE 10-5. View from Fawcett Laboratory, University of California, Riverside, on September 25, 1968. (a) 8:45 a.m., (b) noon,

FIGURE 10-5 (Continued)
(c) 3:30 p.m., and (d) 4:30 p.m. Courtesy of the Statewide Air Pollution Research Center, University of California, Riverside, James N. Pitts, Jr., Director.

and this has led to the state and federally mandated reductions in maximum allowed automobile emissions given in Table 4-2. However, as more data are collected and interpreted on reactions occuring under real-world polluted conditions, the photochemical smog mechanism appears to become enormously more complex, involving many separate reactions and processes to arrive at a model capable of predicting air quality. A procedure that has become extremely valuable in generating a model for photochemical smog is to use a digital computer to attempt to reproduce the reactant and product concentration/time profiles by solving a set of first-order differential equations that simulate the proposed mechanism. Often it is possible to lump reactants such as olefins or aromatics into classes of reactions, represented by generalized species. The advantage of this "lumped mechanism" procedure is that the kinetic mechanism can be expressed as a series of feasible kinetically simple steps for which reasonable rate constants can be estimated, while at the same time digital solution of the simulated rate equations does not require a tremendous amount of computation time. In addition to the need for adequate terms to describe transport, source, and sink reactions of pollutants, it is now obvious that good smog chamber data (obtained at the very low polluted atmospheric concentrations) are necessary, and that *in situ* analyses will undoubtedly be required in many cases. Nevertheless, certain reactions are well established as primary steps in photochemical smog generation, and these in turn suggest additional reactions needed to explain observed time/concentration behavior.

10.3.2 Nitrogen Oxides in the Absence of Hydrocarbons

As pointed out, the largest sources of nitrogen oxides in a polluted atmosphere are emissions from petroleum and coal combustion chambers where temperatures and pressures are high enough for the fixation of nitrogen

$$N_2 + XO_2 \longrightarrow 2NO_X \tag{10-47}$$

and where quenching to low temperatures is rapid enough to prevent the thermodynamically favored back reaction (dissociation) from occurring.

The major oxide of nitrogen produced is nitric oxide NO, a relatively nontoxic gas. Its concentration in a polluted atmosphere is quite sensitive to automobile traffic patterns, in general being a maximum during the peak morning traffic hours (Fig. 10-4). However, NO does not absorb radiation above 230 nm, and therefore it cannot be the primary initiator of photochemical reactions in a polluted lower atmosphere. On the other hand, nitrogen dioxide NO_2 is a brown gas with a broad, intense absorption band absorbing over most of the visible and ultraviolet regions with a maximum at 400 nm ($\epsilon_{max} \cong 7.6$ atm cm) as shown in Fig. 10-6, and undoubtedly this gas is a major atmospheric absorber leading to photochemical smog. As we shall see later, however, there are other light-absorbing species present at very low concentrations in a polluted atmosphere that can also initiate the complex photochemical reactions.

The dissociation energy of NO_2 is 3.1 eV, which corresponds to a photon of wavelength λ = 394.5 nm, and absorption below this wavelength leads to almost complete photodissociation ($\phi \simeq 1$):

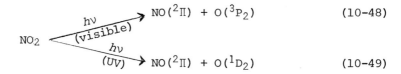

$$NO_2 \xrightarrow[\text{(visible)}]{h\nu} NO(^2\Pi) + O(^3P_2) \quad (10\text{-}48)$$

$$NO_2 \xrightarrow[\text{(UV)}]{h\nu} NO(^2\Pi) + O(^1D_2) \quad (10\text{-}49)$$

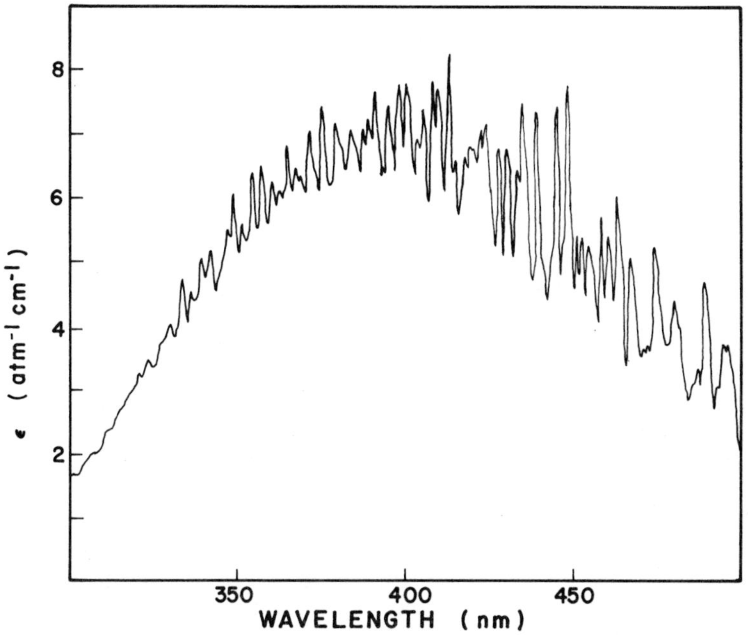

FIGURE 10-6. Absorption spectrum of NO_2 at 25°C. Drawn from data of T. C. Hall, Jr., and F. E. Blacet, J. Chem. Phys. 20, 1745 (1952).

In the absence of other species, molecular oxygen may be formed by a reaction of an O atom with NC_2 [reaction (10-41)], and indeed the quantum yield of O_2 formation approaches unity below 380 nm. [Some dissociation even occurs at wavelengths greater than 394.5 nm, for example, $\Phi_{O_2} = 0.36$ at $\lambda = 404.7$ nm, and thus energetically below the dissociation energy, but this dissociation is strongly temperature-dependent and presumably occurs through internal (rotational and vibrational) energy distribution.] In the presence of molecular oxygen, $O(^3P_2)$ atoms can combine with groundstate O_2 [reaction (10-21)] to produce ozone, and combination of (10-48), (10-21), and (10-40) leads to the quasi-equilibrium

$$NO_2 + O_2 \rightleftarrows O_3 + NO \qquad (10\text{-}50)$$

and hence a steady-state ozone concentration.

As pointed out, NO_2 is a primary light-absorbing species in the atmosphere leading to photochemical smog. A major difficulty until recently with this system has been to explain the very rapid oxidation of NO to NO_2 required to prevent complete depletion of NO_2 in a very short time in sunlight. Thermal oxidation is thermodynamically feasible, but if the reaction is kinetically simple, it must be trimolecular

$$2NO + O_2 \longrightarrow 2NO_2$$
$$k_{298} = 2.1 \times 10^{-38} \text{ cm}^6/\text{molecule}^2 \text{ sec} \qquad (10\text{-}51)$$

requiring two molecules of NO and one molecule of O_2 to come together in a single encounter. This may actually occur at relatively high NO pressures in automobile exhaust systems, but it should not be a significant reaction at normal atmospheric NO pressures, especially since the rate constant for (10-51) is well below that expected if the reaction went at every three-body collision. Nitric oxide is also oxidized to NO_2 by O atoms.

$$NO + O + M \longrightarrow NO_2 + M \qquad (10\text{-}52)$$

However, even though the rate constant is close to triple collisional efficiency, $k = 1.2 \times 10^{-31}$ cm^6/molecule^2 sec when $M = N_2$, this oxidation reaction is unimportant at low-altitude O atom concentrations.

It now appears that the hydroxyl (HO·) and hydroperoxyl (HO_2·) radicals play a key role in the rapid oxidation of NO to NO_2.[6,7] The HO_2· radical is directly involved:

$$NO + HO_2\cdot \longrightarrow NO_2 + HO\cdot \qquad (10\text{-}53)$$

and the HO· radical serves as a chain carrier in several possible free-radical chain mechanisms. The HO· radical has been detected in an urban atmosphere.[8]

Any source of H atoms will produce HO_2· in the atmosphere by the very fast three-body recombination reaction

$$H + O_2 + M \longrightarrow HO_2\cdot + M$$
$$k_{298} = 2 \times 10^{-32}\ \text{cm}^6/\text{molecule}^2\ \text{sec} \qquad (10\text{-}54)$$

Carbon monoxide is a major contributor to a polluted atmosphere, and although it was generally considered to be inert in photochemical smog, it has now been shown to react rapidly (with very low activation energy) with HO·:

$$CO + HO\cdot \longrightarrow CO_2 + H$$
$$k_{298} = 1.7 \times 10^{-13}\ \text{cm}^3/\text{molecule sec} \qquad (10\text{-}55)$$

this reaction also serving as a sink for atmospheric CO. Coupling this reaction with reactions (10-53) and (10-54) gives the CO chain mechanism

$$CO + HO\cdot \longrightarrow CO_2 + H$$

$$H + O_2 + M \longrightarrow HO_2\cdot + M \qquad (10\text{-}56)$$

$$NO + HO_2\cdot \longrightarrow HO\cdot + NO_2$$

A hydroxyl radical leads to the oxidation of a NO molecule to NO_2 and is itself regenerated in the process, so that in principle a single HO· radical can oxidize many NO molecules before the chain is broken by various chain-terminating steps. It is likely, however, that this CO chain does not contribute

[6] *K. J. Demerjian, J. A. Kerr, and J. G. Calvert, The mechanism of photochemical smog formation, in "Advances in Environmental Science and Technology" (J. N. Pitts, Jr., and R. L. Metcalf, eds.), Vol. 4, p. 1. Wiley, New York, 1974.*

[7] *B. J. Finlayson and J. N. Pitts, Jr., Science 192, 111 (1976).*

[8] *C. C. Wang, L. I. Davis, Jr., C. H. Wu, S. Japar, H. Niki, and B. Weinstock, Science 189, 797 (1975).*

to the overall NO oxidation to the extent that similar hydrocarbon chains (discussed below) do in a hydrocarbon-polluted atmosphere.

The CO and hydrocarbon chain reactions can be started by introducing into the system either a HO· radical or a hydrogen atom, and several reactions have been proposed for chain-initiating processes. For example, in a moist atmosphere, nitrous acid (HONO) is formed from NO and NO_2

$$NO + NO_2 + H_2O \longrightarrow 2\ HONO \qquad (10\text{-}57)$$

Like nitrogen dioxide, nitrous acid also absorbs light in the lower visible region, although λ_{max} is in the ultraviolet region (λ_{max} = 360 nm, $\epsilon_{max} \cong$ 2/atm cm). Two photodissociation reactions are proposed leading to either H or HO· chain-carriers:

$$HONO \xrightarrow{h\nu} HO\cdot + NO \qquad (10\text{-}58)$$

$$HONO \xrightarrow{h\nu} H + NO_2 \qquad (10\text{-}59)$$

One or the other of these has a quantum yield close to unity, but either reaction serves as a possible CO chain-initiating step.

A further possible chain-initiating process is the reaction of $O(^1D_2)$ atoms, produced photochemically by reactions such as (10-24), with water

$$O(^1D_2) + H_2O \longrightarrow 2\ HO\cdot$$

$$k_{298} = 1.6 \times 10^{-10}\ cm^3/\text{molecule sec} \qquad (10\text{-}60)$$

The rate of this reaction is approximately one-half that of the ozone-destroying reaction (10-42) for normal concentrations in an urban atmosphere, and therefore needs to be considered in any complete photochemical smog model.

Another $HO_2\cdot$ chain oxidizing process involves the three-body formation of hydrogen peroxide H_2O_2:

$$M + 2HO\cdot \longrightarrow H_2O_2 + M \qquad (10\text{-}61)$$

$$O\ (^1D_2) + H_2O_2 \longrightarrow HO\cdot^* + HO_2\cdot \qquad (10\text{-}62)$$

$$HO\cdot^* + O_3 \longrightarrow HO\cdot + O + O_2 \qquad (10\text{-}63)$$

where $HO\cdot^*$ is an excited hydroxyl radical.

Methane is a natural component of the atmosphere, and a possible chain mechanism with $CH_3O_2\cdot$ as a chain carrier (rather than $HO_2\cdot$) also leading to NO oxidation is

$$HO\cdot + CH_4 \longrightarrow CH_3\cdot + H_2O \qquad (10\text{-}64)$$

$$CH_3\cdot + O_2 + M \longrightarrow CH_3O_2\cdot + M \qquad (10\text{-}65)$$

$$CH_3O_2\cdot + NO \longrightarrow NO_2 + CH_3O\cdot \qquad (10\text{-}66)$$

$$CH_3O\cdot + O_2 \longrightarrow H_2CO + HO_2\cdot \qquad (10\text{-}67)$$

Note that reaction (10-67) is a source of formaldehyde as well as generating the hydroperoxyl radical. It has been pointed out in Section 5.5 that aldehydes are partial hydrocarbon oxidation products from internal combustion engines, furnaces, and power plants, and they are eye irritants in photochemical smog. It is now believed that the direct photolysis of formaldehyde and higher aldehydes are important reactions in initiating formation of photochemical smog in a polluted atmosphere. The gas-phase absorption spectrum of formaldehyde is strongly banded with λ_{max} = 295 nm (ϵ_{max} = 0.4/atm cm), with only very weak absorption near the visible region (for example, $\epsilon \simeq 0.02$/atm cm at 370 nm). This absorption band corresponds to excitation to the $^1(n, \pi^*)$ state, a symmetry-forbidden transition (see Section 9.5.4) accounting for the low extinction coefficient, which then undergoes primary dissociation in two possible manners:

$$CH_2O \xrightarrow{h\nu} {}^1(n, \pi^*) \nearrow HCO\cdot + H \qquad (10\text{-}68)$$

$$\searrow H_2 + CO \qquad (10\text{-}69)$$

Reaction (10-69) is the predominant mode of dissociation at $\lambda > 320$ nm; nevertheless, sufficient radicals are produced by (10-68) to make this reaction competitive with nitrous acid photolysis for CO chain initiation. In addition, reaction (10-68) has the added advantage of providing an additional source of $HO_2\cdot$ from the formyl radical,

$$HCO\cdot + O_2 \longrightarrow HO_2\cdot + CO$$

$$k_{298} = 2 \times 10^{-13} \text{ cm}^3\text{/molecule sec} \qquad (10\text{-}70)$$

which then participates directly in the nitric oxide oxidation reaction (10-53).

If the CO chain mechanism (10-56) is operative, it is clear that enhanced levels of ozone can result from the quasi-equilibrium (10-50) even in a hydrocarbon-free atmosphere as long as it contains significant amounts of CO and NO. It is therefore important that air quality standards be under continual review to reflect the best photochemical smog model.

10.3.3 Reactions in Urban Atmospheres

In an urban atmosphere containing hydrocarbon pollutants, the reaction system becomes much more complex. For example, a mechanism involving over 200 kinetically simple steps is needed to account for the laboratory observed rate of photochemical disappearance of only a single olefinic hydrocarbon in the presence of NO, NO_2, and air. The lifetimes of the different groups of hydrocarbons differ greatly in an irradiated atmosphere, so that the hydrocarbon pollutant concentration varies appreciably from early morning through midday to the late afternoon hours. The olefinic hydrocarbons are found to be the most reactive groups of hydrocarbons in irradiated automobile exhaust gases, reflecting the greater reactivity of these species to chemical oxidants such as ozone, oxygen atoms, and hydroxyl radicals relative to other hydrocarbons. Highly substituted aromatic compounds are also very reactive to these oxidants, however, and this may become a more serious problem in the future if aromatic content is increased in lead-free gasoline to maintain octane number. Under prolonged irradiation, such as occurs on long summer days, even the higher alkanes exhibit some reactivity.

A typical smog chamber run for a single olefin (propene)-nitric oxide-nitrogen dioxide mixture is shown in Fig. 10-7. It is seen that the quasi-equilibrium represented by (10-50) is destroyed on addition of the hydrocarbon, the nitric oxide concentration is greatly reduced by enhanced oxidation to NO_2, the hydrocarbons are oxidized, and ozone is produced (although with a long induction period prior to its buildup). The major secondary pollutants formed from the hydrocarbon photooxidation are aldehydes, ketones, carbon monoxide, carbon dioxide, organic nitrates, and organic oxidants.

The mechanism for the very efficient oxidation of nitric oxide in the presence of hydrocarbons is not clear. Probably alkyl, oxy, and peroxy radicals, generated in the hydrocarbon oxidation by species already described as present in the lower atmosphere (such as ozone, atomic oxygen, and hydroxyl radicals) are involved, although these species alone cannot account for the total hydrocarbon oxidation and nitric oxide conversion. Singlet $O_2(^1\Delta_g)$ has also been proposed as a precursor for such radicals, but at present the evidence

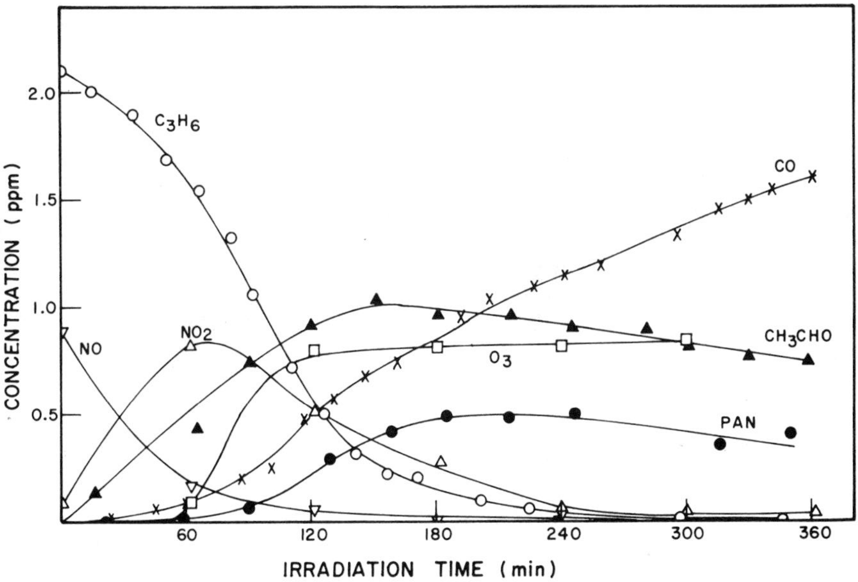

FIGURE 10-7. Typical photochemical smog chamber results for a hydrocarbon-NO-air mixture. Drawn from data of A. P. Altshuller, S. L. Kopczynski, W. A. Lonneman, T. L. Baker, and R. L. Slater, Environ. Sci. Technol. 1, *899 (1967); and in K. J. Demergian, J. A. Kerr, and J. G. Calvert, "The mechanism of photochemical smog formation," in "Advances in Environmental Science and Technology" (J. N. Pitts, Jr., and R. L. Metcalf, eds.), Vol. 4, p. 1. Wiley, New York, 1974.*

seems to indicate that this is not the case, and that singlet oxygen is not an important factor in photochemical smog. However, this conclusion is based on relative $O(^3P_2)/O_2(^1\Delta_g)$ rates and concentrations, and only a change of about one order of magnitude in this ratio, as could result from additional $O_2(^1\Delta_g)$ sources, could greatly modify the relative importance of these two oxidants.

The ozonolysis of olefinic hydrocarbons has been discussed in Section 5.5.2. It was seen that in the gas phase, the reaction produces a molozonide intermediate (5-42) that decomposes into a diradical [reaction (5-44)]. The same diradical as in (5-44) can be formed by addition of oxygen atoms to olefins in the presence of O_2:

$$O + R\text{-}CH = CHR' \longrightarrow [R\text{-}\dot{C}H\text{-}CH(\dot{O})R'] \quad (10\text{-}71)$$

$$[R\text{-}\dot{C}H\text{-}CH(\cdot O)R'] + O_2 \longrightarrow R\text{-}CH(OO\cdot)\text{—}CH(\cdot O)R' \quad (10\text{-}72)$$

In either case, the diradical decomposes into aldehydes and smaller oxygenated diradicals [reaction (5-45)]. The diradicals will react further with oxygen in the atmosphere [reactions (5-46)-(5-61)], and the products of these reactions such as the diradical in (5-47) or the peroxyalkyl radical in (5-49), may oxidize NO:

$$RCH(O\cdot)(O\cdot) + NO \longrightarrow RCHO + NO_2 \quad (10\text{-}73)$$

$$RO_2\cdot + NO \longrightarrow RO\cdot + NO_2 \quad (10\text{-}74)$$

Note that reaction (10-74) is analogous to (10-66) in the chain reaction with methane, and therefore (10-74) is also a source of the hydroperoxyl radical $HO_2\cdot$, thus leading to further oxidation of NO by reaction (10-53). The alkoxy radical ($RO\cdot$) formed in (10-74) may also react with CO to form a nitric oxide oxidation chain mechanism similar to (10-56):

$$RO\cdot + CO \longrightarrow R\cdot + CO_2$$

$$R\cdot + O_2 + M \longrightarrow RO_2\cdot + M \quad (10\text{-}75)$$

$$RO_2\cdot + NO \longrightarrow RO\cdot + NO_2$$

The alkoxy radical is also a source of $HO_2\cdot$ by O_2 hydrogen atom abstraction, comparable to (10-67).

These chain mechanisms are important because they do not remove ozone, and therefore the latter will build up in the reacting system. However, this can only occur after the NO concentration has reached a very low value; thus reaction (10-40), the oxidation of NO by O_3, is unimportant, and this is what is found experimentally as shown in Fig. 10-7. Thus, the reactions given here involving initial olefin oxidation by oxygen atoms and by ozone are able, at least qualitatively, to describe the role of these hydrocarbons in the oxidation of NO to NO_2 in a polluted atmosphere.

Unfortunately, computer simulations of olefinic systems based on the best available rate data indicate that the combined O plus O_3 oxidation reactions are appreciably less than the experimentally observed rates of olefin disappearance. Thus, a further mode of hydrocarbon involvement in nitric oxide oxidation is required, and that role is now considered to be attack by hydroxyl ($HO\cdot$) and hydroperoxyl ($HO_2\cdot$) radicals. With propylene, the reaction is primarily addition to the terminal carbon

$$HO\cdot + CH_3CH=CH_2 \longrightarrow CH_3\dot{C}HCH_2(OH) \qquad (10\text{-}76)$$

$$HO_2\cdot + CH_3CH=CH_2 \longrightarrow CH_3\dot{C}HCH_2(O_2H) \qquad (10\text{-}77)$$

followed by further oxidation by O_2 and subsequent oxidation of NO:

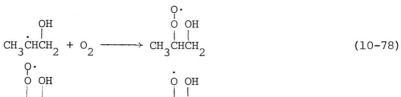

$$CH_3\dot{C}HCH_2(OH) + O_2 \longrightarrow CH_3CH(O\text{-}O\cdot)CH_2(OH) \qquad (10\text{-}78)$$

$$CH_3CH(O\text{-}O\cdot)CH_2(OH) + NO \longrightarrow CH_3CH(\dot{O})CH_2(OH) + NO_2 \qquad (10\text{-}79)$$

and

$$CH_3\dot{C}HCH_2(O_2H) + O_2 \longrightarrow CH_3CH(O\text{-}O\cdot)CH_2(O_2H) \qquad (10\text{-}80)$$

$$CH_3CH(O\text{-}O\cdot)CH_2(O_2H) + NO \longrightarrow CH_3CH(\dot{O})CH_2(O_2H) + NO_2 \qquad (10\text{-}81)$$

There is a complex interaction among the various reactants in a moist polluted atmosphere containing other hydrocarbons (beside olefins) and aldehydes, along with NO, NO_2, and CO. Computer simulations[9] on a model system containing oxides of nitrogen plus CO with *trans*-2-butene representing total olefin, formaldehyde, and acetaldehyde giving the total aldehyde contribution, and methane at natural background level, now appear to suggest that attack by the $HO\cdot$ radical is in fact the most important transient species to initiate

[9] *K. R. Darnall, A. C. Lloyd, A. M. Winer, and J. N. Pitts, Jr., Environ. Sci. Technol. 10, 692 (1976).*

olefin oxidation, with attack by $O(^3P_2)$ atoms being significant only at short times and attack by O_3 or $HO_2\cdot$ contributing only at relatively long times.

We have seen in Section 10.3.2 that hydroxyl radicals are formed in the absence of hydrocarbons from the oxidation of NO by hydroperoxyl radicals [reaction (10-53)], and this oxidation process is also a major generator of $HO\cdot$ radicals in a hydrocarbon atmosphere. Indicative of the large amount of interplay in these systems is the observation in the computer simulation that the olefin also is the major contributor to conversion of $HO\cdot$ to $HO_2\cdot$ initiated either by abstraction of hydrogen from the olefin (for example, propylene)

$$HO\cdot + CH_3CH = CH_2 \longrightarrow \cdot CH_2CH = CH_2 + H_2O \qquad (10\text{-}82)$$

or by addition to the olefin [reaction (10-76)], in which at least one molecule of NO is also oxidized to NO_2 by intermediate $RO_2\cdot$ radicals. It thus appears that the major reactions that oxidize NO to NO_2 in a polluted atmosphere are

$$O_3 + NO \longrightarrow NO_2 + O_2 \qquad (10\text{-}40)$$

$$HO_2\cdot + NO \longrightarrow HO\cdot + NO_2 \qquad (10\text{-}53)$$

$$RO_2\cdot + NO \longrightarrow RO\cdot + NO_2 \qquad (10\text{-}74)$$

We have shown in 10.3.2 that nitrous acid HONO can be photolyzed into H or $HO\cdot$ radicals by absorption in the lower visible region. Although it is not necessary to include this photolysis step to describe the qualitative features of smog formation, inclusion of it does significantly increase the rate of NO_2 production, and hence decrease the time required for NO_2 buildup (as shown in Fig. 10-7). Although HONO has been detected in automobile exhausts and in photolyzed moist air smog chamber mixtures of propylene and nitrogen oxides, at present there are not sufficient data on its concentration levels in polluted atmospheres to determine the full extent of the contribution of nitrous acid photolysis to photochemical smog generation.

Finally, it was also seen in Section 10.3.2 that aldehyde photolysis [typified by reaction (10-68)] is a source of radicals for initiation of the CO chain mechanism (10-56) of nitric oxide oxidation and for direct formation of the hydroperoxyl radical [reaction (10-70)]. Although the photolysis of NO_2 [reactions (10-48) and (10-49)] is the major primary photochemical reaction occurring in the atmosphere, the oxygen atoms produced by it rapidly react with O_2 in the lower atmosphere to form O_3 (10-21), and hence rapidly regenerate NO_2 by (10-40). It would thus appear that NO_2 photolysis should not be a good initiator of photochemical smog, and it may be that the triggering of NO and olefin oxidation follows primarily the aldehyde and/or nitrous acid photolysis. Again, valid photokinetic and thermodynamic data under real polluted atmospheric conditions are needed in order to establish the significance of aldehyde and nitrous acid photolyses in photochemical smog generation.

Only recently have there been detailed studies made of reactions of hydrocarbons other than olefins with the major atmospheric oxidants such as ozone, oxygen atoms, and hydroxyl and hydroperoxyl radicals. Neither the aromatic nor the aliphatic hydrocarbons react significantly with ozone, but they are oxidized by the other species. In fact, it is now believed that highly substituted aromatic compounds may react almost as fast as olefins with O atoms and HO· radicals. Presumably the reaction in either case is one of addition, although the reaction mechanism particularly with the more reactive HO· radical is not clear. Some ring opening may also occur, leading to new conjugated systems that presumably absorb visible light and hence serve as additional photochemical initiators. A reactivity scale for many hydrocarbons found in the atmosphere based on reaction with the HO· radical has recently been proposed.[10]

A major secondary atmospheric pollutant resulting from hydrocarbon oxidation is peroxyacetyl nitrate (PAN). This compound, perhaps the major eye irritant in photochemical smog, is not produced until the nitric oxide is virtually depleted and the ozone induction period is well past (Fig. 10-7). It is easily formed by the direct addition of NO_2 to the peroxyacetyl radical generated by the combination of an acetyl radical and oxygen:

$$CH_3\dot{C}O + O_2 \longrightarrow CH_3C(=O)\text{-}O\text{-}O\cdot \tag{10-83}$$

$$CH_3C(=O)\text{-}O\text{-}O\cdot + NO_2 \longrightarrow CH_3C(=O)\text{-}O\text{-}O\text{-}NO_2\ \text{(PAN)} \tag{10-84}$$

The acetyl radicals may in turn be formed by reactions such as hydrogen abstraction from acetaldehyde,

$$R\cdot + CH_3CHO \longrightarrow RH + CH_3\dot{C}O \tag{10-85}$$

[10] *T. A. Hecht, J. H. Seinfeld, and M. C. Dodge, Environ. Sci. Technol. 8, 327 (1974).*

where R· may be an oxygenated free radical or $NO_3\cdot$ formed by reaction of ozone with NO_2,

$$O_3 + NO_2 \longrightarrow NO_3\cdot + O_2 \qquad (10\text{-}86)$$

or by direct photolysis of acetaldehyde

$$CH_3CHO \xrightarrow{h\nu} CH_3\dot{C}O + H \qquad (10\text{-}87)$$

The slow buildup of PAN may be due in part to the competitive reaction of the peroxyacetyl radical with NO:

$$CH_3\overset{\overset{\displaystyle O}{\|}}{C}\text{-O-O}\cdot + NO \longrightarrow CH_3CO_2\cdot + NO_2 \qquad (10\text{-}88)$$

and in part to reaction (10-86) occurring at relatively high O_3 and NO_2 concentrations.

10.3.4 Summary of Photochemical Smog

Only a very small portion of the reactions possible in photochemical smog have been presented here, but these are representative of the types of mechanisms being suggested to account for photoinitiation, nitric oxide and hydrocarbon oxidation, and for overall product formation in these complex systems. Much success has resulted in recent years through computer modeling, but a clear understanding of photochemical smog can only result from computer simulation of real atmospheric smog conditions based on valid kinetic and thermodynamic data. We can, however, make the following general observations:

(a) Although NO_2 is the major light absorber in a polluted atmosphere, giving photochemical smog its characteristic brownish color, light absorption by nitrous acid, formaldehyde, and aldehydes are now considered to be important photoinitiating reactions in photochemical smog.

(b) The major oxidants in photochemical smog are O_3, O, HO·, and $HO_2\cdot$, with the last two species probably playing the major role in hydrocarbon oxidation and nitric oxide conversion to nitrogen dioxide in the atmosphere; however, there are complex interactions among all the reactive species.

(c) Atmospheric hydrocarbons from automobile emissions and stationary furnaces are oxidized via free-radical mechanisms, and thus they contribute to atmospheric oxidation reactions and lead to formation of eye irritants such as peroxyacetyl nitrate (PAN) and aldehydes.

(d) Although the alkenes are most rapidly photooxidized under atmospheric conditions, O and HO· oxidants also rapidly attack aromatics and (to a lesser extent) alkanes.

(e) Carbon monoxide and water may be involved in chain reactions leading to hydrocarbon and NO oxidation, and therefore these relative major constituents of the atmosphere can no longer be considered inert species.

10.3.5 Inhibition of Photochemical Smog

We have seen in Sections 10.3.2 and 10.3.3 that many of the processes leading to the major components in photochemical smog such as O_3, NO_2, and PAN involve long free-radical chains. A possible mechanism for photochemical smog inhibition therefore might be to add free radical scavengers to the atmosphere, thereby terminating or breaking the chain processes leading to the photochemical oxidants. Such schemes have been considered for many years. One particularly promising free-radical scavenger is diethylhydroxylamine $(C_2H_5)_2NOH$, or DEHA.[11] For example, addition of only 3 mTorr of this compound to a propylene-NO-O_2 mixture (total pressure, 100 Torr) completely inhibits the oxidation of NO to NO_2, while addition of 2.5 mTorr results in 50% inhibition. Apparently inhibition results from radical abstraction of hydrogen:

$$(C_2H_5)_2\,NOH + \cdot OH \longrightarrow H_2O + (C_2H_5)_2NO\cdot \qquad (10\text{-}89)$$

The compound DEHA is a clear, volatile liquid with only a mild odor. So far it has proven to be nontoxic and nonmutagenic, although some urinary animal tests suggest that it may be metabolized into a mutagen. Presumably if it is found to be completely safe, it will be tested by dispersing it into the atmosphere in the early morning hours either by evaporation from cannisters or by spraying it from moving vehicles or airplanes.

There are several arguments to be used against use of such a scheme for inhibiting photochemical smog until much more is known about its long-term effects on humans and on the environment. For example, Pitts[7] has pointed out that structurally DEHA is similar to diethylnitrosoamine $(C_2H_5)_2N\text{-}NO$, a potent carcinogen. It may also be that introduction of a free-radical scavenger at one point will simply delay smog formation until later in the day when the polluted atmosphere has moved downwind, thereby giving the problem to suburbs,

[11] *R. K. M. Jayanty, R. Simonaitis, and J. Heicklen, Atmos. Environ. 8, 1283 (1974).*

rural areas, or to another city. We support the position that extreme caution should be used in introducing one artificial compound into the atmosphere in order to suppress formation of other compounds coming from nonnatural sources.

10.3.6 Sulfur Dioxide

Although sulfur dioxide (SO_2) has so far not been considered a primary contributor to photochemical smog because it is not a major component of automobile exhaust gases, it does provide another potential source of primary light absorption. Most of the reactions in a polluted atmosphere leading to oxidation of SO_2 to sulfur trioxide SO_3, are probably heterogeneous, requiring surfaces such as dust particles to occur. Moist atmospheres very rapidly exhibit increased light scattering when SO_3 is generated, indicating formation of sulfuric acid and particulate sulfate aerosols by the surface-catalyzed reaction

$$H_2O + SO_3 \longrightarrow H_2SO_4 \qquad (10\text{-}90)$$

This is probably one of the major sources of sulfate aerosol in the atmosphere.

The absorption spectrum of SO_2 consists of a very weak absorption band in the near ultraviolet (atmospherically accessible) with λ_{max} = 374 nm, ϵ_{max} = 0.004/atm cm, and a strong ultraviolet band starting at 337 nm (λ_{max} = 294 nm, ϵ_{max} = 10/atm cm).

The bond dissociation energy of SO_2 is 135 kcal/mole (corresponding to λ = 218 nm), and therefore absorption in either band cannot lead to photodissociation to SO and O. The very weak band in the 374 nm region suggests that absorption is the spin-forbidden radiative transition to the lowest-lying triplet state, $^3SO_2$, the ground state of SO_2 being a singlet:

$$SO_2 \xrightarrow{h\nu} {}^3SO_2 \qquad (10\text{-}91)$$

while absorption in the 294 nm band is to the first excited singlet state:

$$SO_2 \xrightarrow{h\nu'} {}^1SO_2 \qquad (10\text{-}92)$$

However, $^1SO_2$ is rapidly converted to the triplet state, either by intersystem crossing

$$^1SO_2 \longrightarrow {}^3SO_2; \; k_{298} = 10^3/\text{sec} \qquad (10\text{-}93)$$

or by collisional induction with SO_2

$$^1SO_2 + SO_2 \longrightarrow {}^3SO_2 + SO_2$$

$$k = 1.7 \times 10^{-12}\ \text{cm}^3/\text{molecule sec} \qquad (10\text{-}94)$$

which occurs at about every collision.

Triplet SO_2 has a very long natural (radiative) lifetime, approximately 10^3 sec, and presumably it is this state that is responsible for much of sulfur dioxide photochemistry and homogeneous oxidation. For example, SO_3 is a product of the reaction of $^3SO_2$ with ground-state SO_2,

$$^3SO_2 + SO_2 \longrightarrow SO_3 + SO \qquad (10\text{-}95)$$

with SO also giving SO_3

$$SO + SO_2 \longrightarrow SO_3 + S \qquad (10\text{-}96)$$

Sulfur trioxide also is formed in the homogeneous reaction with O_2:

$$^3SO_2 + O_2 \longrightarrow SO_3 + O \qquad (10\text{-}97)$$

The photooxidation of SO_2 is much faster in the presence of hydrocarbons and the oxides of nitrogen than in a pollution free atmosphere. Triplet $^3SO_2$ is another potential oxidant of nitric oxide

$$^3SO_2 + NO \longrightarrow SO + NO_2 \qquad (10\text{-}98)$$

followed by

$$NO_2 + SO_2 \longrightarrow SO_3 + NO \qquad (10\text{-}99)$$

with the net reaction being the NO-catalyzed oxidation of $^3SO_2$ by ground-state SO_2:

$$^3SO_2 + SO_2 \xrightarrow{NO} SO + SO_3 \qquad (10\text{-}100)$$

The reaction comparable to (10-100) does not occur between ground-state SO_2 molecules. This then is another example of how a molecule in different electronic states reacts differently, and suggests a potentially important photochemical role of SO_2 in that reaction (10-100) will contribute to the observed rapid removal of SO_2 in photochemical smog systems. However, there are still many uncertainties about the relative importance of various photooxidation steps for SO_2 involving the many free-radical intermediates considered in Section 10.3.3. Radical addition reactions such as

$$HO\cdot + SO_2 \longrightarrow HOSO_2^{\cdot} \qquad (10\text{-}101)$$

$$HOSO_2^{\cdot} + O_2 \longrightarrow HOSO_2O_2^{\cdot} \qquad (10\text{-}102)$$

$$HOSO_2O_2^{\cdot} + NO \longrightarrow HOSO_2O + NO_2 \qquad (10\text{-}103)$$

may be significant in the oxidation of NO to NO_2, but the accurate kinetic data and product analyses needed to assess more fully their roles are not yet available.

BIBLIOGRAPHY

Photochemistry in the Upper Atmosphere

M. Nicolet, An overview of aeronomic processes in the stratosphere and mesosphere, *Can. J. Chem. 52,* 1381 (1974).

P. Crutzen, A review of upper atmospheric photochemistry, *Can. J. Chem. 52,* 1569 (1974).

H. Niki, Reaction kinetics involving O and N compounds, *Can. J. Chem. 52,* 1397 (1974).

D. D. Davis, A kinetic review of atmospheric reactions involving H_xO_y compounds, *Can. J. Chem. 52,* 1405 (1974).

H. S. Johnston and R. Graham, Photochemistry of NO_x and HNO_x compounds, *Can. J. Chem. 52,* 1415 (1974).

S. C. Wofsy and M. B. McElroy, HO_x, NO_x, and ClO_x: Their role in atmospheric photochemistry, *Can. J. Chem. 52,* 1582 (1974).

R. H. G. Reid, Number densities of atomic oxygen and molecular nitrogen in the thermosphere, *Planet. Space Sci. 19,* 801 (1971).

R. Gilpin, H. I. Schiff, and K. H. Welge, Photodissociation of O_3 in the Hartley band. Reactions of $O(^1D)$ and O_2 $(^1\Sigma_g^+)$ with O_3 and O_2, *J. Chem. Phys. 55,* 1087 (1971).

T. C. Degges, Vibrationally excited nitric oxide in the upper atmosphere, *Appl. Optics 10,* 1856 (1971).

K. F. Langley and W. D. McGrath, The ultraviolet photolysis of ozone in the presence of water vapor, *Planet. Space Sci. 19,* 413 (1971).

K. F. Langley and W. D. McGrath, The ultraviolet photolysis of ozone in the presence of molecular oxygen, *Planet. Space Sci. 19,* 416 (1971).

R. F. Hampson, Ed., Survey of photochemical and rate data for twenty-eight reactions of interests in atmospheric chemistry, *J. Phys. Chem. Ref. Data 2,* 267 (1973).

Stratospheric Ozone Reduction by Oxides of Nitrogen

H. S. Johnston, Reduction of stratospheric ozone by nitrogen oxide catalysts from supersonic transport exhaust, *Science 173,* 517 (1971).

H. S. Johnston, Catalytic reduction of stratospheric ozone by nitrogen oxides, in "Advances in Environmental Science and Technology" (J. N. Pitts, Jr., and R. L. Metcalf, eds.), Vol. 4, p. 263. Wiley, New York, 1974.

F. N. Alyea, D. M. Cunnold, and R. G. Prinn, Stratospheric ozone destruction by aircraft-induced nitrogen oxides, *Science 188,* 117 (1975).

Stratospheric Ozone Reduction by Chlorofluoromethanes

M. J. Molina and F. S. Rowland, Stratospheric sink for chlorofluoromethanes: Chlorine atom-catalyzed destruction of ozone, *Nature 249,* 810 (1974).

R. P. Turco and R. C. Whitten, Chlorofluoromethanes in the stratosphere and some possible consequences for ozone, *Atmos. Environ. 9,* 1045 (1975).

R. J. Cicerone, R. S. Stolarski, and S. Walters, Stratospheric destruction by man-made chlorofluoromethanes, *Science 185,* 1165 (1974).

P. H. Howard and A. Hanchett, Chlorofluorocarbon sources of environmental contamination, *Science 189*, 217 (1975).

P. E. Wilkniss, J. W. Swinnerton, R. A. Lamontague, and D. J. Bressan, Trichlorofluoromethane in the troposphere, distribution and increase, 1971 to 1974, *Science 187*, 832 (1975).

V. Ramanathan, Greenhouse effect due to chlorofluorocarbons: Climatic implications, *Science 190*, 50 (1975).

"Halocarbons: Effects on Stratospheric Ozone." Report of the National Research Council (NRC) Panel on Atmospheric Chemistry (1976), and "Halocarbons: Environmental Effects of Chlorofluoromethane Release." Report of the NRC Committee on Impacts of Stratospheric Change (1976). Available from National Academy of Sciences, Printing and Publishing Office, 2101 Constitution Avenue, Washington, DC 20418.

Photochemical Smog

P. A. Leighton, "Physical Chemistry, Vol. IX: Photochemical Aspects of Air Pollution." Academic Press, New York, 1961.

C. S. Tuesday, Ed., "Chemical Reactions in Urban Atmosphere." Elsevier, New York, 1971.

A. P. Altshuller and J. J. Bufalini, Photochemical aspects of air pollution: A review, *Photochem. Photobiol. 4*, 97 (1965).

A. P. Altshuller and J. J. Bufalini, Photochemical aspects of air pollution: A review, *Environ. Sci. Technol. 5*, 39 (1971).

A. J. Haagen-Smit, The control of air pollution, in "Chemistry in the Environment, Readings from Scientific American," p.247. W. H. Freeman, San Francisco, California, 1973.

"Photochemical smog and ozone reactions," Advances in Chemistry Series, No. 113, American Chemical Society, Washington, DC, 1972.

K. J. Demergian, J. A. Kerr, and J. G. Calvert, The mechanism of photochemical smog formation, in "Advances in Environmental Science and Technology" (J. N. Pitts, Jr., and R. L. Metcalf, eds.), Vol. 4, p. 1. Wiley, New York, 1974.

H. Levy II, Photochemistry of the troposphere, in "Advances in Photochemistry" (J. N. Pitts, Jr., G. S. Hammond, and K. Gollnick, eds.), Vol. 9, p. 369. Wiley, New York, 1974.

T. A. Hecht, J. H. Seinfeld, and M. C. Dodge, Further development of generalized kinetic mechanism for photochemical smog, *Environ. Sci. Technol. 8,* 327 (1974).

B. J. Finlayson and J. N. Pitts, Jr., Photochemistry of the polluted troposphere, *Science 192,* 111 (1976).

W. E. Wilson, Jr., E. L. Merryman, A. Levy, and H. R. Taliaferro, Aerosol formation in photochemical smog, *J. Air Pollution Contr. Assoc. 21,* 128 (1971).

W. E. Wilson, Jr., A. Levy, and E. H. McDonald, Role of SO_2 and photochemical aerosol in eye irritation from photochemical smog, *Environ. Sci. Technol. 6,* 423 (1972).

W. S. Cleveland, B. Kleiner, J. E. McRae, and J. L. Warner, Photochemical air pollution: Transport from the New York City area into Connecticut and Massachusetts, *Science 191,* 179 (1976).

J. G. Edinger, Vertical distribution of photochemical smog in Los Angeles basin, *Environ. Sci. Technol. 7,* 247 (1973).

T. H. Maugh II, Air pollution: Where do hydrocarbons come from? *Science 189,* 277 (1975).

M. Corn, R. W. Dunlap, L. A. Goldmuntz, and L. H. Rogers, Photochemical oxidants: Sources, sinks, and strategies, *J. Air Pollution Contr. Assoc. 25,* 16 (1975).

S. L. Kopszynski, R. L. Kuntz, and J. J. Bufalini, "Reactivities of complex hydrocarbon mixtures, *Environ. Sci. Technol. 9,* 648 (1975).

J. M. Pierrard, Photochemical decomposition of lead halides from automobile exhaust, *Environ. Sci. Technol. 3,* 48 (1969).

D. M. Snodderly, Jr., Biomedical and social aspects of air pollution, in "Advances in Environmental Science and Technology" (J. N. Pitts, Jr., and R. L. Metcalf, eds.), Vol. 3, p. 157. Wiley, New York, 1974.

11 PHOTOCHEMISTRY IN THE BIOSPHERE

11.1 INTRODUCTION

Conventional fossil fuels (coal, oil, and gas) represent stored, concentrated energy produced photochemically and biologically eons ago and preserved in a convenient transportable form for present-day usage. We have seen in Chapter 2, however, that although at the present time these fuels account for most of the energy used in the U.S., they are nonreplenishable sources and are of limited amount. It was also pointed out in Chapter 2 that solar energy is the most abundant, continuously replenished form of energy available on earth. It therefore must be considered a serious contender in our search for viable energy sources even though at the present time solar power generation does not appear to be an economically feasible alternative to other means such as nuclear fission. It is the purpose of this chapter to consider radiation from the sun as a source of energy through artificial and natural photochemical processes occuring in the biosphere.

11.2 ARTIFICIAL PHOTOCHEMICAL STORAGE OF SOLAR ENERGY

11.2.1 Energy Conversion and Storage

In Section 2.5.2, several physical methods of solar energy utilization were described in which the energy is collected and stored as heat. The conversion efficiency for these thermal uses of solar energy is subject to the basic limitation of the second law of thermodynamics expressed mathematically in Eq. (2.9a). This states that if a quantity of heat is extracted from a reservoir at temperature T_1, then only part of this can be converted into an amount of work;

the remainder of the heat must pass in the same cyclic operation into a reservoir at a lower temperation T_2. On the other hand, photochemical utilization of sunlight involves the *direct* usage of light in photochemical reactions to produce useful products in the form of stored fuels, etc.; a temperature gradient is not required, and the second law restriction no longer applies.

It turns out that most photochemical reactions of interest to photochemists are exothermic or only slightly endothermic, and involve species that absorb only in the ultraviolet region so that there is an overall large loss in photon energy and consequently very little or no energy storage. The reason for this is that in general photochemists have been interested in producing stable reaction products, and therefore have concentrated on reactions in which the back reaction has a large activation energy in order to prevent appreciable product decomposition. In essence, the light photon in such cases is providing the activation energy for the forward reaction rather than leading to large energy storage. If only a small fraction of the photon energy is lost in a highly endothermic reaction, then the back reaction will have a low activation energy and consequently in general will proceed very rapidly leading to an overall low quantum yield unless an effective and rapid product separation is possible. Reactions leading to a practical photochemical sunlight energy storage system must satisfy the following conditions.

(a) If the absorption of light is efficient (i.e., high quantum yield), then by the principle of microscopic reversibility, the reverse radiative process is also efficient; therefore, nonradiative secondary processes (such as energy transfer, dissociation, or chemical reaction) must occur an order of magnitude faster than the spontaneous emission in order for energy-storing products to be formed.

(b) The overall photochemical reaction must be endothermic, so that the back reaction (which would take place in the dark or in the presence of a suitable catalyst) will give off heat and restore the original reactants for efficient recycling.

(c) The reaction quantum yield (Φ) should be at least 0.1 in order to lead to a realistically efficient storage system.

(d) The exothermic back reaction must occur rapidly under controlled conditions to be practical, but it must have an appreciable activation energy so that it cannot occur so fast that it competes significantly with the photochemical forward reaction under the photolysis conditions (unless the forward reaction products can be easily separated before appreciable back reaction has had a chance to take place during photolysis). Ideally, the most desirable system is one requiring a cheap and stable catalyst or higher temperature for the back reaction to occur, with the elevated temperature being maintained by the heat given off by the reaction.

(e) Light absorption must occur in the visible or near ($\lambda > 350$ nm) ultraviolet region, and preferably over a wide range of wavelengths to utilize efficiently the available solar radiation. (However, the range of light absorption can be extended in some cases by the use of appropriate photosensitizers or organic dyes.) Unfortunately, over half of the available radiant energy from the sun is in the infrared region (See Fig. 4.1), and therefore for the most part photochemically inactive.

(f) Reactants need to be inexpensive, and substances involved in the exothermic back reaction should be easily stored and transported.

(g) In a process involving the direct photochemical conversion of solar energy to electrical energy, the potential energy of the electrons must be raised; if the reverse return of the electrons to lower potential is very fast, no energy storage can take place.

11.2.2 Photolysis of Nitrosyl Chloride

One reaction known for many years[1] that meets most of the above criteria is the photolysis of nitrosyl chloride NOCl:

$$2NOCl \underset{\text{dark}}{\overset{\text{light}}{\rightleftarrows}} 2NO + Cl_2; \quad \Delta H^\circ_{298} = +18 \text{ kcal} \tag{11-1}$$

Nitrosyl chloride absorbs throughout most of the visible region (see Fig. 11-1). The quantum yield for NOCl decomposition in the gas phase has been found to be approximately equal to 2 at all wavelengths below 640 nm. (In solution, the quantum yield is undoubtedly less because of formation of a solvent "cage" around the fragments of the primary photodissociation step leading to very rapid recombination before they have a chance to diffuse apart.) The photodecomposition mechanism is

$$NOCl \xrightarrow{h\nu} NO(^3\Pi) + Cl(^2P_{3/2} \text{ or } _{1/2}) \tag{11-2}$$

[1] *O. S. Neuwirth, J. Phys. Chem 63, 19 (1959).*

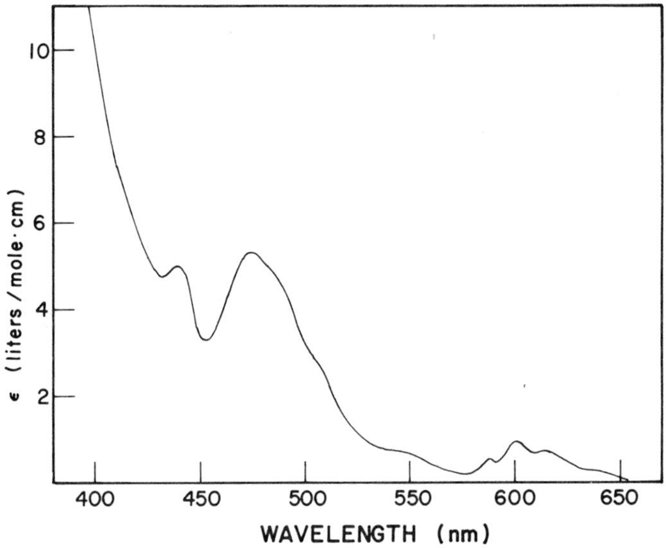

FIGURE 11-1. Absorption spectrum of NOCl. Drawn from data of G. F. Goodeve and S. Katz, Proc. Roy. Soc. (London) A172, *432 (1939).*

$$NOCl + Cl \longrightarrow NO + Cl_2 \qquad (11\text{-}3)$$

This mechanism is consistent with a quantum yield of 2, since two NOCl molecules are decomposed for each photon absorbed. The back reaction to give NOCl takes place rapidly in the dark, so that some means of separating the products and recombining them later is necessary. One method of doing this is to conduct the reaction in a solvent (such as carbon tetrachloride) in which the reactant and only one of the products (in the case of carbon tetrachloride solvent, the chlorine) are soluble, preferably in a flow or dynamic system in which the insoluble product (NO) is continually removed and stored. The back reaction occurs when the NO is bubbled through the chlorine-carbon tetrachloride solution. Unfortunately, much of the solar energy absorbed by this system is lost (or at least is not stored) through the secondary formation of molecular chlorine:

$$2\ Cl \longrightarrow Cl_2; \quad \Delta H_{298} = -58\ \text{kcal} \qquad (11\text{-}4)$$

and the relatively small amount of energy stored in this photoreaction (9 kcal/mole of NOCl) makes it uneconomical at the present time.

11.2.3 Dissociation of Water

The overall dissociation of water into molecular hydrogen and oxygen is endothermic (58 kcal/mole of gaseous water, 68 kcal/mole of liquid water). The reactant water is of course readily available and very cheap, and the molecular hydrogen and oxygen products can be easily stored and recombined in a catalytically controlled 58 kcal/mole exothermic back reaction. Alternately, hydrogen can be separated from oxygen for storage and separate fuel usage. It may very well be that molecular hydrogen will be the most important synthetic fuel of the future, and therefore this splitting of water into hydrogen and oxygen is an extremely important process to consider for potential solar energy utilization.

The dissociation energy of 58 kcal/mole corresponds to a photon of wavelength $\lambda = 493$ nm, while 68 kcal/mole is equivalent to 420 nm. As we have seen in Section 10.2.3, however, the longest possible wavelength of light that can lead to direct water dissociation [into molecular hydrogen and ground-state $O(^2P_3)$ atoms, reaction (10-25)] is 246 nm (116 kcal/mole), which is well below the stratospheric ozone layer cutoff. Thus, direct photodissociation of water is not possible in the biosphere.

The reason for the large energy of dissociation for reaction (10-25) is that the diradical oxygen atom is formed. Since most free radicals cannot be stored, their formation represents energy unavailable for future utilization. A possible solution then is to devise a photosensitized reaction in which intermediate free radicals are not formed.

A reaction along this line that has been extensively studied is the photooxidation of cerous to ceric ions in aqueous perchloric acid solution,

$$Ce^{3+} + H_2O \xrightarrow{h\nu} CeOH^{3+} + H \qquad (11\text{-}5)$$

a reaction that probably involves the hydrated electron as an intermediate species. The hydrogen atom is known to be a strong oxidizing agent for inorganic ions in acid solution, and a possible subsequent thermal reaction that leads to molecular hydrogen formation is

$$H + Ce^{3+} + H_2O \xrightarrow{H^+} CeOH^{3+} + H_2 \qquad (11\text{-}6)$$

Note that one photon absorbed in (11-5) leads in effect to reduction of two water molecules and oxidation of two cerous ions. Ceric ions are known to liberate oxygen from aqueous systems:

$$2CeOH^{3+} \longrightarrow H_2O + 2Ce^{3+} + \tfrac{1}{2} O_2 \tag{11-7}$$

so that combination of (11.5)-(11-7) leads to the net reaction, the Ce^{3+}-photosensitized decomposition of water:

$$H_2O \xrightarrow[Ce^{3+}]{h\nu} H_2 + \tfrac{1}{2} O_2 \tag{11-8}$$

In this reaction, a hydrogen atom is formed instead of an oxygen atom, and the minimum energy necessary for the initiation reaction (11-5) is 86 kcal/mole (333 nm). Actually, cerous solutions absorb well below this wavelength, and the quantum yield of cerous oxidation to ceric is at best only 0.0014 at 254 nm.

Several hypothetical schemes involving no free-radical intermediates have been suggested. In general, they involve forming aqueous binuclear or metal-hydride complexes of specific molecular structures such that the hydrogen and oxygen molecules are formed in the excited state without the free-radical intermediates.[2] It is generally accepted now, however, that effective water photosensitized dissociation will require a two-photon pathway (such as using two sensitizing dyes) in much the same way that photosynthesis utilizes two-photon absorption. This is presented in more detail in Section 11.3.

11.2.4 The Iron-Thionine Photogalvanic System

Another utilization of solar radiation that is not restricted by the second law of thermodynamics is its direct conversion to electrical energy by means of a photogalvanic cell. This type of cell is quite analogous to a thermodynamic galvanic cell in which the chemical energy of a thermodynamically spontaneous reaction is converted directly to electrical energy through oxidation and reduction reactions occurring at electrodes, thus generating the potential for a flow of electrons through an external conducting circuit and the ability to do useful work. For example, in the Daniell cell the spontaneous reduction-oxidation (redox) reaction is

$$Cu^{2+} + Zn \longrightarrow Cu + Zn^{2+} \tag{11.9}$$

[2] *V. Balzani, L. Moggi, M. F. Manfrin, F. Bolletta, and M. Gleria,* Science *189, 852 (1975).*

and the reactions occurring at the two electrodes are $Cu^{2+} + 2e^- \longrightarrow Cu$ at one electrode (the cathode) and $Zn \longrightarrow Zn^{2+} + 2e^-$ at the other electrode (the anode). In a photogalvanic cell, the conditions for a spontaneous redox reaction are produced by absorption of light by a dye molecule D, which in an electronically excited state D* can undergo a one-electron transfer with M^{n+}:

$$D \xrightarrow{h\nu} D^* \tag{11-10}$$

$$D^* + M^{n+} \longrightarrow D^- + M^{(n+1)+} \tag{11-11}$$

If the back reaction (the oxidation of D^-)

$$D^- + M^{(n+1)+} \longrightarrow D + M^{n+} \tag{11-12}$$

is spontaneous, this system constitutes a cyclic coupled photoredox process. The two half-cell reactions

$$D^- \longrightarrow D + e^- \text{ (anode)} \tag{11-13}$$

$$M^{(n+1)+} + e^- \longrightarrow M^{n+} \text{ (cathode)} \tag{11-14}$$

lead to an electrical potential between the two electrodes in the photogalvanic cell and the direct generation of electricity.

One photogalvanic cell that illustrates this technique and has been considered for solar energy-electrical energy conversion involves the photoreduction of the purple dye thionine (absorption spectrum shown in Fig. 11-2) by ferrous ion[3] via the thionine triplet excited state:

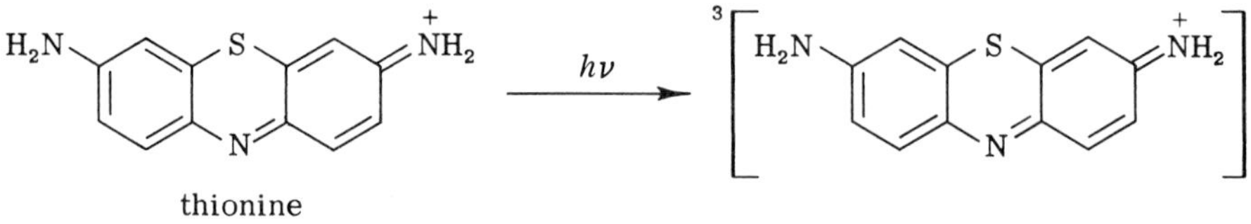

(11-15)

[3] *E. Rabinowitch*, J. Chem. Phys. 8, *551 (1940)*.

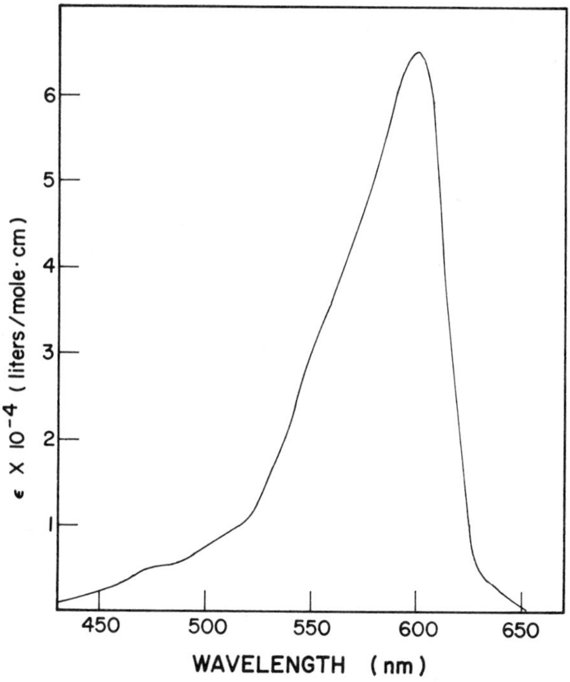

FIGURE 11-2. Absorption spectrum of thionine in 0.1N H_2SO_4. Drawn from data of C. G. Hatchard and C. A. Parker, Trans. Faraday Soc. 57, 1093 (1961).

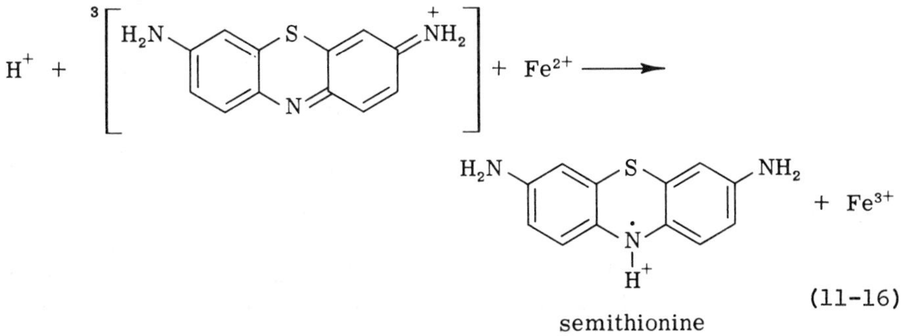

(11-16)

The back reaction between the ferric ion and the half-reduced (semithionine) radical ion is then the photogenerated spontaneous reaction; this back reaction is fast, but slow enough to be observed. If two inert electrodes are placed into a thionine-ferrous ion solution and the region around only one of the electrodes is irradiated, concentration gradients involving the oxidized and reduced forms of thionine and iron are established and an electrical potential of approximately 0.4 V is produced between the two electrodes. The two half-cell reactions ocurring at the electrodes are

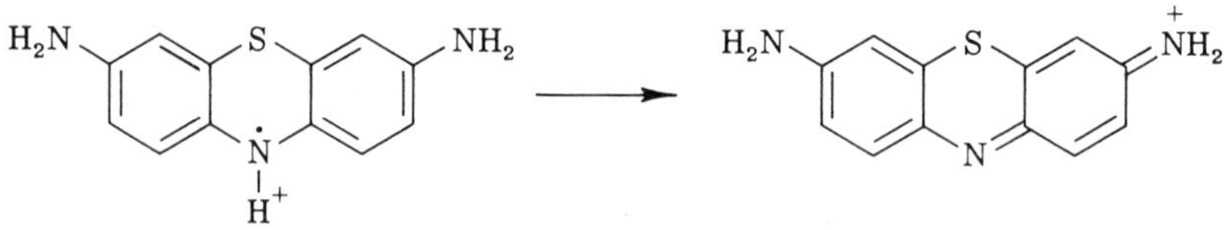

$$+ \; H^+ \; + \; e^- \qquad (11\text{-}17)$$

and

$$Fe^{3+} + e^- \longrightarrow Fe^{2+} \qquad (11\text{-}18)$$

One difficulty with this system is that semithionine reacts with itself in a disproportionation reaction to form leucothionine and thionine:

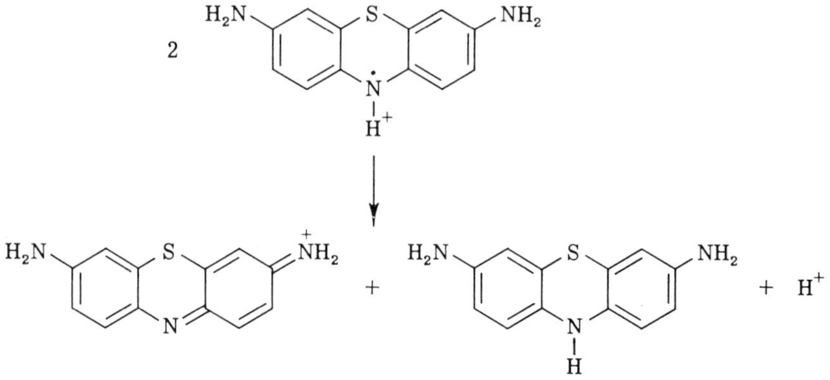

leucothionine (11-19)

The effect of this reaction is to decrease the photostationary-state concentration of semithionine, and hence to reduce the electrical potential of the photogalvanic cell. Some success in retarding the disproportionation reaction apparently has resulted from the use of mixed aqueous-organic solvents.

11.3 PHOTOSYNTHESIS. BIOLOGICAL UTILIZATION OF SOLAR ENERGY

Criteria are given in Section 11.2.1 for a practical photochemical system for energy storage. It turns out that these are met in the natural process of photosynthesis, and it may well be that a system that will provide an effective "artificial" photochemical utilization of solar energy will itself involve some of the biological aspects of this natural process by which growing green plants and certain bacteria store sunlight energy. In this section we shall deal for the most part with the initial act of light absorption by chlorophyll and subsequent primary energy storage and transfer processes.

The overall equation of photosynthesis

$$CO_2 + H_2O \xrightarrow[\text{(chlorophyll)}]{\text{(light)}} (CH_2O) + O_2 + \text{(chemical energy)} \qquad (11\text{-}20)$$

where (CH_2O) represents any carbohydrate or organic matter (in fact, certain rubber plants produce oxygen-free hydrocarbons, and pine trees yield terpenes), has been known for well over 100 years. However, although many aspects of the detailed mechanism are known, we still do not know the full story of all the energy-transfer and photooxidation-reduction interactions. Common to all photosyntheses involving the evolution of oxygen is the presence of the dye molecule chlorophyll a; in addition, many higher plants and algae contain a related molecule, chlorophyll b. Chlorophyll (a or b), shown in Fig. 11-3, is a typical porphyrin derivative with a cyclic structure containing four pyrrole rings I, II, III, and IV (one partially reduced) joined by CH bridges, a fifth five-membered isocyclic ring with a carbonyl group, and a long phytol tail. Within the cyclic structure is a magnesium ion held by four coordinate covalent bonds. In the case of photosynthetic purple bacteria, which do not evolve oxygen, similar dye pigments bacteriochlorophyll a and (in certain cases) b are found (Fig. 11-4).

As would be expected for a molecular species that has evolved into the main light absorber in photosynthesis, chlorophyll absorbs strongly in the long-wavelength red spectral region in which there is a relatively large amount of solar radiation (see Fig. 4-1). The spectra of chlorophyll a and b in ether are given in Fig. 11-5. Each is characterized by two strong absorption bands, one between 400 and 450 nm (the Soret band) that is common to porphyrin derivatives, and one between 650 and 700 nm that is unique to the partially reduced porphine in molecules such as chlorophyll. These bands are red-shifted in the living cell and, in addition, the longer wavelength band becomes broader, suggesting the presence of at least two overlapping bands. For example, very careful analyses and computer simulations of *in vivo* chlorophyll a spectra suggest that the "band" between 650 and 700 nm actually consists of two major bands, at 670 and

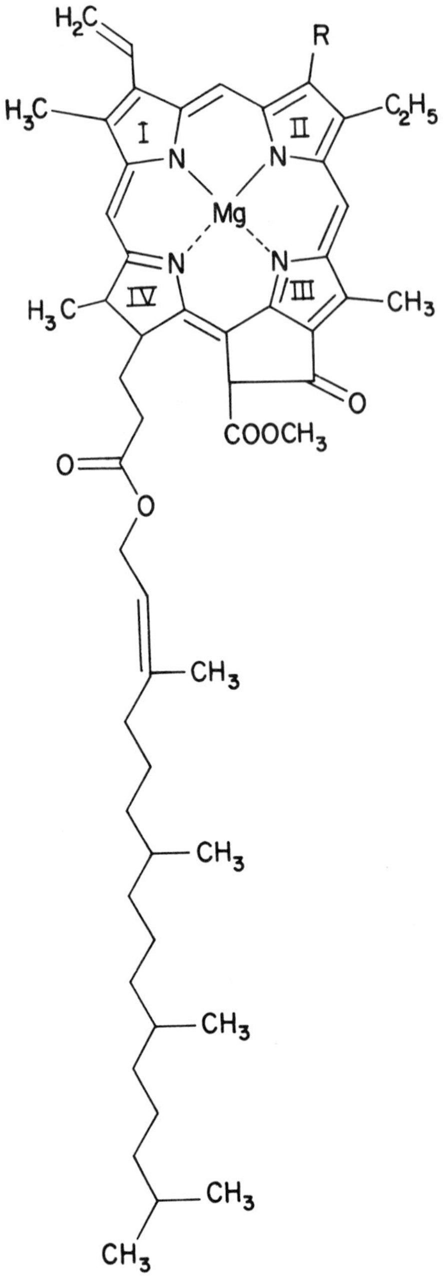

FIGURE 11-3. Molecular structure of chlorophyll: Chlorophyll a: R = CH_3; Chlorophyll b: R = CHO.

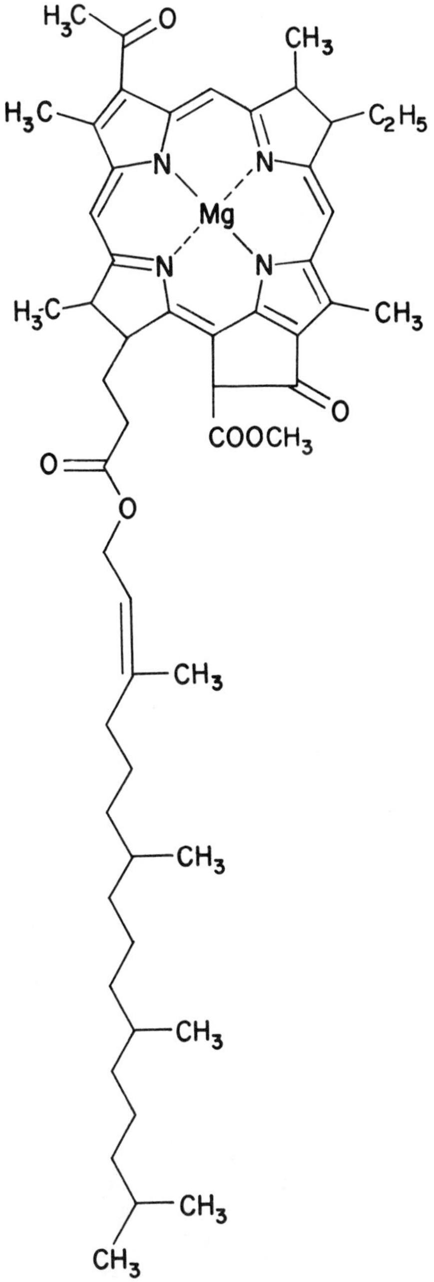

FIGURE 11-4. Molecular structure of bacteriochlorophyll a.

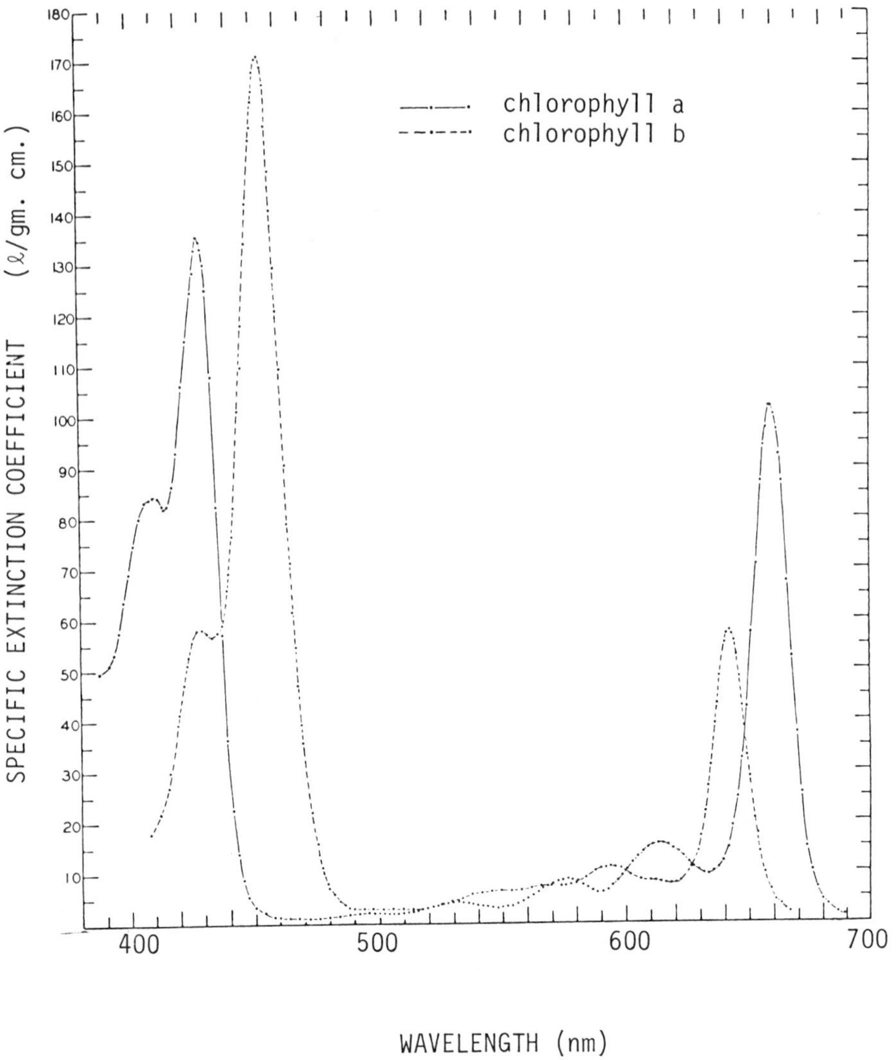

FIGURE 11-5. Absorption spectra of chlorophyll a and b in ether. The molar extinction coefficient ϵ *is obtained by multiplying the specific extinction coefficient by the molecular weight of the chlorophyll. From F. P. Zscheile and C. L. Comar,* Bot. Gaz. 102, *463 (1941)*

680 nm, and at least one minor band at 693 nm. Very sensitive difference spectroscopy measurements, where differences in absorption spectra of plants are determined in darkness and in light, also suggest further very minor components (less than 0.3%) of this red absorption band at 682-690 nm and at 700 nm.

The overall quantum yield of oxygen production in photosynthesis has been the subject of controversy for many years. Figure 11-6 is a typical action spectrum for green algae, which gives the rate of oxygen production as a function of wavelength, compared to the absorption spectrum of the system. This shows that the efficiency of photosynthesis is approximately constant (i.e., oxygen evolution roughly follows the absorption curve) throughout the visible region, falling off however at low wavelengths (presumably due to less efficient absorption at low wavelengths by carotenoid pigments in the living plant) and at wavelengths above ~680 nm (the so-called "red drop"). It is easily shown from bond energies and free energies of formation that the chemical energy stored in reaction (11-20) is approximately 2×10^{-19} cal/molecule of

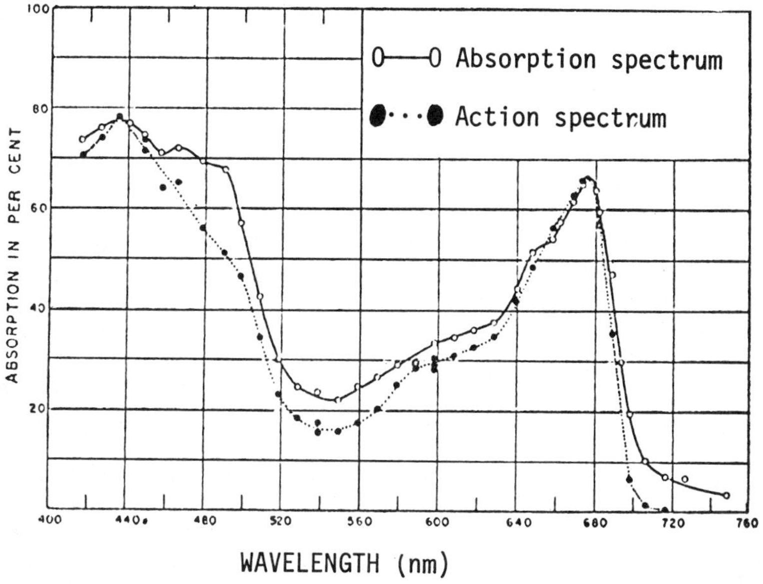

FIGURE 11-6. Typical absorption spectrum and normalized rate of oxygen evolution (action spectrum) for a green alga (Ulva taeniata). *From F. T. Haxo and L. R. Blinks, J. Gen. Physiol. 33, 389 (1950).*

oxygen produced. Since the energy of 680 nm photons [by Eq. (9-3)] is 7×10^{-20} cal/photon, it follows that *at least* 3 photons ($\Phi_{max} = 0.33$) must be absorbed per oxygen molecule evolved. For some time the quantum requirement was thought to be approximately 4, giving a utilization efficiency (that is, energy stored per photon energy absorbed), of about 70-75%. It is now generally accepted, however, that this number of photons is too low by about a factor of 2, and that the maximum quantum yield of oxygen evolution is approximately 0.12.

The absorbing chromophore in chlorophyll is the conjugated tetrapyrrole ring with the complexed Mg ion. Absorption in the 650-700 nm region corresponds (in polar solvents) to (π,π^*) excitation to the lowest excited singlet state, and absorption in the higher energy Soret band results from excitation to a still higher singlet state. Although the fluorescence efficiency of chlorophyll a is quite high in dilute solutions (∿30%), it is only of the order of 5% *in vivo* and competes with photosynthesis, suggesting appreciable singlet-singlet energy transfer. Intersystem crossing to the $^3(\pi,\pi^*)$ state also apparently occurs; this triplet state has been directly detected by flash absorption spectroscopy in dilute solutions in many different solvents, but so far it has not been observed under conditions involving high chlorophyll concentrations in a packed structure as is found in plants.

The absorption and action spectra of photosynthesis (Fig. 11-6) show the presence of photoactive absorbing species in addition to the chlorophylls. Included among these are a variety of molecules called carotenoids [which transfer energy to chlorophyll a with high (20-70%) efficiency, and also prevent deleterious photooxidation reactions sensitized by triplet chlorophyll], and others called phycobilins, which also are involved in accessory light absorption and energy transfer. In many cases, such as in deep ocean water, the presence of these phycobilins permits photosynthesis to occur under very dim blue-green light conditions. The carotenoids belong to the terpenoid group containing an open-chain conjugated polyene system terminating in "ionone" rings, and are either the carotenes (hydrocarbons) or their oxygen derivatives, the xanthophylls (carotenols). The three most common carotenes are the three stereoisomers α, β, and γ carotene. The major carotenoid absorption is in the 400-500 nm region, generally consisting of two or three major bands. The phycobilins are open conjugated four pyrrole rings similar in structure to the bile pigments, and are complexed with proteins forming water-soluble phycobilin-proteins. Two phycobilins found in algae are phycocyanin (blue) and phycoerythrin (red).[4]

Clearly the role of chlorophyll in photosynthesis is quite dependent on its physical environment in the living plant. Photosynthesis occurs in a highly ordered ellipsoidal (4-6 μm long, 1-2 μm thick) structure called the *chloroplast*, which is bounded by a double membrane (the chloroplast envelope) containing all of the light-absorbing pigment molecules. Within this envelope is a complex system of membranes (*lamellae*) in a matrix known as the *stroma* containing a high concentration of lipid and protein. Since the phytol end of the chlorophyll molecule (Fig. 11-3) is soluble in lipid, while the main chlorophyll body complexes with protein, the chlorophyll molecules organize in a closely packed structure.

It has been found that approximately 2500 light-absorbing molecules must collaborate in a group within the chloroplast to evolve one oxygen molecule. Since approximately eight quanta (i.e., $\Phi \cong 0.12$) are required per oxygen produced, this means that a unit of about 300 pigment molecules (the photosynthetic unit) is necessary to absorb and to transmit one quantum of light energy to some "trapping center" where it can join in the initiation of the secondary photosynthetic reactions. The quenching of *in vivo* fluorescence and sensitized fluorescence experiments indicate that energy transmission between like molecules (such as between two chlorophyll a molecules) or between two unlike molecules with overlapping absorption bands--between a carotene molecule and a chlorophyll a molecule, for example--occurs by resonance energy transfer (see Section 9.6). (Similar energy transfer also occurs between molecules without overlapping absorption bands if there is an overlap of the fluorescent band of the donor molecule with the absorption band of the acceptor molecule, leading to a "delayed" resonance between energy states.) In order for this type of energy transfer to take place, the dye molecules must be packed closely together (at least so that their separation is small compared to the wavelength of the absorbed light), a condition satisfied in the chloroplast structure. If all molecules involved in the energy transfer process are identical (or have the same wavelength absorption bands), then transfer is a random process and therefore still

[4]*More details on the chemistry, structure, and biosynthesis of carotenoids and phycobilins may be found in most textbooks on natural products or photosynthesis, such as R. M. Devlin and A. V. Barker, "Photosynthesis." Van Nostrand, New York, 1971.*

very inefficient. But if now there exist in the photosynthetic unit only a few special molecules that absorb at a longer wavelength (lower energy), then most of the pigment molecules will serve as light harvesters or antennas and "funnel" the energy to these few photochemically active species which serve as energy "traps." The reverse process (flow of energy out of the trap) is endothermic, thus requiring an additional input of energy in a temperature-dependent and slow process, so that the photochemically active species is kept almost continually in an electronically excited state. This apparently is what happens in chloroplasts; the photochemically active species, designated P700, is presumably a special form of chlorophyll a (or photochemically special because of its electron-accepting neighbors, thus forming a reaction center) that absorbs at a slightly longer wavelength (700 nm).

So far our discussion of photosynthesis has assumed that only a single photochemical act is involved in initiating the subsequent oxidation-reduction reactions. If this were the case, then the action spectrum for oxygen evolution (Fig.11-6) should follow the absorption spectrum of the plant, modified only in proportion to differences in energy transfer efficiency among the various absorbing molecules. Experimental evidence, such as the red drop in efficiency (or decrease in quantum yield) above 680 nm, the enhancement of oxygen evolution in this region by simultaneous illumination with shorter wavelength light (the Emerson effect), and the fact that the effectiveness of energy transfer from phycobilin absorption is unexpectedly large, suggest, however, that photosynthesis involves the cooperation of two distinct processes with different photochemical functions sensitized by two different photosynthetic units; these are called system I and system II. Although the components of these systems and their roles are not completely known yet, there are several hypotheses about how they function. One hypothesis is as follows: Both systems contain chlorophyll b or phycobilins; system I contains three forms of chlorophyll a (absorbing at ∿670, 680, and 695 nm) and the energy trap P700, while system II also contains chlorophyll a (670 and 680 nm, but not 695 nm) and another energy trap, P680. The two light reactions take place separately but at the same rate (a necessary condition for oxygen evolution to proceed with such a large and uniform quantum efficiency throughout the visible spectrum), and are connected by a relatively slow dark reaction between the products of the two reactions. One difficulty with this scheme is how the two reactions proceed at the same rate regardless of the wavelength of absorbing light even though in some regions the amount of light absorbed by system II is much greater than that by system I. One explanation is that excess energy in system II can "spill over" into system I (but not from I to II because of an energy barrier), which also explains the falloff in quantum yield at $\lambda > 680$ nm (Fig. 11-6) where system I is the more strongly absorbing. Alternately, a "separate package" hypothesis is proposed in which no energy transfer occurs between the two systems, but approximately equal excitation is achieved by different proportions of the same pigment molecules.

Photosynthesis is basically a chemical oxidation-reduction reaction photosensitized by chlorophyll, where carbon dioxide is eventually reduced to carbohydrates and water is oxidized to oxygen. The presence of two distinct but interacting photochemical processes allows the separation of the oxidation and reduction processes in the manner shown in Fig. 11-7. Photoexcitation of system I leads to electron transfer to some as yet unidentified electron acceptor X (reduction) producing the primary photochemical product, which ultimately results in the reduction of CO_2. Similarly, photoexcitation of system II produces an oxidant Z that in some way liberates oxygen from water. The energy difference between the oxidant in system II and the reductant in system I is approximately 1.4 eV. Although this energy difference could be spanned by a single photon (a 680 nm photon corresponds to approximately 1.8 eV), the utilization of the two-photon process allows production of energy-storage products such as adenosine triphosphate ATP in the downhill coupling or linking electron-flow reactions between the two photochemical systems, as shown in Fig. 11-7.

In this very brief treatment of photosynthesis, we have tried to show the various primary photochemical processes of light absorption, electronic excitation, and energy transfer leading up to the initiation of oxidation and reduction by electron transfer, without going into the details of the subsequent or secondary reactions. There are gaps in our understanding of the detailed mechanism of photosynthesis (and one of the major gaps is the process by which oxygen is produced); nevertheless, a great deal has been learned in recent years about the carbon reduction cycle from system I that leads to carbon dioxide reduction, and the techical and economic feasibilities of using the natural or a model photosynthetic cycle as a renewable energy storage and material source are being studied actively. For example, it now appears possible to induce some plants and algae to produce molecular hydrogen instead of reducing carbon dioxide, in essence utilizing solar energy directly to generate hydrogen from water without going through a thermal cycle. Even the well-understood conversion of carbohydrates in the form of such plants as

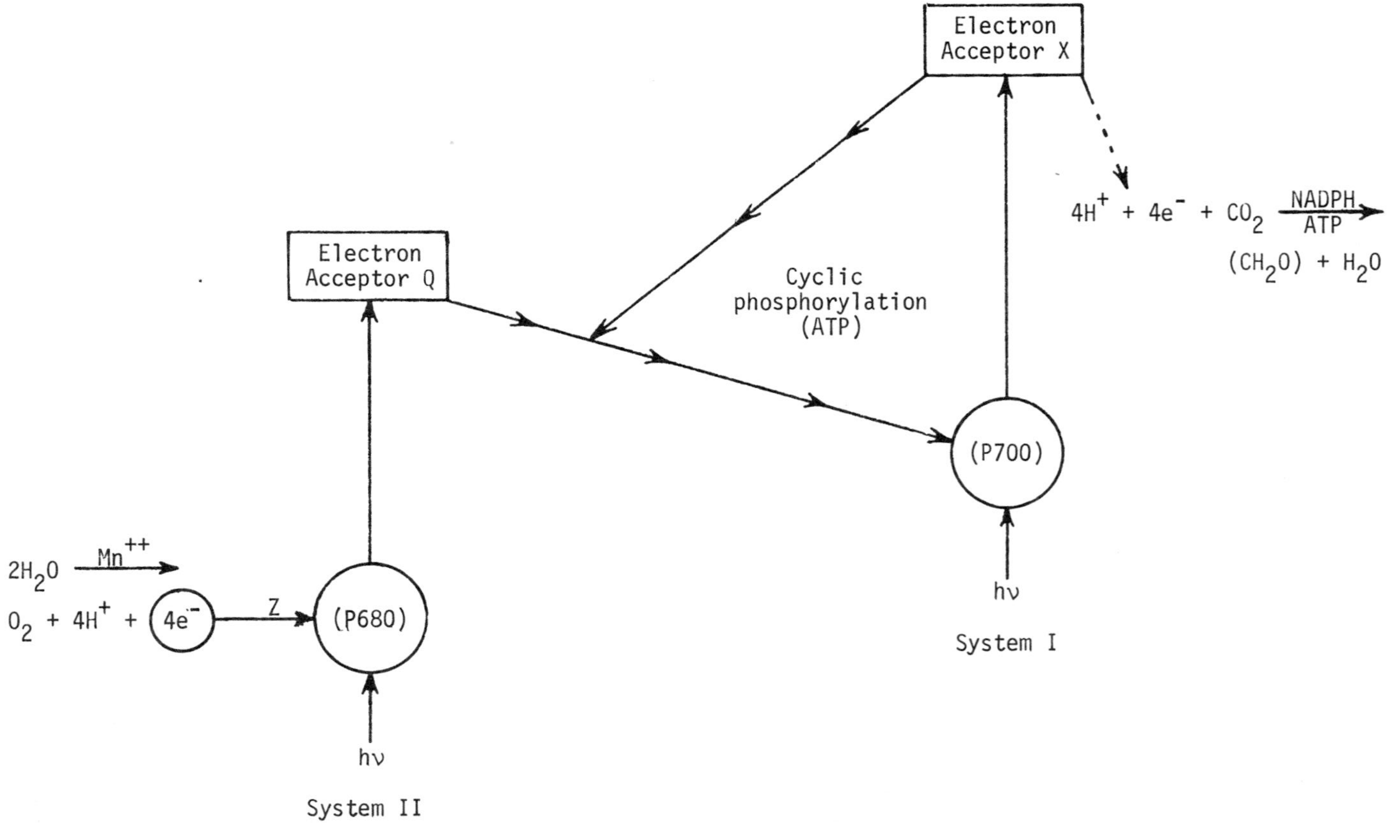

FIGURE 11-7. Schematic representation of cooperation of two photosystems in photosynthesis. Flow of electrons indicated by arrows.

sugar cane and beets into alcohols by fermentation (for use as fuels or fuel supplements) and further into hydrocarbon fuels may now be competitive with fossil fuel usage.[5]

BIBLIOGRAPHY

General

R. K. Clayton, "Light and Living Matter": "The Physical Part," Vol. 1; "The Biological Part," Vol. 2. McGraw-Hill, New York, 1971.

L. J. Heidt, R. S. Livingston, E. Rabinowitch, and F. Daniels, Eds., "Photochemistry in the Liquid and Solid States." Wiley, New York, 1960.

F. Daniels, "Direct Use of the Sun's Energy." Yale University Press, New Haven, Connecticut, 1964.

F. Daniels, Direct use of the sun's energy, *Amer. Scientist 55,* 15, (1967).

M. Wolf, Solar energy utilization by physical methods, *Science 184,* 382 (1974).

B. J. Brinkworth, Solar energy, *Nature 249,* 726 (1974).

J. A. Duffie and N. A. Beckman, Solar heating and cooling, *Science 191,* 143 (1976).

N. N. Lichtin, "The Current State of Knowledge of Photochemical Formation of Fuel." Report of a Workshop Held at Boston University's Osgood Hill Conference Center, North Andover, Massachusetts, Sept. 23-24, 1974.

Artificial Photochemical Utilization of Solar Energy

R. M. Noyes, Kinetic complications associated with photochemical storage of energy, *J. Phys. Chem. 63,* 19 (1959)

[5]*M. Calvin, Science 184, 375 (1974).*

O. S. Neuwirth, The photolysis of nitrosyl chloride and the storage of solar energy, *J. Phys. Chem. 63,* 17 (1959).

V. Balzani, L. Moggi, M. F. Manfrin, F. Bolletta, and M. Gleria, Solar energy conversion by water photodissociation, *Science 189,* 852 (1975).

E. Rabinowitch, The photogalvanic effect. I. The photochemical properties of the thionine-iron system, *J. Chem. Phys. 8,* 551 (1940).

M. D. Archer, Electrochemical aspects of solar energy conversion, *J. Appl. Electrochem. 5,*17 (1975).

Photosynthesis

E. Rabinowitch and Govindjee, "Photosynthesis." Wiley, New York, 1969.

R. M. Devlin and A. V. Barker, "Photosynthesis." Van Nostrand Reinhold, New York, 1971.

R. P. Levine, The mechanism of photosynthesis, in "Chemistry in the Environment, Readings from Scientific American," p. 62. W. H. Freeman, San Francisco, California, 1973. (From *Sci. Amer.,* Dec. 1969.)

Govindjee and R. Govindjee, The primary events of photosynthesis, *Sci. Amer. 231(6),* 68 (1974).

M. Calvin, Solar energy by photosynthesis, *Science 184,* 375 (1974).

M. Calvin, Photosynthesis as a resource for energy and materials, *Photochem. Photobiol. 23,* 425 (1976).

"Synthetic leaf" mimics plants' light conversion, *Chem. Eng. News,* February 16, 1976, page 32.

J. J. Katz and J. R. Norris, Jr., Chlorophyll and light energy transduction in photosynthesis, in "Current Topics in Bioenergetics" (D. R. Sanadi and L. Packer, eds.), Vol. 5, p. 41. Academic Press, New York, 1973.

12
POLYMERS AND PLASTICS

12.1 INTRODUCTION

Synthetic polymers are macromolecules, that is, molecules containing many atoms and of very large size and high molecular weight (ranging from 10,000 to 10,000,000). They are made up of a small number of repeating groups. Many macromolecules are components of living systems; examples are proteins, which are polymers of amino acids, and starch and cellulose, which are polymers of cyclic polyhydroxy compounds (sugars). Since biological polymers are built up in living systems, they can be broken down in these systems (are biodegradable). Biodegradable substances will not cause long-term environmental problems, although the decomposition of large amounts of such materials can be a problem in the short term. In aqueous solution, their degradation will contribute to the biological oxygen demand and use up the dissolved oxygen (see Section 6.6). Under anaerobic conditions, decomposition of proteins will produce undesirable H_2S, NH_3, and CH_4.

Synthetic polymers, on the other hand, are often not biodegradable. In fact, they are chemically tailored for long life in the environment. Their uses are legion, from the manufacture of toy trucks to communication satellites, including house siding, fibers, wire insulation and furniture. Although environmental stability is desirable with these items, there are many instances where a polymeric material is used once and then discarded; e.g., as plastic bottles, food wrappers, and plastic cups. In 1971 over 1 million tons of disposable plastic items were produced in the U.S. and Great Britain. Those polymers designed for long life have often been used to manufacture these disposable items. As these "throw-away" items accumulate in the environment, it has become apparent that biodegradable polymers (coupled with a greater public concern for the environment) are needed to prevent the accumulation of plastic litter on roadsides, in parks, and on beaches. As a consequence, a few polymeric materials recently have been designed that are subject to rapid environmental decay for use in the manufacture of disposable plastic items.

12.2 FORMULATION OF SYNTHETIC POLYMERS

Synthetic polymers are made up of a great number of simple units (monomers) joined together in a regular fashion. Some examples are given in Table 12-1. Most plastic items are not formulated from the polymeric material alone. They usually contain lower molecular weight compounds called plasticizers which increase their flow and processibility. The use of PCBs for this purpose was mentioned in Chapter 7. Esters of phthalic acid are other examples of plasticizers. Fillers, inert materials such as wood flour, cellulose, ground mica, asbestos, or glass fiber, are often added to increase the strength of the plastic or to provide bulk at low cost. Plasticizers may be leached or evaporated from plastic materials in use (the film that deposits on the inside of the windows of new cars in warm weather is caused by plasticizer distilled from the plastic of the car interior), while both plasticizers and fillers may contribute to the environmental disposal and degradation problems. The release of PCBs on incineration was referred to in Chapter 7. Fillers may be inert, but if released as dust during incineration of wastes, they can contribute to atmospheric particulate material and, in the case of asbestos particles, be health hazards in view of the known connection between asbestos dust and lung cancer.

The bulk of the monomers used in polymer synthesis are obtained from petroleum. Therefore, our present dependence on plastic materials is ultimately a dependence on a source of crude oil.

12.3 POLYMER SYNTHESIS

Polymers may be formed from monomers in chain-reaction or step-reaction polymerization. The chain-reaction process is illustrated by the free-radical polymerization of ethylene or its derivatives. The specific example of the polymerization of acrylonitrile to orlon is given in reactions (12-1)-(12-5). In free-radical polymerization, a small amount of initiator, such as a peroxide, is added to the monomer to start the polymerization. The initiator decomposes to form radicals (12-1) which add to the monomer in the chain-initiating step (12-2). Further addition of monomer takes place in the chain-propagating step (12-3). Polymer growth

TABLE 12-1

Some Typical Polymers and the Monomers from which They are Formed

	Repeat Unit	Monomer(s)
Polyethylene (Marlex)	$-CH_2CH_2-$	$CH_2{=}CH_2$
Polytetrafluorethylene (Teflon)	$-CF_2CF_2-$	$CH_2{=}CF_2$
Polypropylene	$-CH_2CH(CH_3)-$	$CH_2{=}CHCH_3$
Poly(vinyl chloride)	$-CH_2CHCl-$	$CH_2{=}CHCl$
Poly(vinylidine chloride) (Saran)	$-CH_2CCl_2-$	$CH_2{=}CCl_2$
Polystyrene	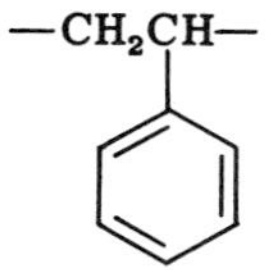	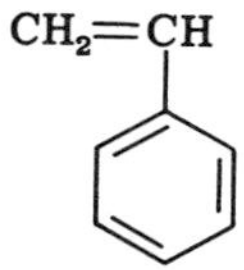
Polycaprolactam (Nylon 6) (Perlon, Caprolon)	$-NH(CH_2)_5-C(=O)-$	$NH_2(CH_2)_5CO_2H$

TABLE 12-1 (Continued)

	Repeat Unit	Monomer(s)
Nylon 66 (copolymer of adipic acid and hexamethylene-diamine)	$-C(=O)(CH_2)_4C(=O)-NH(CH_2)_6NH-$	$HO_2C(CH_2)_4CO_2H$ and $NH_2(CH_2)_6NH_2$
Polychloroprene (Neoprene)	$-CH_2-C(Cl)=CHCH_2-$	$CH_2=C(Cl)-CH=CH_2$
Silicones	$-Si(R)_2-O-$	$HO-Si(R)_2-OH$
Poly(cis-1,4-isoprene) (Natural rubber)	$-CH_2-C(H_3C)=C(H)-CH_2-$	$CH_2=C(CH_3)-CH=CH_2$
Poly(methylmethacrylate) (Plexiglas, lucite)	$-CH_2-C(CH_3)(C(=O)OCH_3)-$	$CH_2=C(C(=O)OCH_3)-CH_3$

TABLE 12-1 (Continued)

	Repeat Unit	Monomer(s)
Poly(ethylene terephthalate) (Dacron, Terylene, Trevira, Mylar)	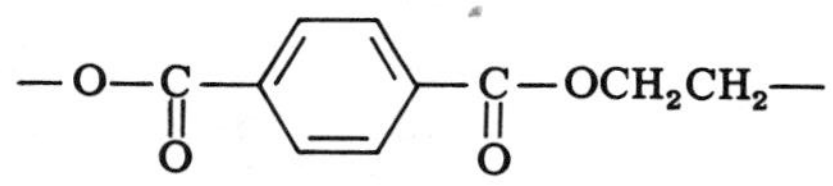	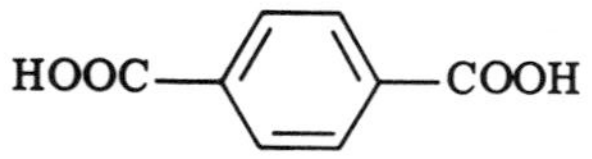 and HO—CH_2CH_2—OH
Lexan (a polycarbonate)	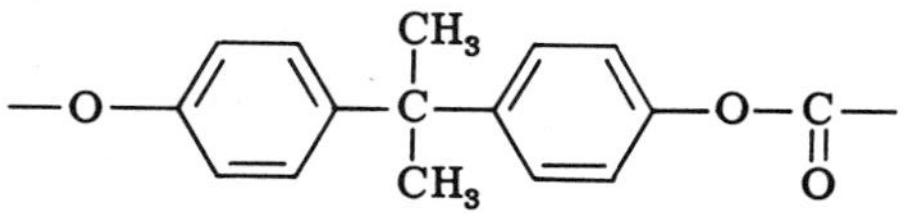	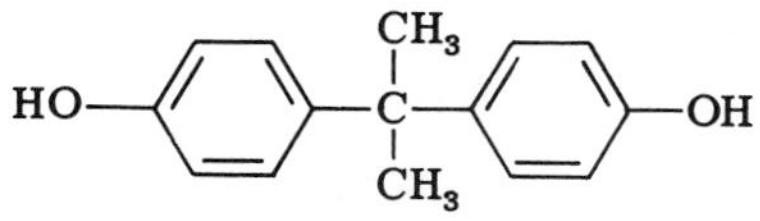 and 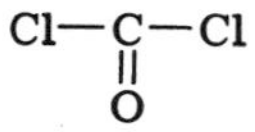
Phenol-formaldehyde resins (Bakelite)	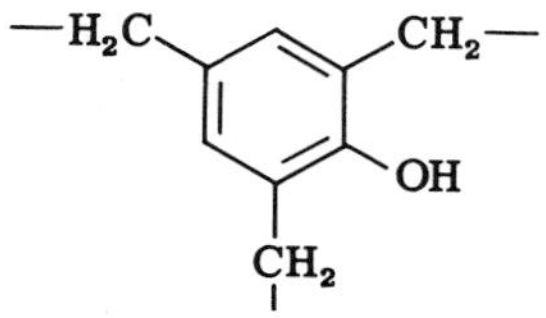	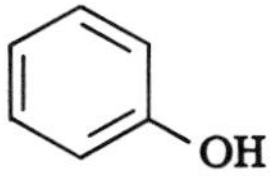 and HCHO

stops in the chain-terminating steps which require the reaction of two radicals such as the disproportionation reaction shown in 12-5.

Initiation

$$ROOR \longrightarrow 2RO\cdot \qquad (12\text{-}1)$$

$$RO\cdot + CH_2 = \underset{\displaystyle CN}{\underset{|}{CH}} \longrightarrow ROCH_2\underset{\displaystyle CN}{\underset{|}{CH}}\cdot \qquad (12\text{-}2)$$

Propagation

$$ROCH_2\underset{\displaystyle CN}{\underset{|}{CH}}\cdot + CH_2 = \underset{\displaystyle CN}{\underset{|}{CH}} \longrightarrow ROCH_2\underset{\displaystyle CN}{\underset{|}{CH}}CH_2\underset{\displaystyle CN}{\underset{|}{CH}}\cdot \qquad (12\text{-}3)$$

Termination

$$2ROCH_2\underset{\displaystyle CN}{\underset{|}{CH}}(CH_2\underset{\displaystyle CN}{\underset{|}{CH}})_nCH_2\underset{\displaystyle CN}{\underset{|}{CH}}\cdot \xrightarrow{\text{combination}}$$

$$ROCH_2\underset{\displaystyle CN}{\underset{|}{CH}}(CH_2\underset{\displaystyle CN}{\underset{|}{CH}})_nCH_2\underset{\displaystyle CN}{\underset{|}{CH}}\text{-}\underset{\displaystyle CN}{\underset{|}{CH}}CH_2(\underset{\displaystyle CN}{\underset{|}{CH}}CH_2)_n\underset{\displaystyle CN}{\underset{|}{CH}}CH_2OR \qquad (12\text{-}4)$$

$$2ROCH_2\underset{\displaystyle CN}{\underset{|}{CH}}(CH_2\underset{\displaystyle CN}{\underset{|}{CH}})_nCH_2\underset{\displaystyle CN}{\underset{|}{CH}}\cdot \xrightarrow{\text{disproportionation}}$$

$$ROCH_2\underset{\displaystyle CN}{\underset{|}{CH}}(CH_2\underset{\displaystyle CN}{\underset{|}{CH}})_nCH{=}\underset{\displaystyle CN}{\underset{|}{CH}} + ROCH_2\underset{\displaystyle CN}{\underset{|}{CH}}(CH_2\underset{\displaystyle CN}{\underset{|}{CH}})_nCH_2\underset{\displaystyle CN}{\underset{|}{CH_2}} \qquad (12\text{-}5)$$

Ionic processes may also be involved in chain reaction polymerization. Cationic polymerization of vinyl monomers is initiated by acids. This process has already been discussed in the synthesis of the ABS detergents from propene (Section 6.4). The molecular weight of the oligomer or polymer will depend on the reaction conditions and the ratio of acid to olefin. Anionic polymerization is initiated by strong bases such as butyl lithium ($CH_3CH_2CH_2CH_2^-Li^+$, which is abbreviated below as *n*-BuLi). A carbanion intermediate is formed (12-6) that continues to add monomer (12-7) until the carbanion is destroyed by the addition of a proton or the elimination of a negatively charged group (11-8).

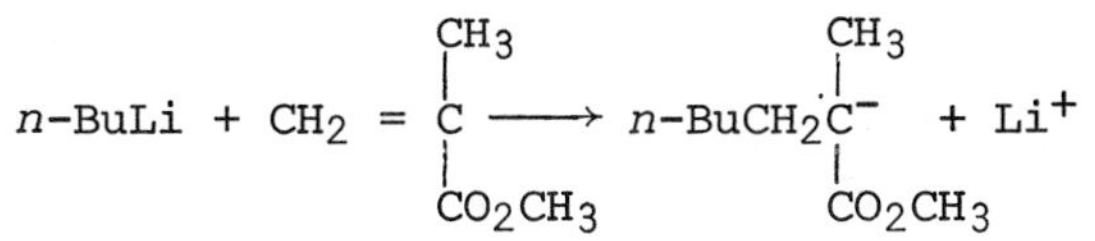

$$n\text{-BuLi} + CH_2 = \overset{\overset{\displaystyle CH_3}{|}}{\underset{\underset{\displaystyle CO_2CH_3}{|}}{C}} \longrightarrow n\text{-BuCH}_2\overset{\overset{\displaystyle CH_3}{|}}{\underset{\underset{\displaystyle CO_2CH_3}{|}}{C^-}} + Li^+ \qquad (12\text{-}6)$$

$$n\text{-BuCH}_2\overset{\overset{\displaystyle CH_3}{|}}{\underset{\underset{\displaystyle CO_2CH_3}{|}}{C^-}} + CH_2 = \overset{\overset{\displaystyle CH_3}{|}}{\underset{\underset{\displaystyle CO_2CH_3}{|}}{C}} \longrightarrow n\text{-BuCH}_2\overset{\overset{\displaystyle CH_3}{|}}{\underset{\underset{\displaystyle CO_2CH_3}{|}}{C}}\text{-CH}_2\overset{\overset{\displaystyle CH_3}{|}}{\underset{\underset{\displaystyle CO_2CH_3}{|}}{C^-}} \qquad (12\text{-}7)$$

$$n\text{-Bu}\{CH_2\overset{\overset{\displaystyle CH_3}{|}}{\underset{\underset{\displaystyle CO_2CH_3}{|}}{C}} \xrightarrow{\quad}_n CH_2\overset{\overset{\displaystyle CH_3}{|}}{\underset{\underset{\displaystyle CO_2CH_3}{|}}{C}}\text{-} + XH$$

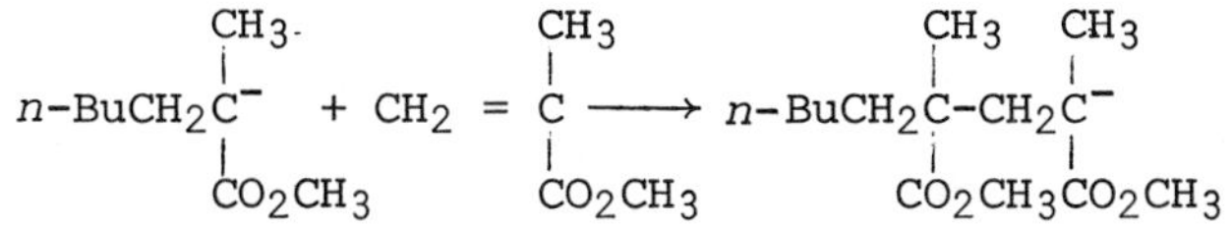

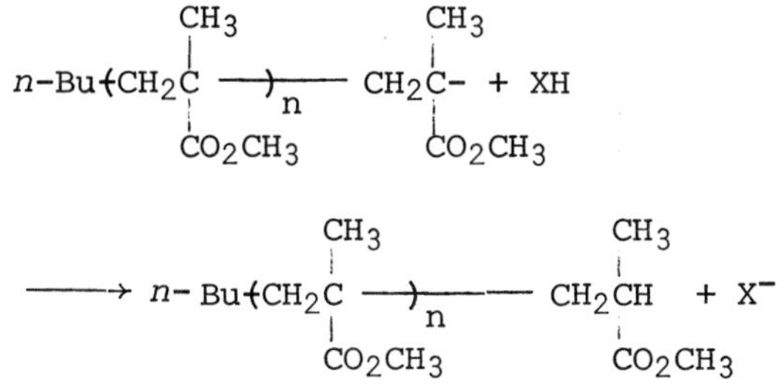

$$\longrightarrow n\text{-Bu}\{CH_2\overset{\overset{\displaystyle CH_3}{|}}{\underset{\underset{\displaystyle CO_2CH_3}{|}}{C}} \xrightarrow{\quad}_n CH_2\overset{\overset{\displaystyle CH_3}{|}}{\underset{\underset{\displaystyle CO_2CH_3}{|}}{CH}} + X^- \qquad (12\text{-}8)$$

In chain-reaction polymerization the molecular weight of the final polymer is dependent on the maintenance of the active species be it the radical, cation, or anion formed initially. In step-reaction polymerization, each reaction that adds a new monomer unit is independent of the preceding one. The formation of nylon 66 illustrates this process:

$$\overset{\overset{\displaystyle O}{\|}}{HOC}(CH_2)_4\overset{\overset{\displaystyle O}{\|}}{COH} + NH_2(CH_2)_6NH_2 \xrightarrow{-H_2O} \overset{\overset{\displaystyle O}{\|}}{HOC}(CH_2)_4\overset{\overset{\displaystyle O}{\|}}{CNH}(CH_2)_6NH_2$$

$$\xrightarrow[-H_2O]{NH_2(CH_2)_6NH} NH_2(CH_2)_6\overset{\overset{\displaystyle O}{\|}}{NHC}(CH_2)_4\overset{\overset{\displaystyle O}{\|}}{CNH}(CH_2)_6NH_2$$

$$\xrightarrow[-H_2O]{\overset{\overset{\displaystyle O}{\|}}{HOC}(CH_2)_4\overset{\overset{\displaystyle O}{\|}}{COH}} NH_2(CH_2)_6\overset{\overset{\displaystyle O}{\|}}{NHC}(CH_2)_4\overset{\overset{\displaystyle O}{\|}}{CNH}(CH_2)_6\overset{\overset{\displaystyle O}{\|}}{NHC}(CH_2)_4\overset{\overset{\displaystyle O}{\|}}{COH}$$

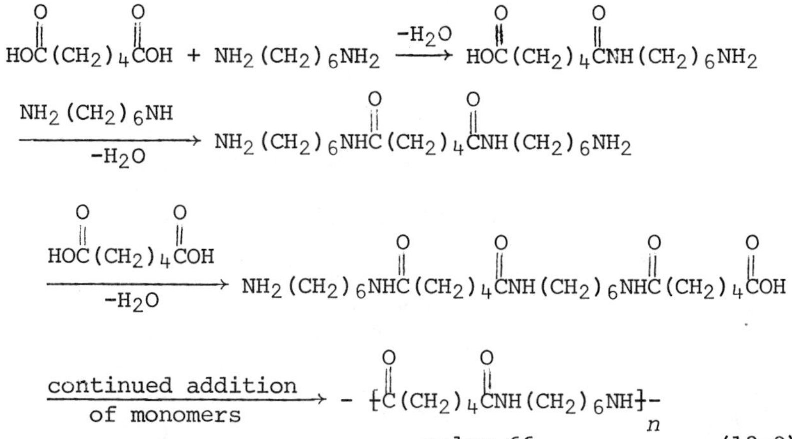

$$\xrightarrow[\text{of monomers}]{\text{continued addition}} - \{\overset{\overset{\displaystyle O}{\|}}{C}(CH_2)_4\overset{\overset{\displaystyle O}{\|}}{CNH}(CH_2)_6NH\}-_n$$
$$\text{nylon 66} \qquad (12\text{-}9)$$

The polyester Dacron is prepared in a similar fashion from ethylene glycol and terephthalic acid.

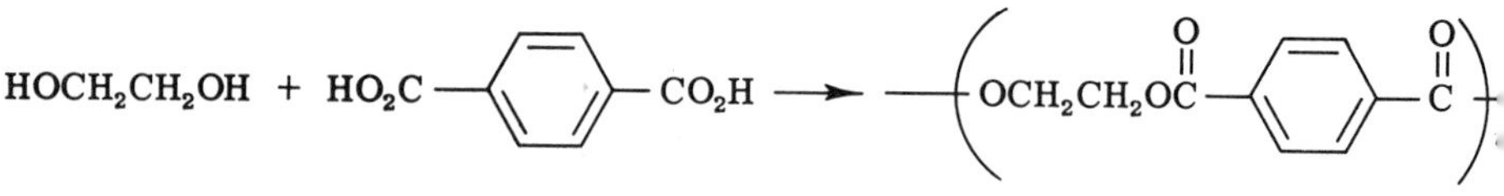

dacron

(12-10)

These stepwise polymerizations are examples of condensation polymerization. A small molecule, water here, is eliminated in each step.

12.4 ENVIRONMENTAL DEGRADATION OF SYNTHETIC POLYMERS

Many hydrocarbon polymers have chemical reactivity similar to that of high boiling petroleum fractions and, as a consequence, these compounds are very persistent in the environment. Of the polymers currently used in large amounts, only the polyesters that contain alphatic ester groupings are susceptible to microbial destruction. Vinyl polymers appear to degrade under environmental conditions but this is often due to the degradation of the phthalate ester plasticizer which results in a change in the mechanical properties of the polymer. It becomes brittle and disintegrates, but the actual polymer molecules are not decomposed.

Some evidence has been accumulated that suggests that polymers are resistant to degradation because they cannot be absorbed into the microbial cell; this is the same reason that the long chain hydrocarbon molecules in petroleum are not degraded (Chapter 5). Polyethylene and linear hydrocarbons are not degraded if they have a molecular weight greater than 450 (about 32 carbon atoms). It has been suggested that polyesters are degraded because extracellular enzymes are produced which cleave the polymer to units that are small enough to be assimilated by the microorganisms.

The successful environmental degradation of most polymers requires a process for cleaving the polymer into units that are sufficiently small for assimilation by microorganisms. Sunlight will effect the partial degradation of some polymers to smaller units that are biodegradable. However, solar radiation is only effective in degrading those polymers that absorb at wavelengths greater than 290 nm since the more energetic *uv* light of shorter wavelength is absorbed by the ozone layer in the atmosphere (see Chapter 4). Consequently, these photochemical reactions are of limited importance in the environmental degradation of polymers. The precise mechanisms of these photochemical reactions are not known but the cleavages may follow one or more of the pathways outlined below. Simple photochemical cleavage is the least likely route for polymer degradation since most polymers do not absorb light at wavelengths as long as 290 nm. Even polystyrene, which contains light-absorbing phenyl rings, undergoes little decomposition by this route. The main pathways for the environmental degradation of commercially important synthetic polymers are:

(a) Light absorption by a complex between the polymer and oxygen: A complex formed between oxygen and the polymer will sometimes shift the absorption of the polymer to wavelengths greater than 290 nm. A hydrogen atom is then abstracted from the photochemically activated polymer by the oxygen molceule to form a radical [R· in reaction (12-11)]. The radical then reacts further with ground-state molecular oxygen as shown in ractions (12-11)-(12-16).

$$R\cdot + O_2 \rightarrow RO_2\cdot \quad (12\text{-}11)$$

$$RO_2\cdot + RH \rightarrow RO_2H + R\cdot \quad (12\text{-}12)$$

$$RO_2H \rightarrow RO\cdot + \cdot OH \quad (12\text{-}13)$$

cleavage products

$$RO\cdot + RH \rightarrow ROH + R\cdot \quad (12\text{-}14)$$

$$HO\cdot + RH \rightarrow HOH + R\cdot \quad (12\text{-}15)$$

$$2\ RO_2\cdot \rightarrow \text{cleavage products} \quad (12\text{-}16)$$

The process outlined in (12-11)(12-16) may also be initiated by the mechanical shearing of the polymer during processing or by attack by ozone (Chapter 5). The oxidative processes initiated by shearing may result in the loss in strength of the polymer, but the extent of the oxidation is not usually large enough to give cleavage products that are sufficiently small for biodegradation.

(b) Photodecomposition triggered by peroxides: Peroxides are formed during the processing of polymers (e.g., mechanical shearing) as outlined in (12-11) and (12-12). The *uv* absorption of peroxides extends to the long wave-length side of 300 nm so they absorb sunlight. The peroxides are cleaved photochemically with a quantum yield close to one to form alkoxyl and hydroxyl radicals (12-13). The radicals formed react further as outlined in reactions (12-14)-(12-16).

(c) Decomposition triggered by carbonyl groups: Carbonyl groups that are present as a result of decomposition of polymer peroxides will absorb sunlight in the vicinity of 300 nm. The excited carbonyl groups can initiate reactions which lead to the cleavage of the polymer. These reactions will be discussed in greater detail in Section 12.5.

12.5 DESIGN OF ENVIRONMENTALLY DEGRADABLE POLYMERS

12.5.1 Introduction

The processes outlined above do not lead to the complete environmental degradation of substances fabricated from polymers, at least not in a reasonable length of time. One has only to walk along the high water mark on almost any beach and it will be apparent from the abundance of items constructed from polystyrene and other plastics that degradation takes place very slowly. Approaches to the design of polymers that undergo a rapid photodegradation have been described recently. Two methods utilized for the design of degradable polymers are outlined below.

12.5.2 Incorporation of Carbonyl Groups in the Polymer

The light energy absorbed by the carbonyl groups of aldehydes and ketones will bring about the cleavage of carbon-carbon bonds. The two processes that can occur are known as the "Norrish type I" and "Norrish type II" reactions. In the Norrish type I reaction (12-17) bond breaking occurs by a radical pathway while a concerted electron shift occurs in the type II reaction (12-18). The net result of either process is the cleavage of a carbon-carbon bond next to the carbonyl group.

$$R\text{-}\overset{\overset{\displaystyle O}{\|}}{C}\text{-}R' \xrightarrow{h\nu} R\cdot + \cdot\overset{\overset{\displaystyle O}{\|}}{C}\text{-}R' \text{ and } R\text{-}\overset{\overset{\displaystyle O}{\|}}{C}\cdot + \cdot R' \qquad (12\text{-}17)$$

$$\cdot\overset{\overset{\displaystyle O}{\|}}{C}\text{-}R' \rightarrow CO + R\cdot' \qquad R\text{-}\overset{\overset{\displaystyle O}{\|}}{C}\cdot \rightarrow R\cdot + CO$$

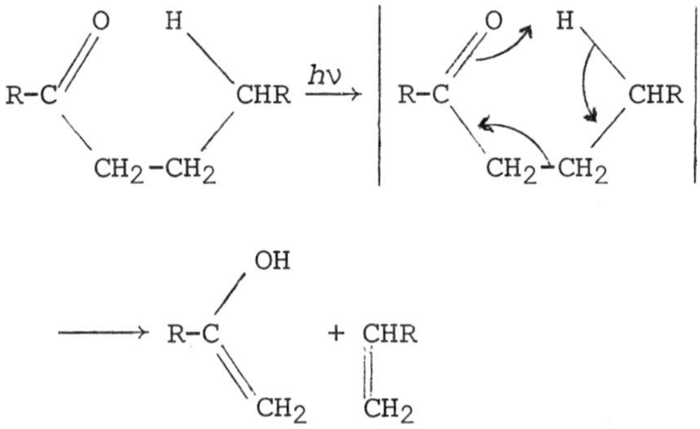

(12-18)

While many polymers contain small amounts of carbonyl groups that may be formed by the slow decomposition of the peroxides produced during the processing of the polymer (Section 12.4), in general there is not a sufficient number of these carbonyl groups to result in photodegradation. It has been found, however, that the copolymerization of small amounts (∿1%) of carbonyl containing monomers together with the normal monomer results in a photodegradable polymer. For example, phenyl vinyl ketone can be polymerized together with the styrene monomer to produce a polystyrene which contains 1% phenyl ketone residues. The small number of phenyl ketone

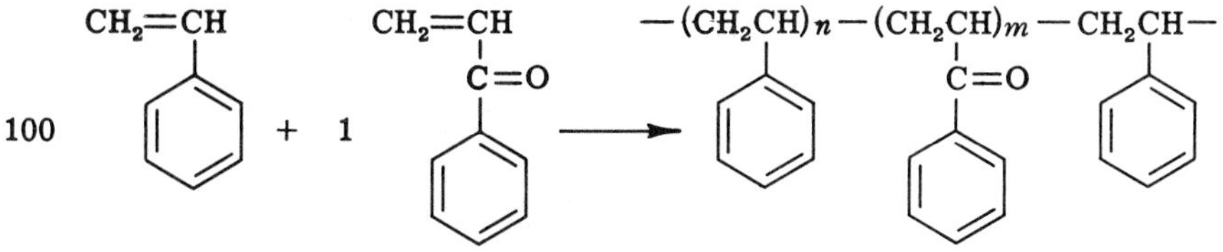

$n + m = 100$ (12-19)

groups does not have much of an effect on any of the physical properties of the polymer except its ability to absorb light; the polymer can be molded, foamed, etc., by the same techniques used for the fabrication of conventional polystyrene. Some photodecomposition may take place before the plastic item is discarded, but this is not usually a problem because fairly extensive chain breaking is required before the physical properties of the polymer change significantly.

The carbonyl group was chosen to effect the polymer degradation because it absorbs light in the 300-325 nm region. This is a critical region for photodegradation because the polymer will not be degraded by the sunlight passing through window glass since the latter only transmits light on the long wavelength side of 325 nm. The polymer will be degraded in the environment because the atmosphere transmits light of wavelength greater than 290 nm. Consequently, materials made from these polymers will be stable before use. e.g., when placed in display windows, but will decompose when discarded.

Cups prepared from polystyrene which contains "built-in" carbonyl groups essentially disappear on standing outside for two weeks (see Fig. 12.1). A wettable powder is produced which is dispersed in the environment. These small particles are eventually degraded to sufficiently small chain length to be attacked by common microorganisms, and hence become biodegradable.

FIGURE 12-1. Disintegration of a polystyrene cup made from plastic containing carbonyl groups: (a) 1-day exposure; (b) 7-day exposure;

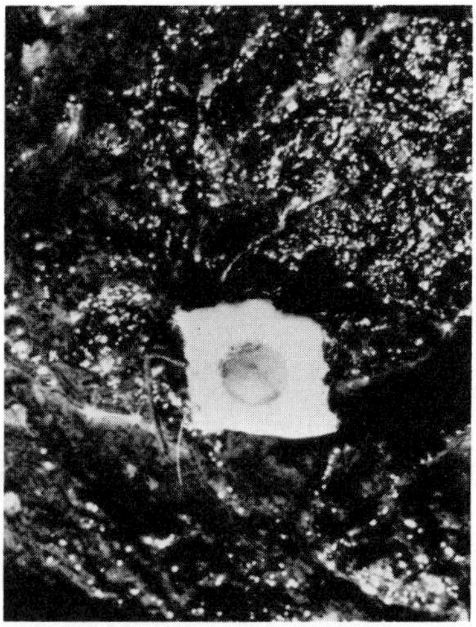

FIGURE 12-1 (Continued)
(c) 14-day exposture; (d) 20-day exposure.
Photographs courtesy of Prof. James E. Guillet, University of Toronto, Canada.

12.5.3 Delayed-Action Photodecomposition Triggered by Metal Chelates

Many compounds have been developed that inhibit the spontaneous decomposition of polymers by reacting with the peroxides formed during processing. Recently, compounds have been devised that provide the same initial stabilization and then accelerate decomposition. The additives are iron (III) or copper (II) complexes of dinonyldithiocarbamate [see reaction (12-20)], which stabilize the polymer by reacting with the peroxy groups generated during processing. However, once the sulfur of the complex is oxidized, the metal ion is released (12-22). The free metal ion accelerates the photochemical decomposition of the peroxides. The following mechanisms of protection and decomposition have been suggested:

Protection of polymer

$$(R_2NCS_2)_2M \xrightarrow{R'OOH} (R_2N\overset{\overset{S}{\|}}{C}-\underset{\underset{O}{\|}}{\overset{\overset{O}{\|}}{S}}-O)_2M \qquad (12\text{-}20)$$

$$(R_2N\overset{\overset{S}{\|}}{C}-\underset{\underset{O}{\|}}{\overset{\overset{O}{\|}}{S}}-O)_2M \xrightarrow{H_2O} RNCS + M^{2+} + SO_4^{2-} + SO_2 + ROH \qquad (12\text{-}21)$$

Accelerated decomposition of polymer

$$ROOH + M^{2+} \xrightarrow{h\nu} ROO\cdot + H^+ + M^+ \qquad (12\text{-}22)$$

$$ROOH + M^+ \xrightarrow{h\nu} RO\cdot + OH^- + M^{2+} \qquad (12\text{-}23)$$

It is possible to formulate polymers of specific lifetime by using varying amounts of the metal complex of the dinonyldithiocarbamates. For example, polyethylene containing 0.05% of the metal complex showed no evidence of degradation after outdoor exposure for seven months, but after 10 months exhibited a 75% decrease in elongation on stretching. Stretching properties are related to chain length; the observed behavior is evidence for extensive chain breaking between the end of the seventh and tenth months of the test. Decomposition was not observed when none of the metal complex was used.

12.5.4 Biodegradable Synthetic Polymers

It was noted in Section 12.4 that polyesters are the only class of polymers used on a large scale that are susceptible to microbial degradation. Since polyesters have limited uses for the fabrication of useful articles, attempts are being made to design other classes of biodegradable polymers.

Some success has been achieved in the synthesis of biodegradable polyamides although these are not in commercial use at present. Polyamides would be expected to be biodegradable if their structures did not differ markedly from that of the proteins (polyamides such as Nylon differ significantly). Recently it has been possible to prepare the acyl azides corresponding to amino acids. These azides are stable as the hydrobromic acid (HBr) salts, but they undergo step reaction polymerization to polyamides when the HBr is removed by treatment with base [reactions (12-25)-(12-27)]. Using this approach, it has been possible to prepare a variety of polymers or copolymers with physical properties reminiscent of nylon, yet are susceptible to microbial degradation.

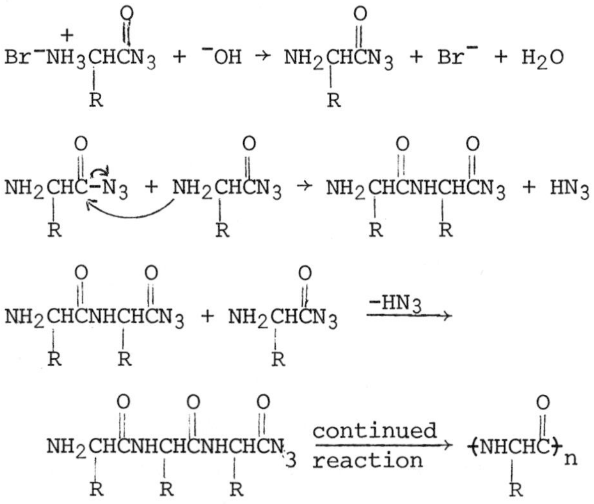

$$Br^-NH_3CHCN_3 + {}^-OH \rightarrow NH_2CHCN_3 + Br^- + H_2O \qquad (12-25)$$

$$NH_2CHC-N_3 + NH_2CHCN_3 \rightarrow NH_2CHCNHCHCN_3 + HN_3 \qquad (12-26)$$

$$NH_2CHCNHCHCN_3 + NH_2CHCN_3 \xrightarrow{-HN_3}$$

$$NH_2CHCNHCHCNHCHCN_3 \xrightarrow[\text{reaction}]{\text{continued}} \{NHCHC\}_n \qquad (12-27)$$

12.5.5 Comparison of the Three Approaches to the Design of Degradable Polymers

Polystyrene containing "built-in" carbonyl groups that accelerate its photodecay is now commercially available. Polymers designed in this way have a major advantage over the use of metal chelates because the carbonyl derivative is covalently bound in the polymer. As a consequence, the polymer can be used to package food without the danger of contamination by the photosensitive compound. On the other hand, the use of metal chelates may have other long-term advantages over the carbonyl containing polymers. First, the chelates offer initial protection and the extent of the protection can be predetermined by the amount that is used. Second, very low concentrations of metal chelates are required ($\sim 10^{-4}$ mol/100 gm polymer) so that contamination of foodstuffs, etc., may not be a significant problem.

Both of these approaches have one major drawback: The resulting polymers will be difficult to color and maintain degradability because most pigments absorb light in the same spectral region as the carbonyl group or metal ion. The pigment will absorb most of the light and no polymer photodecomposition will occur. Biodegradable polyesters and polyamides can be pigmented since pigments should not inhibit the attack of microorganisms. Biodegradable polymers have quite different physical properties than the photodegradable polystyrene because of the presence of a large number of polar functional groups in the polyesters and polyamides. Since the physical properties are so different, the particular commercial application will determine which type of polymer will be used.

BIBLIOGRAPHY

M. Kaufman, "Giant Molecules." Doubleday, New York, 1968.

F. W. Billmeyer, Jr., "Textbook of Polymer Science," 2nd ed. Wiley (Interscience), New York, 1971.

W. Coscarelli, Biological Stability of Polymers in "Polymer Stabilization" (W. L. Hawkins, ed.), p. 377. Wiley, New York, 1972.

Biodegradability: Lofty goal for plastics, *Chem. Eng. News*, Sept. 11, 1972, pp. 37-38.

G. Scott, Improving the environment: Chemistry and plastics waste, *Chem. Britain 9*, 267 (1973).

J. B. Colton, Jr., F. D. Knapp, and B. R. Burns, Plastic particles in surface waters of the Northwestern Atlantic, *Science, 185*, 491 (1974).

J. E. Guillet, Polymers and Ecological Problems, in "Polymer Science and Technology," Vol. 3. Plenum, New York, 1973.

J. E. Guillet, Plastics, energy, and ecology--a harmonius triad, *Plastics Eng. 30* (8), 48 (1974).

N. S. Allen and J. F. McKellar, Photodegradation and stabilization of commercial polyolefins, *Chem. Soc. Rev. 4*, 533 (1975).

13
CHEMISTRY IN AQUEOUS MEDIA

13.1 WATER IN THE ENVIRONMENT

Much of environmental chemistry takes place in aqueous systems. Water itself takes part in a cyclic process of evaporation, precipitation, and transport shown schematically in Fig. 13-1. Approximately 5×10^{14} m^3 of liquid water evaporates and precipitates annually, with the major portion of this being over the oceans. About 0.4×10^{14} m^3 more water falls on land as precipitation than evaporates from it and this is the annual global volume of fresh water run-off. Less than 0.2×10^{14} m^3 is permanently present in the atmosphere as vapor or cloud. The bulk of all water is found in the oceans, where it contains much dissolved inorganic material and small but important amounts of organic substances. Surface fresh water (lakes, streams, etc.) are of highly

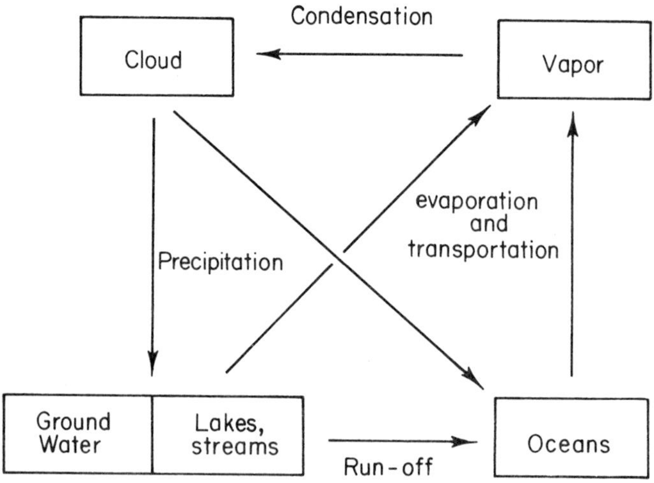

FIGURE 13-1. The water cycle.

variable composition, but usually contain small amounts of dissolved substances, biological materials, and frequently much suspended matter. These water sources are most heavily used by man. Subsurface or ground waters, because of their percolation through soil and rock formations, are normally quite free of suspended and organic materials, but may contain dissolved inorganic material. These waters also are used extensively in some locations. Fresh waters are classified as hard when they contain significant amounts of calcium or magnesium salts in solution. Hardness is easily demonstrated through the precipitate formed with soap.

The fact that liquid water has a maximum density at 4°C, and is more dense than ice, has important consequences in natural water systems. As an unstirred body of water is cooled from the surface, the upper levels sink as their density becomes greater than that of the warmer, deeper water. If cooled below 4°C, the density decreases, and this water remains on the surface. On freezing, the still lower density ice floats. This is of obvious significance in preventing the complete freeze-up of bodies of water, but the density gradients in the liquid are also responsible for stratification effects of considerable importance. Figure 13-2 illustrates stratification in a typical deep lake. In summer, the upper level, the epilimnion, consists of comparatively warm water. Photosynthesis takes place in this region, and the oxygen content is likely to be high. The thermocline represents a region of rapid temperature change. The lower level, the hypolimnion, consists of colder water. Although oxygen is more soluble in cold than in warm water, the only source of oxygen here is diffusion from higher levels. Since oxygen is used up by decomposing organic matter which has settled to the bottom, the oxygen content of the hypolimnion is likely to be low and reducing conditions may exist. A typical lake of this sort will "turn over" in the fall, and often again in the spring, but otherwise there will be little mixing. Turn over refers to displacement of the bottom water by surface water as the latter reaches the temperature of maximum density. Nutrients taken up by organisms in the epilimnion are largely released when the organisms decay in the hypolimnion so that this "turn over" brings nutrients back to the levels where photosynthesis takes place and is important for biological productivity.

Lakes often are described as oligotrophic, eutrophic, or dystrophic. These terms represent stages in the normal development of a lake. An oligotrophic lake has comparatively pure water, is low in nutrient materials and has low biological production. This would be a geologically young lake. A eutrophic lake has a high nutrient content and high biological

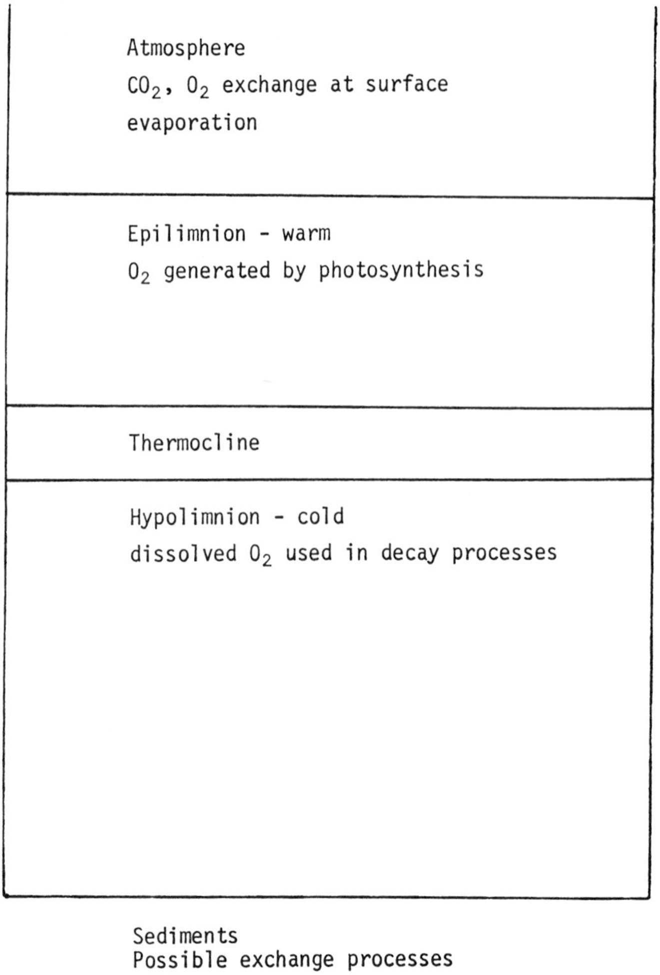

FIGURE 13-2. Schematic representation of stratification in a deep lake.

production. The final stage, a dystrophic lake, will be shallow, marshy, and contain much organic material, and have a high BOD (Section 6.6). Since most of the oxygen is used up by decaying vegetation, actual production is low. These stages form a natural sequence in the history of a lake. The sequence of events can be speeded up artificially by introduction of nutrients to enhance algal growth. The term eutrophication is used commonly to describe this process.

13.2 INTERACTIONS IN WATER

Chemical interactions in aqueous systems make up an important area of environmental chemistry. Water has several properties that are important to the behavior observed in solution. The water molecule is angular and, owing to the large difference in electronegativity of oxygen and hydrogen, quite polar. Consequently, water molecules will have strong electrostatic interactions with one another, and with dissolved ionic materials or other highly polar molecules. The electronegativity of oxygen is large enough that the bonding electrons have little effect in screening the hydrogen nucleus from interaction with centers of negative charge on other molecules, with the result that hydrogen bonding of water to molecules with highly electronegative atoms is important.

Such interactions, to a first approximation, may be regarded as a special case of dipole-dipole attraction which has unusual strength. This leads to molecular association, may determine structure, and in appropriate cases produces stronger hydration of dissolved species than would have been the case on the basis of a simple dipole-dipole interaction. Hydrogen bonding is of greatest importance in aqueous solutions of molecules with oxygen and nitrogen groups, and particularly with OH or NH groups which are not only hydrogen bonded by the hydrogens of the water molecules, but which can themselves hydrogen bond to the water oxygen.

The structure and density properties of water are largely determined by hydrogen bonding. In ice, each water molecule is surrounded tetrahedrally by four others; two H-bonded to it through its own hydrogen atoms, and two more that bond to the oxygen through their hydrogens. This is illustrated in Fig. 13-3. This gives a coordination number of four, which is not very effective in filling space. Consequently the structure is very open with low density. On melting, the regular structure is destroyed and the packing density increases, although the greatest part of the hydrogen bonding remains intact. With further warming, increased hydrogen bond breaking allows the packing efficiency to increase, while greater thermal motion acts to reduce the density. The interplay of these counteracting tendencies results in the variation of density described earlier.

The oxygen atom of a water molecule possesses two pairs of nonbonding electrons (unshared pairs). These may be donated to empty, low-energy orbitals in another atom to form a coordinate bond. This has particular importance with metal ions, and will be discussed in more detail later.

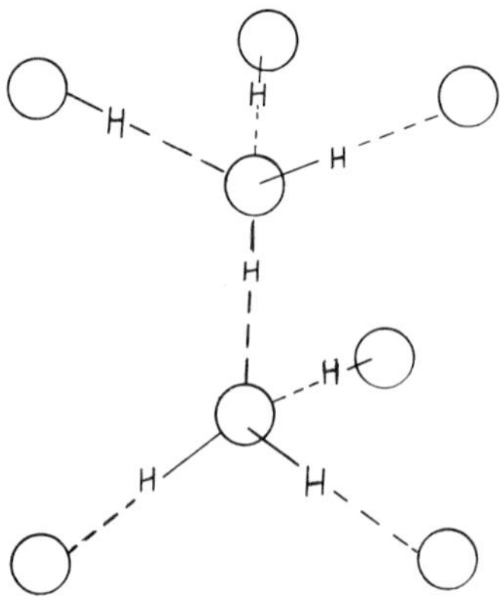

FIGURE 13-3. Hydrogen bonding between the water molecules of ice. (Only two complete water molecules are shown.) Solid lines are covalent bonds, dashed lines are hydrogen bonds.

The solubility of a substance in a liquid is determined by the interactions that take place between the molecules, and by entropy effects. That is, for a process to take place, the free energy of the system must decrease. The free energy change depends upon the enthalpy change, which can be looked at as the change in the energies (heats) of the various interactions involved (bond energies, dipole-dipole interaction energies, etc.), and upon the entropy change which can be regarded as a measure of the change of order, or randomness. The processes involved in dissolving one substance in another are disruption of solute-solute interactions, at least partial disruption of solvent-solvent interactions caused by the presence of solute molecules, and formation of solvent-solute interactions. The first two of these involve an increase in energy (endothermic). Only if the solvent-solute interactions, which are exothermic, are large in magnitude compared to the other interactions will the enthalpy be favorable for solution. In the case of water, large solvent-solute interaction energies can only arise through H-bonding, coordination, ion-dipole attractions, or perhaps dipole-dipole attractions. Solute molecules not able to interact in one of these ways will not interact strongly with the water molecules. As a consequence of these factors, there is a general tendency for like substances to dissolve in one another, and for unlike substances not to do so. Water is generally a good solvent for ionic and highly polar (especially hydrogen bonding) substances, but poor for nonpolar substances such as hydrocarbons. The increase in disorder on mixing is always a driving force toward solubility, although it may be largely outweighed by energy effects in some cases.

These above considerations are valid as a qualitative guide to the type of solubility behavior to be expected. However, a prediction of poor solubility does not necessarily mean no solubility whatsoever. Substances may be important environmentally at very low levels of solubility, so that even if a material is described as "insoluble," its presence in aquatic systems cannot be dismissed. Concentrations in the parts per million (ppm) or even parts per billion (ppb) range may be significant in the case of some pollutants, and solubilities to this level will be found for most substances. Further, the distinction between dissolved and suspended matter is not always clearcut. Very finely divided particulate matter (colloids) may settle or coagulate very slowly, and may pass through most filtering systems. The surface areas per unit mass of such material will be very high, so that reactivity may be much greater than that of the bulk solid.

The solubility of a gas in a liquid depends upon the partial pressure P of the gas. Normally, Henry's law applies, and the equilibrium constant for the process $G_{(\text{gas})} \rightleftarrows G_{(\text{solution})}$ can be written as

$$K = P_{G(\text{gas})}/C_{G(\text{solution})} \qquad (13\text{-}1)$$

Solubility of a gas in most cases is small unless a reaction with the solvent is involved, and generally decreases with increasing temperature.

The equilibrium solubility of a substance that ionizes in solution (e.g., a salt) is expressed in terms of the solubility product. For the solution reaction of the pure solid M_aX_b,

$$M_aX_b \text{ (s)} \rightleftarrows bX^{-a} \text{ (sol)} + aM^{+b} \text{ (sol)} \qquad (13\text{-}2)$$

the equilibrium expression[1] is

$$K_{sp} = [M^{+b}]^a[X^{-a}]^b \qquad (13\text{-}3)$$

[1] *It will be the convention in this chapter to use square brackets to represent the equilibrium concentration of a species, while C, with a suitable subscript, will be used to represent analytical or total concentrations of substances that exist in solution as two or more species.*

This is valid for slightly soluble substances. The equilibrium solubility is dependent upon the concentration of the ions from all sources.

Nonionic solids or liquids will also have an equilibrium solubility, but these are not generally treated by the equilibrium constant approach. Solubilities in general may range from very small to infinite. In environmental situations normally one is dealing with relatively low concentrations, and consequently, equilibrium considerations will be of concern only with slightly soluble substances.

Another property of water is that of autoionization; that is the process

$$2H_2O \rightleftarrows H_3O^+ + OH^- \tag{13-4}$$

The species H_3O^+ represents the hydrated proton. Because of its small size and consequent high charge density, a proton will not exist as such in solution, but will interact with the electron clouds of solvent molecules. In water, it will share one of the lone pairs from an oxygen atom. The actual species present probably is more complex than H_3O^+ (perhaps $H_9O_4^+$), but this is not particularly important for our purposes. This proton transfer process, however, is a very important type of reaction in aqueous chemistry, and acid-base behavior is widespread. Typical waters have a hydrogen ion concentration between 10^{-6} and 10^{-9} M (pH 6 to 9). This is a comparatively narrow range, which is determined by the balance of acidic and basic substances dissolved in the waters. Absolutely pure water at 25°C would have a hydrogen ion concentration of 10^{-7} M, from the equilibrium expression

$$K_w = [H_3O^+][OH^-] = 10^{-14} \tag{13-5}$$

Biological processes, as well as purely inorganic processes such as dissolution or precipitation of minerals, are dependent upon pH. We are interested in what determines the pH of natural waters, and equally important, how resistant the pH value is to natural or man-made influences.

The equations used in equilibrium expressions in this book are written in terms of concentrations rather than activities. The activity a_i of species i in solution of concentrations C_i is

$$a_i = \gamma_i C_i \tag{13-6}$$

where γ_i is the activity coefficient of the substance. In effect, γ_i is the factor by which the actual concentration must be multiplied in order to maintain the validity of the ideal equations. The activity coefficient γ_i may be greater or less than unity, depending on both the concentration C_i and on the nature and concentration of other species in the solution. Because $\gamma_i \to 1$ as $C_i \to 0$, use of concentrations in dilute solutions is acceptable for most purposes, and is necessary in most environmental problems where activities generally are not known. Some care must be exercised, however, as in the case of sea water, which contains roughly 0.5 *M* NaCl in addition to other components and cannot be considered a dilute solution in this context. The presence of a relatively large concentration of an "inert" salt such as NaCl produces nearly constant values for the activity coefficients of ions present in small amounts, such as H_3O^+, although these values may be far from unity. Equilibrium constant expressions such as those for K_w or K_a will give correct results in this case, but only for an equilibrium constant having a value different from that applicable to the dilute solution case. Equilibrium constants valid for a solution of a particular overall composition are sometimes called stoichiometric equilibrium constants.

13.3 ACID-BASE PROPERTIES

The most useful definition of acids and bases in aqueous chemistry is the Brönsted-Lowry definition: An acid is a proton donor; a base is a proton acceptor. This definition may be illustrated by an equation,

$$HCl + H_2O \rightleftarrows H_3O^+ + Cl^- \qquad (13\text{-}7)$$

Here, HCl is acting as the acid (proton donor) and H_2O as the base (proton acceptor) as the reaction goes from left to right. Such reactions are reversible, although, as in this case, the equilibrium may lie very far to the right. If this reaction were reversed, it may be noted that then H_3O^+ would be the acid, and Cl^- the base. $HCl\text{-}Cl^-$ and $H_3O^+\text{-}H_2O$ make two sets of acid-base pairs (conjugates).

Another example of an acid-base-reaction is

$$NH_3 + H_2O \rightleftarrows NH_4{}^+ + OH^- \qquad (13\text{-}8)$$

In this case, H_2O acts as the acid, and this illustrates the amphoteric character of water; that is, its ability to act either as acid or base, depending on the substance with which it is reacting. It is inherent in the Brönstead approach that the terms acid or base have meaning only in terms of a reaction such as given above.

The strength of an acid or a base is given by the equilibrium constant for its ionization process. That is, for the reaction of a weak acid HA with water,

$$HA + H_2O \rightleftarrows H_3O^+ + A^- \qquad (13\text{-}9)$$

$$K_a = [H_3O^+]\,[A^-]\,/\,[HA] \qquad (13\text{-}10)$$

An analogous expression could be set up for a base, with the constant usually called K_b. In place of K_a and K_b, we often find pK_a or pK_b, where $pK = -\log K$.

The larger the value of the equilibrium constant K_a, the stronger is the acid. A so-called strong acid is one which is completely dissociated (equilibrium far to the right) and, in fact, all the available protons exist as H_3O^+. All acids above a certain strength appear equally strong in water, since the acid really present is solely H_3O^+ in those cases. Strong acids may not be equally strong in other solvents. Similar comments hold for bases, with OH^- being the strongest base in water.

Brönsted acids usually fall into three major classes.

Binary acids contain hydrogen and one other element. Naturally occurring examples are HCl (volcanic, may be formed on incineration of chlorine-containing organic substances), and H_2S (volcanic, decomposition product of organic material such as proteins under oxygen-poor conditions). Strengths of binary acids increase with increasing bond polarity and with decreasing bond strength. *Oxoacids* are those in which the ionizable hydrogens exist as OH groups attached to a not too electropositive third element. Environmental examples include H_2CO_3 (formed from CO_2 in water), H_2SO_4 (a result of sulfur oxide emissions) and H_3PO_4 (originating from phosphates in detergents and fertilizers). These particular examples are polyprotic acids. Strengths of oxoacids increase with the number of nonhydroxo oxygen atoms attached to the third element. *Hydrated metal ions* form a third class of acidic substance. Many metal salts on solution in water become strongly hydrated to form complexes such as $[Fe(H_2O)_6]^{3+}$, and such species may be ionized through reaction such as

$$[Fe(H_2O)_6]^{3+} + H_2O \rightleftarrows [Fe(H_2O)_5(OH)]^{2+} + H_3O^+ \qquad (13\text{-}11)$$

More than one proton may be lost. Processes of this sort are important in the behavior of metal salts in water and in the hydrolysis of these salts to form insoluble products, and will receive further treatment later.

Bases, like acids, may be weak or strong. Some examples of bases are NaOH, which in solution ionizes to give OH^- and is a strong base; NH_3, a weak base whose action depends on taking a proton to form NH_4^+; and Na_2CO_3, a salt that also accepts protons via reactions

$$CO_3^{-2} + H_2O \rightleftarrows HCO_3^- + OH^- \tag{13-12}$$

and

$$HCO_3^- + H_2O \rightleftarrows H_2CO_3 + OH^- \tag{13-13}$$

All Brönsted bases have unshared pairs of electrons that can be donated to the 1s orbital of a proton. This picture of acid-base behavior leads to the Lewis definition of acids and bases. A base is an electron pair donor; an acid is an electron pair acceptor. This definition considerably expands the acid-base concept.

Acid-base equilibrium constants (ionization constants) may be used for calculations, such as that of hydrogen ion concentration or pH. Simple examples are discussed in most freshman chemistry and basic analytical texts, and the details will not be dealt with here. However, as an alternative to individual calculations, the variation with pH of the concentration of the various species in equilibrium in a weak acid or base solution may conveniently be expressed graphically in a way that gives a rapid approximation of behavior. For a given total concentration of dissolved acid, $C_T = [HA] + [A^-]$, one can rewrite Eq. (13-10) for K_a in the form

$$[HA] = C_T[H_3O^+] / (K_a + [H_3O^+]) \tag{13-14}$$

or

$$[A^-] = C_T K_a / (K_a + [H_3O^+]) \tag{13-15}$$

In the limit that $K_a << [H_3O^+]$ (pH<<pK), $[HA] = C_T$ and $\log[HA] = \log C_T$; that is, log[HA] is independent of pH. Thus, a plot of log[HA] versus pH has zero slope as long as the pH is small. In this same region, $\log[A^-] = \log C_T + \log K_a + pH$. The slope $d \log[A^-] / d\ pH = 1$. If pH>>pK_a, similar arguments give the slopes of the plots of $\log[A^-]$ versus pH and log[HA] versus pH as 0 and -1, respectively. When pH = pK_a, $[HA] = [A^-]$.

Using these approximations, Fig. 13-4 can be drawn. This refers to a hypothetical monobasic acid of pK_a = 6 at $C_T = 10^{-3}$. The scale of the ordinate will depend on the value of C_T. The pH and pOH lines are plotted from pH + pOH = 14. For a solution of only HA in water, the

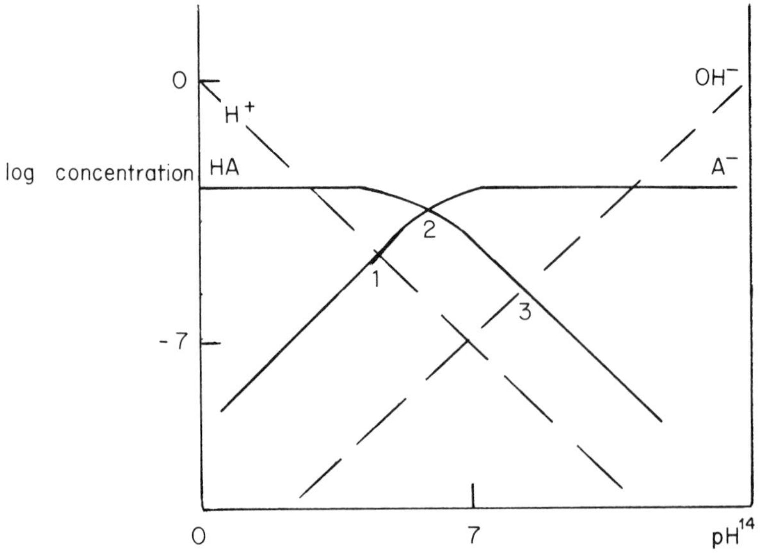

FIGURE 13-4. Diagram of log Concentration versus pH for a weak acid HA for which pK = 6 and $C_T = 10^{-3}$. Point 1 corresponds to a solution of HA in water, point 3 to a solution of a salt (e.g., NaA) in water, and point 2 to equal concentrations of HA and A^-.

electroneutrality requirement gives $[H_3O^+] = [A^-] + [OH^-]$, and pH thus is easily evaluated. In fact, since $[OH^-]$ is so small, $[H_3O^+] = [A^-]$ here.

A solution of the salt NaA of the same total concentration would be partially hydrolyzed to form HA and OH^- in equal concentrations. The pH is given where these lines cross. (Electroneutrality requires $[Na^+] + [H_3O^+] = [A^-] + [OH^-]$.)

A titration curve showing the variation of pH with equivalents of base added to the HA solution discussed can be calculated, and is illustrated in Fig. 13-5. The pH values given by pure HA and the completely neutralized system (equivalent to the NaA solution) have been discussed (points 1 and 3). The half neutralized pH (point 2) also follows easily, since $[A^-] = [HA] = \frac{1}{2} C_T$ here. Values of pH outside of points 1 and 3 are achieved if additional strong acid or base is present. It can be seen that in certain regions, the pH is comparatively insensitive to added acid or base, while in others it is quite sensitive. The term "buffer capacity" is used to denote the rate of change of pH with added strong base or acid. (Buffer capacity is defined as the number of

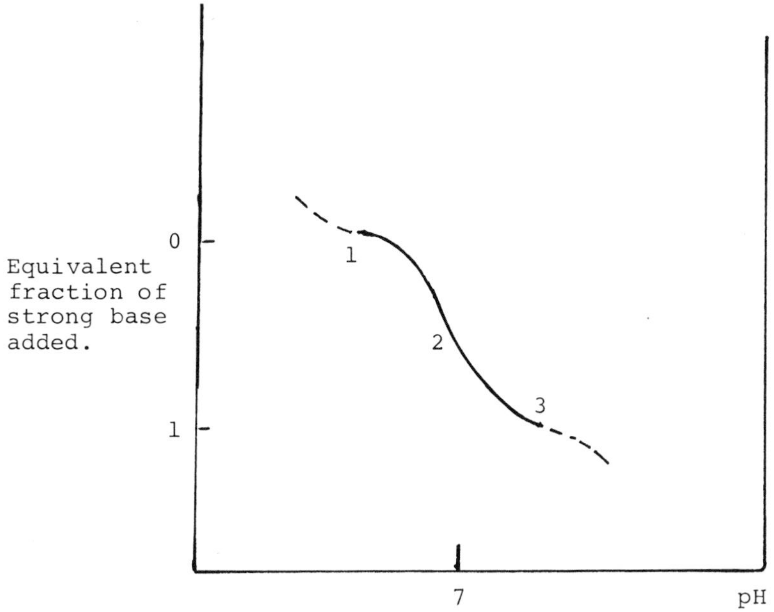

FIGURE 13-5. Titration curve for a weak acid HA.

equivalents per liter of strong acid or base required to change the pH by 1 unit.) The pH changes rapidly in regions of equivalent or neutralization points.

The most important environmental acid is carbonic acid, which is produced in natural waters by the dissolution of CO_2 gas. The dissolution of gaseous CO_2 is comparatively slow; the average lifetime of a CO_2 molecule in the atmosphere is about 7 years. The air-solution transfer process may be catalyzed by the enzyme carbonic anhydrase in waters containing biological materials. When CO_2 enters solution, two steps are actually involved:

$$CO_2\ (g) \rightleftharpoons CO_2\ (aq) \tag{13-16}$$

$$CO_2\ (aq) + H_2O \rightleftharpoons H_2CO_3\ (aq) \tag{13-17}$$

The species $CO_2(aq)$ (molecular CO_2 associated with water of hydration) and $H_2CO_3(aq)$ are not normally distinquished, so that an overall reaction can be written:

$$CO_2(g) + H_2O \rightleftharpoons H_2CO_3^* \tag{13-18}$$

where

$$[H_2CO_3^*] = [CO_2\ (aq)] + [H_2CO_3\ (aq)] \tag{13-19}$$

$$K_{H_2CO_3^*} = [H_2CO_3^*]/P_{CO_2} = 2.8 \times 10^{-2} \tag{13-20}$$

while

$$K_{H_2CO_3} = [H_2CO_3]/P_{CO_2} = 1.58 \times 10^{-3} \tag{13-21}$$

The ionization equilibria, which involve fast reactions, are

$$H_2CO_3^* + H_2O \rightleftharpoons H_3O^+ + HCO_3^- \tag{13-22}$$

$$K_{a_1} = [H_3O^+][HCO_3^-]/ [H_2CO_3^*] = 4.5 \times 10^{-7} \tag{13-23}$$

(the ionization constant for H_2CO_3 alone is about $10^{-3.5}$ but the value above is smaller because only a part of the dissolved CO_2 is in the form of H_2CO_3) and

$$HCO_3^- + H_2O \rightleftharpoons H_3O^+ + CO_3^{-2} \tag{13-24}$$

$$K_{a_2} = [H_3O^+][CO_3^{-2}]/ [HCO_3^-] = 4.8 \times 10^{-11} \tag{13-25}$$

The concentration of species in a CO_2 solution as a function of pH is illustrated in Fig. 13-6 for a fixed total carbonate concentration C_T of 1×10^{-2} moles/liter ($C_T = [H_2CO_3^*] + [HCO_3^-] + [CO_3^{-2}]$).

We can obtain the following expressions from the equilibrium constants and C_T:

$$[H_2CO_3^*] = \frac{C_T}{1 + (K_{a_1}/[H_3O^+]) + (K_{a_1}K_{a_2}/[H_3O^+])} \tag{13-26}$$

$$[HCO_3^-] = \frac{C_T}{([H_3O^+]/K_{a_1}) + 1 + (K_{a_2}/[H_3O^+])} \tag{13-27}$$

$$[CO_3^{-2}] = \frac{C_T}{([H_3O^+]/K_{a_1}K_{a_2}) + [H_3O^+]/K_{a_2}) + 1} \tag{13-28}$$

We can find slopes of the log concentration plots in various regions as illustrated above. Figure 13-6 is plotted from such considerations. The various equilibrium constants on which this figure is based are for pure water containing only CO_2. In sea water, the much higher ionic strength

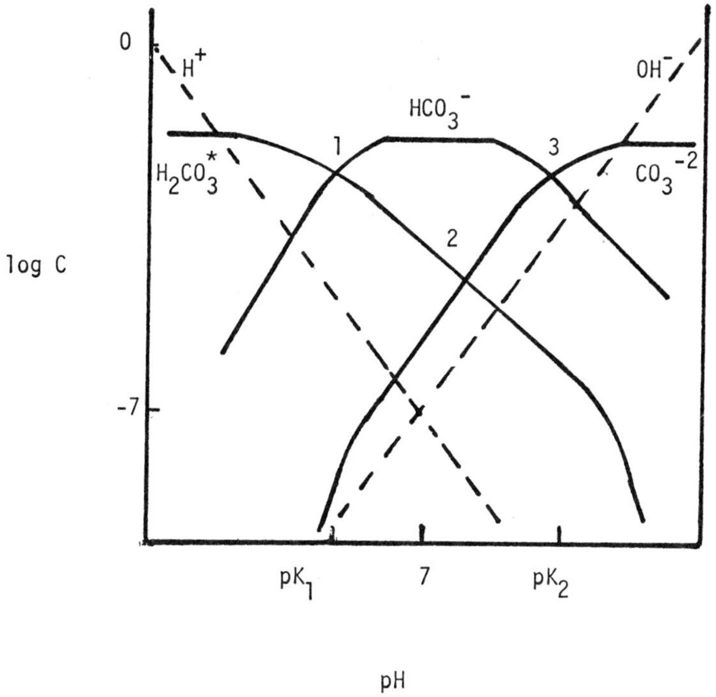

FIGURE 13-6. Log concentration versus pH for the species in equilibrium in a solution of CO_2 in water, at a fixed total concentration C_T of CO_2 and carbonate species. The actual position of the carbonate curves will depend on C_T; this diagram is drawn for a value of approximately 10^{-2} M.

affects activities, and if concentration expressions are used, the corresponding stoichiometric equilibrium constants are different (see list of values in Table 13-1). The net effect is that in sea water all of the curves in Fig. 13-6 would be shifted by ½ to 1 pH unit to the left, for the same C_T.

The pH of a solution containing only CO_2 as the carbonate source would be given from $[H_3O^+] = [HCO_3^-] + 2[CO_3^{-2}] + [OH^-] \simeq [HCO_3^-]$. This has a pH of about 5.6. Natural waters typically are more basic; for example, sea water has a value near 8.1. This is the result of acid-base reactions with bases from rocks. Indeed, a pH of 8 is remarkably near the equivalence point for HCO_3^-; point 2 on the titration curve for this system given in Fig. 13-7. It is evident from the figure that an isolated sample of sea water is not buffered since it is near an equivalence point, and holds its pH of about 8.1 rather precariously. The buffer capacity is very small.

TABLE 13-1

Equilibrium Constants of Some Carbonate Reactions

Reaction	"Pure" H_2O		Sea water	
	0°C	25°C	0°C	25°C
$CO_2(g) + H_2O \rightleftarrows H_2CO_3^*$	8×10^{-2}	2.8×10^{-2}	6.5×10^{-2}	3×10^{-2}
$H_2CO_3^* + H_2O \rightleftarrows HCO_3^- + H_3O^+$	2.6×10^{-7}	4.5×10^{-7}	7.2×10^{-7}	9×10^{-7}
$HCO_3^- + H_2O \rightleftarrows H_3O^+ CO_3^{-2}$	3×10^{-11}	4.8×10^{-11}	4×10^{-10}	8×10^{-10}
$CaCO_3 \rightleftarrows Ca^{+2} + CO_3^{-2}$ (calcite)		4.2×10^{-9}		6.4×10^{-7}

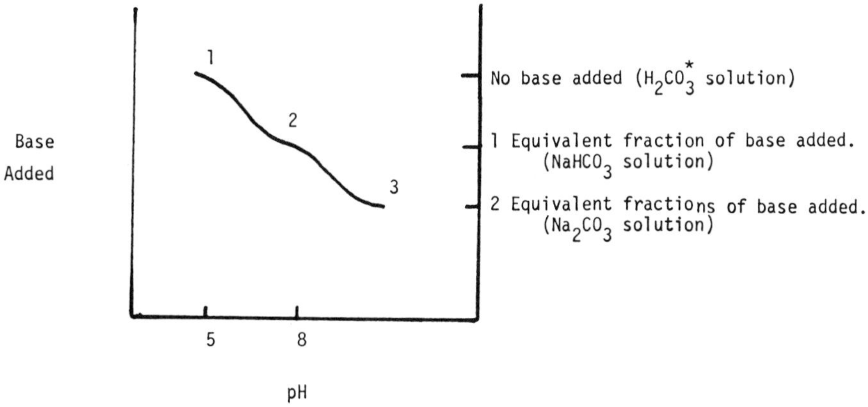

FIGURE 13-7. Titration curve for the CO_2 system.

Natural water not only can react with sediments, but it is in equilibrium with atmospheric CO_2 which will serve to control the concentration of $H_2CO_3^*$ according to the solubility relationship. If $P_{CO_2(g)}$ is constant, then $[H_2CO_3^*]$ also is constant within the time scale permitted by the rate of the gas-solution exchange reaction. Figure 13-8 gives the log concentration versus pH curves for this case, with $H_2CO_3^*$ fixed by $P_{CO_2} = 3 \times 10^{-4}$ atm. The pH of a solution of only CO_2 in water is not much different from the constant C_T situation, but the titration curve is very different and is shown in Fig. 13-9. This system at pH = 8 is strongly buffered toward added base, much less so toward acid. At this pH, most of the dissolved CO_2 is present as HCO_3^-, and the amount of this in the oceans exceeds the amount of gaseous CO_2 in the atmosphere.

It is clear that the pH of the ocean and most other natural waters is far from that predicted by CO_2 alone, and additional base must be present. This may be provided by soluble carbonate salts. In addition, equilibria with insoluble carbonates may contribute to CO_3^{2-} control. The most obvious of these carbonates is $CaCO_3$, since calcium is a very common element in aqueous systems, although others such as $MgCO_3$ are important also. There are a number of complications in the treatment of calcium carbonate equilibria, such as the existence of different crystalline forms (calcite and aragonite) with slightly different solubility constants, and the

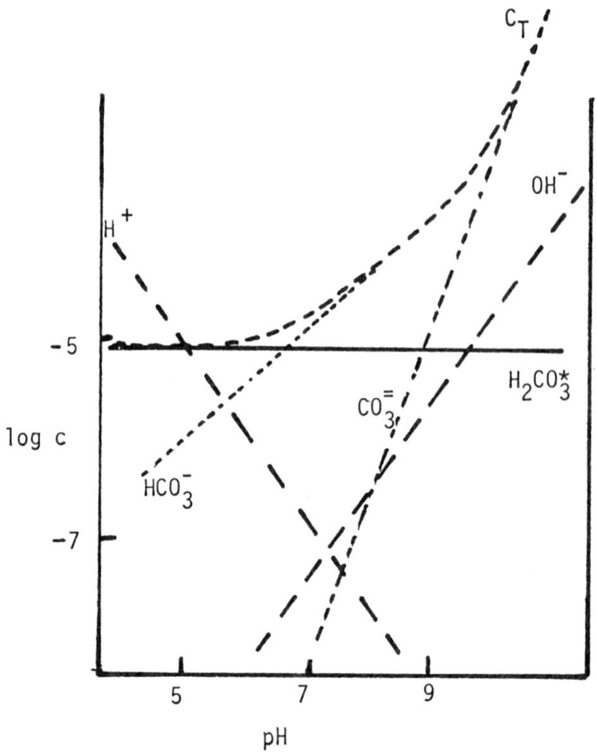

FIGURE 13-8. Log concentration versus pH for the species in a solution of CO_2 in water at a fixed atmospheric pressure of CO_2. The actual position of the carbonate curves will depend on P_{CO_2}; this diagram is drawn for a value of 3×10^{-4} atm, about the normal partial pressure of CO_2 in the atmosphere.

problem of supersaturation; that is, the solubility product concentrations may be exceeded without precipitation taking place.

If $CaCO_3$ precipitation controls the carbonate ion concentration through the Ca^{2+} concentration of the water, it in turn controls the pH and acts as a buffer system for changes in carbon dioxide pressure. Increased $[H_2CO_3{}^*]$ arising from a higher P_{CO_2} would first of all cause the pH to drop as $HCO_3{}^-$ and H_3O^+ form. This lower pH will also lower $[CO_3{}^{2-}]$, since the reaction $CO_3{}^{2-} + H^+ \longrightarrow HCO_3{}^-$ is favored at lower pH values. The carbonate ion in turn is replaced in the solution as $CaCO_3$ dissolves, i.e., the extra H_2CO_3 is titrated by the basic carbonate. The sum of these reaction is

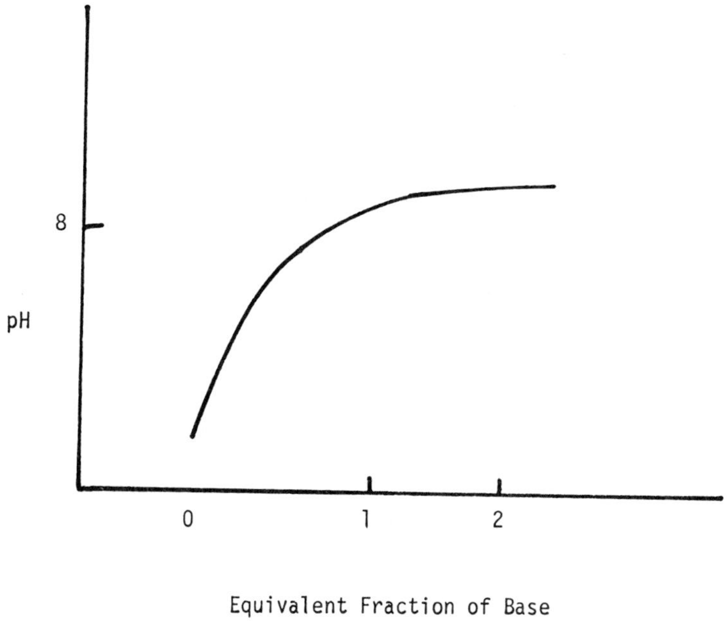

FIGURE 13-9. The titration curve for a solution of CO_2 in equilibrium with a fixed atmospheric pressure of CO_2.

$$CO_2 + H_2O + CaCO_3 \rightleftharpoons Ca^{2+} + 2HCO_3^- \quad (13\text{-}29)$$

The pH and calcium ion values in many fresh water systems are consistent with $CaCO_3$ acting as a CO_2 buffering system. The situation in oceans is more complex, and waters supersaturated and undersaturated with $CaCO_3$ can be found. There is evidence that supersaturation may be encouraged by organic materials in the oceans, which hinder the nucleation process that is essential for initiating formation of the solid phase. Supersaturation is a metastable condition. That is, it is not one of thermodynamic equilibrium, but the rate at which the system moves toward equilibrium is immeasurably slow.

Variation of carbonate solubility with water CO_2 content is responsible for various deposits encountered in fresh water systems. Many soil waters have higher than normal concentrations of dissolved CO_2, as they absorb this as it is formed from reactions of microorganisms in soils at partial pressures exceeding the normal atmospheric value. $CaCO_3$ is then more soluble (it is also more soluble under pressure, see Section 13.4), but when the water comes to the surface, the pressure is released, excess CO_2 leaves the solution, and $CaCO_3$ can deposit. This result is illustrated by stalactite and stalagmite formation in caves in limestone regions. Waters containing Ca^{2+} (or Mg^{2+}) and HCO_3^- are referred to as temporarily hard waters, since heating will drive CO_2 from solution and cause the soluble calcium bicarbonate to convert to the insoluble carbonate in the reverse of reaction (13-29). Permanent hardness is caused by other Ca^{2+} or Mg^{2+} salts, such as sulfate or chloride. It is not eliminated by heating.

An important property often measured for waters to be used industrially or domestically is alkalinity, which is a measure of water quality. Titration of a water with acid to a pH of 8 (the phenolphthalein end point) converts all carbonate to bicarbonate, and the amount of acid required for this is a measure of the carbonate or hydroxide content (carbonate and caustic alkalinity, respectively). Large amounts of carbonate alkalinity, and any caustic alkalinity are undesirable, since such waters will be corrosive and unpalatable. Titration to pH 5 (methyl orange end point) converts bicarbonate to CO_2, and is taken as a measure of bicarbonate content. Caustic alkalinity is present whenever the amount of acid required to reach the phenolphthalein end point exceeds that needed to go from the former end point to the methyl orange end point. Waters with excess acidity, rather than alkalinity, are encountered occasionally, usually in polluted systems such as associated with acid mine drainage (Section 14.6.3.).

13.4 EFFECTS OF TEMPERATURE AND PRESSURE ON EQUILIBRIUM

Equilibrium constants vary with temperature and pressure in a manner that can be predicted thermodynamically. The equilibrium constant is related to the standard free energy change of the process:

$$-RT \ln K = \Delta G^\circ = \Delta H^\circ - T\,\Delta S^\circ \qquad (13\text{-}30)$$

where T is the absolute temperature, R the gas constant, ΔH° the standard enthalpy change, and ΔS° the standard enthropy change. In simple terms, ΔH° is a measure of the change in energy (e.g., bond energies and other interaction energies), and ΔS° a measure of the change in the degrees of freedom, or disorder. The change in ln K with temperature at constant pressure is easily seen by differentiating the expression for ln K (assuming that ΔH° and ΔS° are independent of T, as is normally a good approximation over reasonably small temperature ranges). That is, at constant pressure,

$$\left[\frac{d \ln k}{dT}\right] = \frac{d}{dT}\left[\frac{-\Delta H^\circ}{RT} + \frac{\Delta S^\circ}{R}\right] = \frac{\Delta H^\circ}{RT^2} \qquad (13\text{-}31)$$

The temperature variation depends on whether the enthalpy change of the process is exothermic or endothermic. For solubility, an exothermic heat of solution results in a decrease in solubility with increasing temperature. This is the case with $CaCO_3$, for example, which is less soluble in hot than in cold water. This has consequences for carbonate solubility behavior in lakes and oceans. Values of the equilibrium constants used in this book refer to 25°C unless otherwise indicated.

The pressure variation can be derived from the effect of pressure on ΔG° at constant temperature:

$$d/dP\ \Delta G^\circ = \Delta V^\circ \qquad (13\text{-}32)$$

and

$$d/dP\ \ln k = -\Delta V^\circ/RT \qquad (13\text{-}33)$$

where ΔV° is the standard volume change on reaction.[2] For condensed (solid or liquid) systems, ΔV° is small, so that pressure effects are also small. However, pressures at the ocean's floor may exceed 200 atm, and this can have significant effects. For the dissolution of $CaCO_3$, $\Delta V^\circ \approx -60\ cm^3/mole$ (i.e., the volume decreases because of improved packing of water molecules around the ions) so that $CaCO_3$ should be more soluble in deep water than at the surface even if the temperature were constant.

These effects can be predicted qualitatively from Le Chatelier's principle; when a system at equilibrium is perturbed by a change in concentration, temperature, or pressure, the equilibrium will shift in a way that tends to counteract the change.

[2] *Equations (13-32) and (13-33) may be derived from basic thermodynamic equalities:*

$$G = E + PV - TS$$

so that

$$dG = dE + P\,dV + V\,dP - T\,dS - S\,dT;$$

and

$$dE = T\,dS - P\,dV$$

When T is constant, this reduces to

$$dG = V\,dP \quad \text{or} \quad dG/dP = V$$

13.5 OXIDATION-REDUCTION PROCESSES

The oxidation state of an atom in a molecule is a concept helpful in keeping account of electrons. In many cases it is an indication of the number of electrons involved in bonding, but this is not always true and the oxidation state (or oxidation number) need not have any physical meaning. It is equivalent to the charge on a positive or negative ion in an ionic substance, while for a covalent compound the oxidation state is arrived at arbitrarily by assigning the electrons in a bond to the most electronegative atom. Electrons bonding two atoms of the same electronegativity are shared equally in the assignment.

Reactions that involve changes of oxidation numbers are oxidation-reduction (redox) reactions, and are governed by the oxidation potentials of the systems involved. The potential E of the redox couple or half cell M/M^{n+} is given a numerical value based in principle on an electrochemical cell in which M is in equilibrium with its ions. The potential is measured against a reference couple, the standard being H_2/H^+, which is defined to have a value of exactly zero volts. Values are tabulated as the electromotive series, usually as reduction potentials, i.e., for the process $M^{n+} + ne^- \rightarrow M$. The potential is related to the free energy change of the process as given in Eq. (2-20). Positive potentials and negative ΔG values are characteristic of spontaneous processes.

Standard potentials $E°$ refer to values where the reactants are at unit concentration. The potential varies with concentration according to the Nernst equation. For the process $M^{n+} + ne^- \rightarrow M$,

$$E = E° - \frac{RT}{nF} \ln \frac{1}{[M^{n+}]} = E° + \frac{0.059}{n} \log [M^{n+}] \qquad (13\text{-}34)$$

at 25°C. (R is the gas constant, F is Faraday, 96,500 coulombs, and T is the absolute temperature.)

It is often desirable to express the oxidizing or reducing properties of a system in terms of the concept of pE. In a formal sense, pE = $-\log C_{e^-}$; entirely analogous to pH. Although there are no free electrons in typical chemical systems, the value of pE provides a measure of the oxidizing or reducing properties of a solution, or alternatively serves as a quantitative indication of the oxidizing or reducing conditions necessary for a given oxidation state of an element to be stable. The terms oxidizing and reducing when used to describe a set of conditions or an environment are relative terms; that is, they have meaning only with respect to the reaction or compound being considered. Conditions that exhibit low values of pE will favor reductions of many elements and compounds, and the lower the value, the greater the number of materials which can be reduced. High pE values will favor oxidation similarly. In practice, pE is evaluated from the expression

$$pE = \frac{E}{(RT/F)\ \ln 10} = \frac{E}{0.059} \text{ at } 25°C \qquad (13\text{-}35)$$

The pE value for oxygen in aqueous media can be evaluated for illustration. The oxygen half-cell reaction in acid solution is

$$O_2(g) + 4H^+ + 4e^- \rightleftarrows 2H_2O; \qquad E° = +1.229 \text{ V} \qquad (13\text{-}36)$$

$$E = E° + (0.059/4) \log(C_{H^+})^4 P_{O_2} \qquad (13\text{-}37)$$

Under conditions of equilibrium with the atmosphere where the partial pressure of O_2 is 0.21 atm,

$$pE = 1.229/0.059) + \tfrac{1}{4}\ [4(-pH) + \log 0.21] \qquad (13\text{-}38)$$

At a pH value of 8, in the range reasonable for natural waters, pE = 12.6. This will vary with pH and the oxygen partial pressure. In the absence of other oxidizing agents, the oxidizing power in a body of water is in fact determined by the concentration of dissolved oxygen and P_{O_2} in Eq. (13-37) should have the value of the oxygen partial pressure that would be in equilibrium with the actual O_2 concentration according to Henry's law.

Oxidizing or reducing power in aqueous solution is limited by the fact that water itself can be oxidized or reduced according to the reactions

$$2H_2O \rightarrow O_2 + 4H^+ + 4e^- \qquad (13\text{-}39)$$

and

$$2H_2O + 2e^- \rightarrow H_2 + 2OH^- \qquad (13\text{-}40)$$

respectively. The pE values of these reactions set the limits of strength for oxidizing or reducing substances that can exist in water. The values are dependent on the pH and on the oxygen or hydrogen concentrations (or equilibrium partial pressures).

An example of the use of pE is shown by consideration of the equilibrium between sulfate and sulfide [reaction (13-41)], where the conditions under which SO_4^{2-} or sulfide ion predominates in solution can be related to this variable.

$$SO_4^{-2} + 9H_3O^+ + 8e^- \rightleftarrows HS^- + 13H_2O \qquad (13\text{-}41)$$

(Under most conditions of pH which are of interest environmentally, the predominant form of sulfide is HS^-. Other sulfur species, e.g., sulfite SO_3^{2-} and elemental sulfur, will be ignored for simplicity. In much, but not all, of the aqueous chemistry of sulfur compounds, they are not important.) $E°$ for reaction (13-41) is 0.25 V. Hence,

$$pE = \frac{E°}{0.059} + \frac{1}{8}\log\frac{[SO_4^{2-}][H_3O^+]^9}{[HS^-]} \qquad (13\text{-}42)$$

$$= 4.24 - \frac{9}{8}\,pH + \frac{1}{8}\log\frac{[SO_4^{2-}]}{[HS^-]} \qquad (13\text{-}43)$$

One can thus consider the $[SO_4^{2-}]/[HS^-]$ ratio as a function of pE at a particular pH. At a pH of 8,

$$pE = -4.76 + \frac{1}{8}\log\frac{[SO_4^{2-}]}{[HS^-]} \qquad (13\text{-}44)$$

From this, it can be seen that the concentrations of SO_4^{2-} and HS^- will be equal at pE = -4.76, while at the pE of oxygen-saturated water, 12.6, the ratio $[SO_4^{2-}]/[HS^-]$ is very large ($\log [SO_4^{2-}]/[HS^-] = 138.9$), and hence only SO_4^{2-} is in the stable form. Plots can be made for $[SO_4^{2-}]$ and $[HS^-]$ as a function of pE (the slopes of the plots are easily established as was the pH-log concentration plot for weak acids) at a particular total concentration of SO_4^{2-} plus HS^- to show the behavior graphically.

13.6 COORDINATION CHEMISTRY

13.6.1 General Aspects of Coordination Chemistry

A metal ion invariably is a species with vacant, low energy orbitals. The H^+ ion may be taken as a simple analog. Just as this species never exists in solution as such but always acquires some electron density by sharing an electron pair from another atom, so metal ions generally make some use of their empty orbitals by means of coordinate bonding of this type. The extent to which this is done depends on the energy of the outer orbitals, and, for representative metals, is roughly parallel to the electronegativity. Coordinate bond formation is very slight for the alkali metals, and increases across a period in the periodic table. This type of bonding is also greater for metal ions with small sizes and with high oxidation states. Transition metal ions that have empty inner orbitals are usually particularly strong complex formers. The picture of complex formation, i.e., donation of a nonbonding pair of electrons on an ion or molecule (the ligand) into an empty orbital on the metal, falls into the Lewis classification of acids and bases. The metal ion-donor atom bond may be highly polar, but is essentially a covalent interaction in typical systems. In extremes, however, such as with the alkali metals, it may equally well be regarded as an ion-dipole interaction.

The common donor atoms are the halide ions, the atoms O, N, S, and P in their compounds and C in a few examples such as CO and CN^-. The overall ligand may be neutral or anionic (cationic in a few rare cases), and the complex as a whole may be anionic, cationic, or neutral depending on the sum of the charges on the metal ion and ligands. While there is no general correlation between the strength of ligand behavior and charge, many of the common ligand molecules function this way as anions; for example, the carboxylate group $-COO^-$, inorganic oxoanions, and the halide ions. Coordination tends to facilitate proton loss by neutral ligands through changes in the electron density of the bonds, as has already been illustrated by the aqua complexes. Since the ligands are functioning as bases in the Lewis sense, it is reasonable to expect some relationship between base strength toward a metal ion with that toward a proton, although not a very close one since the electronic characteristics are much different. For strong Brönsted bases (i.e., ligands which form weak acids), there will be a competition between protons and the metal ion for the ligand, and complexing will be influenced by pH. Even with weak Brönsted bases (e.g., Cl^-) there may be a pH dependence, since in basic media competition from OH^- ligands for the metal ions may become important.

The number of donor atoms linked to a metal ion represents the coordination number of the complex. The most common coordination number is 6, with the donor atoms lying at the corners of an octahedron as illustrated in Fig. 13-10. A 4-coordinate, tetrahedral structure also is common, and a variety of other structures are found in particular examples.

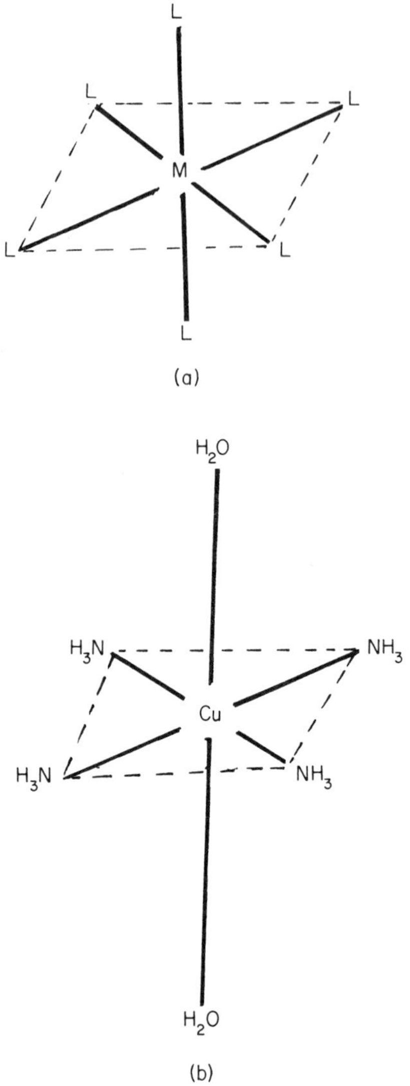

FIGURE 13-10. (a) An octahedral complex $[ML_6]$. *(b) A 6-coordinate complex with two bonds longer than the other four; a common distortion of the basic octahedral structure.*

A donor atom with more than one available nonbonding pair may donate these simultaneously to two or more metal ions, acting to bridge them together. A water molecule can do this, although it is more common with OH^- ligands.

A ligand with two or more separate donor atoms may bridge, but often these ligands are found attached to one metal ion through both donor atoms. This behavior is called chelation, and leads to more stable complexes. An example of a chelating ligand with two donor atoms (bidentate) is ethylenediamine, $H_2NCH_2CH_2NH_2$. Each N possesses one nonbonding electron pair, as in NH_3. The molecule is flexible and can bend to fit two adjacent coordination sites in most geometries. Ligands also are known with 3 (tridentate), 4 (tetradentate), 5 (pentadentate), and 6 (hexadentate) donor groups. Chelation obviously results in a ring structure.

The various metals show differing complexing tendencies and preferences for donor atoms. Ions of the alkali metals (Li, Na, K, Rb, Cs) have little electron pair acceptor tendency, although in aqueous solution they are hydrated by predominately ion-dipole interactions. The tendency for this is greater for the smaller ions, which also have the lowest energy orbitals and greatest covalency to their interactions. Complexing is minimal for this family, but by no means absent. The alkaline earth ions, especially Mg(II) and Ca(II), are very common in the environment. Owing largely to the higher charge densities which they possess, they show a rather larger tendency to attract electrons than the alkali metals, and complexing is important in their behavior. Much of the interaction can be ascribed to coulombic forces, and since the metals are quite electropositive, build-up of electron density on them is unfavorable. Their strongest complexes are formed with strongly electronegative donors such as F^- and oxygen ligands, which form highly polar bonds. Be(II) is much more covalent in its bonding than the other alkaline earths. It forms stronger complexes than the other members of its family, and in biological systems it is believed to act as a poison because it preferentially ties up coordination groups which normally are involved more weakly with Mg(II) or Ca(II) ions. Chelation is very important for the formation of stable complexes with this family. The greater stability of chelate complexes as compared to those with analogous unidentate ligands lies not so much with stronger metal-ligand bonding, but rather with an entropy effect. The number of degrees of translational freedom (disorder) is increased by replacing several unidentate ligands (e.g., water molecules) by a polydentate one.

Ions of charge greater than 2+, and transition metal ions generally, are strongly complexed by a variety of ligands. Oxygen and F^- donors are favored for electropositive metals such as Al and those of the early transition families, where build-up of negative charge on the metal is unfavorable. Nitrogen donors become favored by the midtransition element families in their lower oxidation states [e.g., Co(II), Fe(II)] although higher states still prefer F^- and oxygen donors. Some of the very heaviest and the least electropositive metals (Pt, Au, Hg) favor larger, more polarizable (more covalent) ligands such as I^- or Cl^- in preference to F^-, and sulfur or phosphorus ligands in preference to oxygen or nitrogen donors. For example, the gold in sea water is believed to be present as $AuCl_4^-$. It may be noted that the large amount of chloride in sea water for the most part is not involved in complexing because the predominant cations prefer oxygen donors such as water or carbonate.

Metal-ligand interactions are examples of Lewis acid-base behavior, and the preference of a metal ion for a particular ligand can be correlated with the concept of hard and soft acids and bases. Lewis acids and bases can be ranked according to an empirical property of hardness or softness, which is determined by such factors as size, charge, polarizability, and the nature of the electrons available for interaction. A hard acid typically is small, with a high positive charge and tightly held, chemically inactive electrons, while a soft acid is large, with a small or no positive charge and electrons in orbitals which are readily influenced by other atoms. Hard bases are small, and of high electronegativity, while soft bases are large, with easily distorted electron clouds. (Primarily, these are properties of the acceptor or donor atoms in polyatomic acids and bases. However, the properties of an individual atom in a molecule are often strongly influenced by the rest of the molecule, especially when there is extensive electron delocalization.) The examples in Table 13-2 illustrate these classes. There is a continuous range of this property, but no numerical values can be given.

The utility of the hard-soft concept is simply the rule that like prefers like, that is, interaction among species of similar hardness or softness is preferred over interactions involving a large difference in hard-soft properties. Thus, a metal such as Hg(II), which is soft, prefers to complex with sulfur rather than oxygen donors. This is not to say that complexing between hard and soft species will not take place, only that it is normally less strong than if these properties are matched.

The coordination geometry, that is, the arrangement of the donor atoms around the central metal ion, that is favored by a given element may vary with its oxidation state, with

TABLE 13-2

Some Lewis Acids and Bases on the Hard-Soft Scale

Hard acids	Intermediate acids	Soft acids
H^+, Li^+, Na^+, K^+, Be^{+2}, Mg^{+2}, Ca^{+2}, Al^{+3}, Cr^{+3}, Fe^{+3}, Co^{+3}, metals in high oxidation states	Fe^{+2}, Co^{+2}, Ni^{+2}, Cu^{+2}, Zn^{+2} Pb^{+2}, Sn^{+2}	Cd^{+2}, Pd^{+2}, Pt^{+2}, Hg^{+2}, Cu^+, Ag^+, Tl^+
Hard bases	**Intermediate bases**	**Soft bases**
F^- , OH^-, H_2O, NO_3^-, CO_3^{-2}, PO_4^{-3}, NH_3, organic amines	Br^-, NO_2^-	I^-, CN^-, CO organic sulfur, phosphorus, and arsenic compounds

the nature of the donor atoms, and with other factors. Comparatively few ions form complexes exclusively of a single geometry, although most have a limited number of preferences. The chemical behavior of metal ions in systems in which geometry is critical, e.g., biological activity, may be related to this property. As has been stated, an octahedral, six-coordinate geometry is the one most commonly encountered, and the majority of the aqua complexes can be expected to exhibit this. Not all octahedral complexes are regular, even if all ligands are identical. Six coordinate complexes of Cu(II) are somewhat extreme examples of this, since the usual configuration has four "normal" bonds in a plane, with the remaining two ligands attached by long, weak bonds above and below this plane. Indeed, many Cu(II) complexes often are referred to as square planar, although in fact the additional two weak interactions usually are present, especially in solution, where solvent molecules can take up these positions. The striking deep-blue complex formed when ammonia is added to a solution of a Cu(II) salt has this highly distorted octahedral shape, with four NH_3 ligands in a plane, but two H_2O molecules still weakly attached (Fig. 13-10).

With more highly polarizable ligands than water or ammonia, for example, halide ions, the ions of the first transition series metals tend to form tetrahedral, 4-coordinate complexes. Truly 4-coordinate square planar compounds are found for some ions, chiefly those which have eight electrons [e.g., Pt(II), Au(III)]. This is related to the particular stabilization which this unique number of d electrons can lead to for this geometry in favorable cases. A few other ions are encountered commonly but not exclusively in linear, 2-coordinate complexes [e.g., Ag(I), Hg(II)]. Again, this is related to a particular electronic configuration: ten d electrons. Other coordination numbers and geometries are less common in environmental systems.

In aqueous systems, water is the most abundant ligand. Distortion of the electron distribution on a water molecule as a result of coordination causes the protons to be more readily ionizable, as already mentioned. The result may be the formation of hydroxo (HO^-) or oxo (O^{-2}) complexes. That is, the aqua complex[3] $[M(H_2O)_x]^{n+}$ (x is usually 6) may be partially or even completely deprotonated, depending on the

[3]*It is the convention to write the formula for a complex ion or molecule in square brackets in order to define the coordination sphere. This should not be confused with the use of square brackets to denote concentrations.*

nature of M^{n+} and the pH. This behavior is illustrated by Fe^{3+} or Al^{3+}. One has loss of protons; e.g.,

$$[Al(H_2O)_6]^{3+} + H_2O \rightleftarrows H_3O^+ + [Al(H_2O)_5OH]^{2+} \quad (13\text{-}45)$$

$$[Al(H_2O)_5]^{2+} + H_2O \rightleftarrows H_3O^+ + [Al(H_2O)_4(OH)_2]^+ \quad (13\text{-}46)$$

Ultimately this may lead to oxo species through loss of two protons from one O, and finally to oxo-anions such as $[AlO_3]^{3-}$ if the medium is basic enough. There may be polymerization; e.g.,

$$2\ [Al(H_2O)_5OH]^{2+} \rightarrow OH^- + [(H_2O)_5Al\text{-}\overset{\overset{\displaystyle H}{|}}{O}\text{-}Al(H_2O)_5]^{4+} \quad (13\text{-}47)$$

where the hydroxo ligand bridges two metal ions. With further deprotonation and polymerization, repulsion among ligands usually results in a change in coordination number from octahedral 6 to tetrahedral 4. After aging, we have compounds such as FeO(OH) and AlO(OH) involving bridging ligands around each metal ion to give a three-dimensional cross-linked structure. The reactions are all pH sensitive and often reversible, but once the final highly polymerized products are formed, they may be quite resistant to dissolution upon increase of acidity.

The extent of these hydrolysis reactions can be correlated roughly with size and charge of the metal ion, being more extensive for the smallest and most highly charged ion at a given pH. In general, one has the following:

1+ ions: form simple aqua complexes over most of the pH range.

2+ ions: aqua complexes predominate in acidic media, but hydroxo species occur in basic media. Larger alkaline earths remain aqua complexes even in strongly basic media, while very small ions such as Be^{2+} are much more readily hydrolyzed.

3+ ions: chiefly present as hydroxo complexes in the pH range of natural waters.

4+ ions: most of these form hydroxo or oxo complexes in all but highly acidic solutions.

The highly polymerized hydroxo precipitates that frequently form on hydrolysis of the 3+ and 4+ ions are often of a gelatinous nature, resulting from the irregular cross-linked network that is built up. These usually rearrange and become more crystalline with time. The formation of such precipitates can be used in flocculation processes for water purification as discussed in Chapter. 15.

13.6.2 Complex Stability and Lability

The formation of a complex $[ML_6]^{n+}$ in solution is part of an equilibrium process which can be represented by a series of steps, each described by its own equilibrium constant (often called formation constant);

$$[M(H_2O)_6]^{n+} + L \rightleftarrows [M(H_2O)_5L]^{n+} + H_2O \qquad (13\text{-}48)$$

$$K_1 = \frac{[[M(H_2O)_5L]^{n+}]}{[[M(H_2O)_6]^{n+}][L]} \qquad (13\text{-}49)$$

$$[M(H_2O)_5L]^{n+} + L \rightleftarrows [M(H_2O)_4L_2]^{n+} + H_2O \qquad (13\text{-}50)$$

$$K_2 = \frac{[[M(H_2O)_4L_2]^{n+}]}{[[M(H_2O)_5L]^{n+}\ [L]} \qquad (13\text{-}51)$$

The overall process is

$$[M(H_2O)_6]^{n+} + 6L \rightleftarrows [ML_6]^{n+} + 6H_2O \qquad (13\text{-}52)$$

$$K_{overall} = \prod_i K_i = \frac{[[ML_6]^{n+}]}{[[M(H_2O)_6]^{n+}][L]^6} \qquad (13\text{-}53)$$

These equilibria must hold along with ionization or precipitation equilibria.

A large value for the formation constant means that the formation equilibrium lies far to the right. In the presence of excess ligand, for example, the amount of "free" metal ion would be small in this case. However, it is important to distinguish between thermodynamically stable systems in the above sense, and those that are kinetically inert. Complexes in solution normally undergo continuous breaking and remaking of the metal ligand bonds. (This is true of many other systems also. For example, a weak acid that is largely undissociated according to its ionization constant nevertheless exchanges protons at a very fast rate with solvent water molecules.) If the bond breaking step is rapid, and consequently ligand exchange reactions are rapid, the complex is referred to as labile. If the ligand exchange reactions are slow, the complex is said to be inert. Lability or inertness has no direct relation to thermodynamic stability. Many metal ions complexed by unidentate ligands will undergo complete exchange in a time of fractions of a second; some are much slower, hours or days. These inert complexes are associated with particular electronic configurations of the metal ion. Exchange rates are lower with multidentate ligands. Inert complexes in particular have properties different from both the free metal ion and ligands. An example is $[Fe(CN)_6]^{3-}$, in which the cyanide groups do not exhibit the toxicity of free CN^-.

Complex stability can be correlated with several properties of the metal ion and the ligand. The following are most important:

(a) Size and oxidation state of the metal. Smaller size and higher positive oxidation state lead to stronger complexing. This can be understood in terms of the stronger bonding that arises with shorter metal-ligand distances and the greater coulombic contribution from a greater charge.

(b) Chelation. The free energy change on complex formation depends on both the enthalpy and the entropy changes of the reaction. The enthalpy change is made up largely from the metal-ligand bond energies, while the entropy change involves the change in the degrees of freedom of the system. The number of independent molecules increases when unidentate ligands are replaced by a polydentate one, resulting in a positive contribution to the entropy change and a larger negative free energy change than in the case of a one-to-one exchange. This is a widely accepted explanation for the observed greater stability of chelated systems compared to those of analogous unidentate ligands; the greater the degree of chelation, the greater is the stability. The size of the chelate ring formed is of prime importance. Five-membered rings (including the metal atom) are most favored, with six-membered rings a close second, and other sizes of minor importance.

(c) The electronic configuration of the metal ion. The degeneracy of the d orbitals of a transition metal ion is destroyed when the ion is surrounded by ligands. The d orbitals are said to be split by the ligand field. The splitting pattern depends on the geometry of the complex, while the magnitude of the splitting depends on the nature of the ligand and the metal ion. Occupancy of the lower energy orbitals contributes to the stability of the system (the ligand field stabilization energy) although this is counteracted if the higher energy orbitals are also filled. Thus, this contribution to stability will depend on both the electronic configuration and the magnitude of the d orbital splitting. With a given metal ion, the common ligands usually have the following relative splitting effects: $I^- < Br^- < Cl^- < F^- < H_2O$ (and most other oxygen donors) $< NH_3$ (and amines) $< CN^-$. With octahedral complexes, the electronic configurations leading to a large ligand field stabilization energy are d^3, d^6 (with strongly splitting ligands), and d^8, while d^0, d^5 (with weakly splitting ligands), and d^{10} have no stabilization from this source.

(d) The hard-soft match of the metal and ligand, as already discussed.

13.6.3 Some Applications of Complexing

The formation of complexes may tie up a metal ion to such an extent that solubility is increased considerably; that is, the concentration of free metal ions required to achieve saturation according to the solubility product requirements is less readily reached. This is often desirable when waters containing dissolved minerals are employed for industrial or domestic use, and is the reason for the presence of phosphates as "builders" in detergent formulations. The calcium and magnesium ions in hard water form insoluble salts with soaps and interfere with the action of most detergents. The soap precipitates have a curdlike character and are very difficult to wash away. Hard waters also require larger amounts of the surfactant for adequate cleaning. The Ca(II) and Mg(II) ions can be held in solution by complexing, and certain phosphate ions have been widely used for this purpose, most commonly, sodium tripolyphosphate, $Na_5P_3O_{10}$. The simplest form of phosphate, orthophosphate PO_4^{3-} is tetrahedral. Polyphosphates are based on this structure, but with shared oxygens. The tripolyphosphate ion has the structure shown below. Unlike orthophosphate, this

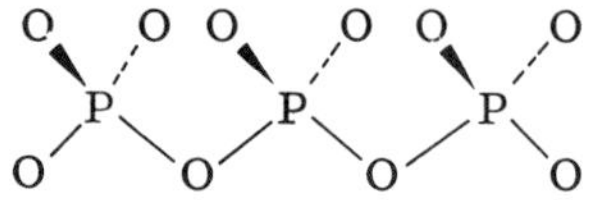

Tripolyphosphate anion

ion can form chelates having six-membered rings, and is potentially tridentate. The extra stability introduced is illustrated by the following equilibrium constants:

$$Ca^{2+} + HPO_4^{\;2-} \rightleftarrows [CaHPO_4]; \quad K = 5.02 \times 10^2 \qquad (13\text{-}54)$$

$$Ca^{2+} + P_3O_{10}^{\;5-} \rightleftarrows [CaP_3O_{10}]^{3-}; \quad K = 1.25 \times 10^8 \qquad (13\text{-}55)$$

In addition to its complexing role, association of protons with the polyphosphate ion produces a basic solution that improves the cleaning action of soaps and detergents by ensuring complete ionization of the hydrophilic anionic functional group. The polyphosphate is slowly hydrolyzed to simple phosphate species.

With low cost and no apparent toxicity properties, phosphates received extensive use in detergent formulations, with typically 35-50% of the detergent materials composed of sodium tripolyphosphate. As this material entered water systems, it was accompanied by extensive increases in the growth of algae, since phosphate is an essential nutrient and an effective fertilizer. The death and decay of these organisms seriously deplete the dissolved oxygen content of a body of water through increased BOD (see Section 6.6). This is the process of eutrophication, which has received much attention. As a result of this effect, phosphates have been banned in detergents in some places (e.g., Canada, New York, Indiana, Michigan). The action of phosphates in water systems is considered further in Chapter 14.

We can predict from the general coordination properties discussed for Ca(II) that a satisfactory complexing agent for replacement of phosphate in detergents would have to be a chelating agent, since only then would adequate stability of the complex be likely. Oxygen would be the chief donor atom, since only this donor atom is sufficiently hard to interact strongly with Ca(II) while at the same time being part of a more elaborate molecule. There are also requirements of non-toxicity, biodegradability, and low cost. The required complexing ability can be achieved with many organic molecules. For example a very effective chelating agent for calcium is ethylenediaminetetraacetic acid EDTA.

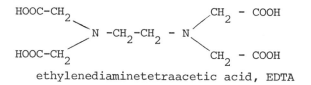

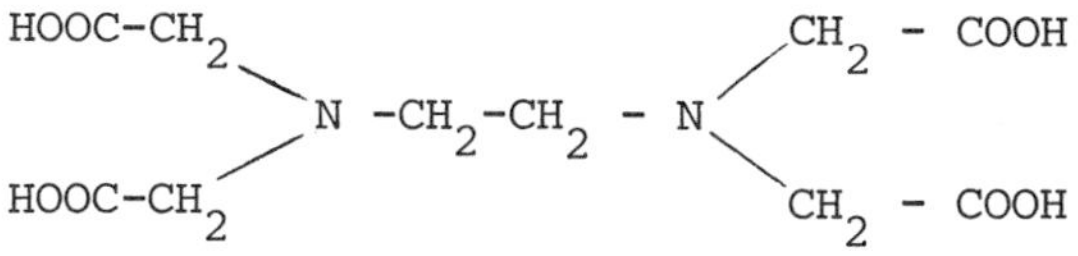

ethylenediaminetetraacetic acid, EDTA

This is potentially hexadentate. It coordinates as the anion, with an oxygen from each ionized carboxyl group and the two nitrogens being the donor atoms. (It may be added parenthetically that while EDTA is potentially hexadentate, it may not be so in all of its complexes. For example, a water molecule may occupy the sixth coordination site leaving a carboxyl group free.) EDTA is widely used in analytical chemistry and for other purposes where it is necessary to have metal ions tied up in soluble form (sequestration). While EDTA itself has not seriously been proposed for use in detergents, a related substance, nitrilotriacetic acid NTA, was suggested as a phosphate replacement, and large scale plans for its introduction were underway in about 1970.

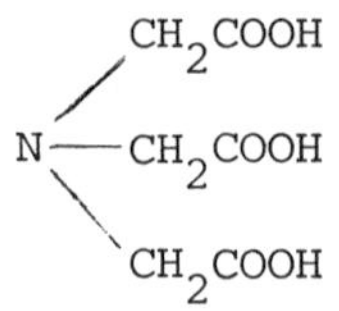

nitrilotriacetic acid, NTA

Again, it is an oxygen from each ionized carboxylate group and the nitrogen which can act as the donor atoms. While this substance does not pollute in the manner of phosphate, and is biodegradable under aerobic conditions, it is not degraded under anaerobic conditions such as occur in septic tanks. There is evidence that NTA enhances the toxic effects of heavy metals (e.g., cadmium), particularly in connection with the production of birth defects. The strong complexes that are formed by NTA with heavy metal ions may permit the metal to be taken up more readily by the cell than under ordinary circumstances. At any rate, NTA has been withdrawn from detergent applications at the present time.

It may be noted here that one alternative to sequestration of the calcium ions is their precipitation in a form that is not obnoxious. Calcium carbonate is one such compound, and this can be precipitated from hard water containing a reasonable carbonate content and high pH. Thus, Na_2CO_3 is used in some detergents for this purpose. Borate (borax) has also been used effectively in detergents for many years. There is some concern that its widespread use could lead to hazards, but as yet there is no proof of this.

Sequestration of metal ions as a means of preventing them from interfering with useful processes or materials is not limited to the laundry. Trace amounts of metals can have undesirable catalytic activity that can be prevented by the presence of complexing agents (e.g., in food). Some forms of metal poisoning have been treated by a strong ligand, e.g., EDTA, and the lack of essential trace metal nutrients in soils has been overcome by addition of complexing agents that extract the needed metal ions from insoluble minerals to produce soluble forms to aid plant growth.

13.6.4 Complexing in Natural Systems

Most natural water systems contain an excess of Ca(II) compared to the available strong ligands. Consequently, although calcium complexes are rather weak, the amount of ligand left over to complex with trace metals is small, perhaps often negligible. However, there are examples where complex formation is believed to have importance in natural waters; for example, higher than expected concentration of dissolved metal ions, solubilization of metal ions for transport in soils and plants, and sometimes unexpected chemical behavior of trace metals. The enrichment of aqueous trace metals by plants tends to follow the normal stability sequence for metal ion complexes [this is, in part, that the complex stability with a given ligand follows the order Mn(II) < Fe(II) < Co(II) < Ni(II) < Cu(II) > Zn(II)]. This suggests that trace metal uptake by plants is related to complex formation. There is evidence that unpleasant outbreaks of algal growth (Florida "red tide," for example) are linked to extraction of ions as complexes of vegetation decomposition products from soils.

Most naturally occurring organic ligands are decomposition products or by-products of living organisms. There are several main types.

(a) Humic acid, fulvic acid, humin, etc. Soils and sediments normally contain organic matter derived from decaying vegetation. Much of this organic matter is made up of fulvic or humic acid materials, or humin. These seem to be a related series distinguished primarily on the basis of solubility differences. Fulvic acid is soluble in both acid and base, humic acid in base only, and humin is insoluble. The molecular weights increase in this same sequence, ranging from 1-2000 to perhaps 100,000. They are not definite simple compounds, but polymeric materials. Their exact origin is uncertain; one theory suggests that they are oxidized and decomposed derivatives of lignin; another that they are formed by polymerization of simple decomposition products from plant materials like phenols and quinones. The structures contain aromatic rings with oxygen, hydroxyl, and carboxyl groups, and some nitrogen groups, all of which have potential coordination abilities. Typical units might be

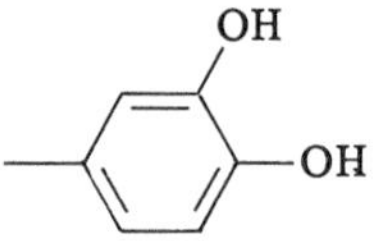

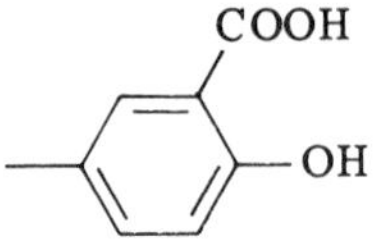

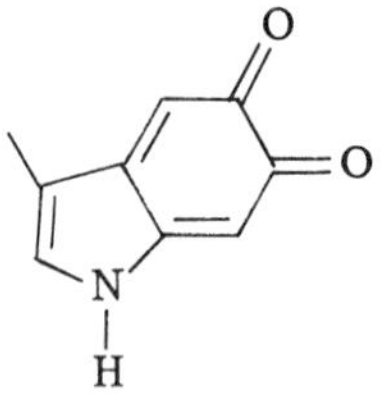

It is well established that strong interactions between these materials and metal ions take place; humic substances have a high capacity for binding metal ions. They form a potential reservoir of trace metals in soils. Humic substances are often associated with color in water, and sometimes are accompanied by a high iron content.

(b) Polyhydroxy compounds: carbonates, sugars. The grouping

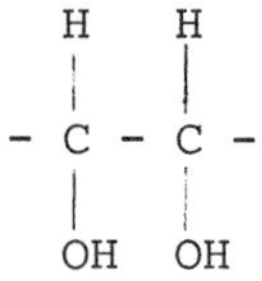

can chelate, although it does not normally form strong complexes. These substances are widespread in products of organisms and in carbohydrate decomposition products.

(c) Amino acids, peptides, and protein decomposition products. A simple amino acid, e.g.,

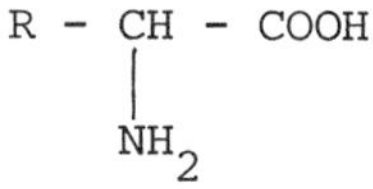

is an effective chelating agent. Such molecules can be products of protein hydrolysis, although there is no definite evidence for their importance in natural water systems. Peptides and related substances are more important in this respect; for example, the ferrichromes are very stable, soluble iron complexes. These ligands, which have high specificity for Fe(III), are common in microorganisms, and are present as decomposition intermediates. They seem to be useful in transporting iron into cells and making it available to iron-containing enzymes. The iron is bound in an octahedron of oxygen atoms from three groups such as

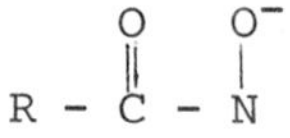

amino acid chain

(d) Porphyrins and related compounds. Compounds based on the tetrapyrole nucleus make up an important class of naturally occurring complexing agents. The parent compound is porphin, the structure for which is given. Considerable scope for electron delocalization is possible in this ring structure (i.e., resonance forms exist) and partly as a result of this, these compounds exhibit a high stability. Substituents on the porphin skeleton, and minor changes to the ring structure, are found in the various naturally occurring compounds of this class.

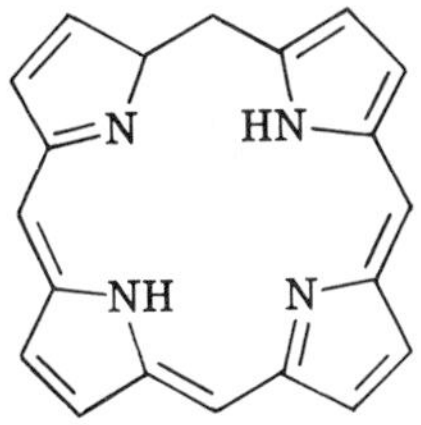

The porphyrin structure is planar, with sufficient room in the center of the ring system for binding a metal ion, as shown below. The electron delocalization referred to above may involve metal orbitals and electrons as well as those of the organic molecule. Metal ions that tend to form octahedral

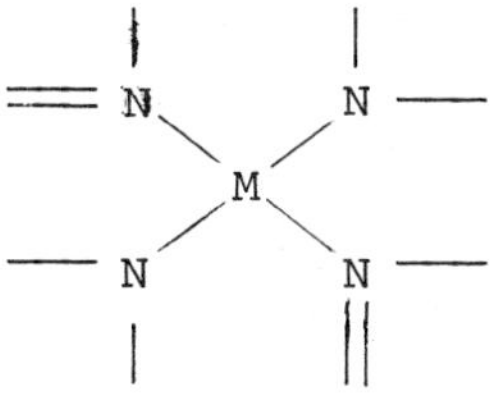

complexes may have ligands above and below this plane; bonds here are also influenced by the electron delocalization in the plane. Examples of compounds of this kind are the iron-containing hemoproteins, including hemoglobin and myoglobin; cobalt-containing vitamin B_{12} and related enzymes (based on the corrin structure); and chlorophyll which has been discussed in Chapter 11 (see Fig. 11-3).

The structure of the metal ion portions of hemoglobin and myoglobin, used in biological oxygen transport and storage, are shown below. In addition to the four porphyrin nitrogen atoms in the coordination sphere of the Fe(II) ion, a fifth nitrogen from a histidine group in the amino acid portion of the molecule is attached to one of the remaining

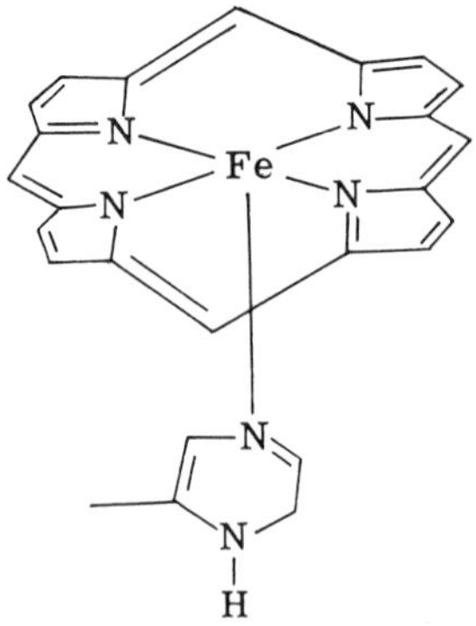

coordination sites. The sixth, active site of the Fe(II) may have the water molecule replaced reversibly by an oxygen molecule. That is, O_2 is taken up when its partial pressure is high, and released when it is low. Other ligands may be attached to the sixth site. If such ligands are not replaced readily, as is the case with CO, which bonds strongly to the Fe(II) in hemoglobin, the oxygen transport function is blocked. This illustrates one process by which the biological function of a metal ion-containing species can be disrupted.

The exact behavior of the metal ion in this and analogous systems is determined not only by the immediate, coordinate environment which influences the electron density on the metal, but also by the extensive protein structure which is part of the molecule. This protein structure may have indirect effects on electron density, and may exhibit important steric effects which control the overall reactions. Thus, iron-containing enzymes with direct coordination structures similar to hemoglobin, but different structures of the protein chains, may not exhibit reversible oxygen uptake, but rather act as electron transfer agents through the Fe(II) $\rightarrow$ Fe(III) conversion.

The availability of trace metals in soil evidently depends on complex formation. Most soil sources of essential metals, which include V, Mo, Mn, Co, Fe, and Cu, are insoluble, and the small amounts that can be made available through acidic attack of the oxides, carbonates, silicates, etc., in which they are present appear to be inadequate for normal plant needs. Large amounts may be available through a reservoir of coordinated metal held in humic acids in the soil, and which may be released on further decomposition, or an attack by agents such as citric or tannic acids, or other substances generated by plant roots.

Complexation may provide a buffering of free metal ion concentration. Excess metal may be tied up as a complex, or released from a complex form, in order to maintain the equilibrium conditions required by the complex stability constant. A regulation of this sort is probable at least in organisms. It is less certain in the nonliving environment, but organic-rich sediments may have some action of this kind in water bodies.

Complex formation with organic materials also may influence mineral weathering reactions, largely through solubilizing a weathering product which otherwise would be insoluble, or through the effect of complexing on the stability of different oxidation states of an element. For example, in the pyrite (FeS_2) oxidation to be discussed later, organic materials and microorganisms can have a considerable catalytic effect. Fe^{2+} is oxidized to Fe^{3+} only slowly; the reaction (which is first order) is 5% complete in 150 days at normal temperatures. This may be faster if it can proceed as follows:

$$Fe^{2+} + \text{organic} + O_2 \rightarrow Fe^{3+}\text{-organic complex} \qquad (13\text{-}56)$$

Ligands such as tannic acid, amino acids with SH groups, or phenol types are particularly effective.

We may also encounter more complicated cycling reactions.

$$[Fe^{3+}\text{-organic complex}] \longrightarrow Fe^{2+} + \text{oxidized organic compound} \qquad (13\text{-}57)$$

$$Fe^{2+} + \text{organic ligand} + O_2 \longrightarrow [Fe^{3+}\text{-organic complex}] \qquad (13\text{-}58)$$

Organic materials in ground water lower pE and result in solubility and stability of Fe^{2+} and Mn^{2+}. (Normally, Mn is present as insoluble oxide, Fe as FeO(OH).) Presence of either of these ions is an indicator of organic contamination of the ground water source.

Complex formation in sea water is significant, but the predominant cations, Ca(II), Mg(II), Na(I), and K(I), are largely present as the aqua species. About 10% of the alkaline earth ions, and 1% of the alkali metal ions in sea water are complexed, mostly by the sulfate ion. Relative concentrations are such (Section 15.2) that this takes roughly half of the sulfate ion concentration. Carbonate and bicarbonate ions are present in much lower concentrations, but the amounts present are extensively complexed with the more abundant cations. This complexing is often taken care of in equilibrium calculations through the use of "stoichiometric" constants valid for sea water conditions.

BIBLIOGRAPHY

General Properties of Water and Water Systems

M. Overman, "Water." Doubleday, Garden City, New Jersey, 1969.

F. Frank, ed., "Water, A Comprehensive Treatise," Vols. I and II. Plenum, New York, 1972.

J. P. Riley and R. Chester, "Introduction to Marine Chemistry." Academic Press, New York, 1971.

R. A. Horne, "Marine Chemistry." Wiley (Interscience), New York, 1969.

H. L. Penman, The water cycle, in "Chemistry in the Environment, Readings from Scientific American, p. 23. W. H. Freeman, San Francisco, California, 1973. (From Sci. Amer., Sept., 1970.)

C. W. Kurse, Our nation's water, its pollution control and management, in "Advances in Environmental Science" (T. N. Pitts and R. L. Metcalf, eds.), Vol 1, p. 41. Wiley (Interscience), New York, 1969.

E. T. Chanlett, "Environmental Protection," Chapter 3. McGraw-Hill, 1973.

Equilibrium in Aqueous Solution

J. N. Butler, "Ionic Equilibrium." Addison-Wesley, Reading, Massachusetts, 1964.

W. Stumm and J. J. Morgan, "Aquatic Chemistry." Wiley, New York, 1970.

L. G. Sillén, Graphic presentation of equilibrium data, in "Treatise on Analytical Chemistry" (P. J. Elving and E. B. Sandell, eds.), Vol. 1, Chapter 8. The Interscience Encyclopedia, Inc., Interscience, New York, 1959.

J. C. Morris and W. Stumm, Redox equilibria and measurement of potentials in the aquatic environment, in "Equilibrium Studies in Natural Water Systems" (R. F. Gould, ed.), Chapter 13. Advances in Chemistry Series No. 67, American Chemical Society, Washington, D.C., 1967.

A. Disteche, "The effect of pressure on dissociation constants and its temperature dependency, in "The Sea" (E. D. Goldberg, ed.), Vol. 5, Chapter 4. Wiley (Interscience), New York, 1974.

W. G. Bresk, Redox levels in the sea, in "The Sea," (E. D. Goldberg, ed.), Vol. 5, Chapter 4. Wiley (Interscience), New York, 1974.

S. Y. Tyree, Jr., The nature of inorganic solute species in water in "Equilibrium Studies in Natural Water Systems" (R. F. Gould, ed.), Chapter 8. Advances in Chemistry Series No. 67, American Chemical Society, Washington, D. C., 1967.

Coordination Chemistry and Natural Ligand Materials

F. Baslo and R. Johnson, "Coordination Chemistry." W. A. Benjamin, New York, 1964.

M. M. Jones, "Elementary Coordination Chemistry." Prentice-Hall, Englewood Cliffs, New Jersey, 1964.

R. J. Barsdate, in "Symposium on Organic Matter in Natural Waters" (D. W. Hood, ed.), p. 485. Institute of Marine Science Occasional Publication No. 1, University of Alaska, 1970.

G. Eglinton and M. T. J. Murphy, eds., "Organic Geochemistry." Springer-Verlag, New York, 1969.

Soap: Some companies' answer to phosphate-containing detergents, *Chem. Eng. News,* Aug. 17, 1970, page 18.

A. L. Hammond, "Phosphate replacements: Problems with the washday miracle, *Science 172,* 361 (1972).

C. F. Baes, Jr. and R. E. Mesmer, "The Hydrolysis of Cations." ORNL-NSF Ecology and Analysis of Trace Contaminants Program Publication 3, Parts I and III, 1974; Parts II and IV, 1975.

D. F. Martin, Coordination chemistry of the oceans, in "Equilibrium Studies in Natural Water Systems" (R. F. Gould, ed.), Chapter 12. Advances in Chemistry Series No. 67, American Chemical Society, Washington, D.C., 1967.

M. N. Hughes, "The Inorganic Chemistry of Biological Processes." Wiley, London, 1972.

W. Stumm, Metal ions in aqueous solutions, in "Principles and Applications of Water Chemistry" (S. D. Faust and J. V. Hunter, eds.), p. 520. Wiley, New York, 1967.

G. L. Eichorn, Coordination compounds in natural products, in "Chemistry of the Coordination Compounds" (J. C. Bailer, ed.). Reinhold, New York, 1956.

14
THE ENVIRONMENTAL CHEMISTRY OF SOME IMPORTANT ELEMENTS

14.1 INTRODUCTION

There are some 90 naturally occurring elements in the environment, each with its own chemistry. Those elements present in high abundance determine the nature of the environment as a whole through the properties and behavior of themselves and their compounds. These, and some of lesser abundance, have important biological roles. This chapter will consider how the chemistry of some of the most important elements relate to the properties of the environment and to biological effects.

The elements that are believed to be essential for the growth and development of organisms are listed in Table 14-1. Many of these elements are present in trace amounts, and it is possible that others will be found that are essential at the trace level. It is extremely difficult to eliminate the ingestion of very low levels of virtually any element by an organism, as must be done to prove that absence of any element is deleterious to the organism's development. Of more immediate concern is the prevention of the ingestion of materials in amounts that are toxic. This includes some of the essential trace elements, since toxic effects may be encountered for some elements if the amounts ingested exceed some limit. Biological toxicity and need both depend upon the organism considered. Effects of an element in both respects may depend upon the chemical form in which it is present.

The chemistry of the elements is best organized around the periodic table (Fig. 14-1). As can be seen, the essential elements occur chiefly among the lighter members of the table. Toxicity, with no known biological use, is encountered among many of their heavier analogs. The general chemical trends in the periodic table are discussed in many textbooks of inorganic chemistry. The properties of a family of representative elements show a general similarity, but with trends governed by the smaller size and higher electronegativity of

TABLE 14-1

Biologically Essential Elements

Elements	General biological function or occurrence
Hydrogen	necessary constituent of virtually all organic compounds
Boron	required by algae and plants; function not known
Carbon	the basic element for organic compounds
Nitrogen	found in amino acids and derivatives, and other organic compounds
Oxygen	present in many organic materials and needed for respiration
Fluorine	small amounts seem to be essential; strengthens teeth
Sodium	a major component of biological electrolytes; nerve action
Magnesium	enzymes, chlorophyll
Silicon	structural material in diatoms
Phosphorus	used in energy transfer processes, DNA and RNA, and bones and teeth
Sulfur	some amino acids, proteins
Chlorine	a major component of biological electrolytes
Potassium	component of biological electrolytes; nerve action
Calcium	bones, enzymes, biological electrolyte component
Vanadium	necessary in plants, marine animals
Chromium	enzymes
Manganese	enzymes
Iron	enzymes, hemoglobin (O_2 transport); biological redox processes
Cobalt	enzymes
Copper	enzymes
Zinc	enzymes
Arsenic	function uncertain; evidence for need in rats
Selenium	liver action, some plants
Bromine	may be necessary for animals
Molybdenum	enzymes
Tin	functions uncertain, but seems to be needed in rats
Iodine	thyroid hormones

Transition Elements $d^{1-10}s^{0-2}$ (usually s^2)

Alkali Metals s^1	Alkaline Earths s^2											s^2p^1	s^2p^2	s^2p^3	s^2p^4	Halogens s^2p^5	Noble gases
													1H				2He
3Li	4Be											5B	6C	7N	8O	9F	10Ne
11Na	12Mg											13Al	14Si	15P	16S	17Cl	18Ar
19K	20Ca	21Sc	22Ti	23V	24Cr	25Mn	26Fe	27Co	28Ni	29Cu	30Zn	31Ga	32Ge	33As	34Se	35Br	36Kr
37Rb	38Sr	39Y	40Zr	41Nb	42Mo	43Tc	44Ru	45Rh	46Pd	47Ag	48Cd	49In	50Sn	51Sb	52Te	53I	54Xe
55Cs	56Ba	57La	72Hf	73Ta	74W	75Re	76Os	77Ir	78Pt	79Au	80Hg	81Tl	82Pb	83Bi	84Po	85At	86Rn
87Fr	88Ra	89Ac															

Inner transition elements

$f^{1-14}d^1$ or $0s^2$

58Ce	59Pr	60Nd	61Pm	62Sm	63Eu	64Gd	65Tb	66Dy	67Ho	68Tm	69Tm	70Tb	71Lu
90Th	91Pa	92U	93Np	94Pu	95Am	96Cm	97Bk	98Cf	99Es	100Fm	101Md	102Nc	103Lw

FIGURE 14-1. A standard form of the perodic table with atomic numbers and symbols, and electronic configurations characteristic of the families. Elements known to be essential in organisms are shown cross hatched.

the lighter elements. Among the representative metals, this is reflected in stronger and more covalent bonding of the lighter members of a family in their typical compounds, but higher coordination numbers and more ionic character for the heavier members. Across a period, the number of valence electrons and the electronegativity increase, with a change taking place from metallic to nonmetallic character. The transition elements owe much of their chemistry to their incompletely filled d orbitals. Complex formation is particularly important among these elements. The heavier transition elements and some of the representative metals immediately following them in the periodic table form highly covalent interactions with many of the nonmetals. They tend to be soft acids.

14.2 THE CARBON CYCLE, CO_2, AND CARBONATES

The most important acid system in nature originates with carbon dioxide CO_2. A large amount of this is present in water, predominantly in the ocean, and the carbonate equilibrium makes up an important buffer system. Carbon dioxide is the beginning and the end product of biological processes, and its presence in the atmosphere influences climate (Chapter 3). Table 14-2 is one estimate of the relative distribution of carbon in nature. Most carbon is present as carbonate rock or as organic sediment material (including oil and coal), the greater portion of which is essentially permanently tied up. The material in the atmosphere and in water is more readily available for environmental and biological activities. The carbon is involved in biological and geological processes that can be represented by the cycle given in Fig. 14-2. Numbers are based on grams of C per square centimeter of earth surface; rates are per year.

Essentially all of the transformations in this cycle involve CO_2 or, in aqueous media, carbonate or bicarbonate ions. Incomplete combustion of reduced forms of carbon produces small amounts of carbon monoxide, especially from internal combustion engines. Much of this also is produced from natural sources. The natural level of CO in the atmosphere is uncertain, but is probably <0.1 ppm. This can be exceeded locally, but the lifetime of CO in the atmosphere is short. Although a variety of chemical and photochemical reactions that would convert CO to CO_2 are possible, in general their rates appear to be too slow to account for the disappearance. The primary sink for CO appears to be organisms in the soil, which effectively absorb this compound.

TABLE 14-2

Relative Distribution of Carbon in Nature

CO_2 in the atmosphere	1
in water as $H_2CO_3^*$ [a]	0.3
HCO_3^-	48.7
CO_3^{-2}	6.0
living organic	0.01
dead organic	4.4
Total in water	59.41
in sediments as organic carbon	10,600
inorganic (carbonates)	28,500
on land as organic carbon	1.2

[a] $H_2CO_3^*$ is both molecular H_2CO_3 and hydrated CO_2 molecules.

Normal soils have been shown to exhibit a rapid uptake of CO in the laboratory, but this uptake is eliminated upon sterilization.

The toxicity of carbon monoxide is well known and is the cause for the concern over high carbon monoxide concentrations which can develop in areas of heavy automobile traffic. The effects of chronic, sublethal CO levels on health are not fully understood. Higher concentrations such as can build up in a closed room through a faulty space heater, or in a closed car through a leaking exhaust system, can quickly be fatal. Carbon monoxide is toxic through attachment to the coordination site of the iron atom in hemoglobin (See Section 13.6.4).

The estimated turnover rates in the carbon cycle indicate that both the biological and geological parts are essentially balanced, with the exception of the step involving the burning of fossil fuels, which increases atmospheric CO_2 (Chapter 3). This will be at least partially offset by an increase in dissolved oceanic CO_2. In evaluating atmospheric CO_2 effects, we are thus concerned with the gas-solution equilibria as well as the acid-base and carbonate solubility equilibria in solution, which will also be influenced by the dissolved CO_2 concentration. The rate at which equilibrium can be achieved is of equal importance. This may take a few years for surface waters (i.e., to depths of less than

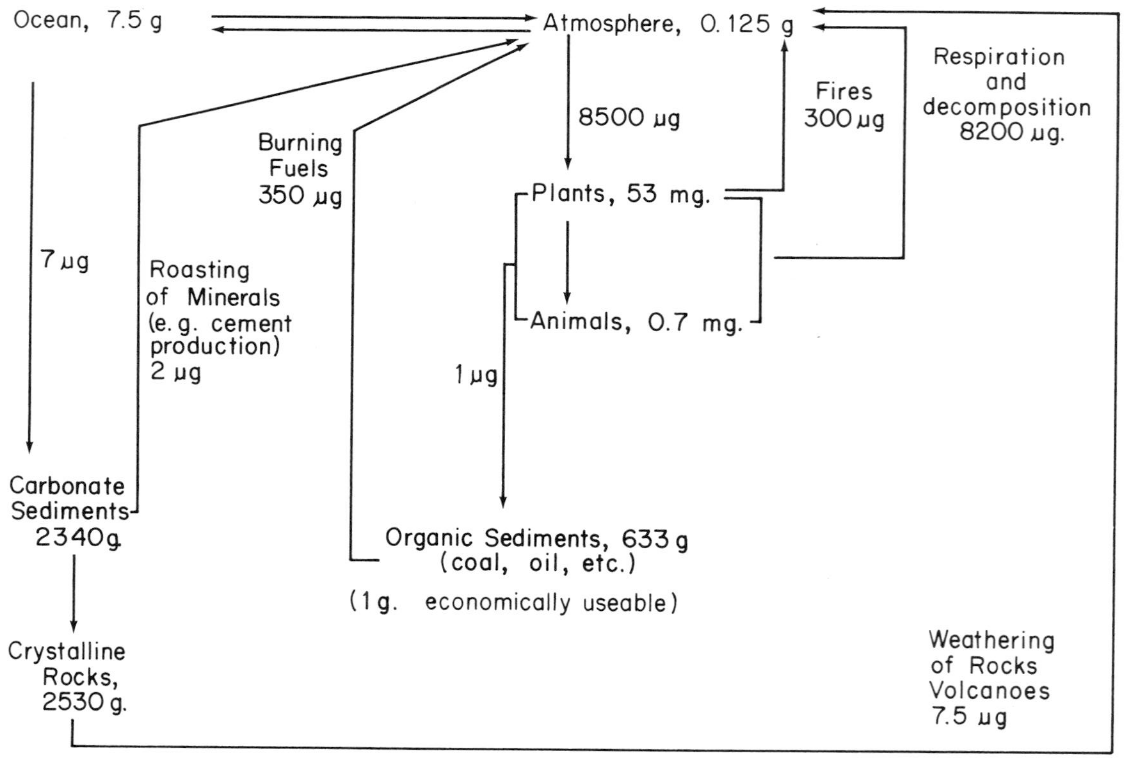

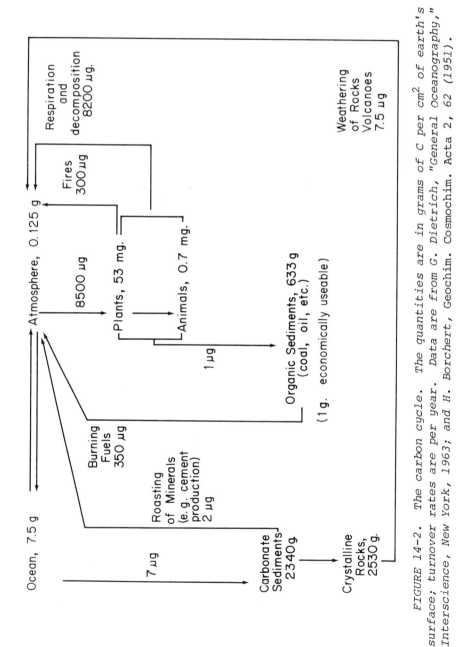

FIGURE 14-2. The carbon cycle. The quantities are in grams of C per cm^2 *of earth's surface; turnover rates are per year. Data are from G. Dietrich, "General Oceanography," Interscience, New York, 1963; and H. Borchert,* Geochim. Cosmochim. Acta 2, *62 (1951).*

100 meters), but the order of 1000 years for deeper ocean layers which make up well over 90% of the total. Carbon dioxide and carbonate equilibria in water are discussed in Section 13.3.

14.3 SULFUR AND THE SULFUR CYCLE

Sulfur is a relatively abundant, essential element of considerable importance. As is true of many elements, it takes part in a biogeochemical cycle (Fig. 14-3). It is a major component of air pollution, particularly in industrialized areas, although natural sources of sulfur also contribute. Several oxidation states are encountered in environmental systems. The most stable under aerobic conditions is S(VI) as in sulfates and SO_3. The reduced form S(-II) is encountered in organic sulfides, including some amino acids, in H_2S, and in metallic sulfides. It is a reduction product of sulfates under anaerobic conditions. Oxidation of sulfides produces chiefly SO_2 [S(IV)] as the immediate product. The most important interrelationships of these states are shown

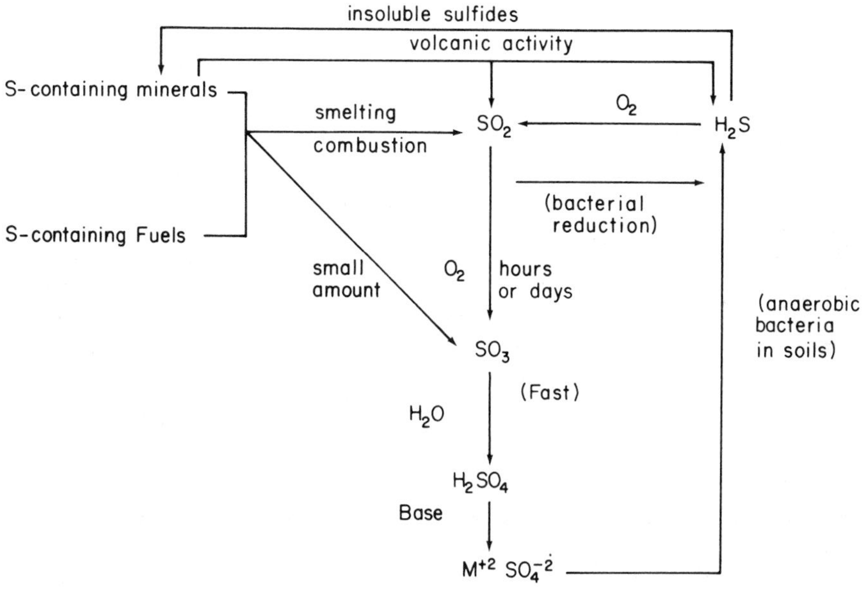

FIGURE 14-3. The sulfur cycle.

in the sulfur cycle. A large amount of sulfur is released to the atmosphere as SO_2 produced from the combustion of sulfur-containing fuels. The sulfur may be present in the fuel as organo-sulfur compounds, or as inorganic sulfide contaminants such as FeS_2. In coal the amounts of each are comparable. Roughly three times as much SO_2 has been produced from coal combustion as from the burning of petroleum in the U.S. in recent years; coal and petroleum together make up the source of about 80% of the total SO_2 emissions, with ore smelting being third. The contribution from coal can be expected to increase as use of high sulfur coals becomes necessary to replace low sulfur petroleum, especially in power generation. A small amount (about 5%) of the sulfur oxides produced in combustion processes is emitted as SO_3 rather than SO_2. However, catalytic exhaust converters on automobiles convert much of the sulfur emissions from automobile exhausts to SO_3. Other sources of sulfur oxides are incineration, chemical processing, and volcanic activity.

Sulfur dioxide is also formed from H_2S, which is released to the atmosphere from volcanoes and from anaerobic decompositon of organic matter in soils and sediments. Hydrogen sulfide is rapidly oxidized in the atmosphere to SO_2 by O_2, O_3, and oxygen atoms. The amount of sulfur released to the atmosphere worldwide in this form (about 2×10^8 tons/year) exceeds the amount released directly as SO_2. However, little H_2S is evolved from localized sources, so that the pollution problem is less serious. Some of the hydrogen sulfide generated in soils and sediments undergoes reoxidation reactions to S, SO_2, or SO_4^{2-}, as well as precipitation reactions to form insoluble metal sulfides. Many of the processes are biochemical.

Sulfur dioxide is thermodynamically unstable with respect to the higher oxide SO_3 under natural conditions. In the atmosphere, the reaction of SO_2 with O_2 is relatively slow. The rate is influenced by photochemical processes and by catalysts (see Section 10.3.5). The most important oxidation process involves catalysis by metal salts present in water droplets or on dust particles. In fog or cloud, SO_2 reacts with water to form sulfurous acid H_2SO_3. This is followed by oxidation:

$$H_2SO_3 + \frac{1}{2} O_2 \longrightarrow H_2SO_4 \qquad (14\text{-}1)$$

Sulfur trioxide itself is extremely hygroscopic and will immediately react with water vapor to form sulfuric acid, and water droplets in air containing sulfur oxides will in fact be a dilute solution of H_2SO_4. This acid may be neutralized to sulfate salts by basic substances such as ammonia. Such materials are present from industrial or natural biological processes. Sulfuric acid is responsible for the corrosive properties of air in industrial localities. The effects can be noticed in the deterioration of some construction materials (H_2SO_4 on carbonate causes decomposition to CO_2) as well as of metals, paint, etc.

Rain water containing H_2SO_4 is acidic and contributes to the acidity of lakes and streams. As has been discussed in Section 3.2, rain water in the northeastern U.S. has shown pH values near 4 in recent years. This is well below the value expected for water in equilibrium with atmospheric CO_2. Not all of this acidity is necessarily caused by sulfur oxides; nitrogen oxides also may play a part. In addition, the pH of rain water depends not only on the acidic oxides, but also upon basic materials emitted to the atmosphere from the same or independent sources. Thus, pH may be influenced by such factors as the nature of the fuels used for heating or power generation (e.g., high or low sulfur content), the combustion temperatures in widespread use (this can influence the formation of nitrogen oxides, see Section 14.4), the nature of particulate emissions, and extensive industrial or natural emissions of other kinds.

Because emission of SO_2 presents a major pollution problem, considerable attention has been given to efforts to reduce these emissions. World-wide combustion of fossil fuels produces the order of 10^8 tons of SO_2 annually, and increasing use of high-sulfur fuels requires some means of reducing the amount of SO_2 evolved in order to control pollution problems. Pretreatment of fuels represents one approach. For example, coal, which contains much of its sulfur in the form of FeS_2 particles, may be separated from the FeS_2 on the basis of different densities. Removal of organic sulfur from coal or oil requires more elaborate and costly chemical treatment. In one promising method of treating high-sulfur coals, extraction of the crushed coal with a heated sodium hydroxide solution under moderate pressure is used to remove nearly all of the inorganic sulfur, a large portion of the organic sulfur, and many of the trace metal components in the coal. Recovery and sale of the sulfur, the metal salts, and the organic materials that are also extracted from the coal in this treatment, along with easy regeneration of the hydroxide solution, give this process a good chance of commercial success.

Removal of SO_2 from exhaust gases is another approach to the pollution problem, and a variety of techniques have been proposed. Since SO_2 is an acid, it can be removed by reaction with a base such as calcium carbonate:

$$2CaCO_3 + 2SO_2 + \longrightarrow 2CaSO_4 + 2CO_2 \quad (14\text{-}2)$$

Calcium carbonate, or calcium magnesium carbonate, are the most likely bases for use in such processes, since they are available cheaply as limestone or dolomite, respectively. In one approach, the crushed carbonate is added with the fuel, and the sulfate salts precipitated by electrostatic precipitators, or washed from the gas stream by a scrubber. Alternatively, the stack gases may be scrubbed by a slurry of $CaCO_3$. This method can be very effective, but produces large amounts of sludge containing $CaSO_4$, $CaSO_3$, unreacted carbonate, and ash, which raises severe disposal problems (Section 2.3).

Another process that has been suggested involves catalytic conversion to sulfur:

$$SO_2 + H_2O \longrightarrow HSO_3^- + H^+ \quad (14\text{-}3)$$

$$HSO_3^- + \text{citric acid} \longrightarrow HSO_3^-\text{citrate complex} \quad (14\text{-}4)$$

$$HSO_3^-\text{citrate complex} + H_2S \longrightarrow 3S + \text{citric acid} + 3H_2O \quad (14\text{-}5)$$

The sulfur can be recovered for industrial use by physical methods.

To be practical, such processes must operate in very large scale systems and must do so reliably with a minimum of maintenance. The volumes of gas to be handled are enormous, for example, the order of 10^8 cubic feet per hour from a 750 MW coal-fired generating plant. Although sulfur is a useful by-product, many other by-products such as $CaSO_4$ have little value and indeed may add considerable expense for their disposal. The capital equipment requirements are likely to be large; economics is a major consideration in applying such treatments. Despite these problems, stack gas scrubbers are expected to work with reasonable operating costs and solvable waste disposal problems.

14.4 NITROGEN

Nitrogen exists in nature in several oxidation states: (-III) as in NH_3, NH_4^+, and various organic compounds; N(III) in nitrates NO_2^-, and N(V) in nitrates NO_3^-. A number of oxides of which nitrous oxide N_2O, nitric oxide NO, and nitrogen dioxide NO_2 are most important environmentally, also occur; in these the nitrogen has other formal oxidation states.

The elemental form N_2 contains a triple bond with a large bond energy. Consequently, reactions that require that the N-N bond be broken are likely to take place with difficulty, even if the overall energy change of the reaction is favorable. As a result, N_2 is relatively inert. Some of its most important environmental reactions are produced by microorganisms, which evidently can provide a reaction mechanism of low activation energy to convert N_2 to ammonia and amines.

The aqueous redox chemistry of nitrogen involves primarily NO_2^-, NO_3^-, and NH_4^+; these take part in oxidation-reduction processes which are expressed in the equations:

$$NO_3^- + 2H^+ + 2e^- \longrightarrow NO_2^- + H_2O \quad (14\text{-}6)$$

$$NO_2^- + 8H^+ + 6e^- \longrightarrow NH_4^+ + 2H_2O \quad (14\text{-}7)$$

$$NO_3^- + 10H^+ + 8e^- \longrightarrow NH_4^+ + 3H_2O \quad (14\text{-}8)$$

These reactions depend on pH, and the equilibrium composition of a nitrogenous system depends on this as well as on the pE of the system. Nitrate has a comparatively narrow range of pE over which it can exist at significant concentrations; most commonly the stable forms are NH_4^+ or ammonia in reducing environments, and NO_3^- in oxidizing ones. At neutral pH ranges, the NO_2^- stability region lies in the vicinity of pE = 6.5.

The essential features of the nitrogen cycle are shown in Fig. 14-4. Table 14-3 gives estimates of amounts of material involved. There are several processes of importance:

(a) Nitrogen fixation refers to the conversion of atmospheric N_2 to N(-III). It is a microbial process, most importantly involving bacteria that are associated with leguminous plants (e.g., clover). It takes place also with some aquatic bacteria, and blue-green algae. The process is complex biochemically, but evidently depends on a metal-containing enzyme system in which both iron and molybdenum take part. This leads to ammonia, which is immediately incorporated into amino acids in the plant. Plants generally, however, cannot absorb nitrogen in any form but the nitrate ion. Other natural fixation processes, such as by lightning, are of negligible importance, although much nitrogen is fixed industrially.

(b) Nitrification is the conversion of amine nitrogen to nitrate. Decay of protein material produces NH_3, which is oxidized through nitrite to nitrate. This is also a bacterial process. Nitrification is reversible, under

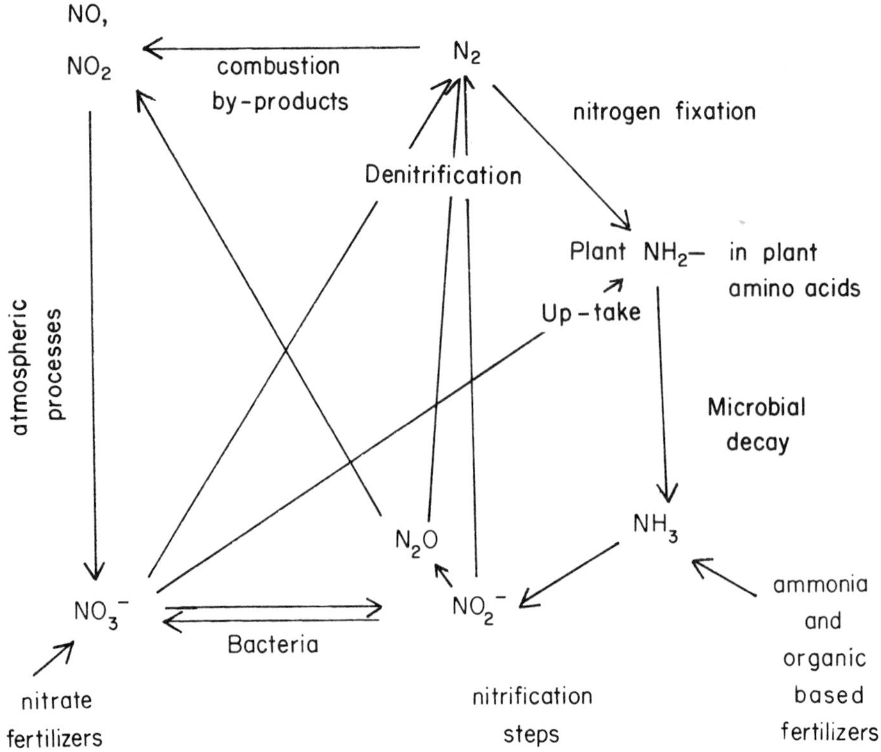

FIGURE 14-4. The nitrogen cycle.

anaerobic conditions, since reduction of nitrate can provide an energy source for bacteria through the net change

$$NO_3^- + \frac{1}{2}\,C_{organic} \longrightarrow NO_2^- + \frac{1}{2}\,CO_2 \qquad (14\text{-}9)$$

($C_{organic}$ refers to the carbon of an organic molecule which is used up in this process.)

The nitrite ion is relatively toxic because of its interaction with hemoglobin. Nitrite in the blood results in the oxidation of the Fe(II) of hemoglobin to form methemoglobin, which has no oxygen-carrying ability; this disease is called methemoglobinemia. While cases of direct nitrite poisoning are rare, the nitrate ion can be reduced to nitrite in the stomach, and for this reason food and water with high nitrate contents are dangerous. This reduction is especially possible in the stomachs of infants, where the low acidity allows the growth of nitrate-reducing microorganisms. There is also some evidence that nitrites in the body can react with organic amines to form possibly carcinogenic nitrosamines.

TABLE 14-3

Quantities Involved in the Nitrogen Cycle[a]

Nitrogen inventory (10^9 metric tons)			
Atmospheric N_2			3,800,000
Plants			13
Animals			1
Dead organic matter:	land		760
	ocean		900
Dissolved N_2			20,000
Inorganic nitrogen (not N_2):	land		140
	ocean		100
Rocks, sediments			18,000,000
Rates of processes in the nitrogen cycle (10^6 metric tons per year):			
Fixation:	Biological	(land)	44
		(oceans)	10
		(atmosphere)	7.6
	Industrial		30
	Total		91.6
Denitrification:		(land)	43
		(oceans)	40
	Total		83
River run-off			30

[a]*Data from Delwiche,* Sci. Amer, 233 *(12), 136 (1970).*

Nitrite (often produced from KNO_3 in situ) is a common additive in cured meats, where it acts to prevent the growth of bacteria, but is also added to produce flavor and an attractive color; the latter is through the formation of nitrosylmyoglobin.

(c) Denitrification is the formation of gaseous N_2 from nitrate to return nitrogen to the atmosphere. Primarily it is carried out by soil bacteria and involves reduction by organic carbon compounds with the production of CO_2. This process can be used to remove nitrates from waste water by addition of a reducing agent such as methanol under anaerobic conditions:

$$5CH_3OH + 6NO_3^- + 6H^+ \longrightarrow 5CO_2 + 3N_2 + 13H_2O \qquad (14\text{-}10)$$

Some nitrogen is returned to the atmosphere as N_2O generated bacterially in soils. Nitrous oxide is comparatively inert, but it is rapidly destroyed by processes that are not entirely clear. It can be photolyzed to N_2 at high altitudes, where radiation exists at a short enough wavelength to be absorbed by N_2O, but it does not seem likely that this reaction can be responsible for the low N_2O levels usual at low altitudes. Other processes for N_2O removal are probable at or near the earth's surface.

Some ammonia is also released naturally from decay of organic materials and is present in air either as NH_3 gas or as an ammonium salt aerosol. These are removed from the atmosphere in rain.

Fertilizers make up an important artificial source of nitrogen input to the environment. Because nitrogen is an essential nutrient for plants, and often a limiting one, large-scale use of nitrogen-containing fertilizers has become commonplace. These may take the form of nitrates, which are immediately available to the plant. Alternatively, fertilizers based on ammonia or organic nitrogen can be used. All these must be converted to inorganic nitrate before used by plants, and since this takes place over a period of time, they can provide a longer lasting source of nitrogen. Nitrates tend to be quite soluble and weakly held by ion exchange force, and hence may easily be leached from the soil and wasted. Organic nitrogen fertilizers may or may not be soluble, but soluble materials can be formulated to resist dissolution (e.g., pellets may be coated with sulfur and wax). However, it should be emphasized that movement of materials through the soil is highly complicated by absorption and ion exchange processes; even readily soluble materials can be retained for long periods in some types of soils. Thus ammonia itself, although a very soluble gas, is an effective fertilizer. It is held in the soil as the ammonium cation by ion exchange. Other widely used compounds are NH_4NO_3 and urea. It is worth noting that since nitrogen is assimilated by plants as the nitrate ion, the original source of the nitrogen has no effect on the plant. Rather, choice of the form a fertilizer should take is a matter of rate of release of NO_3^-, extent of leaching (which results in waste and ultimately water pollution), and other economic factors.

Industrial nitrogen fixation to produce fertilizers (and other nitrogen compounds) is based primarily on the Haber process:

$$\frac{1}{2} N_2 + \frac{3}{2} H_2 \xrightarrow[\text{catalyst}]{\text{elevated } T \text{ and } P} NH_3 \tag{14-11}$$

The source of hydrogen is normally natural gas or petroleum, through reactions such as

$$CH_4 + H_2O \longrightarrow CO + 3H_2 \qquad (14\text{-}12)$$

$$CO + H_2O \longrightarrow CO_2 + H_2 \qquad (14\text{-}13)$$

Consequently, fertilizer production and cost are closely linked to the supply of these fossil fuels. Ammonia is the starting material for most other industrial nitrogen compounds; for example, nitrates are produced through reactions that include oxidation of ammonia by air in the presence of a platinum catalyst.

From the estimates in Table 13-4, nearly 9.1×10^7 tons of nitrogen are fixed annually, while only 8.3×10^7 tons are released to the atmosphere through denitrification. The cycle, then, is not balanced and nitrogen is building up in the biosphere; this should have biological effects as this material acts as a fertilizer. The bulk of the nitrogen compounds enter runoff water and eventually the oceans. However, increased fertilizer use is also believed to lead to increased concentrations of nitrogen oxides in the atmosphere as more N_2O is released in the denitrification step.[1]

Oxides of nitrogen in the atmosphere are a cause of concern with respect to air pollution problems as discussed in Chapter 10. The main compounds of concern are NO and NO_2, which are by-products of combustion processes. They are formed from the reaction of N_2 in the air with O_2. The primary product is NO, but some NO_2 is produced as well. Nitrogen oxide (often called NO_x) production is directly related to the temperature in the combustion zone of the furnace or engine, and can be reduced by operating at lower temperatures. This generally reduces the efficiency of the device, however. Reduction of the amount of excess air in the combustion chamber also will reduce NO_x emission, but at the expense of an increase in the amount of incompletely burned fuel. A two-stage combustion, the first at high temperature with an air deficiency, followed by completion of the combustion at a lower temperature, can be quite effective in reducing NO_x emission.

Absorption of NO_x from the combustion gases is more difficult than is SO_2 removal, although various methods have

[1]*"Halocarbons: Effects on Stratospheric Ozone," p. 171 and references therein. Report of the National Research Council Panel on Atmospheric Chemistry, National Academy of Sciences, Washington, D.C., 1976. (Although primarily concerned with halocarbons, this volume reviews other threats to the ozone layer.)*

been proposed for both acid and alkaline scrubbing of stack gases. Alkaline scrubbers can remove SO_2 and NO_x simultaneously, but neither NO nor NO_2 is efficiently absorbed by a simple basic solution. However, a mixture of NO and NO_2 enters into an equilibrium reaction,

$$NO + NO_2 \rightleftharpoons N_2O_3 \qquad (14\text{-}14)$$

Although this lies far to the left under normal conditions, it is shifted to the right if the N_2O_3 is absorbed into solution; the latter takes place readily in base, since N_2O_3 is the anhydride of nitrous acid,

$$N_2O_3 + H_2O \longrightarrow 2HNO_2 \qquad (14\text{-}15)$$

The nitrite salts formed in a basic solution can be oxidized to nitrate. This process, however, requires equimolar amounts of both NO and NO_2 in the gas stream, or recycling of the excess. Other processes have been proposed, e.g., absorption of NO by alkaline sulfite solutions, which results in production of a compound from SO_3^{2-} and NO_2 with the formula $O_3SN_2O_2^{2-}$, but practical applications are beset by problems of oxidation and stability.

Catalytic reduction of NO_x is possible; reaction with CO, NH_3 or CH_4, for example, can produce N_2:

$$NO + CO \xrightarrow{\text{catalyst}} CO_2 + \frac{1}{2} N_2 \qquad (14\text{-}16)$$

This reaction is favored thermodynamically, but is very slow under normal circumstances. Various metals, e.g., Ag, Cu, Ni, and Pd are effective catalysts, but are subject to poisoning in practical use and in some cases are expensive. Some oxides, e.g., copper chromite and Fe_2O_3, are also effective catalysts for this process. Systems of this kind may be applicable to automobile exhaust, but would have to be built on a very large scale to cope with power generator gases.

14.5 PHOSPHORUS, FERTILIZERS, AND EUTROPHICATION

Phosphorus is an essential nutrient material for plants and animals, being required in biological synthesis and energy transfer processes. The overall photosynthesis reaction in aquatic organisms (70% of all photosynthesis takes place in the ocean) results in the eventual formation of biological material which has an overall C:N:P ratio of approximately 100:16:1. The N:P ratio in ocean water is about 16:1, the same as in biological materials. In fresh water lakes, however, N is usually in excess; typical ranges are N, 10^{-4} to 10^{-5} *M*; P, 3×10^{-5} to 3×10^{-7} *M*. One milligram of phosphorus results in production of 0.1 gm of organic material if N is in excess, and in systems where phosphorus is the limiting nutrient, additional phosphorus input can greatly increase the biological yield. (There is evidence that natural processes act to make phosphorus the most important algal nutrient in any event.[2]) The organic material produced requires 0.14 gm of oxygen for decomposition, and if this is unavailable, the water becomes depleted in dissolved oxygen, with consequent depletion of the higher life forms and production of undesirable products of anaerobic reactions: NH_3, H_2S, and CH_4. In the absence of dissolved O_2, NO_3^-, SO_4^{2-}, and even organic compounds can serve as oxidizing agents in reactions largely controlled by bacterial action. The enhanced growth and decay of algae is the origin of the concern over phosphate pollution and its effects on eutrophication processes in fresh water lakes.

Phosphorus in nature occurs in the form of phosphates, based on the tetrahedral PO_4^{3-} unit as found in orthophosphates. In solution, $H_2PO_4^-$ and HPO_4^{2-} are the predominant forms of phosphate at the usual pH values, and it is in these forms that phosphate is taken up by organisms. The orthophosphate ion PO_4^{3-} will exist in significant concentration in solution only at very high pH values, while orthophosphoric acid is a strong acid. Polyphosphates can be formed from the condensation polymerization of simple orthophosphate units, for example,

$$2H_3PO_4 \xrightarrow{\text{heat}} H_2O + H_4P_2O_7 \qquad (14\text{-}17)$$

The product here is pyrophosphoric acid, which is made up of two tetrahedra with a common corner:

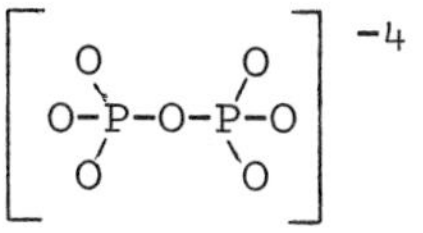

Many polyphosphate compounds can be prepared by heating simpler phosphates in reactions analogous to (14-17). Other polyphosphates may be cyclic such as trimetaphosphate

[2] *D. W. Schindler, Evolution of phosphorus limitation in lakes,* Science 195, *260 (1977).*

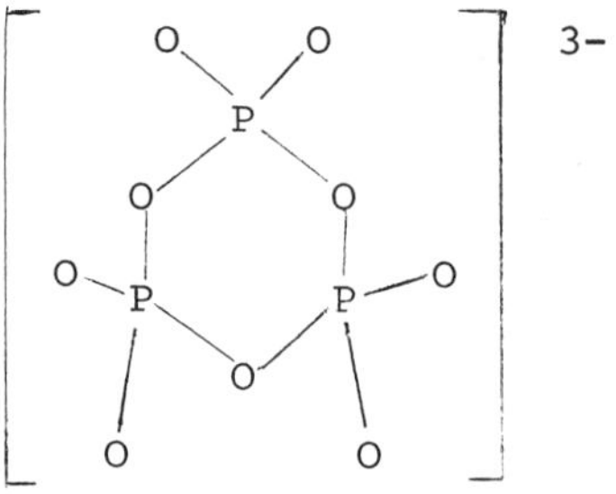

The linear tripolyphosphate used in detergents has already been discussed in Section 13.6. In dilute solutions, polyphosphates are hydrolyzed to the orthophosphate, although the reactions are not always fast.

Natural waters may contain organophosphorus compounds which can make up a significant fraction of the total soluble phosphorus content. These compounds are generally of unknown composition and are derived from biological products or possibly P-containing pesticides. Inorganic phosphate concentration is limited by solubility, since many metal phosphates are insoluble, e.g., $FePO_4$, $K_{sp} = 10^{-23}$. Aluminum phosphates $AlPO_3$ and hydroxyapatite $Ca_5OH(PO_4)_3$ are also very insoluble. Formation of soluble complexes and protonation of the phosphate ions, however, can increase the effective solubility.

Although phosphorus compounds in a variety of other oxidation states are known in chemistry (especially 3- in PH_3 and derivatives, 3+ in phosphorous acid, PCl_3, etc.), the inorganic forms of these oxidation states have no general importance under environmental conditions. The natural phosphorus cycle consists of weathering and leaching of phosphate from rocks and soils, and run-off to the oceans, which serve as a sink. Biological cycling intervenes in this process, but otherwise there are no significant return pathways or atmospheric steps. The low phosphate levels are increased easily by man-made processes, chiefly the use of fertilizers and phosphate-containing detergents.

Phosphate fertilizers are produced commercially mainly from insoluble phosphate rock, which has the formula $Ca_2(PO_4)_2 \cdot CaX$ where X can be CO_3^{2-}, $2(OH)^-$, and others, but is usually $2F^-$. Soluble phosphates are produced by an acid displacement reaction:

$$Ca_3(PO_4)_2 \cdot CaX + 3H_2SO_4 \longrightarrow Ca(H_2PO_4)_2 + 3CaSO_4 + H_2X \qquad (14\text{-}18)$$

If fluoride is present, the HF produced will react with silica in the rock to produce SiF_4 and fluoroscilicate salts (M_2SiF_6). Residual fluoride impurities may remain in the product.

The $Ca(H_2PO_4)_2$-$CaSO_4$ mixture is often sold as such as "superphosphate." The sulfate itself is a useful component, as sulfur is also an important nutrient. A higher phosphate content can be obtained in so-called "triple superphosphate," produced by the process

$$Ca_3(PO_4)_2 \cdot CaX + 6H_3PO_4 \longrightarrow 4Ca(H_2PO_4)_2 + H_2X \qquad (14\text{-}19)$$

Phosphoric acid is formed from excess H_2SO_4 on $Ca_3(PO_4)_2$, and can serve as the source material for other phosphate chemicals through neutralization and condensation reactions. As with nitrogen, phosphate fertilizers are subject to leaching from soils, adding to the nutrient content of run-off water and lakes, and contributing to eutrophication.

In addition to phosphorus and nitrogen, a third major component in lawn and garden fertilizers is a soluble salt of potassium. Most common fertilizers consist of a mixture of these three materials, and the composition specifications (e.g., 6-10-6) refer to the percentage composition in terms of N, P_2O_5, and K_2O. Potassium is not considered a pollution problem. Other fertilizer materials are needed in particular cases; most important examples of these are calcium, magnesium, and sulfur.

14.6 BIOLOGICALLY IMPORTANT METALS AND TRACE ELEMENTS

14.6.1 Introduction

In addition to the elements already discussed, a variety of others have important biological effects. The biological availability of an element, and its ultimate effect on an organism, may depend on the chemical form in which the element is encountered. Thus, solubility properties, the oxidation state, or the bonding characteristics in a particular compound may determine whether it will be inert, poisonous, or detoxified by normal metabolic pathways. Toxic compounds occurring naturally are very often degraded by organisms to nontoxic compounds, but excessive concentrations, or new compounds, may overcome these metabolic capacities.

Table 14-4 lists some of the metallic elements released to the environment by industrial and other processes which may pose concern as health hazards in trace amounts. Strong evidence of toxicity is not present for all of them, and

TABLE 14-4

Trace Metals that May Pose Health Hazards[a]

Element	Sources	Health effects
Nickel	Fuel oil, coal, tobacco smoke, chemicals and catalysts, steel and nonferrous alloys	Lung cancer (as carbonyl)
Beryllium	Coal, industry uses (nuclear power industry, rocket fuel) mine tailings	Acute and chronic system poison, cancer
Boron	Coal, cleaning agents, medicinals, glass making, other industrial	Toxic to plants; slightly toxic to animals in environmentally stable forms
Germanium	Coal	Little innate toxicity
Arsenic	Coal, petroleum, detergents, pesticides, mine tailings	Hazard disputed, may cause cancer
Selenium	Coal, sulfur	May cause dental caries, carcinogenic in rats, essential to mammals in low doses
Yttrium	Coal, petroleum	Carcinogenic in mice on long-term exposure
Mercury	Coal, electrical batteries, other industrial	Nerve damage and death
Vanadium	Petroleum (Venezuela, Iran), chemicals and catalysts, steel and other alloys	Probably no hazard at current levels

TABLE 14-4 (Continued)

Element	Sources	Health effects
Cadmium	Coal, zinc mining, water mains and pipes, tobacco smoke	Cardiovascular disease and hypertension in humans suspected, interferes with zinc and copper metabolism
Antimony	Industry	Shortened life span in rats
Lead	Auto exhaust (from gasoline), paints (prior to about 1948)	Brain damage, convulsions, behavioral disorder, death
Copper	Water pipes, algae control, industrial smoke	Possible liver damage with prolonged exposure; toxic to plants
Manganese	Mining and industrial wastes	Relatively nontoxic
Chromium	Metal plating	Possible carcinogen (as chromate)
Silver	Electroplating, photographic waste	Discoloration of skin
Zinc	Metal plating, mining, industrial smoke	Possible lung effects; low toxicity in solution
Molybdenium	Industrial smoke	Possible lung effects

Antimony, bismuth, tellurium, and zirconium are also encountered as atmospheric and aqueous industrial emissions. Health hazards are not established, but the elements are toxic.

[a]Reprinted, with permission, from Chem. Eng. News, July 19, 1971, p. 30.

for others, evidence that they are harmful is indirect for the environmental forms. Many of the heavy elements not listed here are potentially quite toxic, but either are very rare or very insoluble in forms encountered in the environment.

The importance of the chemical form of an element is illustrated by nickel subsulfide Ni_2S_3 and nickel carbonyl $Ni(CO)_4$, both of which are potent carcinogens (nickel carbonyl is also highly toxic in other ways). Most other nickel compounds, if not noncarcinogenic entirely, certainly have much weaker effects. Both Ni_2S_3 and $Ni(CO)_4$ are possible by-products from the use of nickel catalysts as well as from some of the sources given in Table 14-4. Such catalysts may find very large-scale use in coal gasification processes, for example.

The function of metal ions in biochemistry can be correlated with their coordination tendencies. For example, sodium and potassium, which only enter into weak bonding with typical ligands, chiefly through electrostatic effects, are involved with control processes involving ion transfer. Mg(II) and Ca(II) also are largely electrostatic in their interactions, but are better complex formers. They exhibit rapid ligand exchange, and often are components of enzymes involved in trigger reactions and hydrolysis processes. Electronegative ligands such as oxygen donors are favored by both of these groups. Stronger bonding, preference for nitrogen donors, and stronger interactions with sulfur donors, are found with the ions of Zn, Co, Mn, Fe, and Cu. These are also common in enzyme systems which control a variety of processes. Those that readily form two different oxidation states, especially Cu and Fe, often are involved in redox processes.

The contribution of metals and their ions toward pollution problems is of very serious concern, largely because their physiological effects may be substantial even at concentrations that are extremely low. In some cases effects on behavior and on the nervous system begin with subtle changes and are difficult to detect in their initial stages; some metals may be accumulated in the body so that dangerous levels can build up from long-term ingestion of very small amounts. In many cases, there is little real knowledge about their effects and what constitutes a "safe" level of concentration. Levels that have been set are often based as much on intuition as on fact. At low concentrations, metal pollution is very widespread, with many industrial processes contributing. (In local cases, of course, levels may be quite high.) Natural waters normally contain considerable mineral content; in particular cases levels of harmful inorganic components may be significant quite independent of any man-made pollutants. Finally, because concentrations that are significant may be so small, unexpected sources may be important. For example, although the amounts of metals contained in coal and oil are small, the vast amounts of these materials that are burned result in the liberation of appreciable total quantities of mercury, cadmium, and other metals. Generally, the elements of most concern are found in the environment as compounds of low volatility; nevertheless their contribution to atmospheric problems must not be ignored as they may be present as fine particulate material. In water, they must be considered in suspended forms as well as in true solution.

The actual role played by trace metals in human health is complex and by no means fully understood, and this makes the determination of safe levels very difficult. For example, zinc is known to be an essential element, and a comparatively nontoxic metal by conventional tests. However, there is some evidence that suggests that amounts of zinc very much below present acceptable levels can counteract some of the natural antitumor mechanisms of the body. On the other hand, iron deficiency seems to encourage tumor development.

14.6.2 Alkaline Earth Metals: Beryllium, Magnesium, Calcium, Strontium, and Barium

Magnesium and calcium are extremely common in natural water systems, with calcium carbonate (limestone) and dolomite ($CaMg(CO_3)_2$) being two widespread natural sources. Solubility in water is influenced by pH and CO_2 content. The two ions are concerned with hardness of water, which manifests itself by precipitation with soaps, calcium carbonate deposits when water is heated (boiler scale), etc. For many purposes (washing, waters for certain heat exchange processes), the calcium and magnesium ions are obnoxious, and must be held in solution by chelation or removed by ion exchange processes or precipitation in such a way that the products are not harmful. Both magnesium and calcium are essential elements needed in significant amounts by living organisms. Except in the context of these problems, they are not harmful either in solution or as particulate material.

Chemically, magnesium behaves similarly to calcium, but generally is found in much lower concentrations in water. Both exist solely as the 2+ ions in solution, with interactions that are predominantly coulombic. The ions favor coordination by oxygen donors; simple complexes are weak, but chelation can give moderate stability. Magnesium, being smaller, shows stronger bonding with solvent or ligand molecules and more stable complexes. Both are important in a variety of enzymatic and other biological processes where their weak complexing and rapid exchange determines their function. Calcium is the essential cation in structural features such as bones and shells. In spite of chemical similarities, calcium is not interchangeable with magnesium because of the significant size difference. Calcium metal is extremely reactive and has little industrial use as such; magnesium metal on the other hand, is considerably less reactive and is important structurally since it combines lightness and strength.

The heavier members of the alkaline earth family, strontium and barium, are comparatively rare. Both resemble calcium more than magnesium, and strontium in particular is close enough in size to Ca(II) that its ions frequently can replace those of calcium in solids such as bone. Since one of the radioactive isotopes of strontium is an abundant fission product, this replacement is of considerable importance and is considered in Chapter 17.

The lightest alkaline earth, beryllium, is much different in its properties because its small size, high ionization energy, and greater electronegativity favor much more covalent bonding. It also differs in that it is highly toxic, as has been recognized for many years. Beryllium is employed for a number of high-level technological purposes, such as structural material in nuclear reactors and missiles, as an additive to rocket fuels, and as an alloy with Cu having high corrosion resistance. While these offer at least the potential for industrial pollution, another source is the amount that may be released from burning coal, which usually contains up to 3 ppm Be, but may contain much more in particular cases.

14.6.3 Iron

Iron is a common and important element environmentally. It forms two oxidation states in nature, Fe(II) and Fe(III). The latter is of greater importance. In water solution, Fe(III) forms a strongly hydrated ion in which the water molecules are bonded to the iron by coordinate, largely covalent bonds to give the species $[Fe(H_2O)_6]^{3+}$. The shift of electron density to the iron and out of the OH bonds of the coordinated water molecules makes their protons acidic and results in dissociation equilibria:

$$[Fe(H_2O)_6]^{3+} + H_2O \rightleftharpoons [Fe(H_2O)_5OH]^{2+} + H_3O^+$$

$$K = 8.9 \times 10^{-4} \qquad (14\text{-}20)$$

$$[Fe(H_2O)_5OH]^{2+} + H_2O \rightleftharpoons [Fe(H_2O)_4(OH)_2]^+ + H_3O^+$$

$$K = 5.5 \times 10^{-4} \qquad (14\text{-}21)$$

These are fairly strong acids. Additional steps of proton loss are possible. In addition, one has polymerization reactions such as

$$2[Fe(H_2O)_6]^{2+} \rightleftharpoons [Fe(H_2O)_4(OH)_2Fe(H_2O)_4]^{2+} + 2H_3O^+$$

$$K = 1.2 \times 10^{-3} \qquad (14\text{-}22)$$

The structure involves bridged polynuclear species based on the linkages

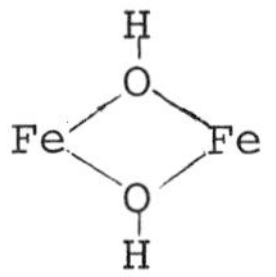

More highly polymerized species also form, and eventually some water molecules as well as H^+ ions are split out. The exact composition of the equilibrium product(s) depends upon pH and total iron content, but frequently a neutral, highly polymerized insoluble substance is produced which is called ferric hydroxide. The formula $Fe(OH)_3$ is often used for this precipitate, but this may not represent a real compound. The fresh material has an uncertain structure, but on aging is mainly FeO(OH). This does not form discrete molecules but involves O and OH groups bridging to other Fe(III) ions. The structurally indistinct nature of the fresh Fe(III) hydrolysis products leads to precipitates of highly gelatinous and often colloidal nature. As a result of hydrolysis reactions such as the above, Fe(III) salts are virtually insoluble in waters with a pH near neutrality; indeed, iron is present mostly as hydroxo complexes at pH = 4. Fe(II), on the other hand, does not form these complexes significantly until pH $\simeq$ 10 and is soluble near neutrality.

The Fe(II)-Fe(III) relationships may be considered. If the solid phase is taken to be FeO(OH), we can write the following reaction and equilibrium constant:

$$FeO(OH)\ (s) + 3H_3O^+ \rightleftharpoons Fe(III)(aq) + 5H_2O; \quad K = 10^4 \qquad (14\text{-}23)$$

[Fe(III) (aq) stands for the ferric aqua complex in solution.] The redox equilibrum is

$$\mathrm{Fe(III)\ (aq) + e^- \rightleftharpoons Fe(II)\ (aq);} \quad K = 10^{13} \tag{14-24}$$

Combining the above,

$$\mathrm{FeO(OH)\ (s) + 3H_3O^+ + e^- \rightleftharpoons Fe(II)\ (aq) + 5H_2O} \qquad K = 10^{17} \tag{14-25}$$

Since

$$K = [\mathrm{Fe(II)}] \,/\, [\mathrm{H_3O^+}]^2[\mathrm{e^-}] \tag{14-26}$$

and from the definition of pE in Section 13-5,

$$\log[\mathrm{Fe(II)}] = \log K - 3\mathrm{pH} - \mathrm{pE} \tag{14-27}$$

At the pE value of air-saturated water, 12.5, log[Fe(II)] = -20, and the concentration of Fe(II) under these conditions is insignificant. Indeed, we can calculate the pE value required for any realistic concentration of Fe(II) to be stable in the presence of solid FeO(OH) and the oxygen partial pressure that would be in equilibrium with this. It will be found on doing so that Fe(II) can be formed only under anaerobic conditions.

The concentration of soluble Fe(III) is expected to be very small. For the solution reaction of FeO(OH), with $[Fe(H_2O)_4(OH)_2]^+$ as the most readily formed soluble species,

$$\mathrm{FeO(OH)(s) + 3H_2O + H_3O^+ \rightleftharpoons [Fe(H_2O)_4(OH)_2]^+} \qquad K = 4.5 \times 10^{-3} \tag{14-28}$$

At pH = 8.1, the concentration of soluble ferric iron is about 3×10^{-11} *M*. The true dissolved Fe(III) concentration in natural waters such as the ocean is difficult to evaluate due to the presence of colloidal material; however, it seems to be about 10^{-6} - 10^{-8} *M*/liter. This value, which is much higher than calculated from solubility equilibria, cannot be accounted for with certainty, but may involve the formation of soluble complexes of unknown kinds, perhaps with natural organic ligands.

The ferric hydroxide precipitate interacts strongly with water; it can hydrogen bond to water molecules effectively and this, coupled with the irregular nature of the chains and cross-linking formed as the material precipitates, leads to a gelatinous floc. On the other hand, if very small aggregates form as might be the case in extremely dilute solutions, the individual particles will be strongly hydrated, and in favorable cases further stabilized against coagulation by adsorbing ions of a particular charge on their surfaces. Under these conditions a more or less stable suspension or colloid results. It is not uncommon for the particles of ferric hydroxide to be extremely small and not easily distinguished from material in true solution.

As previously discussed, Fe(II) may form under anaerobic conditions. Considerable amounts of ferrous sulfide exist, for example, as pyrite FeS_2 (involving the S_2^{2-} species). This is often found in coal areas, where presumably it was formed under oxygen-free conditions. Exposure of pyrite to air and moisture leads to the formation of so-called acid mine drainage, which is highly acidic water and the cause of considerable damage to natural water systems in coal mining areas. The following reactions are involved:

$$FeS_2(s) + \frac{7}{2} O_2 + 3H_2O \rightleftharpoons Fe(II)(aq) + 2SO_4^{2-} + 2H_3O^+ \qquad (14\text{-}29)$$

In this step, the S_2^{2-} is oxidized, forming sulfuric acid, while the now soluble Fe(II)(aq) is free to react further, but does so more slowly:

$$Fe(II)(aq) + \frac{1}{4} O_2 + H_3O^+ \rightleftharpoons Fe(III)(aq) + \frac{3}{2} H_2O \qquad (14\text{-}30)$$

Bacterial action catalyzes the pyrite oxidation, which otherwise is quite slow. Fe(III) hydrolyzes, giving as an overall reaction

$$[Fe(H_2O)_6]^{3+} \rightleftharpoons Fe(OH)_3 + 3H_3O^+ \qquad (14\text{-}31)$$

[$Fe(OH)_3$ may not be a real substance; this formula is used here to refer to the poorly defined precipitate which on drying would be largely FeO(OH).]

One mole of FeS_2 eventually gives 4 moles of H_3O^+: 2 from oxidation of S_2^{2-}, 2 from the oxidation of iron. This process deposits insoluble ferric hydroxide that coats stream beds, as well as contributing to the acidity. Since the formation of Fe(III) from Fe(II) is slow, this may take place over a considerable length of stream. If the Fe(III)(aq) is formed in contact with pyrite, an additional process is involved:

$$FeS_2\ (s) + 14Fe(III)(aq) + 24H_2O \rightleftharpoons 15Fe(II)(aq) + 2SO_4^{2-} + 16H_3O^+ \qquad (14\text{-}32)$$

We may represent the stability relationships in a system such as that of iron as a function of pE and pH diagrammatically. Figure 14-5 illustrates this. Each region represents the conditions of pE or pH necessary for a particular species to be stable. We may see from this that iron metal will be oxidized (rusts) unless the oxygen content is low enough to give pE < -10. The ferrous ion is soluble in slightly acidic reducing solutions, but the carbonate will precipitate near neutrality if CO_2 is present. Fe(III) is soluble only at pH < 4 under oxidizing conditions.

Another major species in the iron system is Fe_3O_4. This oxide, called magnetite, contains iron in both the (II) and (III) states. The exact stability relationships of this compound are unclear, but it has been suggested that the reaction

$$12FeO(OH)(s) \rightleftarrows 4Fe_3O_4\ (s) + 6H_2O + O_2\ (g) \qquad (14\text{-}33)$$

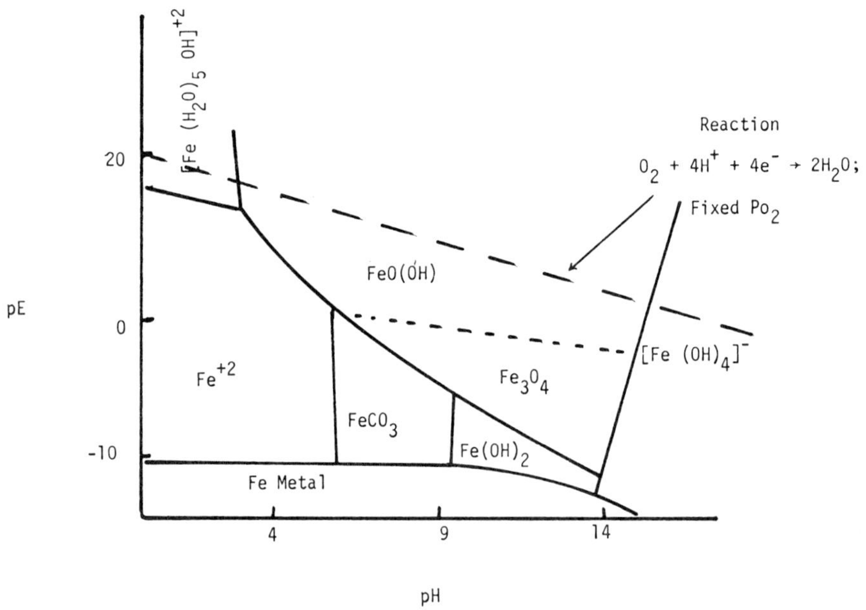

FIGURE 14-5. pE versus pH for some of the components of the iron system in nature. This is highly simplified.

may have acted to regulate the O_2 pressure in the primitive atmosphere.

Iron is a very common constituent of fresh waters, where it sometimes is present in the Fe(III) state either as colloidal material or in solution as a complex with organic material. In large amounts it imparts a characteristic rust color. In some ground waters that are anaerobic, Fe(II) may be present, and since compounds of this oxidation state are soluble, comparatively high concentrations can be reached (often in the 1-10 mg/liter range). This of course is oxidized under ordinary conditions to Fe(III), and ultimately is the source of the colloidal material. Although this oxidation of Fe(II) upon aeration may be moderately slow, significant occurrence of iron in this oxidation state in surface waters requires either the presence of organic material or a high acidity (see Fig. 14-5), and in either case is suggestive of pollution.

Iron in water is not toxic, but it does have a number of harmful effects. It can cause "rust" stains on clothing, porcelain material, etc., as well as giving the water itself an undesirable color and taste. The United States Public Health (USPH) standard for maximum iron content in drinking water is 0.3 mg/liter, and this is based on color and taste effects rather than on toxicity.

14.6.4 Chromium

Although familiar as an inert metal used for decorative and protective plating, large amounts of chromium compounds are used industrially for other purposes, especially in treating steel to resist corrosion. Solutions of the chromate ion CrO_4^{2-} are effective in "passivating" steel, presumably by reacting to form a protective oxide layer on the surface. Wastes from such operations are a prime source of chromium pollution. In acid, chromate condenses to dichromate:

$$2\ CrO_4^{2-} + 2H_3O^+ \rightleftharpoons Cr_2O_7^{2-} + 3H_2O \qquad (14\text{-}34)$$

Chromium in the oxidation state Cr(VI) is highly toxic (chromium limit in water, 0.05 mg/liter) and a possible carcinogen. Chromate dusts have been associated with lung cancer. This form of chromium is also a strong oxidizing agent, and is quickly reduced by the usual organic materials in a natural water system to Cr(III). Chemically, Cr(III) behaves somewhat like Fe(III) in that it is readily hydrolyzed to an insoluble hydroxide. In this form it is comparatively harmless.

14.6.5 Manganese

Manganese is an essential element, and one of low toxicity. In its highest oxidation state as permanganate MnO_4^-, it is a strong oxidizing agent. Its more stable states are Mn(IV), which is encountered in natural systems chiefly as insoluble MnO_2, or as compounds of Mn(II). The chemistry of these states resembles that of Fe(II) and Fe(III); Mn(II) salts are soluble in water, but the dioxide is insoluble, although often producing a colloidal or gelatinous and highly absorbent precipitate. Mn(II) is encountered only under reducing conditions in natural waters, as is Fe(II). Sources of manganese are industrial pollution and acid mine drainage. A manganese compound, methylcyclopentadienyltricarbonylmanganese, has been used for some years as a supplement to tetraalkyl lead as an antiknock agent in gasoline. This compound may be used alone in lead-free gasolines in concentrations of up to 0.13 gm Mn/gal. The USPH limit of 0.05 mg/liter in drinking water again is based largely on staining rather than on toxicity.

14.6.6 Copper

Copper is an essential metal for many organisms. It is used in enzymes that modify redox reactions and also in some oxygen-carrying systems. Its function in these is associated frequently with its ability to exist as both Cu(I) and Cu(II). The most common state is Cu(II). Like many essential metals, large amounts are toxic, and the USPH limit in drinking water is set at 1 mg/liter. It is particularly toxic to lower organisms, and has been used as an algicide in lakes (as $CuSO_4$). Other sources are erosion of copper pipes, industrial wastes, and weathering of rocks. The carbonate and hydroxide are of low solubility, so that the natural level in water is low in the absence of complex-forming substances.

14.6.7 Zinc, Cadmium, and Mercury

These three elements make up one family of the perodic table. They are representative metals possessing two valence electrons. Their position in the perodic classification is immediately following the transition series, and in keeping with the usual periodic trends, they have comparatively high electronegativity values for metallic elements, and form bonds with nonmetals of significant covalent character. The covalent properties are emphasized on going down the family from zinc to mercury.

Zinc is a common metal, and is released into the environment through mining and industrial operations. It is comparatively nontoxic, as suggested by USPHS maximum limits of 5.0 mg/liter in drinking water. Its simple compounds are readily hydrolyzed, while oxygen and nitrogen are the favored donor atoms. It is a constituent of enzymes, for example carboxypeptidase A, which is active in the hydrolysis of the terminal peptide linkage of a peptide chain. The active site in this enzyme is a Zn(II) ion coordinated by N and O from amino acid units in the protein portion of the enzyme. One coordination position on the zinc is free to accept a bond from an oxygen in the substrate. The shift in electron distribution in the peptide resulting from this coordination facilitates the hydrolysis step. Stereochemistry and hydrogen bonding interactions of the protein chain undoubtedly assist in this process.

A primary use of zinc is in metal plating, where it is used extensively in galvanizing iron as a means of preventing rust. Corrosion or rusting of iron is an electrochemical process involving oxygen [reaction (13-36)] and the metal [reaction (14-35)]:

$$\mathrm{Fe} \longrightarrow \mathrm{Fe(II)} + 2e^- \qquad (14\text{-}35)$$

The Fe(II) is further oxidized to Fe(III), which forms a porous coating of hydrated oxide, often increasing further corrosion. A more active metal in contact with the iron will be oxidized preferentially, serving to protect the latter. This is the case with zinc. In contrast to a protective coating of paint which will permit corrosion to start in cracks or pinholes, an electrochemically protective process will function even if the coating is not intact. The zinc on galvanized iron is sacrificed as it performs its function.

Cadmium is the second member of the zinc family, and resembles the latter in its chemistry in many respects. Indeed, it is difficult to obtain zinc free from a cadmium impurity. A major chemical difference is a tendency of cadmium to form more covalent bonding, and more stable complexes. It is not used in natural biochemical processes, and in fact, it is extremely toxic and a very hazardous heavy metal. Its limit in drinking water is 0.01 mg/liter. The reason for the high toxicity evidently lies in its similarity to zinc; it can replace the latter in enzymes for example, but because of stronger bonding and perhaps stereochemical differences, the function of the enzyme is disrupted.

Cadmium also is a frequent material in industrial waste discharges, and has been introduced into water systems through mining operations. It is also employed in metal plating. Cadmium sulfide and selenide are used in some red and orange paint pigments. Although these are very insoluble, this does not guarantee biochemical inertness. It is well known that bacteria can attack solid inorganic sulfides by oxidizing the sulfide to sulfate. In this way, the metal ion can be solubilized. While this process has not been reported for cadmium sulfide pigments, it does suggest the danger of assuming harmless behavior on the basis of *in vitro* chemical properties alone.

The most notable widespread example of cadmium poisoning occurred in Japan, where many people developed a disease called "Itai-Itai" from eating rice grown in paddies watered from a river containing cadmium mine wastes.

Mercury, the third member of the zinc and cadmium family, is a very toxic, cumulative poison, having its chief effects on the nervous sytem. It favors still more covalent bonds than cadmium, especially with relatively heavy elements such as sulfur, which coordinates very strongly with Hg(II) since this is an example of a soft Lewis acid (see Section 13.6). Indeed, mercuric compounds bind strongly to the amino acid sulfur atoms contained in many protein and enzyme structures, thereby disrupting their normal physiological processes. The Hg-C bond has comparatively low polarity and organomercury compounds are quite stable in aqueous media in contrast to the more polar organometallic zinc and cadmium compounds that are easily hydrolyzed. Another feature of mercury is its ability to exist in the 1+ state. This state is stable in solution over a very limited pE range, and under most circumstances would be of importance only as insoluble solid compounds or in the presence of elemental mercury.

The environmental chemistry of this element depends on both organic and inorganic forms, which differ considerably in their toxic effects, although all are hazardous. Elemental mercury is a liquid with a significant vapor pressure at room temperature, and the vapor has severe physiological effects if inhaled. The long term limit is 1 ng (10^{-9} gm)/liter of air. This is easily exceeded in a closed room containing exposed mercury (e.g., a laboratory), although the rate of evaporation usually is slow and reasonable ventilation will keep enclosed areas safe. Oral ingestion of small amounts of the metal is less hazardous (e.g., dental amalgams) since it is relatively inert, as are many of the insoluble inorganic mercury compounds. Some compounds have been used in medicine.

Organomercury compounds are much more toxic. They tend to be lipid soluble, and since higher organisms do not decompose and excrete them effectively, they tend to be concentrated in organic tissues in the food chain. Two general classes need to be considered; organomercurium cations RHg^+, which bond ionically to anions such as nitrate or more covalently to softer anions such as chloride, and species with two organic groups, R_2Hg. Both forms can be found in water systems, and are concentrated in biological materials. In addition to the possibility of natural sources, mercury pollution may come from organomercurials used as fungicides in the treatment of seed grains, and from industrial discharges of organomercurials or of metal. The latter can come from chlorine-alkali plants in which mercury cathodes are used in the electrolysis of NaCl solutions in the productions of Cl_2 and NaOH. Such inorganic mercury discharges were long considered inert and harmless. Unfortunately, bacterial action can convert inorganic mercury in sediments to soluble methylmercury and dimethylmercury species.

The natural volatility of elemental mercury means that there is a mercury cycle in the environment. Mercury ore deposits normally produce a considerable mercury vapor pressure that results in a low level, world-wide distribution of mercury. The element has been found in ice samples that predate industrial use of the metal, although such use has increased the amount available to enter into this cycle. The inorganic forms of mercury undergo interconversions that are mediated by microorganisms; for example, insoluble HgS is converted to soluble Hg(II) by bacterial oxidation and soluble Hg(II) is reduced by some bacterial enzymes to Hg(O). This forms a way in which mercury can be eliminated through volatilization. If Hg(II) and Hg are present together, they will enter into an equilibrium reaction to produce Hg(I); the latter exists as the diatomic ion $Hg_2{}^{2+}$

An alternate method by which Hg(II) is eliminated by some bacteria is through methylation. This appears to proceed through transfer of a $CH_3{}^-$ group. Vitamin B_{12}, a cobalt-containing coenzyme, causes this reaction. Vitamin B_{12} contains cobalt in a porphyrin-like environment of four planar nitrogens, with a fifth nitrogen from a dimethylbenzimidazole group also attached (see Section 13.6.4). A methyl group on the active site of the Co(I) may be transferred as $CH_3{}^-$ to a suitable acceptor, and replaced by H_2O. The cobalt is oxidized from the unusual Co(I) state to the more common Co(II) state in this process. A small amount of methylmercurium ion CH_3Hg^+ can be further methylated to $(CH_3)_2Hg$ in a similar process. It can also be reduced bacterially to Hg(O).

The methylmercurium ion CH_3Hg^+ produced in sediments is partially extracted into the water phase and taken up into the food chain. Dimethylmercury can volatilize to the atmosphere, where it is photolyzed to Hg(O) and methyl radicals. The reactions are summarized below:

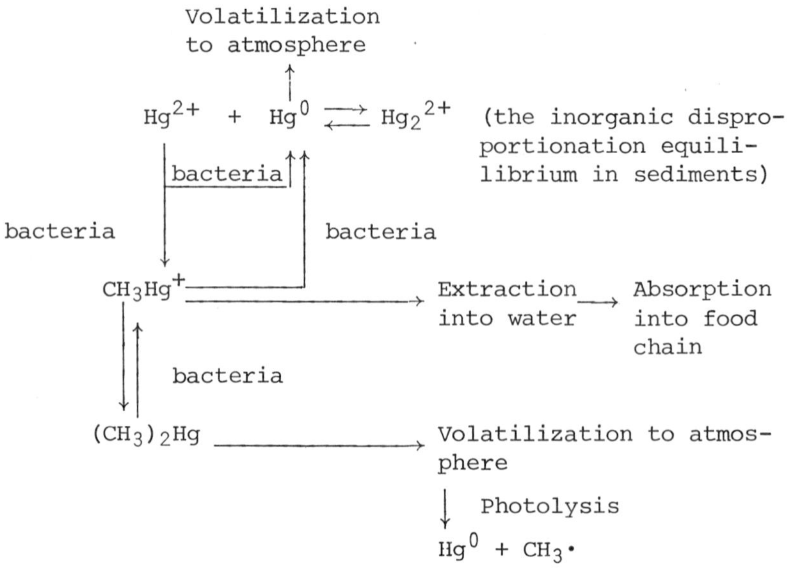

These reactions of mercury undoubtedly occur naturally, although at quite low levels. Industrial input can upset the natural balance by providing additional inorganic starting material, or by direct input of organomercury compounds. There also is evidence that Hg(II) can be methylated by water-soluble methylsilicon, methyltin, and methyllead compounds if these should be introduced from industrial wastes.

The mechanism of formation of methylmercury in the environment is applicable to other elements, e.g., tin, palladium, platinum, and thallium, all of which are heavy metals which form covalent, relatively nonpolar metal-carbon bonds. Some of these elements are employed industrially and could be discharged in wastes. Thallium in particular is toxic in its inorganic as well as organic forms, and bears other resemblances to mercury in its behavior. On the other hand, the more electropositive metals including zinc, cadmium, and lead are not methylated this way, and their alkyls are not stable in aqueous solution.

The environmental hazards of mercury are well documented. Serious and extensive poisoning of the population of Minamata, Japan, which occurred during the 1960s, was caused by consuming mercury-contaminated fish resulting from the organomercury discharges of a local industry. Other cases of mercury poisoning in the U.S. have been traced to seed grain treated with organomercury fungicides that inadvertently were fed to animals shortly before their slaughter.

14.6.8 Boron

Boron in nature is encountered as boric acid or its salts. Boric acid $B(OH)_3$ is quite weak, and this is the form expected in aqueous systems. Boron compounds are widely used industrially. Sodium tetraborate $Na_2B_4O_7$ is used in detergents, and consequently boron is entering water systems. Although necessary for some plants, including algae, it is also known to be toxic to plants at higher concentrations. Toxicity generally is not very well understood, but a maximum limit of 1 mg/liter has been set for drinking water.

14.6.9 Tin and Lead

Tin and lead are the last two members of the carbon family but, in keeping with the general tendency for metallic character to increase with atomic number in a family, they show typical metallic properties. However, they have comparatively weak electropositive characteristics and important electron acceptor properties. Bonds to nonmetals such as carbon have considerable covalent character and make up an important aspect of their chemistry.

Both tin and lead form compounds in which they have the oxidation states (IV) and (II). The former is found in the important organometallic compounds. In general, lead compounds show greater toxicity than those of tin, and environmental lead poses a more significant hazard than does tin.

Organotin compounds are used for a wide variety of purposes, such as wood preservatives, marine antifouling paints, fungicides, and stabilizers for polyvinylchloride. The organotin compounds have the general formula R_nSnX_{4-n}, where X is a suitable anion. The most important systems are those for which $n = 2$ or 3. The toxicity of these is greatest for trialkyl compounds with short carbon chains. Dioctyltin compounds are sufficiently nontoxic to be used in plastics employed for food packaging.

Microbial and photochemical reactions are known to degrade some of the organotin compounds now in use fairly rapidly, although the behavior of many has not been established fully. The final product is an inorganic tin oxide or salt in which form tin is relatively nontoxic. This behavior may be compared to that of lead, which is toxic in its inorganic as well as organic forms.

The toxicity of lead in the environment has caused extensive concern in recent years. The USPHS limit for lead in water is 0.05 mg/liter. Like mercury, Pb(II) forms comparatively covalent bonds with appropriate donor groups in complexes, generally favoring sulfur and nitrogen over oxygen donors, and it may owe some of its physiological reaction to replacement of other metals in some enzymes. Organolead compounds are widely used as the tetraalkyls added to gasolines, but on combustion these are converted to elemental lead, lead oxide, or a halide. These products are not volatile, and atmospheric lead is essentially all particulate in nature. Many common lead salts, such as the carbonate, hydroxide, and sulfide, are extremely insoluble in water; this normally limits the dissolved lead content in a body of water, although in principle complexing agents in the water may considerably increase the effective solubility.

There are various technological sources of lead that can serve as origins of ingested lead in humans and consequent lead poisoning. As with other heavy metals strongly bound by biological complexing agents, lead is a cumulative poison and can act through long-term ingestion of relatively small quantities. Lead poisoning in children from ingestion of paint fragments is well known. This is primarily a hazard with older paints, in which lead pigments were widely used. Basic lead carbonate $(PbCO_3)_2Pb(OH)_2$ is an example of a white pigment, while lead chromate is an orange or yellow one. The oxide, sulfate, and a variety of other lead compounds also were employed. These are largely absent in modern paint formulations, because of both the availability of superior materials and legal restrictions arising from recognition of the hazard. Less common now, but of possible significance in the past is the extraction of lead into drinking water from lead pipes used in the plumbing system. Lead is not now used in drinking water systems. A somewhat more subtle source, however, is pottery utensils. Lead salts have been a common constituent of some of the glazes used since antiquity to decorate pottery, and while apparently inert, some lead in fact may be leached out under acidic conditions particularly if the glaze is improperly fired. Acidic foods (e.g., orange juice) can provide these conditions. One theory for the decline of the Roman empire proposes that the function of much of the population was impaired by chronic lead poisoning originating from wine stored in lead glazed vessels. Another theory attributes the same effect to the use of lead plumbing. Other sources of lead in the human environment have been lead arsenate and lead acetate used as insecticides. Lead also has widespread use in batteries and solder, and a not insignificant amount enters the environment as spent ammunition from hunting. Combustion of coal and oil releases trace amounts of lead. However, roughly half the total environmental lead, and perhaps 95% of the total air-borne lead, comes from gasoline additives.

There has been considerable controversy regarding the hazards associated with lead emissions into the atmosphere. Various evaluations suggest that on the whole, body lead concentrations have not been increased very greatly over what is assumed to be a "normal" level by industrial and automotive emissions. On the other hand, there also are indications that only small increases are necessary before symptoms of lead poisoning can be detected.

14.6.10 Arsenic

Arsenic is a member of the same family as phosphorus, and occurs in the same phosphate rocks from which phosphorus chemicals are obtained. In many industrial phosphates, arsenic remains as an impurity, and thus is found in small amounts in detergents and fertilizers as well as among combustion products, mine tailings, and by-products from the metallurgical processing of copper and other metals. It is also used in pesticides.

Arsenic, like phosphorus, is generally encountered as an oxoanion species, i.e., arsenic acid H_3AsO_4 or a salt of it. In this form its toxicity is well known. A lethal adult human dose is the order of 100 mg but lower doses over a period of time can produce chronic poisoning. Organic arsenic compounds have higher toxicity, and there is evidence that they can be produced from arsenates by the action of anaerobic bacteria, much as methylmercury is formed from inorganic Hg. The following sequence of steps can take place in sediments under the action of the appropriate bacteria:

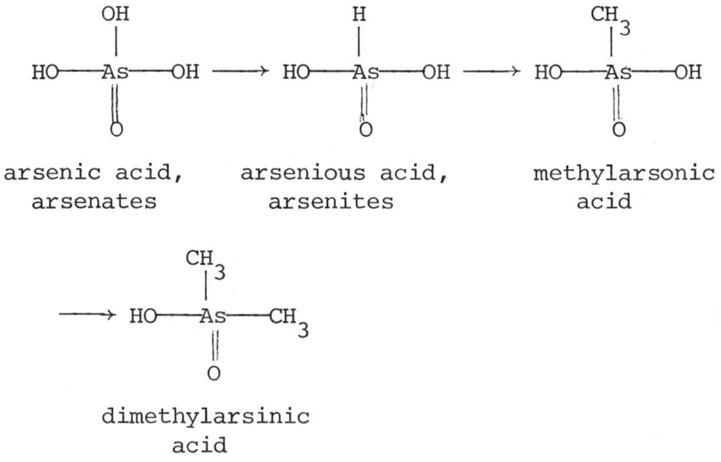

(14-37)

All of these can be extracted into water. Further, additional biological processes can convert the methyl compounds to di- and trimethylarsine $(CH_3)_2AsH$ and $(CH_3)_3As$, which are extremely toxic, volatile compounds. These are readily oxidized, and there is an arsine-arsinic acid biological cycle. As with methylmercury, these methylated arsenic compounds are fat soluble and may concentrate in the food chain.

14.6.11 Selenium

Selenium falls below sulfur in the periodic table, and resembles the latter element in many respects. Metal selenides are found frequently with sulfides, and consequently selenium compounds may appear in the wastes from processing of sulfide ores. Selenates resemble sulfates, and are the most common oxidized form of the element, although selenites are also comparatively stable. Selenium and its compounds are employed in electronic and photoelectric devices, in glass for color, and in photocopying equipment.

The toxic nature of selenium is well known; very small amounts can produce sickness and irritation. Selenium in the soil can be taken up by plants with toxic effects toward animals that consume the material, as has occurred in some areas of the western U.S. Ordinary grasses and cereals can absorb toxic levels, but some plant species can achieve very high degrees of concentration. Availability of soil selenium to plants varies with the total selenium content and soil acidity; the selenium in alkaline soils is chiefly present as soluble selenates that are readily taken up by plants (hence the early name "alkali disease" given to the symptoms of cattle that have eaten high selenium vegetation), while under more acidic conditions the selenium appears to be present as selenites or other less soluble forms. Because of the acute toxicity, and also because of suspected (but at this time unproven) carcinogenic properties, there is a considerable emphasis on reducing selenium emissions.

Bacteria and perhaps other organisms in water sediments and sewage, and some plants, can convert inorganic forms of selenium to volatile organic products such as dimethylselenide $(CH_3)_2Se$, much as inorganic mercury is converted to volatile methylmercury compounds. This may be a means of eliminating toxic levels of selenium in these organisms, and sets up a process of transport and cycling for this element.

In spite of the toxicity of selenium, it is an essential trace element for a number of organisms, possibly including man. It is used in the glutathione peroxidase enzyme system to prevent the formation of peroxides and free radicals from the oxidation of unsaturated fats. Recently, evidence has been presented that indicates that selenium is effective in preventing the onset of cancer; that is, the incidence of cancer can be correlated inversely with selenium levels in the blood. Interestingly, this effect is counteracted by high levels of zinc. If the above observation is valid, it illustrates the difficulty in dealing with trace elements; too thorough elimination of selenium from food and water supplies may do harm, just as too high an intake certainly will.

Tellurium, the next element in the oxygen-sulfur-selenium family, is much less toxic than selenium, because it is much more easily reduced and excreted as organo-tellurium compounds from organisms.

14.6.12 The Halogens

Chlorine is the most abundant halogen element, being widely distributed in water systems as the chloride ion. Sea water is near 0.5 molar in chloride, but some also occurs in fresh waters. The chlorides of some metals are insoluble (e.g., AgCl) but those commonly encountered in nature are soluble. Chlorine is a widely used oxidizing agent, and is added to drinking water to oxidize organic matter. In water, chlorine is hydrolyzed:

$$Cl_2 + 2H_2O \longrightarrow Cl^- + ClOH + H_3O^+ \qquad (14\text{-}38)$$

The hypochlorite ion ClO^- is commonly encountered as sodium hypochlorite used as an oxidizing agent or bleach. Chloride is not harmful, although the total salinity with which it is associated will, if high enough, render water unfit for use. NaCl (and to a lesser extent usually, KCl) are among the final salts to crystallize when fresh water evaporates. This gives rise to natural salt deposits in arid regions, and in some cases a gradual build-up of salt in irrigated soils when irrigation waters containing small amounts of chloride evaporate in the absence of some periodic flushing process. A good deal of chloride as solid NaCl or $CaCl_2$ is used to melt snow and ice, and this also contributes to locally high salinity of soils. Soil salinity is very important for the growth of vegetation, with many plants having very limited tolerance toward this. The Roman act of sowing salt on the ruins of Carthage recognized that a high level of salinity would destroy soil productivity and at least symbolically was intended to prevent regrowth. Chloride ions may accelerate corrosion, for example, by taking part in the mechanism of the metal attack.

The heavier halogens bromine and iodine are much less common than chlorine in natural waters, but they are concentrated from sea water by seaweed, for example. Iodine in particular is essential to health and often is added as an iodide to common salt, because in many areas water and foods are deficient in it.

The first member of the family, fluorine, is an essential element in trace amounts, entering into bones and teeth. It is best known to the public in the latter respect. Its action evidently stems from its ability to replace OH^- in the structures, e.g.,

$$\underset{\text{hydroxyapatite}}{Ca_{10}(PO_4)_6(OH)_2} + 2F^- \longrightarrow \underset{\text{fluoroapatite}}{Ca_{10}(PO_4)_6F_2} + 2OH^- \qquad (14\text{-}39)$$

The size the OH^- and F^- ions are quite similar so that the overall structure is not changed by a limited amount of such replacement, but evidently the hardness and crystallinity of the substances are improved. On the other hand, excess fluoride is harmful. Fluoride is found in minerals, but only rarely does the level in waters become dangerously high from this source. It is of course added to water in fluoridation processes, from industrial processes, and as an impurity in phosphate products (e.g., fertilizers) where it often remains from fluorides present in the original phosphate mineral. Most simple fluoride salts are soluble and will enter the run-off water. Recognition that low levels of fluoride ion in water is effective in reducing dental decay, especially among children, has led many communities to add fluoride to their water systems. Common fluoride materials that may be used in water fluoridation treatments are NaF, which may be obtained readily from reaction (14-40), or silicofluoride salts (Na_2SiF_6 and others) obtained from phosphate purification.

$$\underset{\text{fluorspar}}{CaF_2} + H_2SO_4 \longrightarrow CaSO_4 + 2HF \xrightarrow[Na_2CO_3]{NaOH} 2NaF \qquad (14\text{-}40)$$

The well-known toxicity of excess fluoride has always lead some people to oppose such additions. This is a case where small amounts of a material are beneficial while large amounts are harmful, and failure to recognize that this is a very common situation no doubt influences much of the fluoride opposition. At the controlled levels used in drinking water, fluoride is not harmful, but questions may be raised about the ultimate environmental effects associated with a long-term, low-level injection of fluoride ions into fresh water systems. Other unforeseen effects of fluoride ions in water used for a multitude of purposes may also arise. For example, there have been indications of interference from F^- in kidney dialysis machines.

BIBLIOGRAPHY

General

K. J. Laidler and M. H. Ford-Smith, "The Chemical Elements." Bogden and Quigley, Tarrytown-on-Hudson, New York, 1970.

F. A. Cotton and G. Wilkinson, "Advanced Inorganic Chemistry," 3rd ed. Interscience, New York, 1972.

M. N. Hughes, "The Inorganic Chemistry of Biological Processes." Wiley, New York, 1972.

D. R. Williams, "The Metals of Life." Van Nostrand Reinhold, London, 1971.

"Trace Inorganics in Water." Advances in Chemistry Series No. 73, American Chemical Society, Washington, D.C., 1968.

"Trace Elements in the Environment." Advances in Chemistry Series No. 106, American Chemical Society, Washington, D.C., 1971.

Carbon

G. Skirrow in "Chemical Oceanography." (J. P. Riley and G. Skirrow, eds.), Vol. 2, Chapter 9. Academic Press, New York, 1975.

W. W. Rubey, Geologic history of sea water, *Bull. Geol. Soc. Amer. 62,* 1111 (1951).

V. M. Goldschmidt, "Geochemistry." Oxford University Press, London, 1958.

R. M. Garrels and A. E. Perry, Jr., Cycling of C, S, and O through geologic time, in "The Sea." (E. D. Goldberg, ed.), Vol. 5, Chapter 21. Wiley (Interscience), New York, 1974.

J. M. Gieskes, The alkalinity--total carbon dioxide system in sea water, in "The Sea." (E. D. Goldberg, ed.), Vol. 5, Chapter 3. Wiley (Interscience), New York, 1974.

B. Bolin, The carbon cycle, in "Chemistry in the Environment, Readings from Scientific American," p. 53. W. H. Freeman and Co., San Francisco, California, 1973. (From Sci. Amer., Sept., 1970.)

D. F. Martin, "Marine Chemistry," Vol. 2, Chapter 8. Marcel Dekker, New York, 1970.

Nitrogen

C. C. Delwiche, The nitrogen cycle, in "Chemistry in the Environment, Readings from Scientific American," p. 43. W. H. Freeman and Co., San Francisco, California, 1973. (From Sci. Amer., Sept., 1970.)

I. A. Wolff and A. E. Wasserman, Nitrates, nitrites, and nitrosoamines, *Science 177,* 15 (1972).

E. Robbinson and R. C. Robbins, in "Air Pollution Control" (W. Strauss, ed.), Part II, Chapter 1. Wiley (Interscience), New York, 1972.

D. F. Martin, "Marine Chemistry," Vol. 2, Chapter 7. Marcell Dekker, New York, 1970.

R. F. Vaccaro, in "Chemical Oceanography" (J. P. Riley and G. Skirrow, eds.), Vol. 1, Chapter 9. Academic Press, New York, 1965.

Sulfur

E. Robbinson and R. C. Robbins, in "Air Pollution Control" (W. Strauss, ed.), Part II, Chapter 1. Wiley (Interscience), New York, 1972.

W. W. Kellog, R. D. Cade, E. R. Allen, A. L. Lazrus, and A. E. Martell, The sulfur cycle, *Science 175,* 587 (1972).

G. E. Likens and F. H. Bormann, Acid rain: A serious regional environmental problem, *Science 184,* 1176 (1974). Further comment is provided by L. Newman, *Science 188,* 957 (1975), and G. E. Likens and F. A. Bormann, *Science 188,* 958 (1975).

J. L. Frohliger and R. Kane, Precipitation; Its acidic nature, *Science 189,* 455 (1975).

G. E. Likens, Acid precipitation, *Chem. Eng. News,* Nov. 22, 1976, p. 29.

J. B. Pfeiffer, ed., "Sulfur Removal and Recovery from Industrial Processes." Advances in Chemistry Series No. 139, American Chemical Society, Washington, D.C., 1975.

Chemical desulfurization of coal, in "Modern Energy Technology" (H. Fogiel, ed.), Vol. II, Chapters 48, 49. Research and Education Association, New York, 1975.

Phosphorus

D. F. Martin, "Marine Chemistry," Vol. 2, Chapter 6. Marcel Dekker, New York, 1970.

S. H. Jenkins and K. J. Ives, eds., "Phosphorus in Fresh Water and the Marine Environment." Pergammon Press, Oxford, 1973.

Metals and Trace Elements

E. Frieden, The evolution of metals as essential elements, in "Protein-Metal Interactions" (M. Friedman, ed.), Chapter 1. Plenum, New York, 1974.

E. J. Underwood, "Trace Elements in Human and Animal Nutrition," 3rd ed. Academic Press, New York, 1971.

A. Jernelov, Heavy metals, metalloids, and synthetic organics, in "The Sea" (E.D. Goldberg, ed.), Vol. 5, Chapter 21. Wiley (Interscience), New York, 1975.

Studies firm up some metals' role in cancer, *Chem. Eng. News,* Jan. 17, 1977, p. 35.

"Geochemistry and the Environment, the Relation of Selected Trace Elements to Health and Disease," Vol. 1. National Academy of Sciences, Washington, D.C., 1974.

G. L. Waldbott, "Health Effects of Environmental Pollutants." The C. V. Mosby Co., St. Louis, Missouri, 1973.

S. P. Bapu, ed., "Trace Elements in Fuel." Advances in Chemistry Series No. 141, American Chemical Society, Washington, D.C., 1975.

Z. I. Izreal'son, ed., "Toxicity of the Rare Metals." Israel Program for Scientific Translations, Jerusalem, 1967.

J. O. Leckie and R. D. James, Control mechanisms for trace metals in natural waters, in "Aqueous Environmental Chemistry of Metals" (A. J. Rubin, ed.), Chapter 1. Ann Arbor Science Publishers, Ann Arbor, Michigan, 1974.

S. L. Williams, D. B. Aulenbach, and N. L. Clesceri, Sources and distribution of trace metals in aquatic environments, in "Aqueous Environmental Chemistry of Metals" (A. J. Rubin, ed.), Chapter 2. Ann Arbor Science Publishers, Ann Arbor, Michigan, 1974.

J. Schubert, Heavy metals--toxicity and environmental pollution, in "Metal Ions in Biological Systems" (S. K. Dhar, ed.), p. 239. Plenum, New York, 1974.

Committee on Medical and Biological Effect of Environmental Pollution, "Chromium." National Academy of Sciences, Washington, D.C., 1974.

Committee on Medical and Biological Effect of Environmental Pollution, "Manganese." National Academy of Sciences, Washington, D.C., 1973.

Committee on Medical and Biological Effects of Environmental Pollutants, "Nickel." National Academy of Sciences, Washington, D.C. 1975.

Committee on Medical and Biological Effects of Environmental Pollutants, "Copper." National Academy of Sciences, Washington, D.C., 1977.

L. Friberg, M. Piscator, G. Nordberg, and T. Kjellstrom, "Cadmium in the Environment," 2nd ed. Chemical Rubber Publishing Co., Cleveland, Ohio, 1974.

L. Friberg and J. Vostal, eds., "Mercury in the Environment." The Chemical Rubber Co., Cleveland, Ohio, 1972.

J. W. Huckabee and B. G. Blaylock, Transfer of mercury and cadmium from terrestrial to aquatic ecosystems, in "Metal Ions in Biological Systems" (S. K. Dhar, ed.), p. 125, Plenum, New York, 1973.

E. L. Kothney, The three-phase equilibrium of mercury in nature, in "Trace Elements in the Environment" (R. E. Gould, ed.), Chapter 4. Advances in Chemistry Series No. 106, American Chemical Society, Washington, D.C., 1971.

J. M. Wood, Biological cycles for toxic elements in the environment, *Science 183*, 1049 (1974).

W. E. Smith and A. M. Smith, "Minimata: A Case Study of Mercury Poisoning." Holt, Rinehart, and Winston, New York, 1975.

P. Smith and L. Smith, Organotin compounds and applications, *Chem. Britain 11*, 208 (1975).

B. B. Ewing and J. E. Pearson, Lead in the environment, in "Advances in Environmental Science and Technology" (J. N. Pitts, Jr., R. L. Metcalf, and A. C. Lloyd, eds.), Vol. 3, p. 1. Wiley, New York, 1974.

Committee on Medical and Biological Effects of Atmospheric Pollution, "Lead." National Academy of Sciences, Washington, D.C., 1972.

P.R. Harrison, Air pollution by lead and other trace metals, in "Metal Ions in Biological Systems" (S. K. Dhar, ed.), p. 173. Plenum, New York, 1973.

R. A. Goyer and J. E. Moore, Cellular effects of lead, in "Protein-Metal Interactions" (M. Friedman, ed), Chapter 22. Plenum, New York, 1974.

D. K. Darrow and H. A. Schroeder, Childhood exposure to environmental lead, in "Protein-Metal Interactions" (M. Friedman, ed.), Chapter 20. Plenum, New York, 1974.

D. Bryce-Smith, "Heavy Metals as Contaminants of the Human Environment." Educational Techniques Subject Group, The Chemical Society, London, 1975. (An interesting taped lecture and booklet dealing with Cd, Hg, and Pb, and giving numerous references.)

R. S. Braman and C. C. Foreback. Arsenic in the environment, *Science 182*, 1247 (1973).

A. A. Westing, Plants and salt in the roadside environment, in "Understanding Environmental Pollution" (M. A. Strobbe, ed.), Chapter 43. The C. V. Mosby Co., St. Louis, Missouri, 1971.

H. C. Hodge and F. A. Smith, "Fluorine Chemistry" (J. H. Simons, ed.), Vol. 4. Academic Press, New York, 1965.

Committee on Medical and Biological Effects of Atmospheric Pollution, "Fluorides." National Academy of Sciences, Washington, D.C. 1971.

W. H. Larkin, Selenium in our environment, in "Trace Elements in the Environment" (R. E. Gould, ed.), Chapter 6. Advances in Chemistry Series No. 106, American Chemical Society, Washington, D.C., 1971.

"Selenium in Nutrition." Subcommittee on Selenium, Committee on Animal Nutrition, National Academy of Sciences, Washington, D.C., 1971.

Y. K. Chau, P.T.S. Wong, B. A. Silverberg, P. L. Luxon, and G. A. Bengert, Methylation of selenium in the aquatic environment, *Science 192,* 1130 (1976).

Committee on Medical and Biological Effects of Atmospheric Pollution, "Vanadium." National Academy of Sciences, Washington, D.C., 1974.

W. H. Zoller, G. E. Gordon, E. S. Gladney, and A. G. Jones, Sources and distribution of vanadium in the atmosphere, in "Trace Elements in the Environment" (R. E. Gould, ed.). Advances in Chemistry Series No. 106, American Chemical Society, Washington, D.C., 1971.

15
NATURAL WATER SYSTEMS

15.1 COMPOSITION OF WATER BODIES

The natural composition of natural waters depends on gain and loss of solutes through both chemical reactions and physical processes. For the most part, solutes undergo a geological cycle in which materials entering solution as products of weathering reactions of rocks, volcanism, etc., are carried to the oceans where they undergo further reaction, are deposited in sediments, and eventually are reincorporated into new rocks, which may repeat the cycle. Obviously, these processes have a long time scale. Volatile materials may enter the atmosphere for part of the cycle, and some elements may take part in biochemical, as well as geochemical, processes. Physical transport of particulate material suspended in waters and as dust and sea spray in the atmosphere also have a part in these processes.

The composition of the ocean is not uniform, but variations from average values generally are not excessive. Fresh waters show much greater variation. Table 15-1 compares ocean and average river water compositions for some nonvolatile solute components. Oceans receive solutes from rivers, but the composition of the ocean is not simply that of a concentrated river. A major difference is in the Na-K ratios due to selective incorporation of K^+ in sediments. The natural fresh water sources of the materials in solution are chiefly weathering reactions of minerals. The most important reactions involve decomposition to other insoluble mineral species, with some of the decomposition products such as metallic cations entering solution; some examples are given in Section 16.3.

If a fresh water containing typical amounts of dissolved Ca^{2+}, Na^+, and HCO_3^- in equilibrium with atmospheric CO_2 is allowed to evaporate, the pH will change from an initial slightly basic value as could be calculated from the carbonic acid dissociation equilibria toward greater basicity. The

TABLE 15-1

Average Compositions of Rivers and Ocean Waters--Most Common Elements Only[a]

Element	*Predominant form*	*River (M/liter)*	*Ocean (M/liter)*
C	(HCO_3^-)	1×10^{-4}	2.3×10^{-3}
Ca	(Ca^{+2})	3.8×10^{-5}	0.01
Cl	(Cl^-)	2.3×10^{-4}	0.56
K	(K^+)	5.9×10^{-5}	0.01
Mg	(Mg^{+2})	1.6×10^{-7}	0.05
Na	(Na^+)	3.9×10^{-4}	0.48
S	(SO_4^{-2})	1.2×10^{-4}	0.03
Si	$(Si(OH)_4)$	1.4×10^{-4}	3.6×10^{-5}

[a] *See J. P. Riley and R. Chester, "Introduction to Marine Chemistry." Academic Press, New York, 1971, p. 64 for a more complete list.*

carbonate equilibria will shift to favor CO_3^{2-} as the predominant species. At some point, $CaCO_3$ will precipitate, providing some buffering action. However, the Na^+ will remain in solution since it will form no insoluble salts, and the final solution will be a sodium carbonate solution. The actual behavior of a real fresh water system will be complicated by the other solute material present, but natural soda lakes approximate this situation. In general, magnesium salts and silicates may also be among the precipitates, but F^-, Cl^-, and most SO_4^{2-} stay in solution as Na^+ and K^+ salts. For average fresh water, carbonate species are the predominant anions; thus the solution will become more basic on evaporation as the carbonates become more concentrated.

15.2 MODEL OCEAN SYSTEMS

As was discussed in Section 13.3, the pH of the ocean corresponds to that of a solution of carbonic acid partially neutralized by reaction with basic minerals. That is, the means by which the pH of the oceans is established can be understood only if the interactions of these minerals are considered; the carbonate system alone does not determine the natural pH of the oceans.

The average ocean pH is about 8.1; surface values are a bit higher (∿8.3). The typical variation of pH with depth is shown in Fig. 15-1. The high surface value may relate to biological factors while variation at depth probably relates to pressure and temperature effects.

Physicochemical models of the ocean have been set up to try to understand the pH-controlling reactions, among others. A model devised by Sillén[1] is most notable. In this and related models, it is necessary to consider not only the water and dissolved materials, but also the atmospheric and sediment components that could take part in environmental reactions. Equilibrium is assumed. The amounts of material to be considered per liter of sea water are estimated in Table 15-2. Only the most abundant elements are listed here

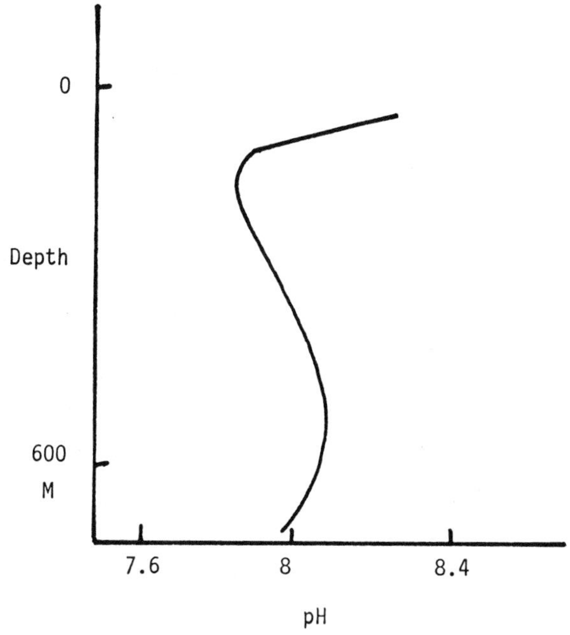

FIGURE 15-1. Variation of the ocean pH with depth.

[1]*L. S. Sillén, The physical chemistry of sea-water, in "Oceanography" (M. Sears, ed.), p. 549. Publication No. 67 of the American Society for the Advancement of Science, Washington, D.C., 1961.*

TABLE 15-2

The Major Components in the Model Ocean, Relative to 1 Liter of Water

Substance	*Amount (moles)*	*Comments*
H_2O	54.9	(1 liter)
Si	6.06	Mostly solid silicates and SiO_2
Al	1.85	Mostly solids
Na	0.76	0.48 moles in solution
Ca	0.56	0.01 moles in solution
Cl	0.55	Mostly in solution
C	0.55	See C cycle
Fe	0.55	Mostly solids
Mg	0.53	0.05 moles in solution
K	0.41	0.01 moles in solution

along with water, since it will be these that will have the most important effects. One may consider the various chemical equilibria that could be set up among these major constituents. This is done most conveniently by imagining the successive addition of components to the water. For charge balance, O^{2-}, OH^-, or H^+ ions may be associated with cations (or anions); these can be regarded as components of the water itself.

After water, the most abundant material is silicon, which can be considered as being added to the model one liter of water as 6.06 moles of silica, SiO_2. Silica will undergo a variety of reactions with water. The solution process is

$$SiO_2 \text{ (crystal)} + 2H_2O \rightleftarrows Si(OH)_4 \text{ (aq.)}$$
$$K = 2 \times 10^{-4} \qquad (15\text{-}1)$$

Solubility of amorphous SiO_2 is higher ($K = 2 \times 10^{-3}$), and consequently this form should dissolve with reprecipitation of the crystal. The fact that amorphous material is found in nature illustrates the problem of rate. Dissolution of the amorphous material is slow. From the equilibrium constant of reaction (15-1), it can be calculated that water should contain about 10^{-4} moles/liter of dissolved silica, and this seems to be the case in bottom ocean waters. The surface concentration of dissolved silica is closer to 10^{-4} because of biological (diatom) action. As a consequence of the use of silica in the skeletons of these organisms, a biogeochemical cycle for silica exists.

The dissolved material undergoes several reactions. One such reaction is

$$Si(OH)_4\ (aq) + H_2O \rightleftarrows SiO(OH)_3^-\ (aq) + H_3O^+$$

$$K = 3.5 \times 10^{-10} \qquad (15\text{-}2)$$

This means that $Si(OH)_4$ is a weak acid. There may be further dissociation steps, but with still smaller equilibrium constants. Another process is

$$4Si(OH)_4\ (aq) \rightleftarrows Si_4O_6(OH)_6^{2-}\ (aq)$$

$$+ 2H_3O^+ + 2H_2O \qquad (15\text{-}3)$$

The same type of polymerization reaction may lead to higher polynuclear species. Since H_3O^+ is involved in this reaction, polymerization is pH dependent, and in most natural waters only mononuclear species will form.

The next most abundant element, aluminum, can be considered to be present as $Al(OH)_3$. The large affinity of this element for oxygen precludes the stable environmental existence of aluminum species other than the hydroxide, oxide, and various minerals in which Al is associated with oxygen. We shall have the reaction

$$Al(OH)_3\ (s) + 2H_2O \rightleftarrows Al(OH)_4^-\ (aq) + H_3O^+ \qquad (15\text{-}4a)$$

which can also be written as

$$Al(OH)_3\ (s) + OH^- \rightleftarrows Al(OH)_4^-\ (aq); \quad (K = 10^{-1}) \qquad (15\text{-}4b)$$

We may note that the low electronegativity of aluminum leads to acid behavior by an indirect process as compared to the silicon case. Other details of the behavior of dissolved aluminum species are unclear, but there can be reactions with silica to produce aluminosilicates. One such reaction is

$$Al(OH)_4^-\ (aq) + Si(OH)_4\ (aq) \rightleftarrows \frac{1}{2}\ Al_2Si_2O_5(OH)_4\ (s)$$

$$+ OH^- + 2\frac{1}{2}\ H_2O;\ K = 10^{-6} \qquad (15\text{-}5)$$

This reaction could produce a basic solution.

Of the remaining elements in Table 15-1, the number of equivalents of the cations Na^+, K^+, Mg^{2+}, and Ca^{2+} exceed the number of equivalents of the listed anions. Hence, some of these cations must be added as the oxides or hydroxides, and omitting the calcium and magnesium associated with 0.55 moles of carbon as $CO_3{}^{2-}$, the basic materials would be in excess. However, the cations may further react with the excess SiO_2 and with the aluminosilicate minerals. One such process could be

$$3Al_2Si_2O_5(OH)_4\ (s) + 4SiO_2\ (s) + 2K^+ + 2Ca^+ + 15H_2O \rightleftarrows 2KCaAl_3Si_5O_{16}(H_2O)_6\ (s) + 6H_3O^+ \qquad (15.6)$$

Such reactions produce H_3O^+ ions, and since the amounts of materials (alumina, silica, cations) are large, they can be the controlling reactions for establishing pH. Addition of carbonate still leaves the aluminosilicates in excess. Various reactions of this type involving known minerals can be proposed. Although accurate equilibrium calculations are difficult to make (partly because equilibrium constants for many of these reactions are not well known) and the hypothesis of equilibrium cannot be checked reliably, it is certain that such reactions can occur. The basic result of Sillén's model is that aluminosilicates control the pH, and the carbonate system merely serves as an indicator and a short term buffer, since reactions involving sediments would presumably reach equilibrium much more slowly. In any real application of buffering action, mixing times are important, as already mentioned.

Equilibria involving cation exchange reactions are important in mineral systems. Bonding of species held by predominantly coulombic interactions will depend largely on packing efficiency, and for cations of the same charge the affinity with which they are held in an aluminosilicate lattice will depend mostly on the relative sizes of the cations and the lattice positions available for them to occupy. For example, relative size (radius ratio) principles would suggest that a small cation would be most stable in a site determined by four oxide ions (tetrahedral site) but would pack less efficiently and be less stable than a large cation if surrounded by six oxygens (octahedral site). Of course, this simplification is often violated when factors other than simple electrostatic interactions become important.

In sediments, K^+ is more abundant than Na^+, and also Ca^{2+} is more abundant than Mg^{2+}. The molar concentration ratios are, in sediments, $K^+/Na^+ = 1.4$, while in sea water $K^+/Na^+ = 0.0026$. That is, K^+ is preferentially bound in the sediments. If these processes are equilibrium reactions, it also follows that the sediments act to control the K^+/Na^+ (and Mg^{2+}/Ca^{2+}) ratios in sea water. Moreoever, the K^+ and Ca^{2+} concentrations must appear in the equilibrium expression for the pH-controlling reactions. For the example given in reaction (15-6), previously,

$$K_{eq} = [H_3O^+]^6 / [K^+]^2 [Ca^{2+}]^2 \qquad (15\text{-}7)$$

The pH of sea water is strongly interrelated to these cation concentrations.

The chloride ions are not taken up significantly by the sediments, and nearly all chloride remains in solution.

If the carbonates are now considered in Sillén's model, as 0.46 moles $CaCO_3$ and 0.09 moles of $MgCO_3$, further reactions with the aluminosilicates become possible, but other new processes available involve precipitation of carbonate solid phases [two crystalline modifications of $CaCO_3$ (calcite and aragonite), $MgCO_3$, and dolomite, $CaMg(CO_3)_2$]. Carbonate will also enter into the pH-controlling reactions, but the significance of the model is that the amount of carbonate is small compared to the proton capacity available from aluminosilicates. For 1 liter of sea water in equilibrium with atmospheric CO_2, 10^{-3} moles of strong acid will change the pH from 8 to less than 6 (biological action may modify this). However, when the minerals are included, the buffering capacity is the order of 1 mole/liter.

The remaining components can be added to this model to complete the picture; oxygen and iron being the most important. Much of the oxygen remains in the atmosphere, but some is in solution according to Henry's law. This controls the pE of the ocean, and determines the states of Fe and Mn to be found in the environment as discussed in Section 14.6. The entire system is subject to the various physicochemical equilibrium equations, and equilibrium conditions can at least in principle be calculated. This model, with some refinements, undoubtedly gives a good description of the equilibrium ocean on a geological time scale.

15.3 RESIDENCE TIMES

For equilibrium models to have any validity, rates of the various reactions must be fast enough to approach equilibrium in the time period in question. Natural reaction rates are complicated by the mixing problem, which has been referred to before.

The length of time a species spends in a given phase (say sea water) is important in terms of how rapidly appropriate cycles can be completed. This leads to steady-state models rather than equilibrium models of the type previously discussed. A steady state assumes that input (for the ocean, weathering products brought in by rivers, volcanic activity, atmospheric input) is balanced by removal (sedimentation, ion exchange, biological production of inert material, etc.). That is,

$$(dC/dt)_{in} = (dC/dt)_{out} \tag{15-8}$$

where dC/dt = rate of change in concentration. Residence time is defined as $T = C/(dC/_{dt})$. Some examples of residence time for the ocean are given in Table 15-3.

In fresh water lakes, we have a similar situation, but output involves outflow as well as the other processes. An inert soluble material should have T about equal to that of the water itself, since it flows in and out of a given region with the water. Shorter T values imply rapid formation of insoluble products (e.g., Al) while larger ones suggest a cycling process that may be inorganic or biological. For example, in upper layers one may have $Fe^{3+} \rightarrow Fe(OH)_3$. However, if much organic material is present in the sediments, giving low pE, the precipitated ferric compound may be reduced to soluble Fe^{2+}, which recycles.

TABLE 15-3

Residence Times of Some Elements in the Oceans

Element	*Residence time (years)*
Al	1×10^2
Ca	8×10^6
Cr	3.5×10^2
Fe	1.4×10^2
K	1.1×10^7
Mg	4.5×10^7
Mn	1.4×10^3
Na	2.6×10^8
Si	8×10^3
Zn	1.8×10^5

Phosphorus and nitrogen are involved in biological cycles; uptake by organisms → death and sedimentation → decomposition and solution. They often have long residence times. The concentrations of substances with T larger than that of water will build up to values larger than that of the input.

15.4 WATER TREATMENT

Large-scale water treatment is necessary in two general circumstances; for water that is taken into distribution systems for household or industrial use, and for waste water that must meet particular standards for pollution control. The most well-known treatment process for drinking water is chlorination. Chlorine is a strong oxidizing agent, and in water produces hypochlorite, which is well known in bleaches (Section 14.6).

$$Cl_2 + 2H_2O \rightarrow Cl^- + HOCl + H_3O^+ \qquad (15\text{-}9)$$

The primary purpose of the chlorine treatment is to oxidize organic materials in the water, especially bacteria. It is necessary to maintain oxidizing conditions throughout the entire distribution system to ensure that septic conditions do not redevelop after treatment, while ensuring that the chlorine content is not excessive near the addition point. The stability of the hypochlorite is dependent upon the pH, which may require adjustment for this reason and to reduce corrosive properties of the water.

Although chlorination oxidizes many organic materials, it does not oxidize all. It can result in chlorination of some molecules, with the possibility of producing chlorinated hydrocarbons of significant hazard to health. This is most dangerous in waters containing much industrial waste, where suitable organic molecules are present in significant amounts. Of course, such waters may already contain chlorinated hydrocarbon wastes and it is not always clear in field situations to what extent chlorination actually contributes to the total. Nevertheless, the hazard has been pointed out especially for waters from the lower Mississippi valley (Section 7.1). While in many localities in the U.S. the law allows no substitute for chlorine treatment of drinking waters, an alternate method is available and receives some use. This is treatment of the water with ozone O_3. This is as effective as chlorine, and avoids the risks of generating chlorinated products.

Water supplies often receive other treatment. For example, filtration through sand beds may be used to remove suspended matter, and coagulation processes using aluminum sulfate or similar materials also may be employed for clarification. Passage through beds of activated carbon is sometimes used to remove organic contaminants that are not oxidized readily by the chlorine treatment. This process has been used on a limited scale for many years. Activated carbon is produced by charring a variety of organic materials, followed by partial oxidation. This material has a high surface area and adsorbs most organic molecules effectively. The activity can be regenerated by oxidation of the adsorbed materials by heating in an atmosphere of air and steam. This method is perhaps the most practical means now available for removing chlorinated hydrocarbons, aromatics, etc., which are among the most dangerous of the organic pollutants likely to be present in a domestic water source.

Clean water is a vital commodity, and usage is now so extensive that waste waters must be repurified in order to avoid destruction of aquatic ecosystems and also because, in many cases, the water will be reused. One very high use of water is for cooling purposes. Such water, if discharged back to fresh water systems, may lead to so-called thermal pollution by increasing lake and stream temperatures, but introduces impurities only if additives are present to reduce fouling of heat exchangers or if substances are dissolved from them. Trace metals extracted as corrosion products would be a possible example of the latter case. The chemistry of thermal pollution lies chiefly in the effects of temperature on physicochemical and biological processes. For example, warm water will reduce the solubility of O_2 which lowers its capacity to support life and to oxidize impurities. Large amounts of artificially warmed water will alter the relative volumes of the epilimnion and hypolimnion in a lake, and may maintain the density gradient when normal conditions would lead to turnover. Since fish and other aquatic organisms often prefer very restricted temperature conditions, the volumes of the habitats available for different species and, hence, the population balance of the water system may be changed greatly.

Treatment of domestic waste water (i.e., sewage), is important, and some of the steps used in common processes are described briefly. Such water normally contains about 0.1% impurities composed mostly of a variety of organic materials. Many of these can be removed by filtration or sedimentation. Large particles (grit, paper, etc.), are removed by mechanical screening or filtration using a variety of techniques depending on the local nature of the material.

Much suspended and fine particulate material remains, which is usually removed by sedimentation processes. In well-designed settling tanks, nearly 90% of the total solids and perhaps 40% of the organic matter can be removed. In order to remove the smaller particles, however, assistance is needed and coagulation may be employed.

Small particles suspended in a liquid are referred to as colloids. Such suspensions often are stabilized by adsorption of ions on the surface of the particles; if ions of a particular charge are adsorbed preferentially, the colloidal particles will repel each other and coagulation will be impeded. Various chemicals can be used that form insoluble precipitates that carry down the suspended materials. Examples are compounds of Fe(III) and Al(III), which on hydrolysis form the insoluble hydrated oxides discussed earlier. Among the more commonly used materials are $Al_2(SO_4)_3$ (aluminum sulfate, or alum) and ferrous sulfate. The latter is oxidized to ferric ion in the water. Hydrolysis produces polynuclear hydroxo species that are readily adsorbed on the surface of the suspended material, and as these hydroxo compounds further condense to insoluble forms, bind the particles with the floc as it settles. If the pH is near neutrality, both Al(III) and Fe(III) are completely hydrolyzed and the added materials completely removed from the water. The pH may be adjusted to ensure this by the addition of a cheap base such as $Ca(OH)_2$ (lime). The reaction can be summarized:

$$Al_2(SO_4)_3 + 3Ca(OH)_2 \rightarrow 3CaSO_4 + 2Al(OH)_3 \qquad (15\text{-}10)$$

In practice, sewage waters often contain considerable bicarbonate ion which also acts as a base; $CaCO_3$ may be part of the precipitate. Lime alone will cause $CaCO_3$ to precipitate and has some coagulant action. An added value to using calcium, iron, or aluminum in this process is that much of the phosphorous present can be removed as the insoluble phosphate salts. Up to 95% removal of phosphate is possible. The solid produced in this process must be disposed of. It may contain toxic industrial wastes, grease, bacteria, etc., and often is buried in landfill, or dumped at sea.

The treatment processes described above are referred to as primary treatment, and a great deal of sewage receives no further processing. A large amount of dissolved material is left behind. Much of this is organic, but there are also nitrate, more or less phosphate, and metal ions. Secondary treatment removes the organic materials. This is important with respect to the overall quality of the discharged water, since a high organic content will use up the dissolved oxygen in the water, cause reducing conditions and the formation of the various undesirable reduction products. Aquatic life also suffers. A practical measure of the reducing impurities in water is the so-called biological oxygen demand (BOD) (Section 6.6).

The most widely used method of secondary treatment is a biochemical one, the activiated sludge process. Bacterial action using special bacterial sludge in an aerated system converts the organic materials to more bacterial sludge. Excess sludge is separated for disposal by sedimentation. Activated sludge residue must be disposed of in landfill or in other ways. A possible use is as a fertilizer, since the material has a high nitrogen and phosphorous content. The feasibility of this use depends on the economics of drying and treating the sludges, and on ensuring the absence of toxic materials.

Alternate sludge processing methods involve further bacterial digestion. Anaerobic digestion of the sludge converts carbon components chiefly to CH_4 and CO_2, while nitrate, sulfate, and phosphate are converted at least in part to NH_3, H_2S, and PH_3. About 70% of the gaseous products are composed of methane. This can be burned to heat the reaction tanks, power pumps, etc., and in fact such degradation of organic wastes is a possible source of energy. The other gaseous products (except CO_2) must be trapped or reacted to less noxious forms.

Aerobic digestion of the sludge with no additional source of nutrients results in autooxidation as the bacterial cells use up their own cell material. The organic materials are converted to CO_2 or carbonates, N, S, and P to the oxoanions, and the solution ultimately produced is potentially useful as a liquid fertilizer. This alternate approach to activated sludge disposal is comparatively new.

Disposal of primary or secondary sludge usually is considered as a problem involving organic material, and N, P, and S nutrients, with the main consideration being their effects on eutrophication processes, BOD, etc. However, toxic materials, including heavy metals, may be present in considerable amounts and represent a hazard that must be considered.

Final clean up of the water effluent from the secondary treatment is referred to as tertiary treatment. Several types of materials must be considered in this respect.

(a) Colloidal organic matter, i.e., the remains of the activiated sludge that do not settle out in the sedimentation tanks. A variety of types of filters can be used, and it is common at this point to use the $Al_2(SO_4)_3$ coagulation process described as the key step in removing these materials.

(b) Phosphate. A phosphate concentration of 1 mg/liter is adequate to support extensive algal growth (blooms). Typical raw wastes contain in the vicinity of 25 mg/liter of phosphate. Thus, a large reduction is necessary to ensure that these effluents do not contribute to eutrophication. Removal of phosphorus by precipitation of an insoluble salt has been mentioned in connection with primary waste treatment. In fact, this precipitation may be performed in the primary, secondary, or tertiary step. Lime is commonly used, with hydroxyapatite being the product:

$$5Ca(OH)_2 + 3HPO_4^{2-} \rightarrow Ca_5OH(PO_4)_3 + 3H_2O + 2OH^- \qquad (15\text{-}11)$$

Lime is cheap, and the efficiency is theoretically very high, although colloid formation by the hydroxyapatite, slow precipitation, and other problems often reduce this efficiency in practice. Polyphosphates, if present, form soluble calcium complexes. Other precipitants are $MgSO_4$ (giving $MgNH_4PO_4$), $FeCl_3$ (giving $FePO_4$), and $Al_2(SO_4)_3$ (giving $AlPO_4$). Phosphate precipitation may occur as part of a coagulation process, especially if iron or aluminum salts are used. An alternate process to precipitation is adsorption, using activiated alumina, i.e., acid washed and dried Al_2O_3. The alumina can be regenerated by washing with NaOH.

(c) Nitrate. The activated sludge process converts some nitrogen to organic forms that are removed with the sludge; much, however, remains behind. It is possible to convert this to nitrate if excess air is used, but this method of operation involves more expense than required for efficient removal of carbon compounds, and usually is not done. The nitrogen remaining in the effluents then is largely some form of ammonia: NH_3 or NH_4^+. This can be stripped from solution by a stream of air if the solution is basic (the pH can be adjusted to be >11 by the addition of lime, which can simultaneously remove phosphate). The chief problem lies in what to do with the ammonia in the air stream; it can be adsorbed by an acidic reagent, but at further cost. If allowed to escape, it is a potential local air pollutant, and eventually enters the run-off water through rain. Alternatively, aerobic bacterial nitrification of ammonia to nitrate, followed by denitrification by other bacteria in the presence of a reducing carbon compound (e.g., methanol) can be employed:

$$NH_3 + 2O_2 \xrightarrow[\text{bacteria}]{\text{nitrifying}} NO_3^- + H^+ + H_2O \qquad (15\text{-}12)$$

$$6NO_3^- + 5CH_3OH + 6H^+ \xrightarrow[\text{bacteria}]{\text{denitrifying}} 3N_2 + 5CO_2 + 13H_2O \quad (15\text{-}13)$$

Purely chemical processes for nitrogen removal are not available; suitable insoluble salts do not exist for a precipitation process, but reduction to N_2 or N_2O by Fe(II) has been suggested for possible development.

(d) Dissolved organic material. Comparatively little dissolved organic material remains after secondary treatment, but what there is, is significant in terms of taste, odor, and toxicity. Chemical oxidation is a possibility, with a variety of oxidants being proposed, including chlorine, hydrogen peroxide, and ozone. Adsorption on activated carbon is effective and perhaps more practical.

(e) Dissolved inorganic ions. In addition to nitrate and phosphate ions, a significant quantity of other inorganic substances are present. Although many of these are not harmful, their accumulation eventually renders water unsuitable for reuse. In addition, toxic heavy metal ions may be present. A variety of methods can be used to remove these substances, although in general the methods are expensive and rarely used in present tertiary waste water treatments. When high quality water is scarce, their use may be practical as it is with brackish or sea water in arid areas. Of main interest are distillation, electrodialysis, reverse osmosis, and ion exchange. Distillation is an old technique, and quite simple if the waters do not contain volatile impurities such as ammonia. It is expensive, but has been used for many years to supply fresh water from sea water, e.g., in Curacao. Electrodialysis is a newer technique. Water is passed between membranes that are alternatively permeable to positive and negative ions. An electrical potential is applied across the system so that the ions migrate, the positive ions passing out through the cation-permeable membrane (which prevents anions from flowing in), while the negative ions move out in the opposite direction through the anion permeable, cation impermeable membrane. The result is a stream of deionized water between the membranes, and this stream can then be separated as clean water. This is critically dependent on the composition of the water; large ions, for example, do not pass through the membranes. It has been applied to tertiary treatment on a pilot plant scale. Reverse osmosis is another new process. If pure water is separated from an aqueous solution by a membrane that is permeable to water (e.g., cellulose), then the solvent will flow from the pure water side into the solution. If the apparatus is designed appropriately, an equilibrium pressure due to the tendency of water to flow into the solution can be developed across the membrane. This is called the osmotic pressure of the solution, and the process is called osmosis. If a pressure exceeding the value of the osmotic pressure is applied to the solution, water will instead flow from the solution to the clean water side. This is the principle of reverse osmosis. This has been shown to be practical, again on a pilot plant scale.

Ion exchange is frequently used for small scale water purification, and often is found in home water softening devices. Ion exchange materials usually are organic resins, but some inorganic materials can also function in this way. The bulk of the resin carries a charge (e.g., a group such as $-OSO_3^-$ or a quarternary amine $\geq N^+$) where an ion of opposite charge is held by coulombic forces to produce electrical neutrality. An ion of a given charge can be replaced by another, and if those present initially are H^+ on the cation exchanger and OH^- on the anion exchanger, ionic impurities can be removed completely if water is passed through one then the other. The exchangers are then regenerated by acid and base solutions, respectively. In water softening, it is often adequate to remove the Ca^{2+} ions, replacing them with Na^+ ions. Regeneration can then use an NaCl solution.

It should be kept in mind that, while there are many tertiary treatment approaches, they are expensive and not widely used. Even secondary treatment is often not applied in many sewage disposal systems. The specific details of waste treatment vary considerably with different facilities.

BIBLIOGRAPHY

Natural Waters

J. P. Riley and G. Skirrow, eds., "Chemical Oceanography," 2nd ed., Vols. 1-7. Academic Press, London, 1975. This is an extensive treatise on many aspects of the environmental chemistry of the oceans.

J. P. Riley and R. Chester, "Introduction to Marine Chemistry." Academic Press, New York, 1971. (Chapter 4 contains an extensive table on river and ocean water compositions and residence times.)

E. G. Goldberg, The oceans as a chemical system, in "The Sea" (M. N. Hill, ed.), Vol. 2, Chapter 1. Interscience, New York, 1963.

G. Sillén, The physical chemistry of sea-water, in "Oceanography" (M. Sears, ed.), p. 549. American Association of Science Publication No. 67, Washington, D.C., 1961.

E.G. Goldberg, Minor elements in sea water, in "Chemical Oceanography" (J. P. Riley and G. Skirrow, eds.), Vol. 1, Chapter 2. Academic Press, New York, 1965.

P.W. Schindler, Heterogeneous equilibria involving oxides, hydroxides, carbonates, and hydroxide carbonates, in "Equilibrium Concepts in Natural Water Systems" (R. F. Gould, ed.), Chapter 9. Advances in Chemistry Series No. 67, American Chemical Society, Washington, D.C., 1967.

W. Broecker, Radioisotopes and large scale ocean mixing, in "The Sea" (M. N. Hill, ed.), Vol. 2, Chapter 4. Interscience, New York, 1963.

R. M. Pytkowicz, The chemical stability of the oceans and the CO_2 system, in "The Changing Chemistry of the Oceans" (D. Dyrssen and D. Jagner, eds.), p. 147. Wiley (Interscience), New York, 1972.

W. W. Rubey, Geologic history of sea water, *Geol. Soc. Amer. 62*, 1111 (1951).

Water Treatment

Metcalf and Eddy, Inc., "Wastewater Engineering: Collection, Treatment, Disposal." McGraw-Hill Book, New York, 1972.

W. W. Eckenfelder and L. K. Cecil, eds., "Applications of New Concepts of Physical Chemical Waste Water Treatment." Pergamon, New York, 1972.

F. L. Evans III, ed., "Ozone in Water and Waste-Water Treatment." Ann Arbor Science Publishers, Ann Arbor, Michigan, 1972.

P. A. Vesilind, ed., "Treatment and Disposal of Waste Water Sludges." Ann Arbor Science Publishers, Ann Arbor, Michigan, 1974.

A. A. Delyannis and E. A. Delyannis, "Water Desalting." Springer-Verlag, New York, 1974.

T. V. Arden, Ion exchange, *Chem. Britain 12,* 285 (1976).

D. McBain, Reverse osmosis, *Chem. Britain 12,* 281 (1976).

K. Dorfner, "Ion Exchangers: Properties and Applications." Ann Arbor Science Publishers, Ann Arbor, Michigan, 1972.

16
THE EARTH'S CRUST

16.1 ROCKS AND MINERALS

Current theories consider the earth to have formed from condensation of dust and gas in space. Gravitational and radiochemical heating melted the aggregated solids, and as the planet cooled, partial solidification and separation of materials took place. The present structure of the earth consists of a molten core composed chiefly of iron and nickel, surrounded by lighter rocks. The outer few miles, the crust, is the only portion of the earth accessible to man, and is the source of most of the substances necessary for a technological society. The crust has undergone a complex evolution as processes such as weathering and erosion have broken up the original rocks, and sedimentation and crystallization have produced new ones. High pressures and temperature and volcanic activity have produced other changes. Indeed, these processes are on-going. These reactions fall into the realm of geology and geochemistry, and are too extensive to be dealt with in detail here, but we shall discuss some of their aspects that most strongly impinge on mankind; in particular, those relating to some of the substances that we use extensively.

Table 16-1 gives the relative abundances of the most common elements in the earth's crust. As can be seen, by far the largest part of this crust is made up of silicon and oxygen, which form the basis of the abundant silicate minerals. The fundamental structure of all silicates consists of four oxygen atoms tetrahedrally arranged around a central silicon atom. In terms of oxidation numbers, these are considered to be Si(+4) and O(-2), but the Si - O bonding is covalent. However, an isolated SiO_4 tetrahedron would carry a charge of -4. Minerals based on the simple SiO_4^{-4} ion in association with cations (salts of silicic acid, Section 15.2) are unusual, and more typically the natural silicates are based on rings, sheets, and chains of linked tetrahedra in which the Si atoms share oxygens. The O:Si ratio is then less

TABLE 16-1

Relative Abundance of Some Elements in the Earth's Crust

Element	Abundance (wt. %)
Oxygen	45
Silicon	27
Aluminum	8
Iron	6
Calcium	5
Magnesium	2.8
Sodium	2.3
Potassium	1.7
Titanium	0.9
Hydrogen	0.14
Magnesium	0.10
Phosphorus	0.10
Cu, Cr, Ni, Pb, Zn	10^{-2}-10^{-3} each
Mn, Sn, U, W	$\sim 10^{-4}$ each
Ag, Hg	10^{-5}-10^{-6} each
Au, Pt	$\sim 10^{-7}$ each

than 4. Some typical structures are given in Fig. 16-1. In aluminosilicates, aluminum, the third most abundant element in the earth's crust, is also present tetrahedrally or octahedrally surrounded by oxygen atoms.

In all of these structures, negative charge on the silicate framework requires the presence of cations for electrical neutrality. Within limits, the nature of the cationic species is of secondary importance. Particular structures are favored by particular cations in order to maximize crystal packing energies, coulombic attractions, and other interactions leading to stabilization of the solid, but small amounts of many cations can be present in a given structure without influencing it in any major way. In particular, ions of similar size and charge may replace one another with little change in mineral structure (isomorphous substitution). If the similarity is great enough, complete replacement of one component by

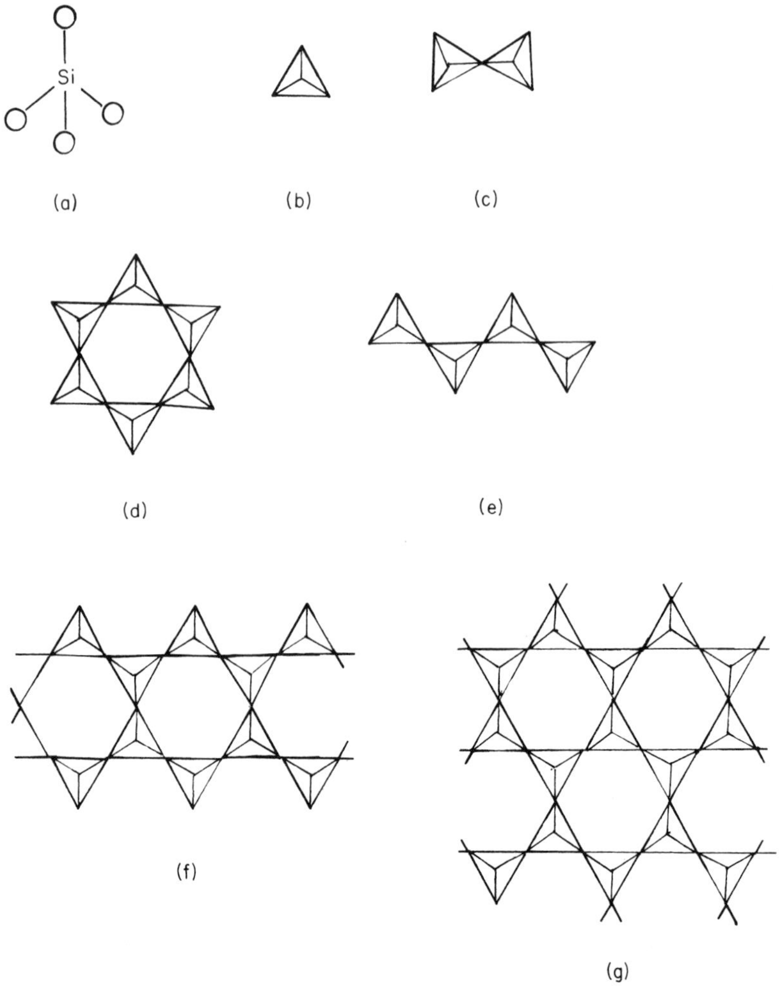

FIGURE 16-1. Examples of silicate structural units: (a) the SiO_4 tetrahedron; (b) a schematic view of a SiO_4 tetrahedron, each apex represents an oxygen; (c) linked tetrahedra, the $Si_2O_7^{6-}$ ion; (d) a cyclic silicate structure; (e) a chain structure; (f) a double chain; and (g) a two-dimensional layer structure.

another can take place without significant structure change. Lesser similarity leads to less replacement that can take place in a given mineral. (Such behavior holds for many solid materials besides silicate minerals.)

From Table 16-1, it is seen that many of the elements essential for modern technology are present in very small amounts in the earth compared to the dozen most abundant elements. The amounts of these less abundant elements in the earth's crust are vast, however, and they are widely distributed. Most common rocks contain a few parts per million of many elements distributed randomly by isomorphous substitution of abundant elements (e.g., Pb replaces some K, and Zn replaces some Mg in most common rocks). For this reason, it has been suggested that mankind will never really run out of mineral resources, and as high concentration ores are exhausted, new technology can be developed to tap the sources in the common rocks. However, the volume of material to be mined and processed, the energy requirements, and the waste disposal problems clearly set economic and environmental restrictions on the minimum concentrations that can be employed for large scale use of any substance. In the past, man has made use of atypical, high concentration sources for most of his mineral needs. Little if any technological development would have been possible if all elements were uniformly distributed in the earth's crust. In the present and future, lower concentration sources are and will be used. Nevertheless, the bulk of the amount of an element actually present in the earth's crust is unlikely to be available for practical exploitation.

For the abundant elements, for example iron and aluminum, availability will not be a problem in the foreseeable future because their overall abundance requires that they be important constituents of many common minerals. As rich deposits are used up, somewhat less rich deposits can be used. To obtain the same amount of metal, more rock must be mined, and of course the cost will increase (and perhaps new methods of processing will be needed) but no sudden shortage need appear. The energy required to produce a fixed amount of metal will rise in a relatively steady manner. For the less abundant metals such as Cu, Zn, and Pb, the common rock sources are much lower in concentration than the ores now used. They represent cases of isomorphous substitution rather than a mixture of an actual compound of the element with common rock as are the present sources. While there may be extensive low grade ore deposits, it does not follow that as the concentration of the element in the ore decreases, the total amount of ore available increases, as seems to be the case with the more abundant elements. Thus, when the ore bodies of the general sort now mined are worked out, a sudden large increase in energy cost per unit of metal recovered can be expected. Compounding this is the fact that many of the important less abundant metals, e.g., Cu, Pb, Sn, Hg, and Au, are being mined relative to their abundance in the crust at a rate that is much larger than that of the abundant metals.

16.2 RECOVERY OF METALS FROM THEIR ORES

Minerals used as sources from which desired elements can be isolated are referred to as ores. The ores of many elements are made up from particular compounds of the element in question, although this may be mixed with much inert rock. High grade ores typically consist of veins of the compound; in low grade ores the particles may be widely disseminated in the rock. Formation of ore bodies in nature may take place in a number of ways, for example, selective crystallization as the molten material (magma) from which the rocks formed cooled, and by hydrothermal processes, which have been important for many widely used metals. In a hydrothermal process, water or brine at high temperature and pressure dissolves metal ions from the bulk rock and redeposits them in concentrated form elsewhere. The extraction process is analogous to a weathering reaction. Exploitable deposits typically form where the liquid flows through a restricted channel where a precipitation reaction can take place. Formation of very insoluble sulfides and arsenides has provided many ore bodies. Oxides, sulfides, arsenides, and carbonates are among the most common ore minerals.

The details of recovery of a metal from its ore depends on the metal and often on the exact composition of the ore used. However, some of the steps employed are common enough to be discussed in general terms. Some of the specific ores and treatment used for individual metals are listed in Table 16-2.

(a) Concentration (beneficiation). Unless the ore is very rich, it is typically concentrated; that is, much of the worthless rock is separated physically from the desired phase. In all such processes the ore is first finely ground. In the common flotation method, the ore particles are suspended in water in the presence of flotation agents. The molecules of the flotation agent may attach themselves preferentially to the desired particles rather than to the worthless rock, and render these particles hydrophobic, so they are not wet by water. They are then carried to the surface in foam when

TABLE 16-2

Summary of Smelting Processes Used for Some Metals

Metal	Chief ores	Common treatment processes	Comments
Aluminum	Bauxite, a combination of aluminosilicate clays	Purification with aqueous NaOH to yield Al_2O_3, followed by electrolysis in molten cryolite, Na_3AlF_6.	Waste sludge of silicates and Fe_2O_3. Requires abundant electrical energy.
Cadmium	Zinc ores	Volatilized from crude Zn and collected in the flue dust, or precipitated from Zn solutions. Separated from Zn by distillation or electrolysis.	A by-product of Zn production
Chromium	Chromite, $Fe(CrO_2)_2$	Reduction of ore with C produces ferrochrome alloy. This is dissolved in H_2SO_4 and Cr obtained by electrolysis. Alternatively, Cr_2O_3 can be formed and reduced with Al, C, or Si.	
Cobalt	Smaltite $CoAs_2$, and cobaltite CoAsS, and other arsenides, sulfides, and oxides	Ore roasted to oxide; sometimes reduced to a metallic mixture, dissolved in H_2SO_4, other metals separated if necessary, and Co precipitated as $Co(OH)_3$ by NaOCl. Heating produces Co_3O_4 which can be reduced by C or H_2. Alternatively, the purified H_2SO_4 solution can be electrolyzed.	Usually a by-product of the production of other metals, e.g., Cu, Pb, Ni.

Copper	Copper pyrite $CuFeS$; cuprite CuO; malachite $Cu_2(OH)_2CO_3$; other arsenide, sulfide, oxide, and carbonate ores	Ore concentrate heated to remove some S, As, Sb, and fused with silica; copper and iron sulfides and oxides separate (copper matte) from an iron silicate slag. $2CuO + 2FeS \rightarrow Cu_2S + FeS + SO_2$ $Cu_2O + FeS \rightarrow Cu_2S + FeO$ The matte is oxidized in the presence of silica; $2FeS + 3O_2 \rightarrow 2FeO + 2SO_2$ $2FeO + SiO_2 \rightarrow$ iron silicate slag $2Cu_2S + 3O_2 \rightarrow 2Cu_2O + SO_2$ $2Cu_2O + Cu_2S \rightarrow 6Cu + SO_2$ Crude Cu usually refined electrolytically; crude Cu anodes dissolve in an H_2SO_4- $CuSO_4$ electrolyte and deposit as pure Cu cathodes.	Much SO_2 released in the melting process, along with As and Sb oxides. Both gas and dust discharges are involved. There are aqueous wastes from electrolysis. Anodic sludge is a source of Ag and Au. It also contains Se, Te, and base metals that must be recovered or disposed of in a controlled manner.
Gold	Free metal	Finely divided ore treated with aqeuous cyanide solution and air; $4Au + 16NaCN + 6H_2O + 3O_2 \rightarrow 4NaAu(CN)_4 + 12NaOH$ $2Au(CN)_4^- + 3Zn + 4CN^- \quad 2Au + 3Zn(CN)_4^{-2}$	
Iron	Magnetite Fe_3O_4; hematite Fe_2O_3; limonite $2Fe_2O_3 \cdot 3H_2O$; siderite $FeCO_3$	The oxide is reduced in a blast furnace with coke and limestone (see Sec. 16.2). Carbonate ores are first roasted; $2FeCO_3 + \frac{1}{2} O_2 \rightarrow Fe_2O_3 + 2CO_2$ Purification and steel-making involve oxidation of C, Si, S, and P with O_2 and the removal as CO_2 and as slag with limestone as flux.	Large volumes of gas (CO_2) produced, and Fe_2O_3 and Fe_3O_4 as dusts.

TABLE 16-2 (continued)

Metal	Chief ores	Common treatment processes	Comments
Lead	Galena Pbs; cerussite $PbCO_3$; anglesite $PbSO_4$; and others	Sulfide ores are oxidized $2PbS + 3O_2 \rightarrow 2PbO + 2SO_2$ $PbS + 2PbO \rightarrow 3Pb + SO_2$	Much SO_2 is released. As, and Sb are often present.
Manganese	Oxides (e.g. pyrolusite MnO_2), and some carbonates and silicates	Ferromanganese (∿80% Mn) is produced in a process similar to that used for iron, but more coke is required. Pure metal is produced electrolytically from MnO_2 dissolved in a sulfate solution	Much high temperature gas and particulate material given off.
Mercury	Cinnabar HgS	Roasted with air and the vapors condensed; $HgS + O_2 \rightarrow Hg + SO_2$ Alternately, roasted with Fe or CaO; metal sulfide and/or sulfates are formed instead of SO_2.	
Molybdenum	Molybdenite MoS_2	Concentrate roasted to MoO_3, reduced with H_2.	SO_2 produced

Nickel	Pentlandite $(Ni,Fe)S$; oxides, silicate, and arsenides	Usually occurs with Cu and Fe. Fractionated by flotation, melted to a matte in a similar process as with Cu and converted to NiO. The oxide may be reduced with C or with water gas, and purified by conversion to $Ni(CO)_4$, volatilized, and decomposed back to the element, or it may be purified electrolytically.	As Cu
Silver	Argentite AgS; cerargyrite $AgCl$ (and other halides); pyrargyrite $AgSbS$; and other sulfides and antimonide ores	By-products of Cu, Pb, and Zn ores, obtained during refining of the base metal. With Cu, the Ag is obtained in the anode "slime," and is recovered by roasting, leaching with H_2SO_4 and smelting to remove Se, Te, and other impurities.	
Tin	Cassiterite SnO_2	Roasted to remove S and As as SO_2 and As_2O_3; metallic oxides removed with an acid leach, and the tin oxide reduced by C.	
Titanium	Ilmenite $FeTiO_3$; rutile TiO_2	The oxide is converted to the chloride, which is reduced by magnesium; $2TiO_2 + 3C + 4Cl_2 \rightarrow 2TiCl_4 + 2CO + CO_2$ $TiCl_4 + 2Mg \rightarrow Ti + 2MgCl_2$	

TABLE 16-2 (continued)

Metal	Chief ores	Common treatment processes	Comments
Tungsten	Scheelite $CaWO_4$; wolframite $(Fe,Mn)WO_4$	The ore is roasted with Na_2CO_3 and $NaNO_3$, leached, and $CaWO_4$ precipitated. HCl converts this to tungstic acid, which is calcined to WO_3. This can be reduced by H_2.	
Zinc	Sphalerite ZnS; carbonate ores	The ore is roasted to ZnO, and this reduced, with the Zn metal vaporizing and being collected; $ZnO + CO \rightarrow Zn + CO_2$ Purification may be by distillation.	SO_2 is evolved in the roasting.

air bubbles are streamed through the water, and can be skimmed off. Typical wetting agents used for CuS ores are alkali xanthate salts,

$$R - O - C\begin{matrix} \nearrow S \\ \searrow S^- \end{matrix} \qquad (R = \text{alkyl group})$$

which are surfactants analogous to detergents, with an affinity for the metal sulfide particle. Other concentration methods involve, for example, density differences, and magnetic separations.

Concentrations methods in general have as their greatest environmental problem the disposal of the waste rock. Dust and water contamination also can be produced.

(b) Roasting. Sulfide and arsenide ores, and some others, are heated with air at high temperature in order to convert them to oxides. This process may also involve the use of limestone or some other flux that will permit removal of silicate materials that react with the flux to form a molten slag. This slag can then be separated from the desired material which is insoluble in it. With sulfide ores, SO_2 is a major gaseous product and must be trapped in order to avoid air pollution problems. Arsenic oxide As_2O_3 is another common product that volatilizes from the furnace but can be condensed to a solid at room temperature. Particulate materials are also given off to the atmosphere unless trapped or precipitated. Possible water pollution problems come from gas scrubbing and slag cooling, and from aqueous treatment of the oxides in some processes.

(c) Reduction. The oxides produced above, or oxide ores, must be reduced to the elemental metal. Carbon, in the form of metallurgical coke made from low sulfur coal, is a common reducing agent. This reduction is carried out at high temperature, such as in blast furnaces, and offers the same waste control problems as (b). Large volumes of very hot gases are involved. Other reduction procedures are in use for some elements; electrochemical reduction of alumina is an obvious example. Some examples of these processes are described below.

(d) Refining or purification. Often the metal obtained initially is too impure for industrial uses. Impurities may be other metals or nonmetals such as S or P that can have very deleterious effects on the physical properties of the metal. Refining techniques may involve remelting of the metal and selective oxidation of the undesirable impurities, or electrolytic purification in which the impure metal is dissolved chemically or electrolytically and plated out in pure form on the cathode of an electrochemical cell.

Many ores are used as sources for several elements, so that actual flow schemes for metal recovery may be elaborate. In addition, unwanted elements are usually present, and if these include toxic materials (Chapter 14), considerable care must be taken in waste disposal practices. Many metal smelters and refineries in the past were noted for their air and water pollution.

Because iron is refined in greater amount than any other metal, some of the processes used will be discussed. The oxide (carbonate ores are first heated to decompose them to the oxide) is reduced directly in a blast furnace with coke, limestone being used as a flux. The solids pass down the furnace against a current of heated air that reacts with part of the coke to heat the charge. The liquid iron produced runs to the bottom of the furnace, while the liquid calcium silicate phase (slag) floats above it. The reactions that take place can be summarized as follows:

$$3Fe_2O_3 + CO \rightarrow 2Fe_3O_4 + CO_2 \qquad (16\text{-}1)$$

$$Fe_3O_4 + 4CO \rightarrow 3Fe + 4CO \qquad (16\text{-}2)$$

$$Fe_2O_3 + CO \rightarrow 2FeO + CO_2 \qquad (16\text{-}3)$$

$$FeO + C \rightarrow Fe + CO \quad \text{[in the lower (hot) part of the furnace]} \qquad (16\text{-}4)$$

The limestone reacts

$$CaCO_3 \rightarrow CaO + CO_2 \qquad (16\text{-}5)$$

$$CaO + SiO_2 \rightarrow CaSiO_3 \text{ (slag)} \qquad (16\text{-}6)$$

Reaction (16-6) is oversimplified; the slag is not pure calcium silicate. The limestone also removes any sulfides that may be present;

$$FeS + CaO + C \rightarrow CaS + Fe + CO \qquad (16\text{-}7)$$

but this reaction is easily reversed by excess sulfide; hence, the requirement that the coke be made from low-sulfur coal.

Some silica is reduced to Si, and phosphate to P. The product of the blast furnace, pig iron, ordinarily contains about 2% P, 2.5% Si; more than 0.1% S, 4% C, and 2.5% Mn. The latter element is present in the iron ore, and goes through reactions similar to those of Fe. Purification to give a steel of improved physical properties is necessary for most purposes. In particular, P, S, Si, and C must be removed or reduced in amount. In the now obsolete open hearth method, the pig iron with perhaps 40% scrap iron, limestone flux, and some Fe_2O_3 and Fe_3O_4 is melted in a furnace heated by gas and air. The P, C, Si, and S are oxidized, and the silicate and phosphate are removed as a slag by reaction with basic CaO.

Modern procedures use the basic oxygen process, in which the oxidation is carried out by a stream of oxygen gas. The process requires much less time (the order of half an hour compared to 12 hr) and less energy than the open hearth process. However, it can use rather less scrap than the latter and has had some depressive effects on recycling of scrap iron and steel.

Special steels, especially those requiring oxidizable components, are made in an electric furnace. This is expensive in terms of energy, but can use a high proportion of scrap, provided the overall composition is suitable for the steel being produced.

Particulate matter in the form of iron oxides are the main emissions from the above processes. Although a great deal of carbon monoxide is involved [Eq. (16-1)-(16-4)], most of it is recycled. Much water is used for cooling and scrubbing, but one of the most significant processes related to water pollution involves removal of oxide scale from the finished steel in a process referred to as pickling. This is achieved in a H_2SO_4 or HCl bath;

$$FeO + 2H^+ \rightarrow Fe^{2+} + H_2O \tag{16-8}$$

The waste liquors are highly acidic and contain ferrous salts; discharge to a water system is reminiscent of acid mine drainage (Section 14.6.3). Injection into deep wells is commonly used for disposal, although this could be dangerous with respect to ground water contamination. A variety of regeneration processes are available if their use is mandated.

Aluminum is also smelted extensively, and the isolation of this metal illustrates the chemistry of an active metal for which chemical reduction is not economically favorable. The aluminum ore used virtually exclusviely is bauxite, a mixture of aluminosilicate clays (Section 16.3). Alumina can be recovered from bauxite by treatment with NaOH solution. Solid impurities, including silica and Fe_2O_3 settle out, and $Al_2O_3 \cdot 3H_2O$ can then be crystallized. Heating produces Al_2O_3. Reduction is performed on the Al_2O_3 dissolved in molten cryolite, Na_3AlF_6, in a cell with carbon electrodes; aluminum metal is produced at the cathode, while the anode is consumed by the oxygen that otherwise would be liberated at an inert anode in an electrochemical process. The net reaction is

$$2Al_2O_3 + 3C \rightarrow 4\ Al + 3CO_2 \qquad (16\text{-}9)$$

(this requires less energy than electrolysis with inert electrodes to produce Al and O_2). It is still an extremely energy-consuming process because of the strong Al-O interaction energy that must be overcome. Alterations to the basic method have been developed to reduce energy requirements.

The electrolysis process produces some fluoride and other particulate emissions that must be trapped to avoid air pollution problems, but the most abundant waste product of aluminum mining is the insoluble materials from the NaOH treatment, the so-called "red mud tailings," and of course, any waste from the generation of the electricity employed.

For all metals it would seem economically preferable to recycle used metal, since this would eliminate the concentration and reduction steps necessary for virgin materials. Unfortunately, collection and sorting are also energy intensive processes. Moreover, under cheap energy conditions, many processes have not been developed with recycling as a major objective. In any reprocessing, the presence of components, metallic or otherwise, that the refining steps are not designed to remove and that can produce deleterious properties in the final product must be prevented. Thus, sorting and separation may have to be extensive. However, as rich ore sources are used up, recycling must be increased for conservation purposes as well as for economic ones.

16.3 SOILS

Surface rocks are subject to break up and chemical change by a variety of processes called weathering. Some of these processed are the following.

(a) Physical disintegration of rocks caused by temperature effects, frost, wind, water, and ice erosion. Some of these factors also are responsible for transporting the weathering products away from their original source.

(b) Chemical reactions with water; for example, hydration or hydrolysis. Hydration is illustrated by reaction (16-10); the change in crystal structure as water molecules are gained or lost will result in disintegration of the material. Hydrolysis involves a more extensive change in the material, for

$$\underset{\text{hematite}}{2Fe_2O_3} + 3H_2O \rightleftarrows \underset{\text{limonite}}{2Fe_2O_3 \cdot 3H_2O} \qquad (16\text{-}10)$$

example, reaction (16-11). Reactions of this sort can liberate cations from the rocks, e.g., K^+ in (16-11).

$$\underset{\text{microcline}}{KAlSi_3O_8} + H_2O \rightarrow HAlSi_3O_8 + K^+ + OH^- \tag{16-11}$$

(c) Attack by acid, especially carbonic acid. Reactions (16-12) and (16-13) are examples of the weathering interactions of acid with minerals. Reaction (16-11) could also have been written as an example of acid attack.

$$\underset{\text{limestone, insoluble}}{CaCO_3} + H_2CO_3 \rightarrow \underset{\text{soluble}}{Ca(HCO_3)_2} \tag{16-12}$$

$$CaAlSi_2O_8 + 2H^+ \rightarrow H_2AlSi_2O_8 + Ca^{2+} \tag{16-13}$$

(d) Oxidation reactions involving substances in low oxidation states. Reaction (16-14) illustrates a hydrolysis reaction accompanied by oxidation of Fe(II):

$$\underset{\text{olivine}}{12MgFeSiO_4} + 8H_2O + 3O_2 \rightarrow \underset{\text{serpentine}}{4H_4Mg_3Si_2O_9} + 4SiO_2 + 6Fe_2O_3 \tag{16-14}$$

(e) Biological effects. These may take the form of physical disruption by plant roots, chemical effects from substances excreted from roots, and bacterial action. Oxidation processes in particular may involve the action of bacteria (e.g., oxidation of FeS_2; see Section 14.6.3).

The eventual result of the breakup and composition changes of weathering reactions is the formation of soil; this, of course, is the fraction of the earth's surface utilized for plant growth. Soil may be regarded as the upper layer of the regolith, the weathered and broken up material overlying bedrock. Soil is the most heavily weathered portion and normally is significantly influenced by plant materials. Soils play a significant role in environmental chemistry in addition to providing a site for plant growth. Much of this activity depends on the microorganisms that are important soil components. The absorption of CO has already been referred to, as have the carbonate equilibria, that in soils are strongly dependent on CO_2 generated from bacterial degradation of organic materials. Nonbiologic processes include rapid SO_2 absorption; soils have been shown to be effective in removing SO_2 from the air. Buffering of pH, and of metal ion concentrations, are also important aspects of soil chemistry.

A typical soil contains stones and gravel, coarse and medium sized material such as sand and silt made up of silica and other primary minerals, and finer weathered material classed as clays. Most soils contain a few per cent organic components such as humic substances (Section 13.6.4) and undecomposed plant materials, although a few organic soils (peats, mucks) may contain up to 95% organic matter. Typical soils exhibit several layers (horizons) of somewhat different composition that are important in soil classification.

The size and nature of the fine particles in the soil are most important with respect to its chemical and physical properties, although the organic constituents can have a major influence as well. Some clays swell when wet, as water molecules enter the crystal lattice. As a consequence of shrinkage on drying, such soils tend to crack. At the same time, the soil may cake and become impervious to moisture. Cohesion of the clay particles, which is also moisture dependent, determines the stickiness of the soil and the agricultural workability of it. The clays also are largely responsible for the very important ion exchange properties of soils, although humic materials contribute significantly.

Typical clays are aluminosilicates present largely as colloidal particles; that is, the size is less than a micrometer. The particles are made up of layers or sheets somewhat analogous to the structure of graphite. The surface area of such small particles is very large and, in addition, internal area between the layers may be available. Thus, surface properties and adsorption are quite important. A number of varieties of clays are recognized, but we shall discuss only two of them here. The two most important classes in soils are represented by kaolinite and montmorillonite, the formulas for which are (idealized and simplified) $Al_2Si_2O_5(OH)_4$ and $NaMgAl_5Si_{12}O_{30}(OH)_6 \cdot nH_2O$, respectively. Both contain sheets of Si atoms in tetrahedral sites surrounded by four oxygens. Kaolinite (Fig. 16-2) contains a layer of an equal number of Al atoms surrounded by octahedra of O and OH groups. The two layers are joined by common oxygen atoms, and these pairs of layers are fairly strongly bound to others by hydrogen bonding. Montmorillonite, on the other hand, has sheets of Al octahedrally surrounded by O and OH, with these sheets associated with two tetrahedral Si sheets (Fig. 16-3). Some of the Al sites are occupied by other ions such as Mg^{2+}. These three layers are only weakly bound to others by electrostatic attraction of cations such as Na^+, which are probably hydrated. The interlayer regions can contain a variable number of water molecules, and this is responsible for the marked swelling of material of this type in the presence of water.

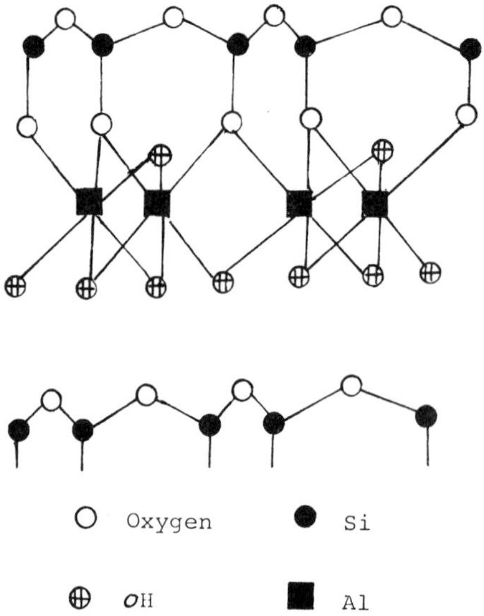

FIGURE 16-2. The structure of kaolinite. This is based on a two-dimensional layer structure of SiO_4 units (Fig.16-1g) perpendicular to the page; only three oxygens are shown around each Si for clarity.

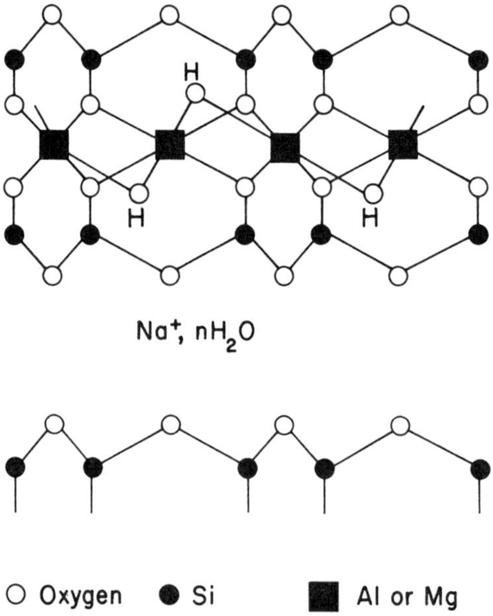

FIGURE 16-3. The structure of montmorillonite. The silicate sheet structure similar to Fig. 16-2g is perpendicular to the page; for clarity only three oxygens are shown around each Si.

Clays are capable of entering into ion exchange reactions in two ways. First, H^+ ions can be lost from the OH groups, and replaced by other cations. This is pH-dependent. Secondly, interlayer cations may be exchanged.

Hydrated metal oxides of Fe and Al also are present in colloidal form in some soils, and dominate those of the tropics. These nonaluminosilicate "clays" have ion exchange properties similar to, but weaker than, those of the aluminosilicates. They tend not to be as sticky and adherent as the silicates, and soils with high proportions of hydrous oxide clays can be cultivated under moisture conditions that make siliceous clays unworkable.

One of the most important natural processes occurring in soils that determines their properties and agricultural characteristics is that of acid leaching, or podzolization. This is a process occurring most markedly in soils of moist, forested, temperate and cold regions, such as the northeastern United States. The top layer of such soils consists of organic materials containing relatively few metal ions. As these materials decompose, acids are formed that are washed into the levels below. As a consequence, the soil beneath the organic layer is leached of most minerals except silica. Aluminum and iron oxides and organic matter are precipitated in the lower layers as soluble material and noncoagulated colloids are washed down. Such soils contain few nutrients except in the upper organic layers, the nutrients having been leached out of the levels beneath the surface. These soils are suitable for deep rooted plants such as trees, but used agriculturally, they soon lose fertility. Besides true podzols,[1] many soils exhibit partial leaching (partial podzolization) when they have less organic matter on the surface and therefore, there is less acid leaching.

In wet, tropical forests, oxidation of the organic material is more rapid, and leaching is less acidic but very extensive. Silica and montmorillonite types of clay are removed, but iron and aluminum oxides remain, giving the soil a red or

[1] *Soils are classified into a large number of types and subtypes. Classification systems have been based on how soils were formed (or thought to be formed) or more recently on soil properties: texture, color, composition, moisture, and others. The presence or absence of certain horizons in the soil profile are important aspects of the classification scheme. Spodosol and oxysol are two examples of soil orders under the newer classification that roughly coincide with some of the podzols and latosols of the older scheme.*

yellow color. Often deposits of these oxides form near the water table. On drying, these may form very hard materials that cannot be cultivated. Such levels are referred to as laterites; the soils are called latosols or oxysols. They are common in much of South America and Africa.

Both podzols and latosols must receive large amounts of fertilization if they are to be useful for large scale agriculture. With the latosols in particular, modern agricultural methods must be applied with great care to ensure that erosion does not expose the hydrous oxide layer with consequent irreversible loss of agricultural use of the soil when this dries. Indeed, primitive agricultural practices on many tropical forest lands, while not highly productive overall, display an excellent adaptation to the chemical nature of the soils. These methods have been the "slash and burn" type. Small fields are cleared of vegetation by burning. This converts the nutrient material in the growing vegetation to inorganic salts that provide a high level of fertilization for crops for a few growing seasons. When these nutrients are used up, the field is allowed to return to forestation for a few years, until the process can be repeated.

Soils with less abundant vegetation and less rainfall have correspondingly less acidic leaching, and in fact may be neutral or basic in nature. Calcium carbonate occurs at modest depths, or even near the surface (calcerous soils), and the nutrient content is often very high. These soils are often highly productive if irrigated.

Slightly acid to neutral soils are favored by many cultivated crops, and although some plants prefer strongly acid soils, it is often necessary to increase the pH of agricultural, garden, and lawn soils by the addition of lime. Agricultural limes are impure forms of calcium oxide, calcium hydroxide, or calcium carbonate, usually with considerable Mg present. Limestone or dolomite are common sources. On the other hand, the acidity of the soil can be increased by the addition of sulfur, which is oxidized to H_2SO_4 by soil organisms. Many fertilizers increase soil acidity also, for example by reactions such as

$$NH_4^+ + 2O_2 \rightarrow 2H^+ + NO_3^- + H_2O \qquad (16\text{-}15)$$

The need for fertilization of soils has been referred to, and several of the major fertilizers that are needed were discussed in Chapter 14. Phosphorus, nitrogen, and potassium are considered macronutrients while many other elements are necessary micronutrients. Of the macronutrients, phosphate may be present in the soil in the form of insoluble phosphate salts that form as a result of the reaction between the hydrogen phosphate ions with metal ions (chiefly, Fe, Al, Ca, either free or held in the clays) or with hydrous oxides. Reversible ion-exchange reactions also may retain phosphate, for example, as in reaction (16-16). Phosphate held in this way is more readily available than that of the insoluble phosphate salts.

$$Al(OH)_3 + H_2PO_4^- \rightleftarrows Al(OH)_2H_2PO_4 + OH^- \qquad (16\text{-}16)$$

Nitrogen may be present in the form of nitrate, which is held only weakly by anion exchange, or as the ammonium ion, which can be held by cation exchange with the clays. The potassium ion is nearly the same size as the NH_4^+ ion, and is held in the same way. Micronutrients, especially metal ions, may be available in clays or other insoluble mineral forms. In some instances one or more such nutrients, although present, may be unavailable to plants, and some metal ion deficiencies have been relieved by treating the soil with complexing agents of the EDTA and NTA types (Chapter 13). Some input of micronutrients such as chloride and sulfate comes from rainwater.

16.4 NONMETALLIC MINERAL PRODUCTS

Many minerals are sources of nonmetallic materials of widespread use. Some of these are employed with little treatment, while others undergo considerable chemical modification before use. The environmental problems associated with these substances are chiefly those associated with mining or quarrying and crushing of large volumes of rock or clay; that is, with destruction of landscape and emissions of particulate materials into water and the air. The common minerals that are involved are not considered toxic, but the physical form in which the particulates can be emitted can render them harmful to health. In what follows, we shall consider briefly a few aspects of some of the more important examples of substances of this class.

(a) Stone. From the beginning of civilization, stone has been used as a material of construction. Marble (a calcium carbonate rock) and granite (a silicate rock) are two examples of widely used building stones. Like other rock, some buildings are subject to weathering processes and particularly to attack by acid constituents of urban airs. Sulfur oxides are the primary acid components involved. Attack is particularly severe on carbonate stones. It is well known that many stone buildings and carvings that have existed for centuries have undergone rapid and extensive deterioration in recent times as industrial and urban air pollution has increased.

While much of the weathering attack of building stone is due to acid attack from H_2SO_4 arising from SO_2 emissions and to carbonic acid, water itself can be destructive by entering the pores in the surface layers of the stone and causing the surface layers to flake off through stresses imposed in freezing and thawing. The most serious cause of surface disintegration of this sort, however, is the strain imposed just beneath the surface by recrystallization of salts as water enters and evaporates from the pores. Any source of a salt having a significant water solubility can contribute to this, e.g., sea spray, but in general calcium and magnesium sulfates formed from the acid attack of industrial air [reaction (16-17)] are probably most important.

$$CaCO_3 + H_2SO_4 \rightarrow CaSO_4 + H_2O + CO_2 \qquad (16\text{-}17)$$

That is, air pollution does not just dissolve the stone from the surface, but can also cause layers a few millimeters thick to disintegrate. In many places, attempts are underway to protect stone monuments and historical buildings from deterioration from these causes.

Protective measures were undertaken as long ago as the first century B.C. utilizing wax rubbed on the warmed stone. Various waxes, resins, silicones, and other materials have been applied more recently. Such treatments rarely achieve deep enough penetration to be effective; water can get behind a thin film and eventually cause the treated layer to flake off. Some use is being made of polymers using processes in which solutions of monomers of good penetrating power are applied to the stone, followed by *in situ* polymerization. Acrylics, epoxies, and alkoxysilanes are among the materials used. Another treatment for carbonate stones involves the deposition of $BaSO_4$ in the stone. This material is much less soluble than $CaSO_4$ or $MgSO_4$, and also forms solid solutions with $CaSO_4$. Consequently, the strains introduced through the recrystallization process are much reduced.

(b) Clay-based materials, brick, etc. Bricks for building purposes have been prepared simply by drying clays or muds. The hydrous oxide layers of lateritic soils are one example that provides a good building material by simple drying, although generally aluminosilicate clays will produce a stronger and more permanent substance by firing. Brick, tile, and pottery are all made in this way. A nonswelling clay is required that can be formed into the desired shape as a paste with water, and does not deform significantly on drying. The firing process causes changes in the chemical nature of the aluminosilicate, and permits the individual particles of the clay to fuse together to produce a hard, strong structure. This is essentially irreversible; water will not reconvert the fired material to soft clay, as it will after simple drying. The fired clay is porous, but a nonporous surface can be achieved by a glaze of material that is melted to a glass.

(c) Glass. Glasses are rigid, noncrystalline materials which are often referred to as supercooled liquids, but this is a serious oversimplification. Common glass is a silicate, but does not correspond to a particular compound. Silica, SiO_2, is a 3-dimensional network based on Si-O-Si linkages. If the atoms are arranged in a regularly repeating pattern, a crystalline form of quartz is produced, but they may be arranged irregularly, forming a glass. Addition of an oxide of a metal with a small tendency to form a covalent bond to oxygen (e.g., Na_2O) to SiO_2 results in breakage of the Si-O-Si links:

$$Na_2O + \text{-Si-O-Si-} \rightarrow \text{-Si-O}^- + {}^-\text{O-Si-} + 2Na^+ \qquad (16\text{-}18)$$

As a result, the softening temperature is lowered and the properties of the polymer are changed. Control over these properties is possible by appropriate selection of the composition of the glass. Addition of B_2O_3 will result in Si-O-B linkages, since boron also has a strong affinity for oxygen. This is the basis of the borosilicate glasses of which Pyrex[2] is an example; they have higher softening points and a smaller coefficient of thermal expansion than the soda glasses. Lead oxide produces a high index of refraction and is used in crystal glasses.

Common glass is made from SiO_2, Na_2O, and some CaO, which are fused at high temperatures. The source of the Na_2O and CaO may be the carbonates or other salts that decompose to oxide on heating; SiO_2 is obtained as sand. There are restrictions on the purity of the starting materials, since many metal ions and collodial metals, sulfides, and selenides produce colors in the glass. Indeed, this is how colored glasses are produced. Scrap glass can easily be used in the melt, but the need to sort the glass from contaminants and to separate the colored glasses is a problem in recycling.

(d) Cements. Use of brick and stone for construction requires a cement or mortar to bind them together. A cement is the binding material, in this context a powder activated by water. If sand is added as a filler, the mixture is

[2]*Trade mark of the Corning Glass Co., Corning, New York.*

called a mortar. Mortars are used in bonding bricks, etc. Mixtures with coarse materials such as gravel or crushed rock are referred to as concrete. Use of cements goes back to antiquity, with various substances having been used.

The earliest cements were based on gypsum, which can be dehydrated at temperatures in the vicinity of 200°C;

$$\underset{\text{gypsum}}{CaSO_4 \cdot 2H_2O} \rightleftarrows \underset{\text{plaster of paris}}{CaSO_4 \cdot \frac{1}{2} H_2O} + \frac{3}{2} H_2O \qquad (16\text{-}19)$$

On mixing with water, the reaction can be reversed and the material sets to a hard mass. Plaster of paris has some uses even today. It evidently was used with sand as a mortar for masonry by the ancient Egyptians, but gypsum is slightly water soluble and unsuited to moist climates.

Lime mortars were also discovered quite early in the development of civilization. These are formed by heating (calcining) calcium carbonate, and slaking the quicklime (calcium oxide) produced with water to give slaked lime, calcium hydroxide.

$$CaCO_3 \xrightarrow{\text{heat}} CO_2 + CaO \xrightarrow{2H_2O} Ca(OH)_2 \qquad (16\text{-}20)$$

A paste of slaked lime, sand, and water can be used as a plaster on walls or as a mortar; it hardens by chemical reaction with atmospheric CO_2:

$$Ca(OH)_2 + CO_2 \rightarrow CaCO_3 + H_2O \qquad (16\text{-}21)$$

Lime mortars have two disadvantages. First, they depend on diffusion of CO_2 from the air into the interior of the mass, and consequently they cure slowly and not always completely. Second, they will not harden on contact with water.

Hydraulic cements, those which will harden in the presence of water, first were made by mixing volcanic ash with lime. These (called lime-pozzolana cements), when mixed with water, harden by a chemical reaction that produces calcium silicates and aluminosilicates from the calcium hydroxide, silica, and alumina. The last two are provided by the volcanic ash. Limestone containing considerable clay also produces hydraulic lime on calcining; similar other reactions can take place. These materials were the forerunners of Portland cement, the material now in general use for construction purposes.

Portland cement is chiefly composed of calcium silicates and calcium aluminosilicates; the primary compounds in it have the empirical formulas $3CaO \cdot SiO_2$, $2CaO \cdot SiO_2$, $3CaO \cdot Al_2O_3$, and

$4CaO \cdot Fe_2O_3 \cdot Al_2O_3$. The CaO is ordinarily obtained from limestone; the SiO_2, Al_2O_3, and Fe_2O_3 from clay or shale, although blast furnace slag, fuel ash from power plants, and industrial calcium carbonate wastes are sometimes employed. The starting materials are ground and heated to 1300-1500°C to incipient fusion. All water and CO_2 is thereby removed, and chemical reactions to form the silicate and aluminate compounds take place. After cooling, the "clinker" is finely ground with gypsum. When water is mixed with the cement, complex reactions take place which involve hydration of the compounds present and the formation of new compounds. It is these chemical hydration reactions that are responsible for the hardening of the cement. These reactions are not completed for weeks, although they are fast enough to cause the paste to become solid in a matter of hours.

(e) Asbestos. Several different aluminosilicate minerals are referred to as asbestos. All are based on shared SiO_4 tetrahedra, but some (amphiboles) are based on double chains of linked tetrahedra, the double chains linked to each other by interactions with cations, while others (chrisotiles) are based on double layers of SiO_4 units. Magnesium and hydroxide ions link the layers together. However, the layers are not flat but curved into cylinders. In the bulk, both forms of asbestos are fibrous as they are built up of overlapping rods or cylinders. Asbestos materials are widely used for thermal insulation, and have been used as inert fillers in plastics and other materials. Asbestos is relatively inert chemically. However, it has now been recognized as having one environmental property that is of great importance for mankind: its dust is carcinogenic, producing lung cancer. Manipulation of the asbestos fibers invariably causes some breakdown and the release of asbestos particles; this can be quite extensive in asbestos mining and processing operations if precautions are not taken. The carcinogenic character evidently is due to the physical irritation caused by the tiny rodlike particles.

Although it is well established that long-term inhalation of high concentrations of asbestos dust is harmful, there is more question about the effects of small amounts such as that to which the general population is exposed. Asbestos can be quite widespread in small amounts; normal building construction and destruction work, wear from vehicle brake linings, asbestos contaminants in other mineral-derived products (for example, talcum powder, plaster) are all sources of asbestos particles in the air, but whether or not they constitute general health hazards is undetermined. There is also uncertanty, but cause for concern, over the presence of asbestos particles in drinking water; in some localities quite appreciable concentrations exist. Evidence is not conclusive as to whether such particles are hazardous on ingestion; their interaction with the gastrointestinal tract need not be similar to their interaction with the lungs, but the risk that it will be similar is certainly real. Such water contamination may arise from discharge of waste waters used in mining or mineral treatment; for example, that used in scrubbers to remove dust from exhaust gases. It may also come from dumping solid mineral wastes or sludges into water systems.

REFERENCES

B. J. Skinner, A second iron age ahead? *Amer. Sci. 64*, 258 (1976).

H. O. Buckman and N. C. Brady, "The Nature and Properties of Soils," 7th ed. MacMillan, New York, 1969.

C. A. Price, Stone, decay and preservation, *Chem. Britain 11*, 350 (1975).

C. Hall, On the history of Portland cement after 150 years, *J. Chem. Ed. 53*, 222 (1976).

M. McNiel, Lateritic soils, in "Man and the Ecosphere, Readings from Scientific American," p. 68. Freeman, San Francisco, California, 1971. (From *Sci. Amer.*, November, 1964.)

F. M. Lea, "The Chemistry of Cement and Concrete," 3rd ed. Chemical Publishing Co., New York, 1971.

A. N. Rohl, A. M. Langer, and I. J. Selikoff, Environmental asbestos pollution related to use of quarried serpentine rock, *Science 196*, 1319 (1977). See also J. T. Hack, and A. N. Rohl, A. M. Langer, and I. J. Selifoff, *Science 197*, 716 (1977).

L. J. Carter, Asbestos, Trouble in the air from Maryland rock quarry, *Science 197*, 237 (1977).

P. M. Cook, G. E. Glass, and J. H. Tucker, Asbestiform amphibole minerals: detection and measurement of high concentrations in municipal water supplies, *Science 185*, 853 (1974).

17
NUCLEAR CHEMISTRY OF THE ENVIRONMENT

17.1 INTRODUCTION

In this chapter we shall attempt to provide an understanding of the nature, the sources and the hazards of ionizing radiation in the environment. As the name signifies, ionizing radiation interacts with matter in ways that cause ionization of atoms and molecules. When such radiation is absorbed by man, it can produce harmful effects.

We shall emphasize naturally occurring and man-made nuclear sources of ionizing radiation. These sources and the characteristics of the radiation they produce will be treated from the viewpoint of nuclear chemistry. In addition, we shall consider other important sources of ionizing radiation such as X-ray machines. Although these are nonnuclear sources, they must be taken into account as significant sources of ionizing radiation to which the general population is exposed. Thus, in considering the potential effects of any one type of source, such as a nuclear power plant, we must define the environment broadly to include the nuclear and non-nuclear sources used for dental and medical diagnosis and for medical therapy.

Ionizing radiation has been a component of the environment ever since the earth was created, i.e., for about 4.5×10^9 years. Thus, atoms of the primordial, long-lived radioisotopes of elements such as potassium, rubidium, and samarium and of primordial long-lived isotopes of the radioelements uranium and thorium together with their short-lived daughters have made the earth itself a complex source of ionizing radiation. In addition, the earth has been bombarded, apparently continuously, by cosmic radiation, which is high energy ionizing radiation originating in outer space. Cosmic radiation also produces some of the radioactivity found in the environment.

Although the presence of a variety of ionizing radiation at the earth's surface greatly predates the presence of man, it was not until the end of the nineteenth century that man devised means to detect and produce ionizing radiation. Specifically, Roentgen's discovery of X rays in 1895 and Becquerel's related discovery of radioactivity in 1896 mark the beginning of man's investigation of the properties, effects, and applications of ionizing radiation. Interestingly, for both discoveries the potential benefit to mankind in the field of medicine was almost immediately recognized. Also, it was not long before luminous "radium dial paint" for clocks and watches brought radium into pockets and homes. "X rays" and "radium" soon became household terms.

During the period 1900-1940, there were many advances in experimental and theoretical nuclear physics. Artificial radioactivity was discovered and particle accelerators for producing man-made radioactivity were devised. In 1939 there was a discovery which was destined to have particular significance for the environment. In that year Hahn and Strassmann detected neutron-induced nuclear fission of uranium. Potential uses were immediately recognized and within a few years, during World War II, the technology for large-scale utilization of the energy released in nuclear fission was developed. Whether the energy is released explosively in a nuclear weapon or more slowly in a nuclear power plant, hazardous radioactive fission products and other radioactive substances are formed. The extent to which these sources of ionizing radiation enter the environment, add to the natural background level of ionizing radiation, and thereby become a threat to the health of this and future generations, is a matter of serious concern now and will continue to be so in the future.

With the signing of international agreements limiting the testing of nuclear weapons in the atmosphere by nations and in the absence of nuclear warfare, this source of atmospheric contamination has been greatly reduced, but has not been eliminated. Atmospheric weapons tests and underground nuclear detonations or tests are still conducted, but recently attention has been focused on the environmental impact of nuclear power plants. The possibility of a direct impact arises because the plants produce large quantities of radioactive material of the same type that is produced in nuclear weapons. Within a nuclear power plant and in subsequent handling steps, the radioactive material is confined to systems and containers that are designed to prevent its release to the environment. The problem of storage of radioactive waste from the nuclear power plants is an unsolved but by no means a new problem. The possibility of indirect environmental impact stems from use and production in the plants of materials needed to fabricate nuclear weapons, even though the materials as used or produced would require extensive purification and processing.

For a given scientific discovery, potential uses that will meet the needs of mankind may or may not be obvious at the time of discovery. The development of applications is expedited, of course, if potential benefits are recognized. Too often the cost has been taken to be simply the development cost in dollars. The true cost may not be measurable in dollars and may not even be recognized until many years have passed. Harmful effects involving new phenomena as well as familiar ones may slowly emerge. After the recognition and evaluation of long-term effects, as in the case of DDT, for example (see Chapter 7), one proceeds cautiously into the future with 20-20 hindsight. The harmful effects of ionizing radiation have been characterized and studied in depth for several decades. Although much is known, much has yet to be learned. Some of the effects do not become evident until many years after exposure and even then one cannot always be certain that an observed effect was caused by ionizing radiation rather than some other agent capable of producing precisely the same effect.

We shall begin the chapter with a brief summary of the properties of radioactive material. Such properties include the radioactive half-life and the type and energy of the radiation emitted. We shall then consider the interaction of ionizing radiation with matter, dosimetry, and the effects of such radiation on the human body. Next, we shall examine the environment with respect to both natural and man-made sources of ionizing radiation. Finally, current levels of population exposure will be summarized.

During the last decade or so, the public, industry, and government have increasingly recognized that cost-benefit (or risk-benefit or cost-effectiveness) analysis is an essential part of the decision-making process. This represents a drastic change from the situation that prevailed during the industrial revolution. It is now commonplace for environmental impact to be a key factor or even a critical factor in the making or delaying of decisions that affect agriculture, the automobile industry, the power industry, etc. Hopefully, this chapter will provide the reader with a basis for thinking objectively when considering the pros and cons of issues that involve ionizing radiation.

17.2 THE KINETICS OF RADIOACTIVITY

17.2.1 Half-Life

Radioactive decay is typically a first-order rate process. For a radioactive substance, the number of atoms decaying per unit time in a sample is, therefore, directly proportional to the number of atoms of the radioactive substance, the radionuclide, present. A given radionuclide is designated by a chemical symbol, a subscript corresponding to the atomic number, and a superscript indicating the mass number. Thus, carbon-14 is written as $^{14}_{6}C$ or simply ^{14}C since the atomic number is implicit in the chemical symbol.

If N is the number of atoms of a single radionuclide present at time t, the rate of change of N is given by

$$dN/dt = -\lambda N \tag{17-1}$$

The proportionality constant λ is the disintegration constant, a characteristic constant for each radionuclide. It is commonly expressed in the units of sec^{-1}. The disintegration rate or the activity for a single substance A, at any instant is equal to λN. Equation (17-1) is analogous to the equation for the rate of a first-order chemical reaction (Section 9.3). The symbols used in this chapter are those traditionally used for a radioactive decay. Thus, the rate is expressed in terms of the number of atoms N, instead of concentration, and the rate constant is written as λ instead of k. The unit of activity is the curie, where one curie, 1 Ci, is 3.700×10^{10} disintegrations per second (dps). Commonly used subunits are mCi (10^{-3} Ci), μCi(10^{-6} Ci) and pCi (10^{-12} Ci). For many purposes it is useful to know the specific activity of a radionuclide. This is the activity per unit weight of element, e.g., mCi per gram of carbon. Another useful quantity is the activity concentration, e.g., pCi per liter of solution.

Integration of Eq. (17-1) between two times t_1 and t_2 leads to

$$N_2 = N_1 \exp\left[-\lambda(t_2 - t_1)\right] \tag{17-2}$$

where N_1 and N_2 are the numbers of atoms of the radionuclide at time t_1 and t_2, respectively. Commonly t_1, the time at the beginning of the decay period of interest, is set equal to zero with the resulting familiar expression

$$N = N_0 \exp(-\lambda t) \tag{17-3}$$

where t is the lapsed time and N_0 is the number of atoms of the radionuclide present in the sample when $t_1 = 0$. (Letting t_1 equal zero is analogous to resetting a laboratory timer to zero before turning it on.)

When N/N_0 is equal to 0.5, t is equal to $t_{½}$ the half-life of the radionuclide. Then $\exp(-\lambda t_{½}) = 0.5 = \exp(-0.693)$ and

$$t_{½} = 0.693/\lambda \tag{17-4}$$

Equation (17-3) may then be written in the form

$$N = N_0 \exp(-0.693t/t_{\frac{1}{2}}) \qquad (17\text{-}5)$$

which is more useful than Eq. (17-3) because it is $t_{\frac{1}{2}}$ rather than λ that is tabulated for radionuclides. When Eq. (17-5) is rewritten as

$$N = N_0 (\tfrac{1}{2})^{t/t_{\frac{1}{2}}} \qquad (17\text{-}6)$$

the resulting equation is useful for quick exact calculations or working approximations. The mean life τ is of interest in certain calculations. It is equal to $1/\lambda$.

By multiplying Eqs. (17-3), (17-5), and (17-6) by λ and replacing λN by A, the activity, we obtain the equations showing the time dependence of the activity of a sample. Thus Eqs. (17-5) and (17-6) become

$$A = A_0 \exp(-0.693t/t_{\frac{1}{2}}) \qquad (17\text{-}7)$$

and

$$A = A_0 (\tfrac{1}{2})^{t/t_{\frac{1}{2}}} \qquad (17\text{-}8)$$

where A_0 is the initial activity. A typical decay curve corresponding to these equations is shown in Fig. 17-1. To illustrate the use of Eq. (17-8), we might ask how long it would take for a sample of a radionuclide to decay to about one thousandth of some specified activity. Recognizing that $(\frac{1}{2})^{10}$ is equal to 1/1024, about ten half-lives will be required to provide the specified reduction in A/A_0 for *any* radionuclide. Thus, for the fission products $^{90}_{38}Sr$ ($t_{\frac{1}{2}}$ = 28.5 yr) and $^{106}_{45}RH$ ($t_{\frac{1}{2}}$ = 30 sec) 285 yr and 300 sec, respectively, would be required for a 1 Ci sample of each to decay to about 1 mCi.

In the laboratory or in the field, we measure a counting rate R, rather than N or A, using a particular detector chosen for its sensitivity to the particular type of radiation emitted by the source. The counting rate is proportional to A and depends on the response efficiency of the detector. Assuming the efficiency remains constant during the measurement, a plot of R instead of A would give a decay curve similar to that in Fig. 17-1. The line would have the same slope, but would be displaced by the ratio of R to A. From a decay curve, we can evaluate $t_{\frac{1}{2}}$ from the slope by simply observing the time required for R to decrease by a factor of ½.

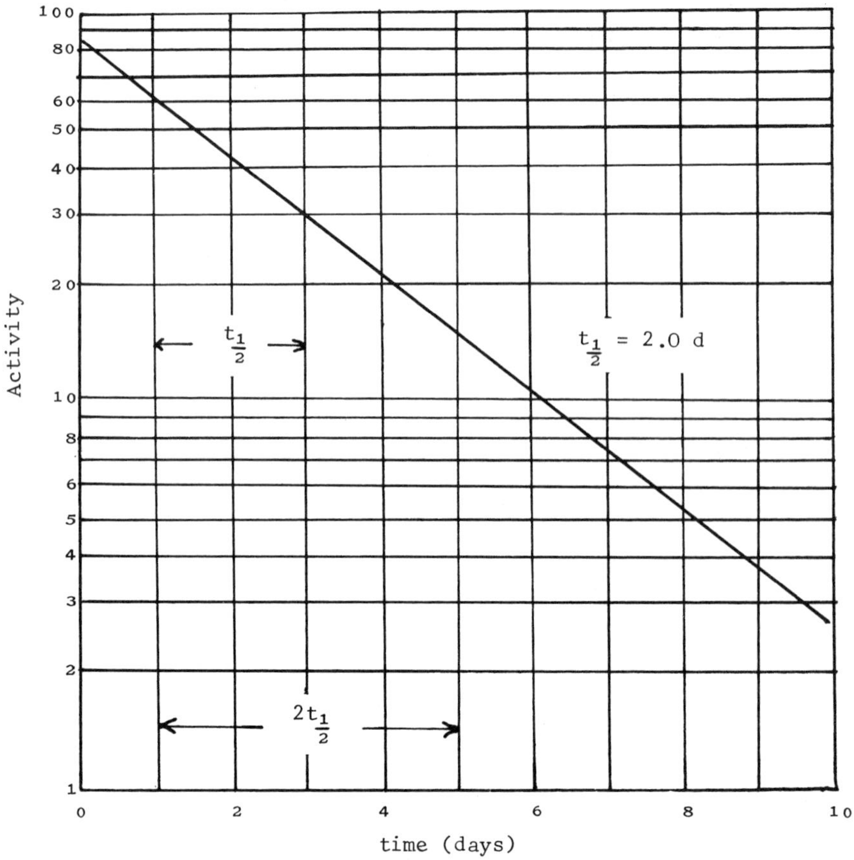

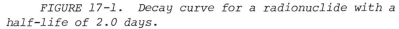

FIGURE 17-1. Decay curve for a radionuclide with a half-life of 2.0 days.

A few examples of half-lives of naturally occurring radionuclides are given in Table 17-1.

The primordial radionuclides have half-life values comparable to or longer than the age of the earth, i.e., about 4.5×10^9 yr. One of the most important single (independent) sources of background radiation is ^{40}K. Other important radionuclides are among the isotopes of elements $_{81}Tl$ through $_{92}U$. Most short-lived radionuclides in nature are formed as daughters of long-lived parents. About 20 short-lived radionuclides are also being produced continuously by cosmic ray interaction in the atmosphere. Two such cosmogenic radionuclides of interest in terms of the environment are ^3H (also written as T) and ^{14}C. These are produced in the atmosphere by nuclear reactions of fast neutrons (energy in the MeV range) with ^{14}N. The reactions are $^{14}N(n,^3H)^{12}C$, and

TABLE 17-1

Examples of Half-lives of Radionuclides in Nature

Radionuclides	*Half-life*
^{3}H	12.33 yr
^{14}C	5730 yr
^{40}K	1.28×10^{9} yr
^{50}V	4×10^{14} yr
^{214}Po	1.64×10^{-4} sec
^{226}Ra	1600 yr
^{232}Th	1.40×10^{10} yr
^{238}U	4.47×10^{9} yr

$^{14}N(n, p)^{14}C$, where the target nuclide is written first, the incident particle is indicated next within the parenthesis, the low mass number reaction product or products are indicated next within the parenthesis and finally the product nuclide is given. The relative yields for the production of ^{3}H and ^{14}C from ^{14}N are about 1:100. Among the other radionuclides of cosmogenic origin in the atmosphere, soil, and sea are ^{7}Be, ^{22}Na, ^{26}Al, ^{32}Si, ^{32}P, ^{33}P, ^{35}S, ^{36}Cl, ^{39}Cl, ^{37}Ar, ^{53}Mn, and ^{60}Co. These are present in smaller amounts, and according to current thinking they are of lesser importance in the environment.

17.2.2 Mixtures of Radionuclides

In nature and in nuclear power reactors there are complex mixtures of radionuclides. Mixtures are of three types: (1) independent, (2) genetically related (parent decays to radioactive daughter), and (3) a combination of (1) and (2). For a nongenetically related mixture of only two radionuclides A and C, such that A → B (stable) and C → D (stable) and C has the longer half-life, the composite decay curve might be as shown in Fig. 17-2, where the component decay curves are also shown.

Most of the naturally occurring radionuclides in the rocks and soil of the earth are members of one of the three series known as the uranium, thorium, and actinium series. The series are named after the longest-lived member, the

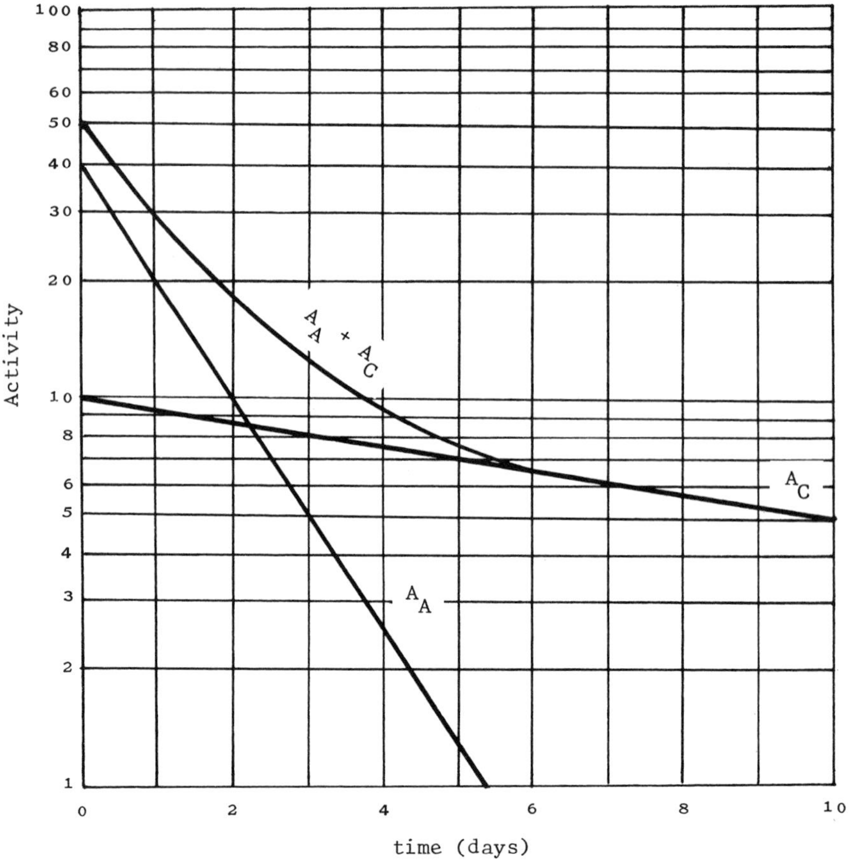

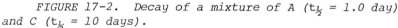

FIGURE 17-2. Decay of a mixture of A ($t_{1/2}$ = 1.0 day) and C ($t_{1/2}$ = 10 days).

parent of the series. Similarly, fission products are commonly members of chains typically having three or four members.

For a given series or chain, one can write A → B → C → D -----------→ to a stable end product. The parent-daughter relationships in the three naturally occurring series are represented in Fig. 17-3 and certain characteristic data are given in Tables 17-2 - 17-4.

The uranium series, for example, begins with $^{238}_{92}U$ and ends with $^{206}_{82}Pb$ (stable). Because all members of the series have mass numbers representable by four times an integer plus two, the series is also known as the $4n+2$ series.

To determine the amount of daughter B present in these or any series at time t, one starts with the rate equation for the net rate of change of N_B atoms:

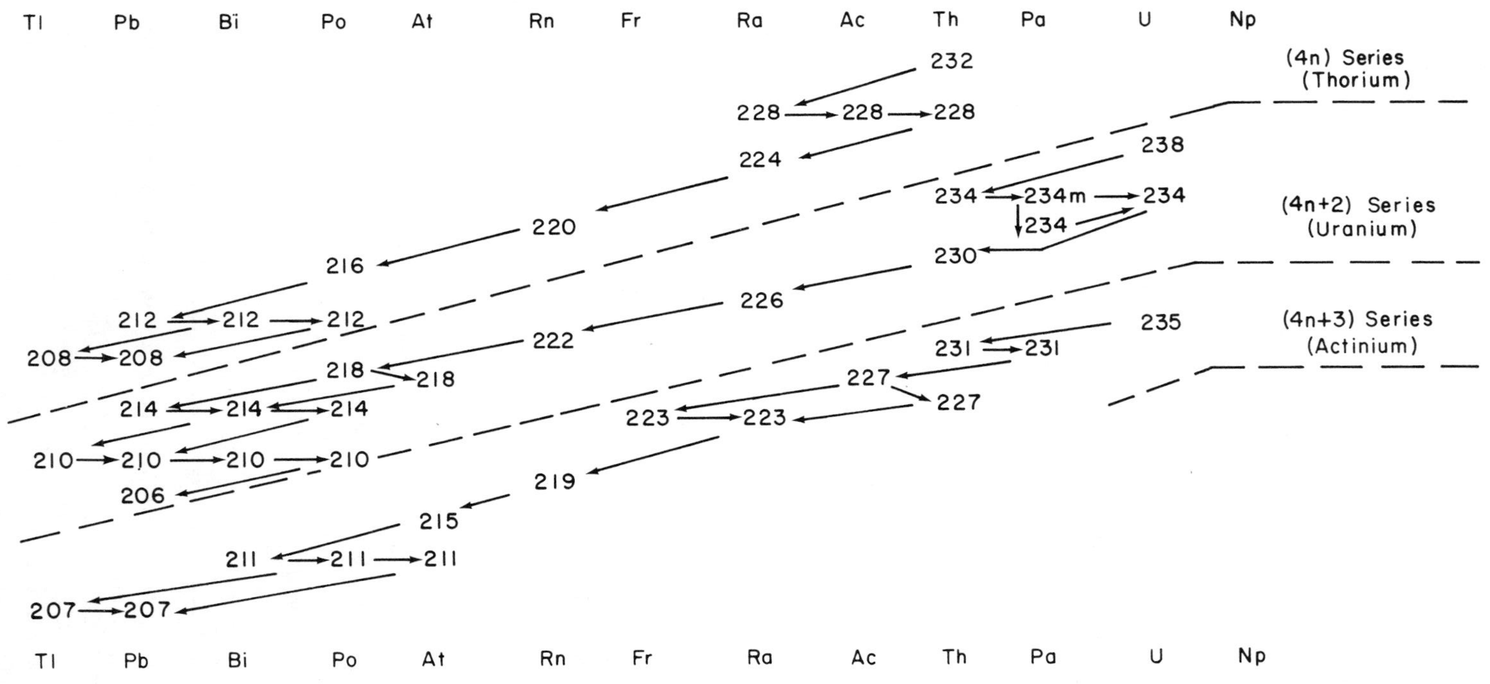

FIGURE 17-3. *Genetic relationships for the thorium, uranium, and actinium series showing only the significant branching. The numbers are mass numbers.*

TABLE 17-2

The Uranium (4n+2) Series

Nuclide[a]	Half-life	Major type of radiation[b]
$^{238}_{92}U$ (uranium I)	4.47×10^9 yr	α
$^{234}_{90}Th$ (uranium X_1)	24.1 day	β^- (γ)
$^{234m}_{91}Pa$ (uranium X_2)	1.17 min	β^- (99.87% to ^{234}U)
$^{234}_{91}Pa$ (uranium Z)	6.67 hr	β^-, γ
$^{234}_{92}U$ (uranium II)	2.44×10^5 yr	α
$^{230}_{90}Th$ (ionium)	7.7×10^4 yr	α
$^{226}_{88}Ra$ (radium)	1600 yr	α, (γ)
$^{222}_{86}Rn$ (radon)	3.824 day	α
$^{218}_{84}Po$ (radium A)	3.05 min	α(99.98%), β^-(0.02%)
$^{214}_{82}Pb$ (radium B)	26.8 min	β^-, γ
$^{218}_{85}At$	∿2 sec	α
$^{214}_{83}Bi$ (radium C)	19.8 min	α(0.02%), β^-(99.98%), γ
$^{214}_{84}Po$ (radium C')	164 μsec	α
$^{210}_{81}Tl$ (radium C")	1.30 min	β^-, γ
$^{210}_{82}Pb$ (radium D)	22.3 yr	β^-, (γ)
$^{210}_{83}Bi$ (radium E)	5.01 day	β^- (γ)
$^{210}_{84}Po$ (radium F)	138.4 day	α
$^{206}_{82}Pb$ (radium G)	stable	

[a]With name used in older literature and still used in some instances.

[b](γ) indicates low intensity (1-5%); photons with intensity less than 1% are not indicated; where both α and β are given, branching occurs.

TABLE 17-3

The Thorium (4n) Series

Nuclide[a]	Half-life	Major type of radiation[b]
$^{232}_{90}Th$ (thorium)	1.40×10^{10} yr	α
$^{228}_{88}Ra$ (mesothorium I)	5.75 yr	β^-
$^{228}_{89}Ac$ (mesothorium II)	6.13 hr	β^-, γ
$^{228}_{90}Th$ (radiothorium)	1.91 yr	α, (γ)
$^{224}_{88}Ra$ (thorium X)	3.64 day	α, (γ)
$^{220}_{86}Rn$ (thoron)	55.6 sec	α
$^{216}_{84}Po$ (thorium A)	0.15 sec	α
$^{212}_{82}Pb$ (thorium B)	10.64 hr	β^-, γ
$^{212}_{83}Bi$ (thorium C)	60.6 min	α(36%), β^-(64%), γ
$^{212}_{84}Po$ (thorium C')	304 nsec	α
$^{208}_{81}Tl$ (thorium C'')	3.054 min	β^-, γ
$^{208}_{82}Pb$ (thorium D)	stable	

[a]*With name used in older literature and still used in some instances.*

[b]*(γ) indicates low intensity (1-5%); photons with intensity less than 1% are not indicated; where both α and β are given, branching occurs.*

TABLE 17-4

The Actinium (4n+3) Series

Nuclide[a]	Half-life	Major type of radiation[b]
$^{235}_{92}U$ (actinouranium)	7.04×10^8 yr	α, γ
$^{231}_{90}Th$ (uranium Y)	25.5 hr	β⁻, γ
$^{231}_{91}Pa$ (protactinium)	3.25×10^4 yr	α
$^{227}_{89}Ac$ (actinium)	21.77 yr	α(1.4%), β⁻(98.6%)
$^{227}_{90}Th$ (radioactinium)	18.72 day	α, γ
$^{223}_{87}Fr$ (actinium K)	22 min	β⁻, γ
$^{223}_{88}Ra$ (actinium X)	11.43 day	α, γ
$^{219}_{86}Rn$ (actinon)	3.96 sec	α, γ
$^{215}_{84}Po$ (actinium A)	1.78 msec	α
$^{211}_{82}Pb$ (actinium B)	36.1 min	β⁻, (γ)
$^{211}_{83}Bi$ (actinium C)	2.14 min	α(99.7%), β⁻(0.28%)
$^{211}_{84}Po$ (actinium C')	0.53 sec	α
$^{207}_{81}Tl$ (actinium C'')	4.77 min	β⁻
$^{207}_{82}Pb$ (actinium D)	stable	

[a]With name used in older literature and still used in some instances.

[b](γ) indicates low intensity (1-5%); photons with intensity less than 1% are not indicated; where both α and β are given, branching occurs.

$$dN_B/dt = \text{(rate of formation of B by decay of A)} - \text{(rate of decay of B)}$$
$$= \text{(activity of A)} - \text{(activity of B)}$$

or

$$dN_B/dt = \lambda_A N_A - \lambda_B N_B \qquad (17\text{-}9)$$

In this equation it is assumed that substance A can decay in only one mode, namely, that in which daughter B is formed. There are many cases in which branching (decay by more than one mode) occurs. When branching occurs, the $\lambda_A N_A$ term must be multiplied by the fraction of decays in which substance B is formed.

The value of A_B, the activity of the daughter at any time t, can be obtained in terms of $(A_A)_o$ and $(A_B)_o$, the activities of the parent and daughter at some reference time $t = 0$, by integrating Eq. (17-9) to obtain N_B, multiplying the result by λ_B, and finally rewriting $\lambda_B N_B$ as A_B. The resulting equation is then

$$A_B = \frac{(A_A)_o \lambda_B}{(\lambda_B - \lambda_A)} [\exp(-\lambda_A t) - \exp(-\lambda_B t)] + (A_B)_o \exp(-\lambda_B t) \qquad (17\text{-}10)$$

The last term represents the decay of the daughter radionuclide present in a sample of A when $t = 0$. Three types of parent-daughter systems can be recognized on the basis of the relative values of λ_A and λ_B. When $\lambda_A <<< \lambda_B$, that is, $(t_{½})_A >>> (t_{½})_B$, the system can attain a steady state called "secular equilibrium." When $\lambda_A < \lambda_B$ or $(t_{½})_A > (t_{½})_B$, the system can reach a steady state called "transient equilibrium." Finally, when $\lambda_A > \lambda_B$ or $(t_{½})_A < (t_{½})_B$, a steady state can not be attained.

For systems that can attain secular equilibrium, Eq. (17-10) reduces to the following when only pure A is present at $t = 0$:

$$A_B = (A_A)_o [1 - \exp(-\lambda_B t)] \qquad (17\text{-}11)$$

In this equation, t, the lapsed time, is short compared to the half-life of the parent; hence $(A_A)_o$ remains essentially constant during time t. The growth of the daughter activity according to Eq. (17-11) is represented by Fig. 17-4. Recalling that $\exp(-\lambda_B t)$ can be replaced by $(½)^{t/t_{½}}$, we see that in one half-life of the daughter, its activity is equal

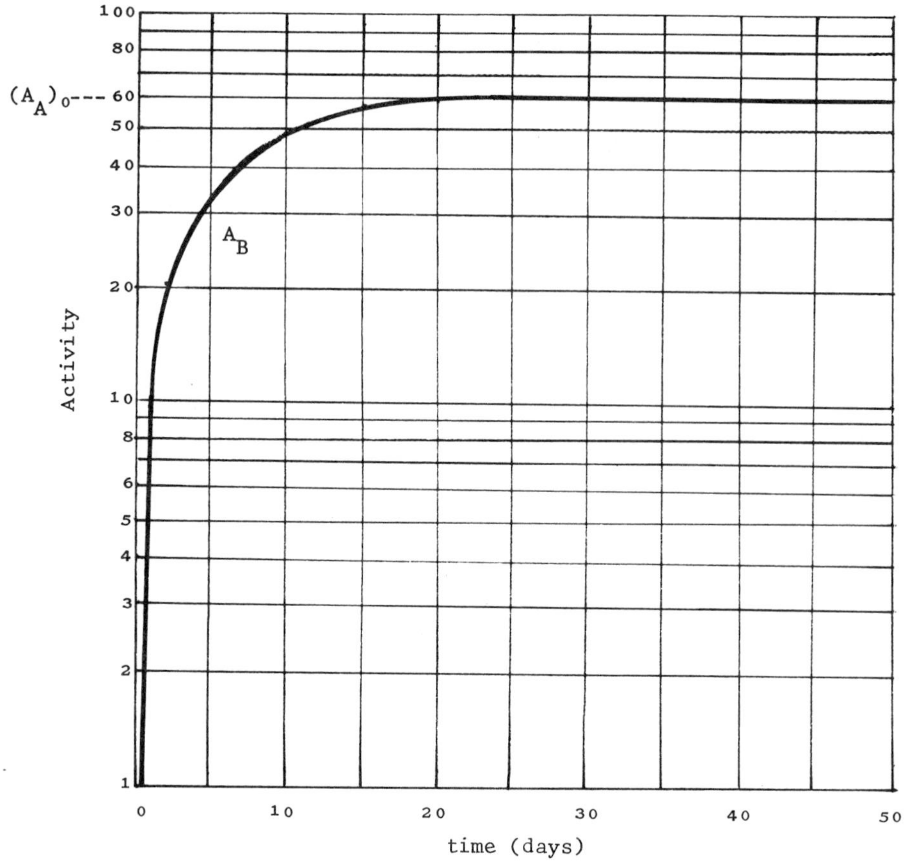

FIGURE 17-4. Growth of 3.824 day $^{222}Rn(A_B)$ in pure ^{226}Ra (1600 yr) having $(A_A)_O$; attainment of secular equilibrium.

to 50% of that of the parent; after two half-lives, 75%; and after ten half-lives, 99.9%. When a system has attained secular equilibrium, the activity of the daughter is equal to that of the parent. The number of atoms and therefore the weight of each is, however, very different.

Two examples of parent-daughter combinations that can attain secular equilibrium are

$$^{226}_{88}Ra \xrightarrow{1600\ yr} {}^{222}_{86}Rn \xrightarrow{3.824\ day} {}^{218}_{84}Po \xrightarrow{etc.}$$

and

$$^{106}_{44}Ru \xrightarrow{369\ day} {}^{106}_{45}Rh \xrightarrow{30\ sec} {}^{106}_{46}Pd\ (stable)$$

The former is important in the uranium series and the latter is an important contributor to fission product wastes. In the uranium series and in the other two naturally occurring series, the parent of each series is a primordial radionuclide. In primary rocks that have not been disturbed by weathering, leaching, or the escape of the radon isotope that constitutes the noble gas member of each series, all members except the last are in secular equilibrium with the parent. This is important to keep in mind when considering the radioactivity background associated with the uranium and thorium content of rocks, soil, and water.

17.2.3 Man-Made Radioactivity

When a radionuclide is produced by means of an accelerator or a nuclear reactor, its rate of formation is dependent upon its half-life. If t is the irradiation period for a target containing atoms of a specific nuclide A which undergoes nuclear reaction to form radionuclide B, the activity of B at the end of the irradiation is given by

$$A_B(\mathrm{dps}) = (R_f)_B\,(1 - \exp[-0.693t/(t_{1/2})_B]) \qquad (17\text{-}12)$$

In this equation it is assumed that $(R_f)_B$, the rate of formation of B, is a constant during the irradiation. The magnitude of the term $(R_f)_B$ is directly proportional to (1) the number of atoms of A in the material being irradiated, (2) the number of reactant projectile particles, e.g., neutrons, striking the target per second, and (3) the probability (called the cross section) that the product B will be formed when the projectile particle reacts with a nucleus of A. Equation (17-12) is similar in form to Eq. (17-11) for the growth of the activity of a daughter that can attain secular equilibrium with its parent. After an irradiation time of ten half-lives of B, A_B is essentially equal to R_f and the product reaches a "saturation level" or upper limit of activity as shown in Fig. 17-5. At "saturation" the radionuclide B is decaying as rapidly as it is being formed. After termination of the irradiation, the sample decays as shown in Fig. 17-5. The decay time t' is sometimes referred to as the "cooling" time because the heat released by a given radioactive source is proportional to the disintegration rate. The appropriate equation becomes

$$A_B = (R_f)_B(1-\exp[-0.693t/(t_{1/2})_B])(\exp[-0.693t'/(t_{1/2})_B]) \qquad (17\text{-}13)$$

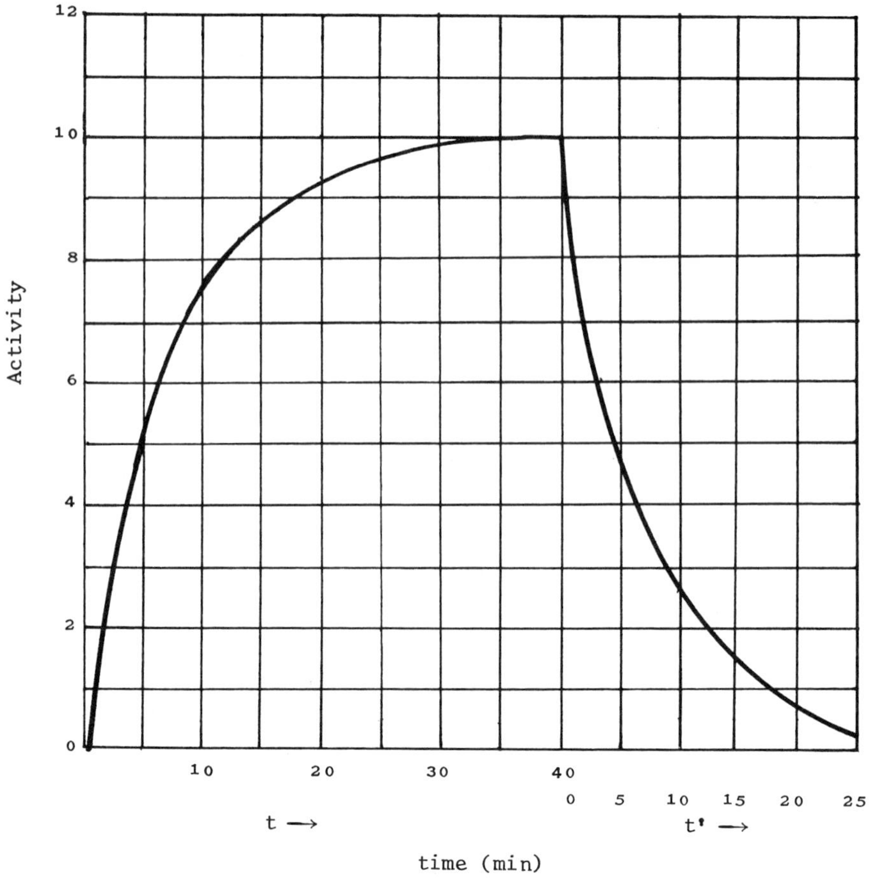

FIGURE 17-5. Production and decay of a radionuclide having $t_{\frac{1}{2}}$ *= 5.0 min; irradiation time t = 40 min, decay time out to t' = 25 min.*

Nuclear fission, the main large-scale source of man-made radioactivity will be discussed in some detail in Section 17.8.3.

17.3 MODES OF RADIACTIVE DECAY

The equations for the time dependence of radioactive decay given in the preceding section do not provide information about decay mechanisms. The mechanisms by which an unstable nucleus may spontaneously disintegrate are called the modes of decay. The type or types and the quantity of radiation emitted by a nucleus in the decay process depend on the mode or modes of decay. Only the more common modes will be considered here.

17.3.1 Beta Decay

There are three types of beta decay. In negatron emission the unstable nucleus emits an electron β^- and is transformed into an isotope of the adjacent element having higher atomic number. The overall reaction may be represented by

$$ {}^{A}_{Z}\mathrm{El} \rightarrow {}^{A}_{Z+1}\mathrm{El} + {}^{0}_{-1}\beta + {}^{0}_{0}\bar{\nu} \qquad (17\text{-}14)$$

For an electron of nuclear origin (a beta ray), the symbol β^- will be used; for one of nonnuclear origin, e^- will be used. In the unstable nucleus a neutron is transformed into a proton at which time the negatron and the antineutrino $\bar{\nu}$ are created and emitted. The negatron, the antineutrino, and the product nuclide share the transition energy. The product nuclide takes a relatively small fraction of the disintegration energy as recoil energy. Both the negatron and the antineutrino have continuous spectral energy distributions, i.e., the number of particles of a given type having a given energy within the maximum available for the transition varies continuously with energy in contrast to a spectrum consisting of discrete energy groups of lines. An example of a beta ray spectrum is illustrated in Fig. 17-6a. The upper limit of the beta ray energy E_{max} is characteristic of the transition energy. Only one of the two particles emitted, the negatron, is readily detectable. Because of its charge, it causes ionization in any material through which it passes and is, therefore, classified as ionizing radiation. By contrast, the antineutrino has no charge. Its rest mass (as opposed to its motion or inertial mass) is essentially zero. This elusive particle plays an essential role in conserving energy and nuclear spin in beta decay. Thus the negatron and the antineutrino share the available energy E_{max} so that the beta ray spectrum contains all values of energy between zero and E_{max}.

If the nuclide produced in negatron emission is formed in its ground state, the transition is represented by an energy level diagram or decay scheme of the type shown in Fig. 17-6b. In such a diagram, energy is plotted as the ordinate and atomic number as the abscissa. Carbon-14 with E_{max} equal to 0.156 MeV has a simple decay scheme. More commonly the product nuclide may be formed in an excited state that may deexcite at once, i.e., within 10^{-12} sec, or after

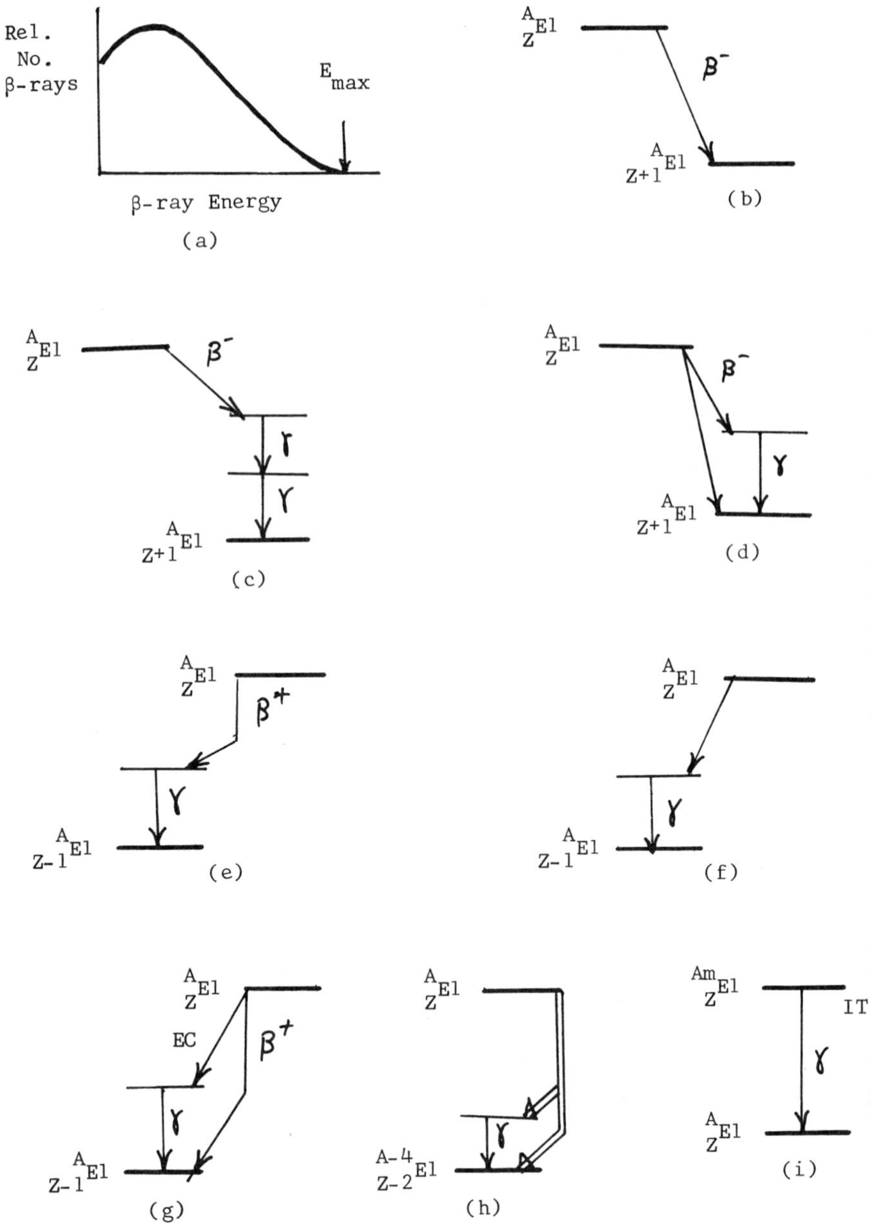

FIGURE 17-6. (a) β - ray spectrum; (b) pure β^- decay; (c) β^-, γ decay; (d) β^-, γ decay; (e) β^+, γ decay; (f) EC, γ decay; (g) β^+, EC, γ decay; (h) α, γ decay; (i) isomeric transition.

some delay, by emission of one or more gamma rays (high energy photons) as shown in Fig. 17-6c,d, where two excited states are formed by beta decay. If the deexcitation occurs with a measurable half-life, the product is called an isomer and is in a metastable state designated by the letter "*m*" after the mass number. The total beta transition energy Q_{β^-} is for decay to the ground state of the product and, therefore, includes any gamma transitions. For a given gamma transition, the gamma rays are monoenergetic.

In the second type of beta decay, positron emission, a positively charged electron β^+ and a neutrino are created and emitted when a proton is transformed into a neutron in the unstable nucleus. For our purposes the neutrino is the same as the antineutrino, although they differ with respect to the directions of their spins. The positron, like the negatron, will produce ionization. The overall change in positron emission is

$$^{A}_{Z}\mathrm{El} \rightarrow {}^{A}_{Z-1}\mathrm{El} + {}^{0}_{+1}\beta + {}^{0}_{0}\nu \qquad (17\text{-}15)$$

In this case the total disintegration energy Q_{β^+} includes 1.02 MeV (the rest energy of two electrons), E_{max} for the positron, recoil energy of the product nuclide, and any gamma radiation emitted if the product nuclide is formed in an excited state. The rest energy of two electrons is included in the energy balance because the atomic mass, which includes the mass of the Z electrons that balance the charge on the nucleus, decreases not only by the mass of the positron emitted, but also by the mass of an outer electron. The outer electron is lost because the nuclear charge of the product atom is one unit less than that of the unstable precursor. The numerical value for the rest energy of an electron, 0.51 MeV, can be calculated from the equivalence between one atomic mass unit (1 amu) and an energy of 931.48 MeV. This conversion factor comes from the equation $E = m_0c^2$, where m_0 is the rest mass and c is the velocity of light. An example of a decay scheme for a positron emitter is shown in Fig. 17-6e. The shape of the positron spectrum would be similar to that in Fig. 17-6a.

When the available energy is less than 1.02 MeV, the minimum (neglecting recoil energy) for β^+ emission, a proton can be transformed into a neutron by electron capture (EC), the third type of beta decay. The process is

$$^{A}_{Z}\mathrm{El} + e^- \rightarrow {}^{A}_{Z-1}\mathrm{El} + {}^{0}_{0}\nu \qquad (17\text{-}16)$$

where the electron, e.g., a K-electron, is in the unstable atom. As for the other types of beta decay, the product nuclide may be formed in an excited state that emits gamma radiation. If there is no gamma ray emission, the detectable ionizing radiation emitted in the electron capture process consists of X rays that are characteristic of the *product* atom, because it is the source of the X rays. An example of a decay scheme illustrating EC is shown in Fig 17-6f. Because the process can compete with positron emission, decay schemes often show both processes as in Fig. 17-6g.

17.3.2 Alpha Decay

The process of alpha decay is represented by

$$^{A}_{Z}El \rightarrow {}^{A-4}_{Z-2}El + {}^{4}_{2}He \qquad (17\text{-}17)$$

The alpha particle $^{4}He^{2+}$ is a highly ionizing type of radiation. For a given transition between specified energy levels, the alpha particles are monoenergetic. As for the other modes of decay, monoenergetic gamma rays may be emitted. An example of a decay scheme is shown in Fig. 17-6h.

17.3.3 Isomeric Transition

Deexcitation of a nuclear isomer or metastable state can be represented simply as

$$^{Am}_{Z}El \rightarrow {}^{A}_{Z}El + \gamma \qquad (17\text{-}18)$$

There is no change in atomic number. In isomeric transition (IT) or in any of the already mentioned gamma emissions processes, a competitive process known as internal conversion may occur and, in fact, may predominate. In such cases the excited nuclide emits one or more groups of monoenergetic electrons, called conversion electrons and represented by e^- to indicate that their origin is outside the nucleus. These electrons have energy equal to that which the photon would have minus the binding energy of the electron, e.g., the binding energy of a K- or L-electron. In either case the photons and the electrons are ionizing radiation. A simple decay scheme for IT is shown in Fig. 17-6i.

17.3.4 Neutron Emissions

A few of the products of nuclear fission, e.g., ^{87}Kr and ^{137}Xe, are short-lived neutron emitters. Fission products are formed with an excess of neutrons relative to the number of protons needed for stability. When a nuclide such as $^{236}_{92}U$ (formed when $^{235}_{92}U$ captures a neutron) undergoes nuclear fission, the fission fragments are isotopes of elements having atomic numbers between 30 and 66, as discussed in Section 17.8.3. The ratio of neutrons to protons for stable isotopes of the elements above calcium increases with increasing Z. This ratio is, therefore, less for stable isotopes of the fission product elements than for $^{236}_{92}U$. The fission fragments formed in a fission event and their daughters, the fission products, have a nuclear composition that is neutron rich and is more nearly characteristic of the fissioned nuclide. Normally the excess neutrons are converted to protons by β^- decay. Usually several successive β^- decay steps are required to reach a stable composition; hence the existence of fission product chains.

17.3.5 Decay Schemes

The decay schemes shown in Fig. 17-6 are schematic. The schemes for two specific, long-lived fission products, ^{137}Cs and ^{90}Sr, are shown in Fig. 17-7. In a sample of ^{137}Cs each disintegration results in the emission of a β^-, but 93.5% of the β^- disintegrations are to the metastable ^{137m}Ba, which has a half-life of 2.55 min. The latter decays by isomeric transition with emission of the 0.662 MeV gamma ray attributed to ^{137}Cs sources. For ^{137}Cs sources having an age of about 25 min, the ^{137m}Ba is in secular equilibrum. Direct decay to the ground state of ^{137}Ba occurs in 6.5% of the cases. The value of Q_{β^-} is 1.176 MeV. When considered as a source of gamma rays (from the daughter), only 93.5% of the disintegrations of ^{137}Cs could result in emission of a gamma ray. Thus, a one curie source of ^{137}Cs would, according to the decay scheme, emit $0.935 \times 3.700 \times 10^{10}$ or 3.460×10^{10} photons/sec. However, the internal conversion ratio e^-/γ is 0.11. This information is not generally included in the energy level diagram. The resultant gamma emission rate for a one curie ^{137}Cs source is, therefore, 3.117×10^{10}/sec. This is the value needed when considering the hazard of a one curie source of ^{137}Cs as a gamma emitter, apart from the hazard as a negatron emitter.

The energy level diagram for $^{90}_{38}Sr$ includes that of the 64 hr daughter $^{90}_{39}Y$. The latter emits beta radiation having

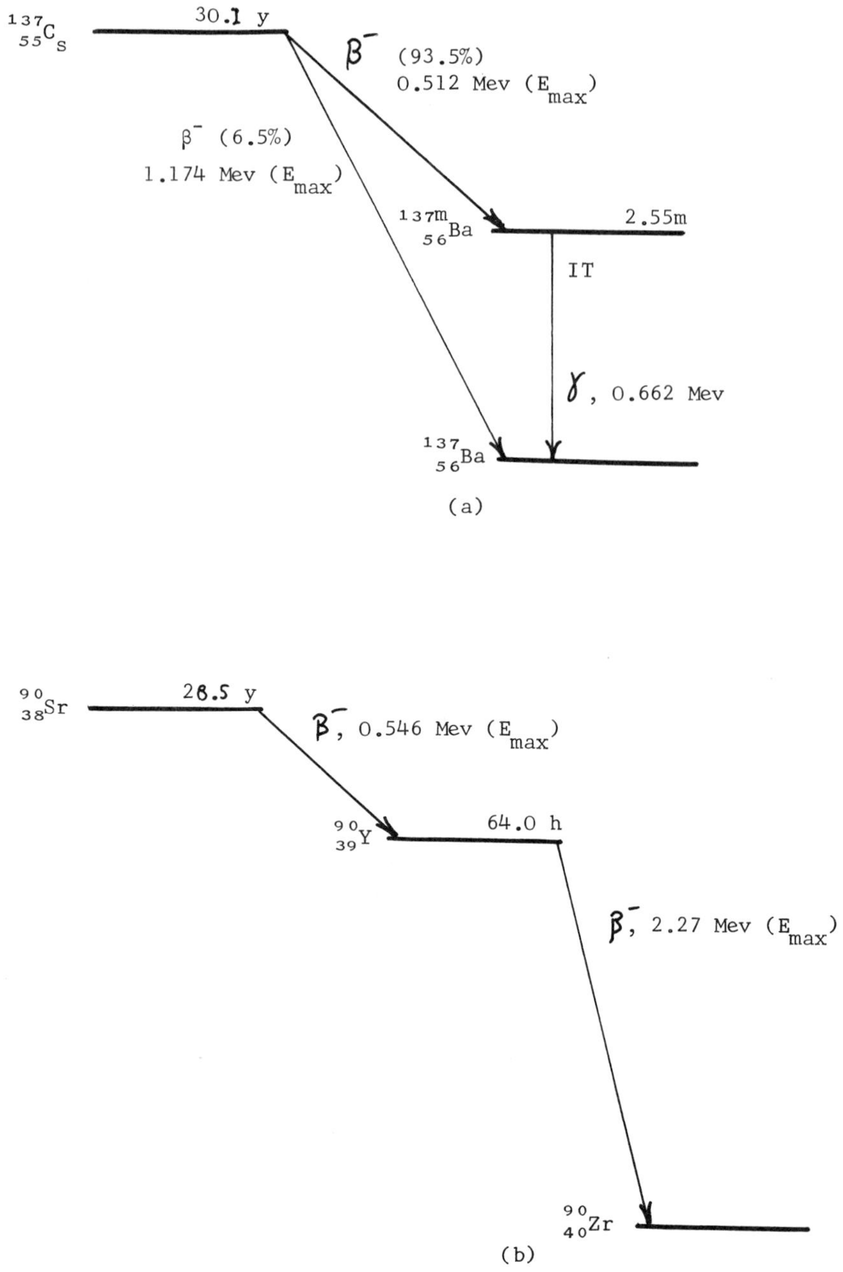

FIGURE 17-7. Decay schemes for (a) ^{137}Cs and (b) ^{90}Sr-^{90}Y.

a high E_{max} and a negligible fraction (0.02%) decays to an excited state of $^{90}_{40}Zr$. In terms of radiation hazard, strontum-90 and its daughter are, therefore, considered to be sources of only negatrons.

17.4 INTERACTIONS OF IONIZING RADIATION WITH MATTER

17.4.1 Charged Particles

Charged particles such as electrons e^- and e^+, beta rays β^- and β^+ (electrons emitted by a radioactive nucleus), and alpha particles $^4He^{2+}$, interact electrostatically with matter, e.g., air, radiation detectors, shielding, body tissue, etc. The interaction is predominately with electrons in atoms, although interaction with atomic nuclei becomes increasingly important at high energies, e.g., 10 MeV or higher. In the electrostatic interaction, the charged particles transfer energy to the absorbing medium in two ways, namely, by producing ion-pairs (an electron and a positively charged ion) and by causing electronic excitation of atoms and molecules. Eventually, the energy transferred to the absorbing medium is degraded to translational (thermal) energy although some may be converted to chemical energy (resulting from the breaking of chemical bonds), depending upon the chemical nature of the medium. High energy electrons can also radiate a fraction of their energy as electromagnetic radiation (bremsstrahlung) having a continuous energy spectrum. This effect is important in connection with the shielding of high-level, beta-emitting radioactive sources, and high-energy electron accelerators.

17.4.2 Gamma Rays and X Rays

Gamma rays (high energy protons originating in a nucleus) interact with matter in four ways: (1) photoelectric absorption, (2) Compton scattering, (3) pair production, and (4) photonuclear reaction. Only the first three modes of interaction of electromagnetic radiation are significant for our purposes. In each of these modes electrons are released or produced with kinetic energy by the gamma ray. In the photoelectric process, the primary photon, i.e., the photon incident upon the absorbing medium, is absorbed and transfers all of its energy to the absorbing atom by interacting with a bound electron in an inner shell. This photoelectron is ejected from the absorbing atom with energy equal to that of the absorbed photon minus an amount equal to the binding energy of the electron. The photoelectron may escape from the absorbing medium or it may lose its energy by causing ionization and electronic excitation within the medium. Because of the vacancy created in a shell of the absorbing atom, the atom emits characteristic X rays as it would if such a vacancy were produced in an atom of the target of an X ray tube.

In the Compton process, some of the primary photon energy is transferred to an electron (not a firmly bound electron) during a scattering interaction in the absorbing medium. The scattered photon is degraded in energy relative to the primary photon by an amount equal to that received by the Compton electron. The distribution of energy during the interaction can be calculated as a function of the scattering angle taken relative to the direction of the primary photon. The scattered photon may undergo additional Compton scattering or it may undergo photoelectric absorption. The latter becomes more probable with decreasing photon energy. For energies of about 1 MeV or so, the Compton effect predominates.

In the third type of γ-ray interaction, the primary photon transfers all of its energy to the absorbing medium in the vicinity of a nucleus and in its place a pair (not a ion-pair) is *created*. The pair consists of a positron e^+ and a negatron e^-. For this process to occur, the primary photon must have at least sufficient energy to create the pair, i.e., 2×0.51 MeV. Any excess above the threshold energy of 1.02 MeV appears as kinetic energy of the e^+ and the e^-. Eventually, the e^+ is annihilated by reaction with an electron to produce two photons, each having energy 0.51 MeV. These photons then behave as primary photons and interact by the Compton and the photoelectric processes.

In one way or another, these interaction processes for γ rays lead to the emission of characteristic X rays that are equivalent to γ rays in being electromagnetic radiation but have much lower energies in the keV range and interact mainly by the photoelectric process. Although the overall degradation of γ-ray energy in an absorber is obviously a complex sequence of events, it can be analyzed in terms of the principle of the conservation of energy and these modes of interaction.

X rays originate when deexcitation involving inner electrons occurs in an atom or when high energy electrons are rapidly decelerated in passing through matter. In the former case, the X rays are called characteristic because their energies are determined by the energy level spacings that are characteristic of the chemical element emitting the X rays. In the latter case, the X rays have a continuous rather than discrete spectrum and are known as "white" X rays when produced in X-ray tubes and as bremsstrahlen when produced in high energy (MeV range) electron accelerators or in the absorption of beta rays in matter. X rays are emitted following the electron capture process, as pointed out in Section 17.3.1, and in most atoms following a nuclear transition. The intensity of the X radiation varies widely from one radionuclide to another. In general, most X rays of interest in this chapter will be in the energy of 10-100 keV. Since they are identical with gamma rays in being electromagnetic radiation, their modes of interaction with matter are those described above for gamma rays. Because of the relatively low energy of characteristic X ray, however, the predominate mechanism will be the photoelectric effect.

17.4.3 Neutrons

Free neutrons are obtained by liberating them from nuclei by nuclear reactions such as (γ,n), (p,n), and fission. When ejected from nuclei, neutrons will generally have kinetic energies in the range of from about 1 MeV to several MeV. Because of their high speed, such neutrons are called fast neutrons. As neutral particles with a mass close to that of a proton, they are scattered by atomic nuclei as they diffuse through matter. Usually the scattering is elastic. The neutron loses energy in a sequence of collisions and slows down. The medium in which the slowing down process occurs is called a moderator. When a neutron has been slowed down so that its energy is about 1 eV, it is called a thermal neutron. Thermal neutrons can be in temperature equilibrum with their surroundings and have a Maxwell-Boltzmann energy distribution. In many ways they resemble the molecules of an ideal gas. Thus, at 300°K their average kinetic energy is 3/2 kT or 0.039 eV.

Fast neutrons can release protons, alpha particles and other charged particles by nuclear reactions such as (n,p) and (n,α). Thermal neutrons are generally captured by nuclei with the release of high energy photons. The (n,γ) reaction is a source of gamma rays in a neutron-absorbing medium.

The nuclear reactions of fast or slow neutrons and the scattering of fast neutrons can impart sufficient recoil energy to the nuclei involved to break chemical bonds and cause dislocations in the lattices of crystalline solids. When H atoms are released as recoil atoms from H_2O or organic compounds, they may have sufficient energy to become protons, which then produce ionization.

Free neutrons are radioactive and decay by negatron emission with a half-life of 10.6 min. This property can be observed for neutrons introduced into a high vacuum, but decay does not normally take place because the free neutrons are quickly captured by nuclei in air, vessel walls, or whatever medium they penetrate.

17.4.4 Penetration of Ionizing Radiation into Matter

In terms of the depth to which ionizing radiation will penetrate into matter, alpha particles have small finite ranges and those from typical alpha emitters travel only a few centimeters in air, e.g., 4-10 cm, before losing all of their kinetic energy in excess of the thermal energy corresponding to the temperature of the absorber. They are stopped by clothing and by the outer layer of a few microns (a few thousand nanometers) of skin. Beta rays also have finite ranges of penetration but will travel up to several meters in air or millimeters in tissue, depending on their energy. Gamma rays have no finite range; they are absorbed exponentially, i.e., their intensity decreases exponentially with distance traveled in an absorbing medium. They can penetrate deeply into the body or even pass through it.

With this interaction behavior in mind, it is helpful to classify sources of ionizing radiation as either external or internal to the body. Thus, external sources of alpha radiation are generally not hazardous unless placed in contact with the skin. It is a simple matter to use air or a thin container to absorb the radiation. Similarly, external sources of beta radiation can generally be shielded rather easily. The external sources of greater concern, then, are those consisting of gamma emitters. As discussed above, commonly, but not always, gamma radiation is emitted along with alpha or beta radiation.

Internal sources of ionizing radiation arise from ingestion, inhalation, or absorption (through a break in the skin) of radioactive material. The radiation is released within the body and the energy is transferred to nearby tissue. Alpha radiation becomes the most dangerous because, as we shall see, it transfers all of its energy and concentrates the radiation damage within a small volume of tissue.

17.4.5 Radiation Chemistry

It is perhaps appropriate at this point to compare briefly the interaction and the effects of the interaction of ionizing radiation with matter with those of light discussed in Chapter 9. It should be recalled that when light in the wavelength range of about 2000-10,000 Å (200-1000 nm) is absorbed, one photon transfers its energy to one molecule in the interaction or primary step. When absorption occurs, the energy of the photon corresponds to the energy needed to excite the molecule from a lower discrete level to a higher discrete level. For light in the visible or ultraviolet region, sufficient energy is available to cause electronic transitions. Ionization requires more energy, e.g., 15.5 eV to form N_2^+ and 12.5 eV to form O_2^+, corresponding to light having a wavelength of about 100 nm.

The electronically excited molecules will undergo one or more of a variety of processes or secondary steps. For example, a molecule may undergo deexcitation by the processes of fluorescence or phosphorescence, it may undergo photolysis, i.e., dissociate into chemically reactive species, or it may collide with other molecules thereby transferring energy and possibly undergoing chemical reaction.

Charged particles and gamma rays with energy in the range of thousands to millions of electron volts can readily produce ionization in the primary interaction step. In fact, a single alpha particle may, depending on its energy, undergo 10^5 or so ionizing events in passing through air before becoming a normal helium atom in thermal equilibrium with its surroundings. Although a gamma ray may disappear in a single primary ionization event, the electrons or the positron-electron pairs released into the absorbing medium will produce many additional ion pairs.

The chemical changes that occur as a consequence of ionization are studied in the branch of chemistry called radiation chemistry. Electronically excited molecules are also produced directly by ionizing radiation and indirectly when an ion is neutralized. Many of the observed chemical changes are the same as those involved in photochemistry. Dissociation into fragments is called radiolysis rather than photolysis. As a point of contrast, ionizing radiation, which essentially transfers its energy by brute force, can populate a wider range of excited states than light can. Furthermore, the reactions of ionic species are dominant in radiation chemistry, but are uncommon in photochemistry.

17.5 DOSIMETRY OF IONIZING RADIATION

The damage (chemical and biological) produced in a system by ionizing radiation is related to the energy absorbed by the system. Historically, the concept of radiation dose was developed to provide a quantitative measure of the dose of X rays or γ rays (e.g., from radium) used for medical diagnosis and therapy. The original unit of dose, the roentgen, has now been designated a unit of exposure, because it is defined in terms of the ionization produced in air at a point in space that is exposed to a source of X or γ rays rather than in terms of energy absorbed by a biological system at the same point.

One roentgen (1 R) is defined as that quantity of X or γ radiation such that the associated corpuscular emission per 0.001293 gm of air produces, in air, ions carrying one electrostatic unit of quantity of electricity of either sign. The associated corpuscular emission refers to the electrons released in air by the X or γ radiation. These energetic electrons produce ion pairs as they travel through air. The quantity of air specified is the weight of dry air in 1 cm^3 at STP. Accordingly, one roentgen is equivalent to 2.08×10^9 ion pairs/cm^3 of air, 1.6×10^{12} ion pairs/gm of air, and 2.58×10^{-4} Coulomb/kg of air. Time does not enter into the definition. Radiation detectors, e.g., Geiger-Müller counters, ion chambers, etc., can easily be used in portable radiation survey instruments that can be calibrated in roentgens or milliroentgens per hour by using a standard source of a gamma emitter.

The absorbed energy equivalent of 1 R can be calculated from w, the average energy lost by ionizing radiation in the process of forming an ion pair in air. An average value for w is obtained by measuring the number of ion pairs formed and dividing this number into the energy of the radiation that produced the ion pairs. A commonly used value for w is 34 eV/ion pair formed in air. Using this value and the conversion factor for electron volts to ergs, we find that 1 R = 87 ergs/gm of air. Because the absorption of ionizing radiation is dependent upon the atomic number of the absorbing atoms, we would expect that the energy absorbed per gram of biological tissue at a point where the exposure dose is 1 R would be somewhat different from 87 ergs. To deal with the energy absorbed in systems other than air, a unit of absorbed dose called the rad was introduced. One rad is equal to 100 ergs/gm (0.01 J/kg) of absorber. Then, the absorbed dose D is calculated from

$$D = (\text{ergs absorbed/gm of matter})/100 \qquad (17\text{-}19)$$

Biological effects depend not only on the energy absorbed per gram, but also on the kind and energy of the incident radiation. In order to compare effects produced for a variety of conditions, the dose equivalent DE was defined. The unit of DE is the rem (roentgen equivalent man). It is calculated from

$$DE = D \times \text{(modifying factors)} \tag{17-20}$$

$$= D \times (QF \times DF \times \text{other factors})$$

where *QF* is the quality factor and depends on the energy deposited per unit track length, i.e., the linear energy transfer LET. Units that have been used for LET include ergs/cm, keV/μm, and eV/nm. The *QF* is similar to the RBE, the relative biological effectiveness, formerly used to calculate rems from rads. The second factor *DF* in Eq. (17-20) is the distribution factor. It takes into account special effects associated with the localization of internal sources of radionuclides. Table 17-5 contains typical values of *QF*. The dependence of *QF* on LET is shown in Table 17-6. The data in these two tables can be related to the range of an ionizing particle and to the variation of specific ionization along the track. Thus alpha particles have a large LET and a short range. LET increases along a track as a charged particle slows down.

An exposure of 1 R results in absorption of about 93 ergs/gm of tissue. Then, 1 R is approximately 0.93 rad and for X and γ rays with *QF* = 1, 1 R is approximately 0.93 rem.

TABLE 17-5

QF for Several Types of Radiation

Radiation	*QF*
γ *rays*	*1*
X rays	*1*
Beta rays[a] (β^-, β^+) *and* e^-	
$< 0.03E_{max}$	*1.7*
$> 0.03E_{max}$	*1*
Alpha particles	*10*
Recoil nuclei	*20*
Neutrons	
thermal	*3*
10 MeV	*6.5*

[a] *Beta rays have a continuous spectrum with an upper energy limit* E_{max} *and an average energy 30-40% of* E_{max}.

TABLE 17-6.

Dependence of QF on LET

Ion pairs/ μm of H_2O	*LET (keV/μm)*	*QF*
100	*3.5 or less*	*1*
100 - 200	*3.5 - 7.0*	*1 - 2*
200 - 650	*7.0 - 23*	*2 - 5*
650 - 1500	*23 - 53*	*5 - 10*
1500 - 5000	*53 - 175*	*10 - 20*

It is often adequate to take 1 R = 1 rem for radiation with unity *QF*. By contrast, the absorption of 100 erg/gm of tissue from alpha radiation would result in a dose equivalent of 10 rem or, considering the absorbed dose required to produce the dose equivalent of 1 rem of γ rays only 0.1 rad of alpha radiation would be required.

Returning to ^{137}Cs as an example, the exposure rate at one meter from a 1 Ci *point* source is 0.346 R/hr. For a point source, the exposure rate falls off inversely as the *square* of the distance from the source if absorption along the path is neglected. This follows from geometrical considerations. The number of photons originating at the center of a sphere that pass through a unit area of the surface of the sphere varies inversely with the square of the radius as the radius of the sphere is changed. We are dealing with a specific example of the "inverse square law." The law applies to sources that can be treated as a mathematical point. That is, the actual size and shape of the source can be neglected when the source is very small and the distance from the source is very great relative to the dimensions of the source. Thus, a 1 cm sphere or a 1 × 1 cm cylinder suspended a few meters from any large object that could backscatter radiation would behave as a point source at distances of a meter or more. On the other hand, for a large plane such as the soil of the earth's surface or a large cylinder such as a nuclear power reactor, exposure would not decrease according to the inverse square law for distances of meters or hundreds of meters. The exposure would decrease only for distances sufficient to result in significant absorption of the radiation in air (assuming there are no objects in the line of sight that could act as shielding).

As an approximation, we can calculate the exposure at 1 ft from a point source of a gamma emitter by the equation

$$\text{R/hr at 1 ft} = 6nCE \tag{17-21}$$

where C is the number of curies, E the energy (MeV) of the gamma ray for which the exposure is calculated, and n the fraction of disintegrations in which the gamma ray is emitted. When several gamma rays are emitted (a rather common situation), the individual gamma-ray contributions are summed. To properly use Eq. (17-21) or a more exact one, it is necessary to know the disintegration scheme for the radionuclide.

Another equation that can be used to estimate the exposure to a point source of γ rays is

$$\text{R/hr per mCi, at 1 meter} = 1.50 \times 10^5 \mu_a E_\gamma \tag{17-22}$$

where μ_a is the absorption coefficient of the γ ray for air and E_γ is the γ ray energy in MeV. For E_γ between 0.1 and 2 MeV, an energy range that brackets the values commonly associated with radioactive sources, μ_a for air is approximately 3.5×10^{-5}/cm. As written, the equation is based on the assumption that the one γ ray for which the calculation is being made is emitted in 100% of the disintegrations.

Reduction in the exposure rate and, therefore, the absorbed dose rate from a given point source can be achieved by increasing the distance from the source, thereby taking advantage of the inverse square law, or by shielding the source. If a *collimated* beam of γ rays having an intensity I_0 is perpendicularly incident on an attenuator (usually called the absorber), e.g., a sheet of lead, of thickness x and if the emerging beam intensity I is measured after passing through another collimator, the attenuation is given by

$$I = I_0 \exp(-\mu_0 x) \tag{17-23}$$

If x is in cm, μ_0 is the linear attenuation coefficient, and has the units of cm^{-1}. Other units used are gm cm^{-2} (density-thickness) for x and $\text{cm}^{-2}\,\text{gm}^{-1}$ for μ_0, which is then called the mass attenuation coefficient. Density thickness is simply the product of the absorber density ρ times the linear thickness of the absorber. The unit is relatively independent of the atomic number of the absorber because for a given thickness in gm cm^{-2}, all materials have about the same number of electrons that can interact with the ionizing radiation. An attenuation coefficient is the sum of three specific coefficients: one each for the photoelectric, Compton, and pair production processes.

Equation (17-23) can be derived by integration of Eq. (9-5) for the absorption of light, but with αc replaced by μ. Equation (17-23) can be expressed in terms of logarithms or, as is more commonly the case, in terms of absorber half-thickness in a manner analogous to the rewriting of Eq. (17-3) in terms of half-life. The half-thickness of an absorber is the thickness in cm or in gm cm^{-2} needed to reduce the beam intensity by a factor of two. Then, $x_{1/2} = 0.693/\mu$, where μ is μ_o or μ_o/ρ. Values for μ and $x_{1/2}$ are available in handbooks in tabular or graphical form as a function of energy for a given absorber material.

As an example, $x_{1/2}$ for lead for the 0.662 MeV γ ray emitted by a source containing ^{137}Cs is 6.5 gm cm^{-2}. Therefore, if a lead shield having a thickness of 13 gm cm^{-2} were placed around the source, 75% of the primary γ rays would be absorbed or scattered from a collimated beam of the 0.662 MeV γ rays by the shield. Taking the density of lead to be 11.3 gm cm^{-3}, we find the linear thickness of the shield to be 1.18 cm. In an exact calculation of the effectiveness of such a shield, a correction would be required for the scattered radiation and secondary radiation, e.g., characteristic X rays of lead, escaping from the outer surface of the shield. The correction, called a buildup factor, is needed because for attenuation the data for μ or $x_{1/2}$ assume collimation that would eliminate scattered radiation. (For a given absorber and a given E_γ, the energy absorption coefficient μ_a is less than the attenuation coefficient.)

17.6 EFFECTS OF IONIZING RADIATION ON THE HUMAN BODY

17.6.1 Introduction

We shall very briefly summarize here some of the effects of ionizing radiation on the human body. These can be divided into two major categories: (1) somatic effects (produced in the body receiving the radiation), and (2) genetic effects (transferred to the next or later generations). When ionizing radiation transfers energy to a biological cell, the effect (damage) to the complex organic molecules such as DNA, RNA, etc., may result from two modes of interaction. *Direct* interaction with such molecules leads to breaking of covalent bonds and the formation of ions and radicals. The complex components may be highly fragmented or may be structurally altered but still functional. In the *indirect* mode of interaction, the ionizing radiation transfers its energy to those molecules, e.g., H_2O, most likely to be in its path. Water undergoes radiolysis to form highly reactive species such as H, $OH^\bullet$, $HO_2^\bullet$, H_2O^- (or $e^-_{H_2O}$, the hydrated electron), and H_2O_2. These species can react with the organic components of the cell to produce changes of the same type as are produced in a direct interaction. Indirect interaction is probably the more likely cause of biological effects.

17.6.2 Acute Somatic Effects

Acute effects result from total body exposure during a very short period of time, e.g., seconds or minutes, as in an accident.

One cannot specify a precise dose at which a particular effect will occur. It varies from person to person and depends somewhat on the health of the individual. However, the following are typical:

(1) About 25 rems: Approximate lower limit of detection by blood analysis. There is a decrease in the number of leukocytes (white blood cells). In order to determine accurately the change, a base blood count must be obtained before exposure to ionizing radiation. Blood cell count returns to normal after a few months.

(2) About 200 rems: Nausea, fatigue, death possible within six weeks or so. Susceptibility to infection begins to increase rapidly beyond this level.

(3) About 400 rems: LD-50 (lethal dose for 50% of those receiving this dose). Blood-forming organs, e.g., bone marrow and spleen, are damaged. Higher doses are generally needed for LD-50 when the exposure rate is low.

17.6.3 Chronic (Long-Term) Somatic Effects

Localized doses of 700-1500 rems of X rays to careless users of X ray machines can produce skin burns and skin cancer. Similarly, doses of 600 rems or more of any kind of ionizing radiation to the eyes can cause cataracts. The chronic whole body external dose needed to cause leukemia is not established. A dose of about 200 rems can produce temporary sterility. Whole body doses sufficient to produce permanent sterility are also in the lethal range.

In general, rapidly dividing cells are the most radiation sensitive. The bone of children (and especially of a fetus) is, therefore, relatively sensitive to ionizing radiation. Growth can be inhibited.

Chronic effects from internal sources of ionizing radiation arise mainly from radionuclides having a long effective half-life T_{eff} in the body. The effective half-life T_{eff} of a radionuclide in a biological system is defined in terms of the radioactivity half-life (customarily written as T_{rad} in this context) and the biological half-life T_{biol}. Thus,

$$T_{eff} = T_{rad}\, T_{biol} / (T_{rad} + T_{biol}) \qquad (17\text{-}24)$$

The biological half-life is the time for the body to eliminate one-half of an administered or otherwise acquired dosage of any substance by regular processes of elimination. It is taken to be the same for all isotopes of a given element. A few examples of T_{biol} are given in Table 17-7. The effective half-life is determined mainly by the shorter half-life when the two half-lives are very different. Thus for ^{32}P with a 14.3 day radioactivity half-life, and a 3.16 year biological half-life in bone, T_{eff} in bone is 14.1 days. A radionuclide that deposits selectively in an organ or in bone and has a long radioactivity half-life is particularly hazardous. Iodine, which concentrates in the thyroid and can produce nodules and cancer, is a particularly hazardous fission product element. Among the isotopes of iodine produced in fission are ^{129}I(1.59 × 10^7 yr), ^{131}I(8.04 day), ^{132}I(2.29 hr), ^{133}I(20.8 hr), and ^{135}I(6.59 hr). The elements Ca, Sr, Ra, and Pu are examples of bone seekers, i.e., substances that deposit preferentially in bone. Radioisotopes of these elements can cause bone tumors and bone cancer and leukemia. Strontium-90 ($t_{1/2}$ = 28.5 yr) is an example of a fission product in this category. As additional examples, the two naturally occurring isotopes of radium, ^{226}Ra ($t_{1/2}$ = 1600 yr and a member of the uranium series) and ^{228}Ra ($t_{1/2}$ = 5.75 yr and a member of the thorium series) must be handled very carefully to avoid ingestion. The longer-lived alpha-emitting isotope, the one that is implied when we refer to radium, is the relatively more dangerous isotope of the two. On an equal weight basis, however, the beta-emitting ^{228}Ra has a much higher disintegration rate. In some instances the two isotopes have been used as a mixture. Finally, ^{239}Pu has been used in nuclear weapons and it has an essential role in the fuel cycle of one of the proposed breeder power reactors (see Section 17.8.6).

17.6.4 Genetic Effects

Ionizing radiation is a mutagen, i.e., it is an agent that can cause gene mutations and chromosome aberrations for any exposure.

As an estimate, perhaps 2-10% of the spontaneous mutations that occur can be associated with exposure to the natural radiation background for about 50 yr. The genetic effects produced by radiation are not unique and, therefore, are not distinguishable from effects produced by other mutagens.

TABLE 17-7

Biological Half-Life of Selected Elements

Element	Biological half-life
H	12 day (total body)
C	10 day (total body)
P	257 day (total body) 3.16 yr (bone)
Ca	about 50 yr (bone)
Fe	2.2 yr (total body)
Zn	2.5 yr (total body) 5.4 yr (muscle) 3.6 yr (bone)
Sr	35 yr (total body) 50 yr (bone)
I	138 day (total body) 138 day (thyroid)
Cs	70 day (total body) 140 day (muscle)
Rare earths	about 2 yr (total body)
Po	30 day (total body) 70 day (kidneys)
Ra	22 yr (total body) 44 yr (bone)
Th	150 yr (total body) 200 yr (bone) 150 yr (liver)
U	100 day (total body) 300 day (bone)
Pu	180 yr (total body) 200 yr (bone) 80 yr (kidneys) 80 yr (bone)

17.6.5 Correlations between Dose and Effect

Most of the information on the relationship between dose and health effects for man caused by internal or external exposure to various types of ionizing radiation has been obtained indirectly by means of laboratory experiments with insects and with animals such as mice and rats. In a number of instances, data have been obtained directly in connection with medical exposure, X ray exposure received by radiologists, radiation exposure accidents, and population studies involving statistically significant numbers of people exposed in known ways that can be documented. In some cases the population studies have provided bases for maximum permissible levels of exposure or of body burden for radioactive material deposited in the body. It may be of interest to note a few examples of population studies.

The potential medical value of radium was recognized in the early 1900s. It was used successfully, for example, to cure carcinoma. Unfortunately, it was also administered as a tonic and a cure-all before the hazards were recognized. "Radium water," both as a commercial preparation and as mineral water with above average radium content, was promoted for its alleged medicinal value. Some of the people who acquired a body burden of radium developed bone tumors after a period of 10-20 yr. It was possible in some instances to relate the incidence of bone tumor with the radium content of the bone.

Even before the discovery of radioactivity, it had been observed that uranium miners in Europe had an unusually high incidence of lung cancer. The cause was not known except that it was related to the material being mined. The uranium was mined because its compounds are highly colored and were useful as pigments. Even until recently its compounds were used to produce orange and green glaze on dinnerware. Today uranium is mined in the United States and in many parts of the world because of its value as a source of the energy of nuclear fission. Studies have shown that the lung cancer problem is caused by both radon gas ^{222}Rn and its daughters ^{218}Po, ^{214}Pb, ^{214}Bi, and ^{214}Po. The daughters are more dangerous than the radon because they are not gases and attach themselves to particulate matter that is retained by the lung after inhalation. Although the concentrations of these substances vary widely from mine to mine, they are generally about 4000 pCi/liter of air for radon and about 1000 pCi/liter of air for the daughters. Standards have been set for safe operation of uranium mines.

Another population study that has been underway for many years involves a group of workers who were employed during the period of about 1915 to 1925 to paint clock and watch dials with luminous paint containing radium salts. Ingestion of radium resulted from the practice of the workers to use their lips to form a fine tip on a paint brush. Some of the workers developed ill effects such as bone tumors, bone cancer, and anemia after a latent period of several years. There have been several fatalities attributable to radium. Studies have related the various types of injury to the radium level in the bones of those having the injury.

For a period of about 15 yr prior to 1945, several radiologists administered "Thoratrast" (a colloidal suspension of ThO_2 containing the alpha emitter ^{232}Th) to patients in order to improve the images obtained in the diagnostic use of X rays. These patients have been studied because after 10-20 yr some of them developed malignancies at the site of injection or malignancies of the liver.

Finally, mention should be made of population studies involving nuclear weapons. One is an evaluation of the effects of the exposure of the populations of Hiroshima and Nagasaki to gamma rays and neutrons in 1945. A second is an evaluation of the effects of external and internal exposure on the people on the Marshall Islands who were accidentally exposed to radioactive fallout from a hydrogen bomb in 1954. A third is concerned with possible long-term effects for persons who were observers of weapons tests.

17.7 THE NATURAL BACKGROUND OF IONIZING RADIATION

17.7.1 Sources of Natural Background Radiation

We are constantly exposed to a natural background of ionizing radiation. The external portion of this background consists of cosmic radiation and radiation from radionuclides present in all materials of construction, soil, rocks, air, water, and food. In addition, an internal contribution arises from several naturally occurring radionuclides that are normal constituents of our bodies. These sources are discussed individually below.

There are some 75 radionuclides that occur in nature. These are scattered throughout the periodic table from hydrogen to uranium but most have an atomic number greater than 82. Only a relatively small number contribute significantly to the natural background. Others can be neglected because of a very low abundance in nature, i.e., a low abundance of the element or the radioisotope or both, or because of an exceedingly long half-life, e.g., 10^{15} yr, and, therefore, a very low specific activity.

17.7.2 Cosmic Ray Background

Most of the cosmic rays to which man is exposed reach the surface of the earth as secondary radiation produced in the earth's atmosphere by the interaction of primary cosmic rays with nuclei of nitrogen and oxygen atoms. Except for a small contribution from the sun, the primary cosmic rays (primaries) are of galactic origin. About 77% of the primaries are protons and about 20% are ^{4}He. The remainder consists mainly of nuclei of elements heavier than helium plus a small contribution from electrons and positrons. The energy spectrum of the primaries is from about 1 MeV to about 10^{14} MeV (10^{5} GeV). At the top of the earth's atmosphere, the incoming flux is about 1 primary cosmic ray/cm^2/sec with an average energy of about 2 GeV. When these high energy particles collide with nuclei in the earth's atmosphere, they produce showers of nuclear debris. Thus, protons with energy over about 8 MeV can eject neutrons from nuclei by the (p,n) reaction and if the energy is over about 300 MeV, pions can be created. Pions may have positive, negative, or zero charge as indicated by the symbols π^+, π^-, and π^0. For a charged pion, the rest mass is 273 times greater than the rest mass of an electron; for a neutral one, 264 times greater. Pions may interact with nuclei or they may decay into muons (rest mass 206.7 times that of an electron). A positive pion can decay into a positive muon μ^+ plus a neutrino; a negative pion into a negative muon μ^- plus a neutrino; and a neutral one, into two photons. A positive muon can decay into a positron plus two neutrinos; a negative muon into an electron and two neutrinos.

Neutrons released in the atmosphere by primary cosmic rays may escape into space, may reach the surface of the earth, or may produce ^{3}H and ^{14}C in the atmosphere by the previously mentioned (n,t) and (n,p) reactions on ^{14}N, respectively. The ^{14}C, in the form of CO_2, exchanges with carbon in the other reservoirs of the earth and reaches an equilibrium concentration with living matter. The average equilibrium value, 13.6 disintegrations of ^{14}C per min per gram of carbon, varies a few percent from reservoir to reservoir of carbon. Assuming that the primary cosmic ray intensity has remained constant for tens of thousands of years, one can use the ^{14}C content of an object to determine the time at which it ceased to be in equilibrium with its surroundings. This is the basis for the ^{14}C dating method for objects having an age of up to about 50,000 yr. Some relatively short-term fluctuations in the ^{14}C equilibrium value have been observed for objects whose age could be determined by other methods. On the long-term average, however, the primary cosmic ray flux seems to have been constant.

The secondary cosmic radiation at sea level is commonly characterized as having two components. The "hard" (penetrating) component consists of muons having energy in the GeV range, fast neutrons, neutrinos, and a small percentage of primaries. The "soft" component consists of electrons, positrons and photons, all having energies from a fraction of an MeV to a few MeV. The intensity of cosmic rays at sea level varies with latitude. It increases about 10% in going from zero to 50° geomagnetic latitude because of the interaction of the earth's magnetic field with the various charged particles contained in primary and secondary cosmic radiation. The intensity decreases with increasing barometric pressure because of increased absorption of the secondary radiation in the atmosphere with increasing density.

At sea level the dose rate from cosmic-ray produced ionizing radiation is about 30 - 50 mrems/yr. The value is, of course, subject to some variation with latitude. An additional contribution from slow neutrons (about 8 slow neutrons/ cm^2 / sec) adds about 5 mrem per year. The dose rate from cosmic rays approximately doubles in going from sea level to an altitude of 10,000 ft. This effect results from a combination of a decrease in absorption and an increase in the relative number of high energy primaries with an increase in altitude. Passengers in jet planes and residents of cities located at relatively high elevations receive greater than sea-level doses. Cosmic radiation must be considered as one of the hazards of space travel.

17.7.3 Distribution of Naturally Occurring Radionuclides in the Environment

(a) Rocks and Soils. Granitic rocks of the upper part of the earth's crust contain average concentrations of 3-4 ppm of uranium (as the element), 10-15 ppm of thorium, and 2-6% of potassium (0.0118 atom % ^{40}K). The concentrations are generally less in basic igneous works, sedimentary rocks, and limestones, but can, of course, be much higher in ores. Based on the average concentration in the crust, the ratio Th/U is about 4 and, for many rocks, this ratio varies roughly in the same way as the potassium content.

Natural uranium consists of three long-lived isotopes, namely, ^{238}U (99.274%), ^{235}U (0.7205%), and ^{234}U (0.0056%). The ^{238}U is the parent of the uranium series in which ^{234}U is one of the daughters. The ^{235}U is the parent of the actinium series, which terminates in stable ^{207}Pb. For thorium the long-lived isotope is ^{232}Th, the parent of the thorium series, which terminates in stable ^{208}Pb.

Each of the three radioactive series contains an isotope of radon. Specifically,

$$^{238}U \rightarrow \dashrightarrow \; ^{222}Rn \; \underset{3.824\ day}{\overset{\alpha}{\rightarrow}} \; ^{218}Po \rightarrow \text{---} \; \text{products}$$

$$^{232}Th \rightarrow \dashrightarrow \; ^{220}Rn \; \underset{55\ sec}{\overset{\alpha}{\rightarrow}} \; ^{216}Po \rightarrow \text{---} \; \text{products}$$

$$^{235}U \rightarrow \dashrightarrow \; ^{219}Rn \; \underset{4.0\ sec}{\overset{\alpha}{\rightarrow}} \; ^{215}Po \rightarrow \text{---} \; \text{products}$$

As a noble gas, radon can diffuse out of rocks, especially porous rocks, and out of soil. Diffusion is limited by the half life for the shorter-lived isotopes of Rn. Radon-222 migrates into the atmosphere and into water supplies in which the source water comes into contact with rocks or sediments containing uranium. For some soils, essentially all of the radon can escape (can be emanated).

Soil is a source of γ rays emitted by members of the three series. Generally the γ rays that escape are emitted in the top half-meter depth of soil. Escape of radon from soil results in loss of γ-emitting daughters of radon, such as 19.8 min ^{214}Bi from ^{222}Rn; 60.6 min ^{212}Bi and 3.05 min ^{208}Tl from ^{220}Rn; and 36.1 min ^{211}Pb from ^{219}Rn. If there is no wind, these daughters are formed from the escaped Rn isotopes just above ground level and cause an increase in the γ ray intensity near the ground. Air, of course, provides much less attenuation of γ radiation for a given linear thickness than does soil.

Various factors can affect the radon levels in soil. In the winter when the soil is frozen and covered with snow, the γ ray intensity in the soil is increased. On the other hand, when the soil is warm, heavy rain can wash the radon to lower levels. Variations in atmospheric pressure can alter the concentration of radon to depths of several feet.

The dose at 1 m from granitic rocks and soils due to γ rays from thorium, uranium, and potassium is commonly taken to be in the range of 75-200 mrad/yr. (In sections of Brazil and India the value may be 100 times greater because of monazite sands consisting of phosphates of rare earths and thorium.) The uranium series contributes about one-fourth of the total and ^{40}K contributes about as much as the thorium series. Uranium tends to migrate to lower depths of soil than thorium because it forms water-soluble compounds more readily, while significant amounts of ^{222}Rn from the uranium series can be lost by emanation. Both factors tend to lower the uranium series contribution to the background from soil. Although spontaneous fission of ^{232}Th, ^{235}U, and ^{238}U does occur, the contribution of this process to the background is neglibible.

(b) Materials Used for Building Construction. Buildings can provide some shielding from extraterrestrial radiation and from γ radiation from rocks and soil. Certain materials of construction, e.g., stone containing uranium or a high potassium content, can, however, raise the dose received by the occupants of a building. Dosages vary depending on the location of the building, but approximate values might be 50 mrad/yr for a wooden house, 70 mrad/yr for a brick or concrete house, and 100 mrad/yr for a stone (granite) house.

Granite, marble, sandstone, and other stones typically used for building construction contain insufficient uranium to make them practical sources of this element. In recent years, however, a problem has arisen from the use of the tailings from uranium ore processing plants in Colorado, Utah, and seven other Western States. The tailings (in a total amount approaching 100 million tons) have the texture of fine sand and are suitable as base and backfill material under and around concrete floors and basements. The tailings have been used in the construction of thousands of buildings that include private homes, schools, and public and commercial buildings. Although the tailings that remain after the leaching of uranium do not contain sufficient residual uranium to be of interest commercially, they do constitute a relatively concentrated source of ^{226}Ra, ^{222}Rn, and the remaining daughters of the uranium series. The radon can escape into the air in the buildings constructed using the tailings. In addition, the γ-ray background is raised by the decay products remaining in the tailings under or near the buildings. In a sense, the type or radiation hazard that the uranium miner faces (Section 17.6.5) is being shared with others, albeit at a lower level. For tailings, however, long-term genetic effects in later generations are probably more likely than lung cancer. A simple, inexpensive solution to the problem is not apparent. Only removal of the tailings from the buildings will eliminate the hazard.

Potentially, similar problems could arise from the use of residues obtained in the mining and milling of phosphate rock for the production of fertilizer. Some phosphate rock, particularly that found in Florida, contains more than just a trace of uranium. When the rock is treated with acid to form phosphoric acid, the uranium remains with the fertilizer, but the radium precipitates as the sulfate with the calcium sulfate (gypsum) by-product. The gypsum may contain about 25 pCi of ^{226}Ra per gram. Because gypsum is used in the fabrication of wallboard, concern has been expressed that the contaminated gypsum might find its way into buildings.

(c) Air. Radon is the important naturally occurring radioactive component of the atmosphere. Its concentration varies from place to place on the earth and with atmospheric conditions. A mean value for the world-wide concentration of ^{222}Rn, the predominant isotope, is 0.5 pCi/liter. For ^{220}Rn, from the thorium series, the value is 0.02 pCi/liter. The daughters of ^{222}Rn (^{210}Pb and ^{210}Po) attach themselves as atoms to dust particles and are washed out of the air onto vegetation during a rainstorm.

Natural gas used in space heaters and unvented kitchen ranges contains ^{222}Rn that is released into the air in homes using such devices. Natural gas used in furnaces for home heating and used in industry or for power generation is also a direct source of ^{222}Rn in the atmosphere. The radon content of natural gas varies from source to source, but the average for most of the U.S. is about 20 pCi/liter. Natural gas from gas fields in Kansas, Colorado, New Mexico, and the Texas panhandle contain higher levels of about 50 pCi/liter.

It is also of interest to note that coal-fired power plants add not only products of combustion but also radon to the atmosphere. Uranium and its daughters are present in the inorganic fraction of coal that becomes ash.

(d) Water. The important naturally occurring radionuclides in water are ^{226}Ra and its α-emitting daughters ^{222}Rn, ^{218}Po, and ^{214}Po. There is a wide variation in the ^{226}Ra content of water in public water supplies. It may be almost undetectably low or it may be as high as about 5 pCi/liter. Supplies that depend on wells often have somewhat elevated levels. Tap water generally has less radium than raw water, but this is not always so. A typical value for tap water is 0.04 pCi/liter. The maximum permissible level for a large population is considered to be 3.3 pCi/liter. At spas the level can go to values in excess of 10 pCi/liter and "artificial radium waters" have been used with over 2×10^6 pCi/liter. At a given spa, the radium content can vary widely from spring to spring. Although the existence of ^{226}Ra in the waters of certain spas has been known for over six decades, recent concern has led to attempts to limit the daily intake of such waters by individuals. We should also expect that hot water and steam that is tapped from the earth for geothermal generation of power might also have appreciable natural radioactivity. For example, a geothermal power plant such as that in operation in Wairakei, New Zealand might release to the environment 80 mCi/hr of ^{222}Rn in water vapor discharge and another 470 mCi/hr of ^{222}Rn in the liquid discharge. The liquid discharge would also contain ^{226}Ra.

Accidental spillage or leakage of radium-contaminated process water from mining and milling operations such as the previously mentioned case of phosphate rock have also recently become of concern as a potential source of environmental contamination.

The main source of radioactivity in the ocean is ^{40}K. Secular equilibrium is not obtained for the members of the uranium and thorium series in the ocean because of adsorption by sediment of some of the elements, especially thorium, e.g., ^{230}Th in the uranium series.

(e) Food. The naturally occurring alpha activity of food is mainly associated with ^{226}Ra, its α-emitting daughters (especially ^{210}Po) and the α-emitting daughters of ^{228}Ra. With respect to ^{210}Po, it should be recalled that it is the daughter of ^{210}Bi, which is the daughter of the 22-yr β-emitter ^{210}Pb. Sources of lead are sources of ^{210}Pb and are, therefore, sources of ^{210}Po in the body. The intake of lead in a normal diet may be as high as 400 μg/day. Both ^{210}Pb and ^{210}Po can enter food not only from the soil, but also from the atmosphere, where they are present as the daughters of ^{222}Rn. Examples of these concentrations at ground level are 4×10^{-6} pCi/liter of ^{210}Pb and 1×10^{-6} pCi/liter of ^{210}Po. For leafy vegetables, the ^{210}Pb and ^{210}Po content are more directly related to rainfall than to the composition of the soil.

The concentrations of naturally occurring radionuclides in food vary so widely, e.g., by factors of several hundred, for a given type of food that we shall settle for a few examples of typical values. Examples of the total α activity of foods in pCi/kg are fruits, 1-20; milk and milk products, 1-20; meat, 1-20; fish, 2-300; and cereals, 10-600. Anyone with a craving for Brazil nuts may be interested to know that values in excess of 16,000 have been reported.

Most of the naturally occurring β activity in food can be attributed to ^{40}K. The ^{40}K activity can be calculated readily from the potassium content of the food. For example, fish have, on the average, about 0.3% by weight of potassium, which corresponds to 2500 pCi of ^{40}K per kilogram. In passing, we note that above average values of ^{40}K activity have been reported for caviar.

For growing plants, the availability of potassium is subject to large variation. Depending on the nature of the soil, only a small percentage of the potassium in the soil may be exchangeable and, therefore, available. Potassium may be a significant constituent of fertilizers. Some fertilizers are also sources of some of the members of the uranium series that were present in the phosphate rock used to manufacture the fertilizers.

Food accounts for 80-90% of the intake of ^{226}Ra. The remainder comes from water. The average intake of ^{226}Ra is about 3 pCi per day in food having 1-10 pCi of ^{226}Ra/kg. The daily intake of ^{40}K and ^{210}Pb in food is about 3400 pCi and 2 pCi, respectively. The total intake of any radionuclide in food is, of course, subject to variation in the concentration of the radionuclide in the food and to the amount of food consumed. An "average" diet in the U.S. is about 3.5 kg of food.

(f) Natural Radioactivity in the Human Body. The "standard man" weighs 70 kg and contains 140 gm of potassium and between 10^{-10} and 10^{-11} gm of radium. The latter is in the skeleton and may provide a dose of about 10 mrem/yr. In soft tissue the chief α emitter is ^{210}Po where the dose may be about 38 mrem/yr. A dose of about 20 mrem/yr to soft tissue comes from the beta-gamma radiation ^{40}K. The contribution from ^{14}C (a low-energy beta emitter) to the internal dose rate is negligible.

17.8 MAN-MADE SOURCES OF IONIZING RADIATION IN THE ENVIRONMENT

17.8.1 X-Ray Machines

Because of the obvious benefits, the use of X rays for medical diagnosis is continually increasing. It has been estimated that more than 90% of the population exposure to man-made sources of ionizing radiation in the United States comes from the diagnostic use of X rays. In general, the risks are considered to be acceptable by the population. Government regulatory agencies however, are becoming increasingly concerned about the risks. For example, restrictions in the use of X rays during pregnancy have been recommended. Moreover, as the percentage of the population receiving X rays increases, so does the probability of future genetic effects. There are regulations intended to reduce the genetically significant exposure to both patient and operator by limiting the X-ray beam area to the minimum needed, by reducing leakage of X rays from the source, and by improving the sensitivity and reliability of the measuring methods. Examples of diagnostic exposure values are chest X ray, 10-200 mR; GI tract, about 20 R; and examination during pregnancy, about 50 R.

To a much lesser extent, X rays are also used for medical therapy. Here the cost-benefit analysis is different from that for diagnosis. As an example, treatment of tumors can result in a localized exposure of 3×10^3 to 7×10^3 R.

The use of X-ray machines for industrial radiography is increasing. In this case the hazard is to the operator of the equipment. With proper shielding of the equipment and with the use of safe procedures, the radiographic equipment should not be a source of exposure.

Routine fitting of shoes by means of X rays is an example of an application of X rays for which the benefit does not justify the risk. The practice of using X-ray shoe fitting machines has been largely discontinued, but a few years ago people, especially children, trying on shoes were sometimes exposed to X rays repeatedly while on a shopping expedition. Exposures of about 10 R per fitting were not uncommon. The shoe salesmen were also exposed repeatedly.

A potential source of X rays in the home today is the television receiver. The high voltages used in these receivers, especially in those designed for color television, can generate X rays. Shielding, e.g., glass for the picture tube and metal for the high voltage supply, are necessary.

Although X-ray machines are not strictly within the realm of nuclear chemistry, they constitute a major source of the ionizing radiation to which the population is exposed. It is with reference to this source and natural background that one must evaluate effects of man-made sources that are in the realm of nuclear chemistry.

17.8.2 Radionuclides Produced in Accelerators and Reactors

The main exposure of the public on an individual basis to these man-made radionuclides is by way of the diagnostic and therapeutic uses in nuclear medicine. The levels of activity used vary from μCi to multi-mCi for oral and intravenous administration. This branch of medicine is developing very rapidly. Some hospitals have their own accelerators for producing short-lived radionuclides. Certain short-lived radionuclides used to prepare radiopharmaceuticals are available from "generators" that are commercially available. These generators contain a relatively large amount of the longer-lived parent of the radionuclide of interest. The shorter-lived daughter is removed chemically as needed. An example is the separation of 6.0 hr ^{99m}Tc from the 66.7 hr ^{99}Mo parent by simple chromatographic elution.

Nuclear reactors are used to produce most of the common radionuclides needed for research in industry, universities, etc., although accelerators designed for use in nuclear medicine are becoming more numerous. Similarly, reactors produce most of the radionuclides used in nuclear medicine. Research reactors also provide the means for applying the methods of nuclear chemistry to the analysis of environmental samples for many nonradioactive constituents at trace concentrations. Whether a reactor is used as a source of neutrons to produce radionuclides by activation of target materials or is used as a research facility in its own right, it produces radioactive fission products (Section 17.8.3) during its normal operation. In most reactors, these are contained within the sealed fuel rods containing the fissionable material, e.g., ^{235}U, and are not released unless there is a failure of the cladding, i.e., the metallic tubular container part of the rod. Eventually, as for power reactors discussed below, the fission products are released when the fuel rods are sent to a chemical processing plant for recovery of the residual fissionable material.

In the case of large air-cooled reactors, the cooling air is exhausted through a high stack so that the radioactivity induced in the air by neutrons in the reactor core is diluted to an acceptable level before it reaches the ground. The neutron-induced short-lived radionuclides include ^{16}N (7.14 sec), ^{17}N (4.16 sec), and ^{19}O (29.1 sec) and the somewhat longer-lived ^{41}Ar (1.83 hr). The three short-lived radionuclides are produced from oxygen and are, therefore, also present in the water-cooled reactors. They are also produced in air used for ventilation of buildings containing certain types of accelerators.

A variety of devices known as SNAP (systems for nuclear auxiliary power) have been developed and put into service. In general, these systems use thermoelectric conversion of nuclear energy to electrical energy. The larger ones use small reactors as the heat source and are intended for use on satellites. Their electrical power output may be the order of a kilowatt. Smaller units having electrical power output of the order of a few watts are isotopic sources utilizing the heat from either long-lived β emitters such as ^{90}Sr or ^{137}Cs or α emitters such as ^{238}Pu or ^{244}Cm having half-lives of 87.8 and 17.9 yr, respectively. The systems are attractive for applications where reliability is important and routine servicing is undesirable or even impossible. Large isotopic power generators have been used to provide power for navigational buoys, remote weather stations, communications satellites, and lunar surface experiments in the Apollo program. They may contain kilocuries of radioactive material, but they are not used in locations readily accessible to the general population. Small generators using ^{238}Pu, which does not require bulky shielding, are incorporated in one type of implanted cardiac pacemaker.

Direct energy conversion has been used to obtain electrical energy from radioisotopes. Electrical power output of the order of microwatts is obtained by absorbing the ionizing radiation in a semiconductor material such as silicon, thereby generating a direct current.

In general, man-made radionuclides are not incorporated into household items in as many ways as naturally occurring ones were in the past. As pointed out in Section 17.6.5, the oxides of uranium were used for many years to produce a beautiful orange or green glaze for dinner ware and other ceramic items. Similarly, radium has been used in luminous paint for clocks and watches. On occasion watches have been observed to be sources of alarmingly high exposure to γ rays from radium and its daughters. Some shielding is provided by the case and works of a wrist watch. Man-made tritium is a low-energy pure negatron emitter (E_{max} = 18.6 keV) with a half-life of 12.3 yr and would seem to be an attractive substitute for radium in luminous paint. It is of interest to note, however, that it has been found that wearers of watches containing tritium (bound in some form in paint) may have tritium levels in the body above the natural level. This is consistent with the relative ease with which hydrogen diffuses through metal and, in this case, continues to diffuse through the skin and become incorporated as HTO (tritiated water) in the body by isotope exchange.

Relatively recently, man-made radioactivity has been used as the source of ionization of air in one type of smoke detector designed for use in any type of building, but especially suited for early detection of residential fires. Such devices, either battery- or ac-operated, are small and can be attached to a ceiling. They incorporate a small amount of americium-241, an α emitter that also emits low-energy γ radiation and has a half-life of 433 yr. Smoke particles reduce the steady α-produced ionization current reaching the detector and trigger an audible alarm.

It should be pointed out that appropriate licenses are required for the manufacture of devices containing radioactive material, and for the possession and use of radioactive materials in excess of amounts specified by law for each radionuclide. Licenses are obtained from the U.S. Nuclear Regulatory Commission (NRC) or through state health departments for those states having an agreement arrangement with the NRC. Beginning with the Atomic Energy Act of 1946 and continuing until 1975, the federal regulatory agency was the U.S. Atomic Energy Commission.

17.8.3 Nuclear Fission

Before considering the production of fission products on a large scale and the ways by which they may enter the environment, we shall briefly examine the nature of the fission reaction. When the nucleus of a $^{235}_{92}U$ atom captures a thermal neutron, the resulting compound nucleus $\left[^{236}_{92}U\right]$ may deexcite by emitting γ radiation to form the α-emitting $^{236}_{92}U$ ($t_{1/2}$ = 2.34 × 10^7 yr), or it may undergo fission after deforming into the shape of a dumbbell. Fission is the more likely process by a factor of 5.46. In other words, the cross section for fission, which is the probability that a thermal neutron will cause fission, is 5.46 times the cross section for the radiative capture (n,γ) reaction. The two highly charged, primary nuclear fragements produced in fission have collectively, as kinetic energy, about 80% of the 200 MeV released per fission. The remainder of the energy is distributed among neutrons and γ rays released by the fragments and by negatrons, γ rays and antineutrinos emitted later by the fission products. Very soon after formation, the fission fragments evaporate one or more neutrons and emit γ rays to become secondary fission fragments that are still highly unstable and decay by negatron emission (see Section 17.3.7). When the half-life of a fragment becomes sufficiently long so that the nuclide can be chemically identified, it is called a primary fission product. Unless stable, the primary fission product will decay by negatron emission into a daughter, which in turn may decay into a daughter, etc., until a stable nuclide is formed and the decay chain is terminated. The following is an example of such a chain.

$$^{137}_{53}I \xrightarrow[\beta^-]{24.6\ sec} {}^{137}_{54}Xe \xrightarrow[\beta^-]{3.84\ min} {}^{137}_{55}Cs \xrightarrow[\beta^-]{30.1\ yr} \begin{cases} 92\% \nearrow {}^{137m}_{56}Ba \\ \qquad \downarrow I.T.\ 2.55\ min \\ 8\% \searrow {}^{137}_{56}Ba\ (s) \end{cases} \qquad (17\text{-}25)$$

There are over 75 known fission chains in the mass number range of 72 - 161. In addition, there are five cases in which a stable nuclide is formed as a primary fission product. As a group, the fission products include the 37 elements from zinc through dysprosium. By examining a periodic table (Fig. 14-1), we can see that the fission product elements cover a wide range of chemical behavior. Also to be noted are the facts that ten are members of the lanthanide series, and two, technetium and prometheum, are radioelements, i.e., all isotopes are radioactive.

The total number of radioactive fission products is over 200. Half-lives fall between a fraction of a second and 2×10^5 yr. The radioactive species are distributed among the 75 chains, which have, on the average, three members. The large number of chains reflects the fact that symmetrical fission into two fragments of equal mass number is relatively uncommon when fission is induced by thermal neutrons. Instead, a light and a heavy fragment usually are produced. This gives rise to the double-peaked fission yield curve shown in Fig. 17-8. The fission yield for a given mass number is the total (maximum) yield for the chain of that mass number. It

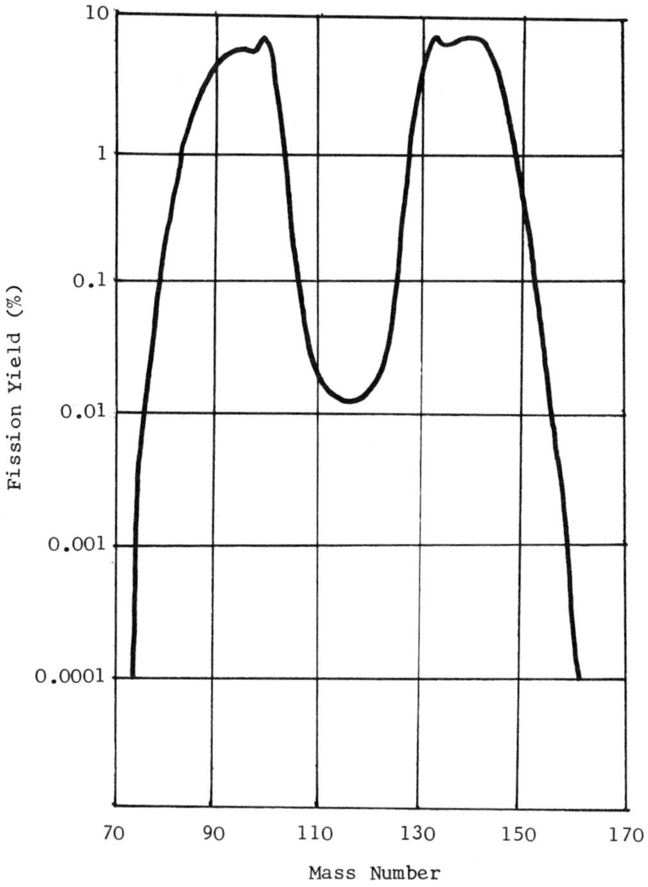

FIGURE 17-8. Yields for fission product chains; fission of ^{235}U with slow neutrons.

is expressed as the percent of fissions in which fission products of the given mass number are formed. Individual fission products may or may not have the same yield as the chain. One reason for the latter is that there may be direct formation of members later in the chain. Selected fission products having moderately long or very long half-lives and produced with relatively high yield are listed in Table 17-8.

When fission of $\left[{}^{236}_{92}U\right]$ occurs, the sum of the masses of the pair of one light and one heavy fragment must equal 236, since mass number is conserved. Similarly, the resulting fission product masses plus the masses of the neutrons evaporated by the fragments must also equal 236. In about two out of 1000 fission events a rare type of fission known as ternary fission occurs and ${}^{4}He$ or ${}^{3}H$ is released.

Although a small integer number of prompt neutrons is released immediately following fission, the number varies in accordance with the extent of the neutron excess of the fragments. The average number of such prompt neutrons released per thermal fission of ${}^{235}U$ is 2.46. There are also a few neutron emitters among the fission products. These produce the delayed neutrons that are important in the control of nuclear reactors. An example of such a fission product is

TABLE 17-8

Selected Fission Products and Their Half-Lives

${}^{85}Kr$	10.73 yr	${}^{133}Xe$	5.29 day
${}^{89}Sr$	50.5 day	${}^{137}Cs$	30.1 yr
${}^{90}Sr$-${}^{90}Y$	28.5 yr, 64 hr	${}^{140}Ba$	12.79 day
${}^{91}Y$	58.6 day	${}^{141}Ce$	32.53 day
${}^{95}Zr$-${}^{95}Nb$	65.5 day, 35.1 day	${}^{143}Pr$	13.58 day
${}^{99}Tc$	2.13×10^{5} yr	${}^{144}Ce$	284.4 day
${}^{103}Ru$	39.6 day	${}^{147}Nd$	10.99 day
${}^{106}Ru$-${}^{106}Rh$	369 day, 29.9 sec	${}^{147}Pm$	2.623 yr
${}^{131}I$	8.041 day		

^{137}I [Eq.(17-25)], which has a branched decay to the extent of about 0.03% to form stable ^{136}Xe.

Although fission has been discussed with reference to ^{235}U, fission can be induced in many other nuclides. Thus ^{233}U and ^{239}Pu, like ^{235}U, can be fissioned by thermal or fast neutrons. The cross section for thermal neutron fission is much higher, however, than that for fast neutron fission for all three nuclides. Fast neutrons are required to fission ^{238}U and ^{232}Th, the abundant isotopes of these two elements. Commonly ^{233}U, ^{235}U, and ^{239}Pu are referred to as fissile and ^{232}Th and ^{238}U as fissionable. Very high energy (several hundred MeV) neutrons, protons, and other charged particles will induce fission not only in these five nuclides, but also in isotopes of several elements including Bi, Pb, Tl, Au, Pt, Re, and W.

When a compact mass containing a few kilograms of fissionable material is caused to fission at once, a nuclear explosion can result. On the other hand, if the same quantity of fissionable material is fissioned slowly in a nuclear reactor, the rate of energy release is controlled and the energy can be used to produce steam and electricity at a steady rate. Water-cooled nuclear power reactors obtain their energy from thermal neutron induced fission under conditions that are not those required for a nuclear explosion.

Quantities of fission products sufficient to be of concern if released into the environment are produced by military nuclear weapons, nonmilitary nuclear explosives, large nuclear research reactors, and nuclear power reactors.

17.8.4 Nuclear Weapons

Since 1945 nuclear weapons have been altering the ionizing radiation component of the environment. In the fission-type weapon using ^{235}U or ^{239}Pu, the fission products listed in Table 17-8 and many others having short half-lives are released and are distributed on the surface of the earth as fallout and rainout. The size of a weapon is expressed in terms of its TNT equivalent. One ton of TNT is considered to have an explosive energy of 4×10^6 kJ. In nuclear fission of ^{235}U about 165 MeV of the total fission energy of 200 MeV is immediately available as kinetic energy of the fission fragments. Thus if 1 kgm of ^{235}U were completely fissioned in an explosion, the energy release would be equivalent to 17 kilotons of TNT. Depending on the efficiency of the device, a weapon would be expected to release only a fraction (perhaps one-tenth to one-fifth) of the maximum TNT equivalent per kilogram.

Thermonuclear weapons or H-bombs can release much greater explosive energy (in the megaton TNT equivalent range) than fission weapons, but have a smaller release of fission products per unit explosive energy because fission products are produced only in the fission reaction that triggers the main thermonuclear reaction. The thermonuclear reaction is the main source of energy. Thermonuclear reactions are fusion reactions, such as the so-called "D-T reaction,"

$$ {}^{2}_{1}H + {}^{3}_{1}H \rightarrow {}^{4}_{2}He + {}^{1}_{0}n \qquad (17\text{-}26)$$

which is exoergic by 17.6 MeV. The tritium can be made in situ by the reaction

$$ {}^{6}_{3}Li + {}^{1}_{0}n \rightarrow {}^{3}_{1}H + {}^{4}_{2}He \qquad (17\text{-}27)$$

where the neutrons are initially supplied by the fission of ${}^{235}U$ or ${}^{239}Pu$. The energy released in reaction (17-27) is 4.8 MeV. The deuterium needed for the fusion reaction can be introduced in the form of lithium-6 deuteride. The ${}^{6}_{3}Li$ can also undergo a thermonuclear reaction with ${}^{2}_{1}H$ as the temperature rises to form $2{}^{4}_{2}He$ with the release of 22 MeV per event.

The rather unfortunate word "clean" has been used to characterize the fission-fusion type of the H-bomb described because the quantity of fission products released per kiloton energy release is less than that for a fission weapon the same kiloton-size as the trigger device. By adding an outer layer of ${}^{238}U$, a relatively inexpensive material, neutrons released in the D-T reaction and other fusion reactions will cause fast neutron fission of ${}^{238}U$. Now, the weapon can be characterized as "dirty" and becomes more so as more ${}^{238}U$ is incorporated.

If the ${}^{238}U$ were omitted and the weapon were designed as a warhead for a missile or artillery shell for tactical use, it could be conceived as an enhanced-radiation weapon (sometimes called a neutron bomb). The blast effect would be reduced and the fast neutrons from the fusion reaction could provide a very high dose of incapacitating radiation.

Weapons of a megaton or larger generally release the radioactive debris in the stratosphere, where the residence time of the long-lived fission products may be from one to five years with an average of about two years. When the debris falls into the troposphere, it can be carried to the earth's surface by rain. Weapons under 1 megaton release the debris in the troposphere or below (perhaps a few thousand feet or so). Tropospheric fallout takes place within about a month and may occur at great distances from the point of origin. The debris may be carried along rapidly in a jet stream with less dispersion than might otherwise occur. For lower level detonations the bomb debris is adsorbed on soil, etc., which is carried upward and settles out in a matter of minutes. In the latter case, the fallout is deposited locally, relative to the detonation, and can include radioactivity induced by neutron capture reactions in the soil (or seawater if the detonation is near or over an ocean). When the fission products are in the stratosphere they receive world wide distribution. (See Section 3.6 on the circulation of the atmosphere.) For underground detonations, most of the radioactivity remains underground, but there is usually some venting and localized fallout.

Three other radionuclides should be mentioned in connection with nuclear weapons. These are ^{239}Pu, ^{3}H, and ^{14}C. In a weapon in which ^{239}Pu is used, any unfissioned ^{239}Pu would be a component in the residual radioactivity. Tritium is produced in the rather rare type of fission known as ternary fission. It is produced in large quantities in H-bomb detonations as is ^{14}C, which is formed by the (n,p) reaction of ^{14}N in the atmosphere.

Before 1954, when atmospheric testing of H-bombs began, the ^{3}H and ^{14}C on earth were very largely cosmogenic. The natural ^{3}H, mainly as HTO in the hydrosphere, was estimated to be about 900 gm (about 9 MCi). Prior to 1954, rainwater had about 0.5 to 5.0 tritium units of ^{3}H. One tritium unit (1 T.U.) is equal to one ^{3}H per 10^{18} atoms of ^{1}H. For water, 1 T.U. corresponds to 3.3 pCi/liter. After 1954, levels as high as 500 T.U. were found in rainwater.

The long term value for the cosmogenic or natural level of ^{14}C (13.6 disintegrations/min/gm of carbon) is based on measurements of the ^{14}C content of wood that was in living equilibrium with the environment before 1890. With the beginning of the industrial era at the end of the nineteenth century, the great increase in the combustion of fossil fuels (low in ^{14}C) led to a decrease of about 2% in the level of ^{14}C in the atmosphere. After the main flurry of weapons testing, the ^{14}C level in the atmosphere rose to about double the natural value. Without further detonations of nuclear weapons in the atmosphere, the ^{14}C level should slowly decrease to a new, constant value somewhat higher than the pretest value.

Data on the distribution and change in distribution of ^{3}H from weapons testing have been used to study the movement of water in the atmosphere and in the oceans. For example, the residence time of ^{3}H in the stratosphere has been estimated to be 2-4 yr. Tritium is washed from the troposphere within about two months. For ^{14}C as CO_2 the residence time in the stratosphere is about 4 yr and for the atmosphere as a whole, the residence time is about 10 yr.

Of the fission products in fallout listed in Table 17-8 three have been long recognized to be of especially serious ecological importance, namely, ^{90}Sr, ^{131}I, and ^{137}Cs. In evaluating the hazard of such radionuclides to man, attention is focused on them as an internal hazard. It then becomes necessary to examine their mode and extent of entry into man and effect of retention of each in tissue. This matter will be considered in Section 17.9.1.

Quite apart from disastrous physical effects and radiological consequences on a global scale that would result from nuclear warefare, another potential hazard has been receiving attention. As pointed out in discussions of the atmosphere in preceding chapters, ozone in the lower stratosphere plays a vital role in protecting the various forms of life on the earth from short-wave ultraviolet radiation. Nuclear weapons produce oxides of nitrogen in the atmosphere and these can react with ozone to deplete the protective layer as discussed in Section 10.3.3.

17.8.5 Nonmilitary Nuclear Explosives

In the United States, the use of nuclear explosives for nonmilitary purposes has been known as Project Plowshare. The name of the project refers to the objective of using the technology of nuclear weaponry for the benefit of mankind, thereby "beating swords into plowshares." Various uses have been proposed; a few have been tested. One of the proposed uses is that of earth-moving on a large scale such as that required to dig a major canal. The calculated cost in dollars per mile for various methods of such a major excavation may, in some cases, show that the nuclear method is the most favorable. Evalution of the problem of residual radioactivity, a negative feature, is not so easy to achieve.

A second type of peaceful application is that of releasing gas or oil from rock strata by underground explosions. It is the intent that such underground explosions should not release fission products into the atmosphere. Tests have not always been free of atmospheric release. Another disadvantage of the method is the contamination of the released gas or oil with fission products and other radionuclides produced in the matrix materials.

17.8.6 Nuclear Fission Power Plants

Because the only type of nuclear power plant in commercial operation in the U.S. today utilizes nuclear fission as its source of energy, we shall simply refer to such plants as nuclear power plants. Later we shall consider briefly certain aspects of proposed nuclear power plants that would use nuclear fusion as the source of energy. Fusion power plants are only in the conceptual stage.

The first central station nuclear power plant operated by a public utility in the U.S. for the generation of electricity was built in Shippingport, Pennsylvania in 1957. Its net electrical capacity was 90 MW. Today some plants are being built and designed for capacity in excess of 1200 MW. Many commercially operated plants have been or are being built with capacities in the range of 500-1100 MW.

By early 1978, about 69 nuclear power plants were operating in the U.S. These provided about 13% of the electrical power generated in the U.S. at that time. Most were located east of the Mississippi River and these, in turn, were concentrated along the eastern seaboard and in the Great Lakes area. The remainder were mostly on or near the west coast. Similarly, most of the plants under construction or planned were to be located east of the Mississippi, but placed to provide a somewhat more even distribution within this region. It should be pointed out that once a site for a nuclear power plant has been selected, it is rather common for more than one unit, e.g., three or four, to be built. The units, each having a capacity in the 500 MW to 1100 MW range, may be spaced a few years apart in their construction. The trend has been toward the higher capacity. It usually takes at least five years to build a nuclear power plant and make it operational. Up to five additional years may now be required for site selection, site approval, etc., before construction can begin.

In, say, 1970 estimates of the number of new nuclear power plants needed by about 1975 or so were based on several assumptions, two of which were (1) that the annual increase in demand for electrical power would be the same as or even slightly higher than that in the past, and (2) that the relative costs of nuclear versus fossil fuels in various geographical areas could be predicted. Thus, in regions where the demand for power is high because of heavy industry and relatively high population density but where there is no nearby source of fossil fuel, the economics of nuclear power plants has been favorable. By 1975, however, the process of planning and committing funds for new nuclear power plants had become more complex because of (1) greater uncertainties in the future demand for power; (2) unpredicted increases in the costs of both uranium and fossil fuels; (3) the development of other energy sources, e.g., solar energy, to meet some of the needs such as heating of buildings; (4) increased time required for licensing; and (5) opposition to the building of nuclear power plants because of concern about their potential impact on the environment in general, and questions about safety, waste disposal, and security related to potential use of reactor-produced plutonium for nuclear weapons. A drop in the annual increase in the demand for energy and increased costs also reduced the rate of construction of fossil-fueled power plants.

(a) Types of Nuclear Power Plants. There are several distinct types of nuclear power reactors. We shall consider only a few and limit the types to those in use or under development in the U.S., where water (light water) is the most commonly used coolant. The first commercial plant mentioned was a pressurized water reactor (PWR). This type of light-water reactor (LWR) is still popular for new plants. As we should expect, details of design, materials of construction, fabrication methods, the physical and chemical composition of the fuel rods, safety features, etc., have evolved over the years as operational experience has been gained and as the challenge of building larger plants has been faced.

Currently, the fuel used in LWRs is uranium slightly enriched to perhaps 3-5% in ^{235}U and is commonly in the form of UO_2 pellets. Enrichment of the ^{235}U is necessary to compensate for the fact that the hydrogen in water, which is used as both moderator and coolant, captures a significant fraction of the thermal neutrons needed to maintain the chain reaction in which at least one of the neutrons produced in each fission must be available to induce another fission. Very briefly the series of steps from mine to fuel rod is as follows: (1) ore is converted to pure U_3O_8, which is converted to UF_6 containing 0.720% ^{235}U; (2) the UF_6 is enriched to about 3% ^{235}U in a gaseous diffusion plant; and (3) the enriched UF_6 is converted to UO_2, which is used to prepare the fuel rods. Expansion of the ^{235}U enrichment facilities will be required to support the nuclear power industry if it grows and is based on the burning of enriched uranium.

Central station nuclear power plants almost qualify as closed systems with respect to radioactive material when operating normally. A simplified block diagram for a PWR system is shown in Fig. 17-9.

In a PWR the UO_2 pellets, as cylinders about 0.5 in. in diameter and about the same length, are sealed in a fuel rod consisting of a cylindrical tube (called the cladding) made from an alloy of zirconium and having a diameter slightly

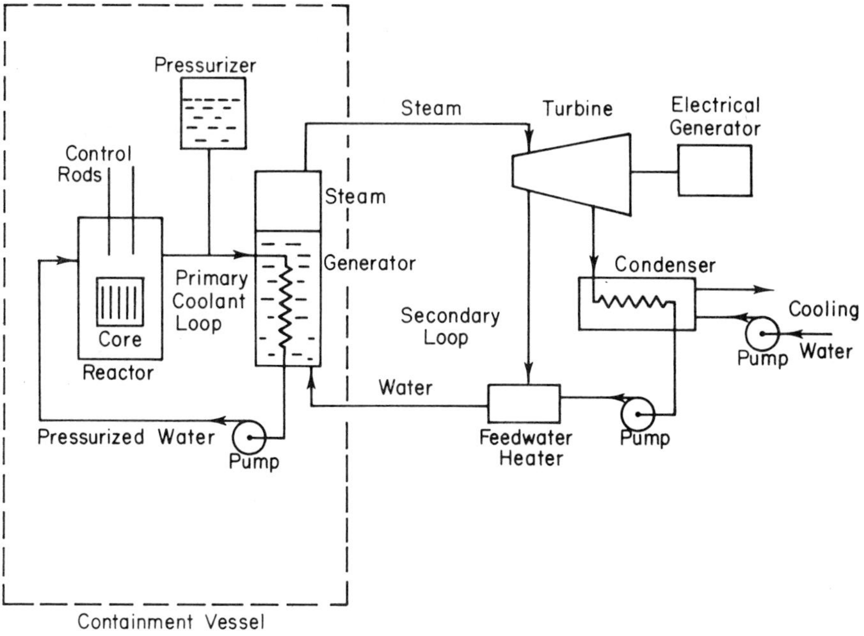

FIGURE 17-9. Simplified schematic for pressurized water reactor (PWR) power plant.

larger than that of the pellets and thickness of about 0.03 in. Zirconium has a low probability for capture of thermal neutrons and some of its alloys (containing mostly Zr) have both a high resistance to corrosion and good mechanical strength. The rods, which are about 13 ft. long, are held together in bundle-like assemblies of about 200 rods each. There are about 200 assemblies in a reactor core, which then contains about 90 metric tons of uranium.

Water under pressure, e.g., 2000-2500 psi to prevent boiling, is pumped through the core where it flows between the rods to remove the heat. After leaving the reactor vessel at about 600°F, the water passes through one of several primary closed loops that include heat exchangers and steam generators outside the reactor vessel. Water in the secondary loop is converted to steam at perhaps 500°F and 800 psi to drive a steam turbine connected to an electrical generator. Another heat exchanger is used to condense the steam from the turbine and reject waste heat to the surroundings, e.g., a river or a lake or an ocean. The region of the plant containing radioactive material (reactor, primary coolant loops with heat exchangers, and the steam generators) is confined within a structurally strong, steel-lined, reinforced concrete containment vessel to contain the radioactivity in the event of an accident (nonnuclear) to the reactor that could release radioactive material. The latter consists mainly of fission products but also includes uranium, plutonium, and other transuranic elements formed from uranium, and radionuclides produced by radioactivation of corrosion and erosion products from fuel rods, piping, reactor vessel, control rods, welding alloys, and impurities or anticorrosion additives in the water. Table 17-9 contains a partial list of radioactive material (consisting chiefly of neutron-activated corrosion and erosion products in the absence of appreciable leakage of fission products from the fuel rods) that could be contained in the primary coolant of a PWR power plant under normal operating conditions.

The buildup of fission products in the fuel rods generates several problems. These include (1) expansion within the fuel-element as two fission product atoms are formed for each atom fissioned; (2) buildup of pressure caused by noble gas fission products; (3) buildup of neutron poisons, i.e., fission products having very high cross sections for the

TABLE 17-9

Radioactive Contaminants that may be in Reactor Primary Water Collant

^{3}H (T)	12.33 yr	^{58}Co	71.3 day
^{16}N	7.11 sec	^{60}Co	5.27 yr
^{17}N	4.16 sec	^{64}Cu	12.74 hr
^{19}O	26.9 sec	^{95}Zr	65.5 day
^{24}Na	15.02 hr	^{181}Hf	42.4 day
^{51}Cr	27.71 day	^{187}W	23.9 hr
^{54}Mn	312.5 day	*Fission products*[a]	
^{55}Fe	2.7 yr	*Fissionable material*[a]	
^{59}Fe	44.6 day		

[a] *From ruptured fuel rods, fuel rods with tiny leaks, or new fuel rods with surface contamination of fissionable material.*

capture of neutrons needed to maintain the fission chain reaction; (4) release of decay energy as heat after the reactor has shut down; and (5) the need for removal and disposal of megacurie amounts of β-γ emitters. A power reactor "burns" and converts to fission products about 1 gm of ^{235}U per day for each megawatt of thermal power, i.e., per megawatt(th), produced. The 1 gm fissioned per day would produce an electrical output of about 0.3 MeW(el). The basis for the relationship between the thermal power output and quantity of ^{235}U used is that each fission event releases about 200 MeV and, therefore, 1 W of power, as heat, requires 3.1×10^{10} fissions per second. In general, a reactor can burn only a fraction of the fissile material in the core before the partially spent fuel must be replaced with fresh fuel.

The composition of the fission product mixture in a reactor is very complex because of the large number of different primary fission products formed, the decay chains, and the very great range of half-lives. The short-lived fission products reach saturation or equilibrium rather rapidly, as discussed in Section 17.2.3. Long-lived fission products become increasingly important as the reactor continues to operate with a given set of fuel rods. After about one year the core might, for example, contain about 10^{10} Ci of fission products.

A second popular type of light-water-cooled, thermal power reactor in the U.S. is the boiling water reactor (BWR). As the name implies, in this type of reactor the water used to remove heat from the reactor core is allowed to boil within the reactor. The steam thus generated has a temperature of about 550°F and a pressure of about 1000 psi and is used directly to drive the turbine. Steam is delivered to the turbine through a steam separator that separates the steam from the water-steam mixture leaving the reactor. A simplified block diagram for a BWR is shown in Fig. 17-10.

Typically, the fuel and fuel rods are similar to those used in the PWR except for differences in dimensions and in the number and arrangement of the rods. About 150 metric tons of uranium are contained in the core. In both the BWR and the PWR there are provisions for preventing dangerous buildup of hydrogen in the water circulated through the core. The hydrogen is produced by radiolysis of water.

Two systems are used to contain the radioactivity in the event of an accident in a BWR plant. The primary containment is achieved by a steel pressure vessel that is surrounded by reinforced concrete. The pressure vessel surrounds the cylindrical reactor vessel. The building that houses the reactor and the primary containment vessel is constructed of reinforced concrete and constitutes a secondary containment. Radioactive material that escapes from the primary containment

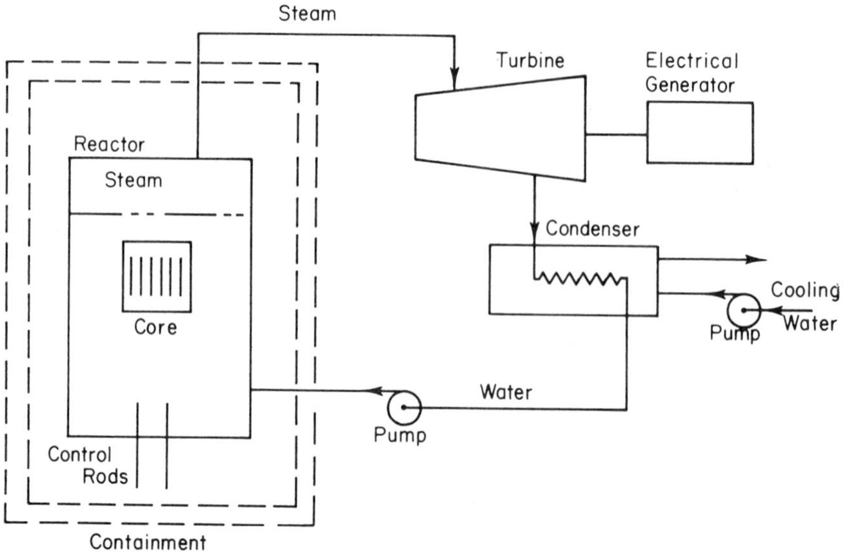

FIGURE 17-10. Simplified schematic for a boiling water reactor (BWR) power plant.

vessel is confined within the secondary containment. Air from the latter building is treated to remove radioactive contaminants before being exhausted through a tall stack, or at least at a level well above ground.

A third type of power reactor that has been used in the U.S. to a very limited extent is the gas-cooled reactor using helium as the coolant. A simplified block diagram is shown in Fig. 17-11. With helium as coolant the only radioactivity expected in the gas leaving the reactor would be activation products of impurities such as argon and oxygen and any fission products that leak from the fuel-elements.

Canadian reactors use heavy water D_2O as the moderator of fission neutrons to provide the thermal neutrons needed to fission ^{235}U. Because deuterium has a much lower cross section than hydrogen for absorbing thermal neutrons, the neutron economy is improved, i.e., fewer thermal neutrons are absorbed in the moderator and coolant, and natural uranium can be used as the fuel. The cost of extracting D_2O from water is traded for the cost of enriching the fuel with respect to ^{235}U. When the heavy water is used to remove heat from the core as well as to moderate the neutrons, it must be circulated through a

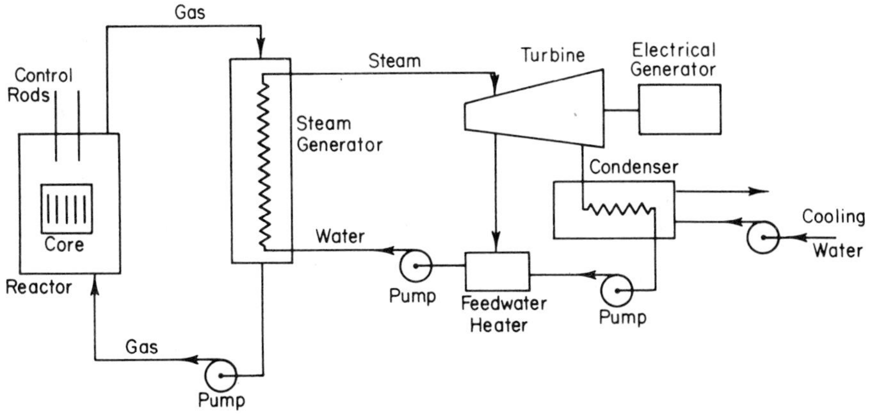

FIGURE 17-11. Simplified schematic for gas-cooled reactor (GCR) power plant.

heat exchanger in which light water is converted to the steam used to drive the turbines.

Various types of breeder reactors have been proposed and are under study in anticipation of the time when the limited supply of ^{235}U will be exhausted. Recall that natural uranium contains only 0.720 atom % ^{235}U. The remainder, almost entirely ^{238}U, can undergo several kinds of nuclear reactions with neutrons. One of these, the capture reaction, is the first step in the conversion of ^{238}U to ^{239}Pu, which is fissionable with thermal neutrons. The process is as follows:

$$^{238}_{92}U\ (n,\gamma)\ ^{239}_{92}U \xrightarrow[23.5\ \text{min}]{\beta^-} {}^{239}_{93}Np \xrightarrow[2.35\ \text{days}]{\beta^-} {}^{239}_{94}Pu \qquad (17\text{-}28)$$

The ^{239} Pu is an α emitter with a half-life of 24,390 yr and is a particularly hazardous substance, as pointed out in Section 17.6.3. Special handling techniques must be used in working with large or small quantities. In thermal power reactors of the types currently in use, some ^{239}Pu is formed but the amount is less than the amount of ^{235}U fissioned because of nonfission absorption of thermal neutrons in the water and in materials of construction. Nonfission absorption can be decreased by utilizing the neutrons before they become thermal. Although more fuel is required when fast neutrons are used to induce fission because the fission cross section decreases with neutron energy, more neutrons are available to make ^{239}Pu by reaction (17-28). By using fast neutrons in a reactor, which is then called a fast reactor, it is possible to use ^{239}Pu as the fuel and at the same time make more ^{239}Pu than is consumed. A reactor operating in this manner is called a breeder and the ^{238}U is referred to as the fertile material in the fuel cycle. Breeding ^{239}Pu in this fuel cycle would permit the expansion of the nuclear power industry. Initially ^{235}U would be used to make sufficient ^{239}Pu from ^{238}U to serve as the fuel in breeder reactors. Also, the ^{239}Pu produced in thermal LWR systems can be used most efficiently in a fast breeder reactor.

Another possible fuel cycle for breeder reactors utilizes ^{233}U as the fuel and ^{232}Th as the fertile material to make ^{233}U by the following process:

$$^{232}_{90}Th\ (n,\gamma)\ ^{233}_{90}Th \xrightarrow[22.2\ min]{\beta^-} {}^{233}_{91}Pa \xrightarrow[27.0\ day]{\beta^-} {}^{233}_{92}U \qquad (17\text{-}29)$$

Uranium-233 is an α emitter with a half-life of 1.58×10^5 yr. By means of this cycle the relatively abundant ^{232}Th, the predominate isotope of thorium in nature, could be used to generate power. In this case breeding seems to be theoretically possible with either thermal or fast reactors.

In 1972, the decision was made to build a large scale [400 MW(el)] fast breeder power plant near Oak Ridge, Tennessee. It is known as the CRBR (Clinch River Breeder Reactor). The type of reactor selected for this demonstration plant breeds plutonium and is an LMFBR, liquid metal fast breeder reactor. Various aspects of the technology of this type of reactor have been studied over the years by means of experimental reactors, but construction of the demonstration plant has been delayed and its future is in doubt.

In the proposed reactor, the fuel would be about 20% PuO_2 and about 80% UO_2. The UO_2 would be made from depleted uranium, i.e., uranium whose ^{235}U content has been reduced below that in natural uranium in order to provide enriched uranium for LWR systems. Depleted uranium is now stockpiled. Cladding material for the fuel rods would probably be a stainless steel.

The coolant in an LMFBR type of reactor is liquid sodium, which is much less effective than water in slowing down neutrons. In passing through the core the sodium captures neutrons to form 15.0 hr ^{24}Na and 2.6 yr ^{22}Na from the stable ^{23}Na. The coolant is, therefore, highly radioactive quite apart from contamination by fission product leakage from fuel elements. As shown in the simplified block diagram in Fig. 17-12, an intermediate-heat-exchange loop containing liquid sodium transfers heat from the primary loop to the

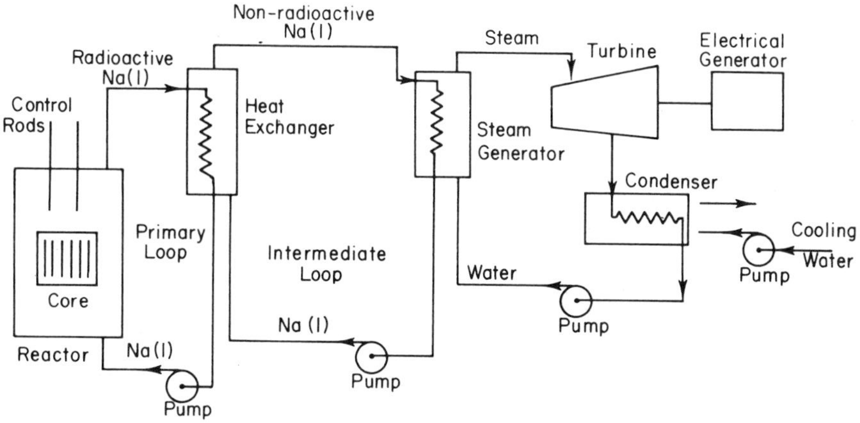

FIGURE 17-12. Simplified schematic for a liquid metal fast breeder reactor (LMFBR) power plant.

steam-generating loop. In such reactors the primary system can be operated at much higher temperatures than for water-cooled cores. Specially designed heat exchangers are required for the transfer of heat from liquid sodium to water in order to prevent accidental contact of the sodium with the water. On the basis of chemical considerations, liquid sodium is not a pleasant substance to handle on a large scale. It is spontaneously flammable in air and must be handled in an inert atmosphere. It reacts violently with water to liberate hydrogen, which generally ignites.

(b) Efficiency. In general, nuclear power plants of the LWR type operate with thermal efficiencies of about 30-35%. Fossil fuel power plants can be designed to operate at somewhat higher thermal efficiencies, e.g., about 40%. Thus, both types of power plants "throw away" about 2/3 of the energy available. In other words, only about 1/3 of the heat released in nuclear fission or in combustion is converted to electrical energy. If we make the oversimplified assumption that a power plant can be treated thermodynamically, to a first approximation, as an ideal heat engine, then we can see in a qualitative manner the way in which the temperature of the steam delivered to the turbine influences the thermal

efficiency of the plant. It should be recalled (Section 2.2.2) that η, the maximum fraction of heat that can be converted to mechanical energy by an engine operating reversibly in a Carnot cycle, can be calculated on the basis of the second law of thermodynamics. If the higher temperature at which heat is supplied to the engine is T_2 and the lower temperature at which the engine rejects heat to the surroundings is T_1, η is given by

$$\eta = (T_2 - T_1)/T_2 \tag{17-30}$$

where T_1 and T_2 are absolute temperatures. The actual efficiency will be less than the maximum for a real system because of irreversibility and heat losses. The efficiency for converting mechanical to electrical energy is generally very high. By inspection of Eq. (17-30), we can see that the thermal efficiency can be increased by increasing T_2 for a fixed T_1 that is determined by the environment. Nuclear power plants of the LWR type operate at temperatures lower than those of the most efficient fossil fuel plants and, therefore, for plants of the same thermal capacity they reject more heat to the surroundings than do fossil fuel plants. It is expected that the LMFBR will operate at a higher temperature and, therefore, with a higher efficiency than either the PWR or the BWR. All power plants transfer waste heat to the surroundings by heating water from a source of cooling water such as a river, a lake, or an ocean. The source of cooling water is warmed especially near the point of discharge of water from the condensers, where the temperature of the water may have been raised by about 20°F. This is sometimes referred to as "thermal pollution." Both deleterious and beneficial effects on aquatic life may occur. Fossil fuel power plants generally cause less thermal pollution than nuclear plants, not only because they may operate more efficiently, but also because about 1/6 - 1/10 of the waste heat is released directly into the atmosphere through the stack that releases the products of combustion into the atmosphere. In the U.S. the LWR nuclear power plants transfer essentially all of the waste heat to the cooling water. Cooling towers could be used to divert some of the waste heat into the atmosphere without adding any chemical or radioactive pollutants to the atmosphere. In the past, their use has been ruled out mainly on the basis of the increase in cost of electricity that would result from adding them to the power plant, i.e., increase in the capital cost that is already relatively high for a nuclear plant.

(c) Siting. Before any type of central station power plant can be built at a proposed site, its potential impact on the environment must be assessed. This is a major undertaking involving predictions of the extent to which the plant could affect the quality and temperature of the cooling water it uses, and the extent to which it could cause radioactivity pollution of the atmosphere, vegetation, aquatic life, etc., in the surroundings. For nuclear power plants there has always been a lengthy procedure required to assure public safety. A preliminary hazards report evaluating the environmental impact of credible accidents is needed in order to obtain a construction permit from the Nuclear Regulatory Commission. Site selection must take into account nearby population density, availability of cooling water, geological stability of the area, e.g., the presence or absence of a fault, etc. A final hazards report must be approved before an operating license is issued.

(d) Surveillance. A nuclear power plant may release small amounts of radioactive materials, such as gaseous fission products, tritium, and activated corrosion products into the environment. Therefore, prior to operation of a nuclear power plant, an environmental radioactivity survey is made to establish a background baseline that can be used for reference in later surveillance measurements when the plant is in operation. A preoperational survey is conducted over a period ranging from several months to about a year. Details of a survey program depend on the nature of the site, but we shall note the types of measurements that have been made in surveys in the past. Radioactivity is measured for samples of soil, sediment in the cooling water supply, well water, surface water, air (both particulates and moisture), and precipitation (rain and snow, if applicable). Often the radioactivity can be sufficiently characterized by low resolution γ ray spectrometry. By such spectrometry we can identify and quantify some of the members of the uranium and thorium series, ^{40}K, and fallout ^{137}Cs. Field measurements are made and samples for laboratory analysis are collected in such a way that they can be repeated after the plant is in operation. In the field measurement of soil, a detector such as a sodium iodide scintillation detector shielded on the top and sides by a cylinder of lead is placed close to the soil in a reproducible manner. A γ ray spectrum is obtained by means of a portable pulse height analyzer. The same detector and analyzer can be placed in the bottom of a boat to measure the γ ray emitters in water below the boat.

For laboratory analysis of soil, core samples, perhaps cylindrical and 6 × 6 in., are taken at selected locations. After being screened and dried, the samples are analyzed by γ ray spectrometry.

When water samples are taken for laboratory analysis, the water is filtered through a series of filters differing in porosity so that the radioactive contaminants are separated into two components: the dissolved and the undissolved that remains on the filter. Both components are analyzed for γ emitters and, in addition, the filtered water is analyzed for tritium by liquid scintillation counting.

Samples of the particulate matter in air are obtained by pulling a large volume, e.g., 10^5 liters, of air through a filter. The filter, as for those used for laboratory determination of particulates in water, is dissolved in a suitable solvent and after dilution to a suitable volume, the radioactivity is measured. The activity observed is mainly that of the daughters of ^{222}Rn. These, but not the minor components from ^{220}Rn can also be detected in the field by using the sodium iodide detector.

Tritium, as HTO, in air can be determined by drawing air through a desiccant and then removing the water from the desiccant by heating it in a closed system containing a suitable trap for condensing the water. Filtered rain and melted snow samples are also analyzed for tritium.

In preoperational surveys additional samples such as samples of local meat, milk, vegetables, fish, etc., may be collected and analyzed for radioactivity. On the other hand, the nature of the site, e.g., type of soil, types of plant and animal life, may be such that the survey may not require more than confirmation of expected gross α, gross β, and gross γ measurements without spectrometry.

Under NRC regulations, an operating nuclear power plant (or other installation in which large quantities of radioactive material are handled) must have a surveillance program and must file periodic reports detailing the release of radionuclides in gaseous or liquid form. From such reports and a knowledge of the pathways that the released radionuclides can follow to reach man, an estimate of population exposure can be made. The pathways are discussed in Section 17.9.1. The following is indicative of the types of samples taken as part of a surveillance program: (1) air: particulates and iodine, (2) surface water, e.g., river water above and below the power plant, (3) ground water, (4) drinking water, (5) sediment and aquatic plants in surface water, (6) milk, (7) fish, (8) shellfish, (9) fruits, (10) vegetables, (11) meat, and (12) poultry. Analysis of surface water, drinking water, and milk would provide data on γ-emitting radionuclides, gross β emission, and radiostrontium content. Tritium would also be determined in the water samples. In addition, so called "direct radiation" is measured at various locations off site, e.g., in the nearest communities by means of strategically located integrating dosimeters. Over the years, much experience has been gained with respect to surveillance sampling procedures, location and method of sampling, radiochemical methods of analysis, and statistical analysis of the data.

Within a nuclear power plant, radiation detectors continously monitor the level of radioactivity in coolant loops, condenser effluent, etc., in order to detect any sudden increase. Water samples are also withdrawn on a regular schedule for radiochemical analysis to detect fission products, activated corrosion products, and for measurement of pH and conductivity.

How do nuclear power plants release radioactivity into the environment and how much is released? Both nuclear and fossil-fuel power plants discharge gaseous and liquid effluents into the environment. The radioactivity in gaseous effluents from nuclear power plants includes ^{3}H, ^{16}N [from the reaction $^{16}O(n,p)^{16}N$] and radioisotopes of the fission product elements Kr, I, and Xe.

The tritium is produced in ternary fission in both PWR and BWR plants and to a greater extent from boron in boric acid used for control purposes in the primary coolant for PWR plants. Tritium from fission can diffuse from the UO_2 pellets into the gap or free space in the fuel rod and can then escape along with other volatile fission products into the core cooling water if the cladding develops leaks. In addition, tritium gas can diffuse through most metals, particularly at elevated temperatures, and can reach the cooling water even if there are no leaks in the cladding. When the primary cooling water in a PWR or the cooling water in a BWR is processed for purification and removal of gases, some gaseous effluent containing tritium as HTO is generated.

If the gases are stored in hold-up systems, the many short-lived components will decay leaving ^{85}Kr as the major source of radioactivity along with significant amounts of ^{3}H and ^{131}I. For a PWR plant, the volume of gas to be stored is less than that for a BWR plant and a hold-up time of about 30 days is feasible. Most of the ^{131}I is removed by adsorption before the ^{85}Kr is released to the environment. For a BWR plant the off-gases are diluted by noncondensable gases that increase the volume of gas so much that practicable hold-up times are reduced and have been as short as less than 1 hr. One diluent is air, which can leak into the system at the turbine where the steam pressure on the low pressure side is less than 1 atm. The second diluent is a mixture of radiolytic H_2 and O_2. In a PWR, H_2 is introduced into the primary loop to reduce radiolysis. In principle, the H_2 and O_2 mixture can be removed by careful recombination.

For a 1000 MW(el) plant, the annual release might be up to 50 Ci of ^{3}H, up to 10^4 Ci of ^{85}Kr and up to 1 Ci of ^{131}I. These values are for normal operation and do not take into account accidental releases of radioactive gases.

Core cooling water is a source of liquid radioactive waste in a nuclear power plant. The waste accumulates from coolant purification processes, from leaks at pump seals, and from samples taken for analytical control. Most of the dissolved nonvolatile cationic species of fission products (entering the coolant because of cladding failure) and of activated corrosion products are removed by chemical means, such as ion exchange, and are retained for waste disposal. The resulting low-level waste water containing residual dissolved radioactivity and the original amount of HTO is commonly mixed with and diluted by the condenser cooling water being returned to the environment.

Annual releases as liquid effluent from a 1000 MW(el) plant might be 10^2-10^4 Ci of ^{3}H and a few curies of other radionuclides. Perhaps 10^4 Ci/yr of radioactive waste might be collected and converted to solid form for disposal. Again, the values do not include accidental releases that can occur if leakage occurs between "closed" cooling systems or if an operator opens the wrong valve.

Radioactivity limits for effluents from a nuclear power plant are set by the NRC. These are specified in terms of maximum permissible concentrations (MPC) (see Section 17.9.2), such that an individual outside the restricted area of the plant would not receive a yearly dose in excess of 0.5 rem if all his or her intake of air or water were at the MPC levels. The regulations are also intended to limit the effluent radioactivity to values as far below the MPC values as is practicable. Suggested lower levels may be 10 to 10^4 times less than MPC values as improved technology makes these levels practicable.

Even though the MPC values may be low, e.g., 10^{-2} - 10^{-6} μCi/ml, the extremely large dilution factors that apply when plant effluents are discharged into the atmosphere or into a river, for example, make it possible to discharge seemingly large total amounts of radioactivity. The rate of discharge of raw water used in a single pass through a 1000 MW(el) plant might be as high as 10^6 gal/min.

(e) Safety. Beginning with the construction of the first reactor, nuclear safety has been an important aspect of reactor design. Safety features are a significant factor in making the capital cost of a nuclear power plant relatively high compared with that of a fossil-fuel plant. The

importance given to safety and the measures taken have resulted in a good safety record. Through 1975 there were no nuclear accidents in nuclear power plants in the U.S. that affected the public. In fact, there were no major reactor incidents. Perhaps the most serious accident was a fire in electrical cables near the control room at the Browns Ferry Nuclear Plant near Decatur, Alabama in March 1975. The fire burned for several hours and caused extensive damage, including destruction of some of the electrical cables needed to operate safety devices. Even so, the two reactors were shut down and there was no hazardous release of radioactivity. The incident emphasized the need for improvement in certain equipment and operating procedures.

It is reasonable and proper to ask how great the risk to the public would be following a major accident directly involving a nuclear power reactor. In the absence of data on an observed frequency of occurrence of such accidents, it is necessary to estimate the risk by special methods. We must start by describing the types of accidents that could affect the health of the public. At the outset, it is important to recognize that thermal power reactors, the general type in use in the U.S., do not have proper composition in terms of fissionable material and other materials to suddenly undergo a nuclear explosion and thereby become an "atom bomb." This is only one of several reasons to discount a nuclear explosion. The most serious type of accident that could occur is one in which the radioactivity in the core of a power reactor is released from the fuel rods and then a fraction of it escapes from the containment and is dispersed in the environment around the plant site. Depending upon the burn-up history of the fuel in the core, the radioactivity in a 1000 MW (el) reactor would be on the order of 10^9 Ci of fission products, about half as much ^{239}Np, and very much smaller amounts of other radioactive material such as isotopes of uranium, plutonium, americium, and curium. For several years it has been recognized that such an accident could be initiated by a sudden loss of core coolant because of, say, a break in a large pipe in the core cooling loop of an operating power reactor. A loss of coolant accident (LOCA) could result in melting of the core, if the safety systems failed and the temperature rose to above 5000°F. It is the core melting process that would release the radioactivity into the containment area. If the metallic cladding of the fuel rods but not the UO_2 fuel pellets melted, only a fraction of the radioactivity that is present in a volatile form in the gap or space within the fuel rod not occupied by the pellets would be released. It is conceivable that the molten core could penetrate into the earth and some of the radioactivity could escape from the confinement. Additionally, core melting could be followed by a steam explosion caused by sudden contact of molten core material with water or by a combination of a steam explosion and an explosion of hydrogen produced by decomposition of water or by reaction of water with the zirconium of the cladding. Parts of the reactor vessel, for example, might then become projectiles that could penetrate the containment and provide an escape route for radioactivity. Alternatively, the explosion pressure might cause venting below the containment and release of radioactivity from the molten core material.

As a safeguard against melt-down following a LOCA, power reactors are equipped with a reserve supply of water that can be used in an emergency cooling system (ECCS) designed to flood the core or spray it with water in a fraction of a minute. At the time of a LOCA the reactor would be automatically shut down. Even if there were no failures of the control systems and the reactor were safely shut down, heat would still be released as fission product decay heat. Thus, 10 sec after shutdown, the decay heat release rate might be about 5% of the operating power level of the reactor just before shutdown. This decay heat would be sufficient to melt the core in the absence of emergency cooling.

In the summer of 1972, an in-depth study of nuclear reactor safety[1] was initiated by the U.S. Atomic Energy Commission. The results of an earlier and much more limited study had been published in 1957. The new study was completed under the sponsorship of the U.S. Nuclear Regulatory Commission and the final report titled "An Assessment of Accident Risks in U.S. Commercial Nuclear Power Plants," WASH-1400, was issued in October 1975. The report, which is also known as the Rasmussen Report after the name of the director of the study group, was first issued in draft form. Comments on the draft report were used in the preparation of the final report and are included in one of the 11 appendices that amount to over 2000 pages. Because this is a landmark report on a controversial issue concerning the environment, we shall summarize the methods used in the study and the conclusions reached.

The overall risk to the public depends on the consequences of radioactive releases and on the probability and magnitude of radioactive releases. Risk can be expressed in several ways. Consider first a familiar example, namely, the risk of being killed in an automobile accident. This is a

[1]*"Reactor Safety Study, An Assessment of Accident Risks in U.S. Commercial Nuclear Power Plants." U.S. Nuclear Regulatory Commission, WASH-1400 (NUREG 75/014), NTIS, Springfield, Virginia, 1975.*

nonnuclear risk often used for comparison. The risk can be calculated easily because, unfortunately, there are statistically significant data. In 1968, for example, there were 55,200 deaths caused by the automobiles in the U.S. The societal risk that year was 55,200 deaths per year. The risk for an individual in a population of 2×10^8 people is then 2.8×10^{-4}/person/yr or a chance of about 1 in 3500/yr.

In the Reactor Safety Study (RSS) it was necessary to determine the probability that each of many possible initiating events would occur and would be followed by a sequence of events, e.g., failures of safety systems such as the ECCS after a LOCA, leading to melting of the core. The accident sequences following an initiating event were described by event trees. An event tree is a means for identifying possible outcomes of an initiating event. Event trees were prepared for accidents initiated by LOCAs and by transient events other than LOCAs that require reactor shutdown or occur in a reactor already in shutdown condition, e.g., failure of the system used to remove decay heat. Event trees were also prepared for the sequence of events following containment failure.

For each system identified in an event tree, a fault tree was developed to determine the probability of failure of the system. Failure probability was determined from failure data on identical equipment, if available, from the probability of operation errors and from the probability of errors in equipment maintenance. Two specific operating nuclear power plants, one PWR and one BWR, were used as typical of the only two types of reactors considered in the RSS.

For an accident leading to release of radioactivity, the magnitude of the release is controlled by (1) the amount and composition of the radioactivity released from the core into the containment, (2) the fraction of this radioactivity that is retained by adsorption, etc., within the containment, and (3) the way in which the containment fails. In the RSS, the released radioactivity was considered to contain 54 biologically significant radionuclides. The potential for release of fission product elements (in elemental form or as oxides) from the core during an accident sequence is greatest for the elements Xe, I, Cs, and Te, about ten times less for Sr and Ba, and for Ru and other noble metals, and about a hundred times less for La and the other rare earths.

The health and economic consequence of a release of radioactivity depend on factors such as (1) the manner in which the radioactivity is dispersed in the environment, (2) the size of the population and the amount and type of property exposed, and (3) the effects that the exposure would have on the population and the property. Physical damage to property would be limited to the power plant itself. Beyond the plant site, damage would arise from contamination possibly covering a few hundred to a few thousand square miles. Exposure to the population would be reduced by evacuating people living within 25 mi of the plant and located in the direction of the wind. It was considered that exposure could lead to (1) external dose from the released activity as a passing cloud or from radioactivity deposited as contamination from the passing cloud, or (2) internal dose from inhaled or ingested contaminated material. The probabilities were then calculated for the following health effects: early fatalities (within 1 yr of an accident), early illness (acute somatic effects), thyroid illness, latent cancer fatalities (after 10 - 40 yr, or so), and genetic effects. For the latent cancer fatality risk, an upper limit was obtained by assuming no threshold and using linear extrapolation of observed data at high exposures down to predicted exposures.

The probability of population exposure was calculated by multiplying together the probability that a core melt will occur, the probability of release of a specified fraction of the radioactivity per core melt, the probability of occurrence of specified weather conditions and the probability of a population being located in the path of the released radioactivity. For the accident having the largest consequences in the RSS, the largest release was calculated to occur in less than 10% of the core melts and the worst weather in less than 10% of the time and for high population areas the probability was less than 1%. Thus, the estimated probability of an accident having the largest consequence is 10^{-4} times the probability of a core melt.

The report includes extrapolated probabilities for an accident and for various consequences for the case of 100 nuclear power reactors in operation. For this number of plants a few of the conclusions in terms of probabilities are as follows: (1) core melt down: 1 in 200 per yr or 1 in 20,000 per reactor per yr; (2) 100 or more early fatalities: 1 in 10^5 yr (compared to 1 in 2 yr for an airplane accident and 1 in 7 yr for fires); (3) 300 cases of early illness: 1 in 10,000 per yr; (4) 170 cases of latent cancer deaths per yr: 1 in 10,000 per yr (normal incidence per yr is 17,000 for a population of 10^7 people); (5) 1400 cases per yr of thyroid illness: 1 in 10,000 per yr (normal incidence per yr per 10^7 people is 8000); (6) 25 cases per yr of genetic effects: 1 in 10,000 per yr (normal incidence rate per yr per 10^7 people is 8000); (7) contamination of 2000 sq. mi.: 1 in 10,000 per yr; (8) property damage of $\$0.9 \times 10^9$: 1 in 10,000 per yr (compared with an annual rate of property loss due to fires of $\$2.2 \times 10^9$ in 1970). For increasing magnitude of a given consequence, the calculated probability decreases.

For comparison with the seven risks given above, the following is pointed out in WASH-1400: The average individual risk of fatality in the U.S. is about 1 in 4,000 for motor vehicle accidents, about 1 in 10^5 for air travel accidents, and about 1 in 2×10^6 for tornadoes. Expressed in terms of frequency of occurrence for 100 or more fatalities, the values are 1 in 100 yr for release of a toxic gas, e.g., rupture of a railroad tank car containing 90 tons of liquid chlorine; 1 in 5 yr for a tornado; 1 in 20 yr for an earthquake; and 1 in 10^5 yr for a meteorite impact. A fire causing property damage in the range of $50 million to $100 million occurs about once every 2 yr.

The RSS did not attempt to answer the question, "Are the nuclear accident risks acceptable?" When does an individual or society consider that a risk is unacceptable? Apparently in the case of risk of death in an automobile accident, the annual loss of population equal to that of a city of 50,000 is acceptable when considered in terms of an average individual risk of death as 1 in 4000 per yr. The individual also tends to consider that the statistics do not apply to him or to her and that he or she can manipulate the chance of a fatal accident. Also the individual tends to be inconsistent in demanding risk levels and tends to interpret risk differently according to the degree of familiarity or understanding of the nature of the accident.

The results of another study[2] of the safety of LWRs, sponsored by the American Physical Society (APS), were published in 1975. Population doses were calculated and risk estimates were made. Comparisons of results with those in the Draft WASH-1400 were made. Some of the predicted consequences are greater in the APS report. In both the RSS and the APS study many assumptions had to be made about accident sequence, population exposure, and consequences. The two study groups made several different assumptions about exposure and consequences. Some of the assumptions originally made by the RSS group were modified before the final report was prepared. The result was an increase in certain consequences, e.g., the number of latent cancer fatalities.

The APS report contains not only an analysis of the risks of a nuclear accident, but also descriptions of the PWR and the BWR plants, discussions of safety problems, and recommendations for the reduction of the uncertainties of risk estimation, and for reactor safety research.

[2] *Report to the American Physical Society by the study group on light-water reactor safety,* Rev. Mod. Phys. *47, Suppl. No. 1 (1975).*

The safety aspects of LMFBRs have been studied in less depth than those of the LWRs and they are even more controversial. The limited experience gained so far in the U.S. with test reactors has not been very encouraging. One possible source of trouble has already been mentioned, namely, the use of sodium as the coolant. Additionally, fast reactors have different control characteristics than LWRs and the consequences of a core melt-down from a LOCA might lead to a far more serious uncontrolled nuclear reaction that could in turn lead to the release of larger amounts of plutonium and fission products than for an LWR accident. Also, the materials of construction in the core region of fast reactors are exposed to higher neutron fluxes than those in LWRs and, as a consequence, the extent of radiation damage to the materials is expected to be higher. Again, these problems are considered to be soluble by the proponents of fast breeder reactors and such reactors are being built in Europe. The use of safety research experiment facilities for conducting safety-related experiments on LMFBRs should answer many of the questions about safety and performance.

17.8.7 Nuclear Fuel Reprocessing

Reprocessing of fuel from nuclear reactors consists of opening the fuel rods, discarding the cladding, dissolving the UO_2, extracting the uranium and any plutonium away from the fission products, and preparing the fission products for disposal. The UO_2 in the spent fuel is converted to $UO_2(NO_3)_2$ from which the fission products and plutonium are removed separately. The purified uranyl nitrate is converted to UF_6 (containing about 1% ^{235}U) that is returned to the enrichment plant for use in the preparation of UO_2 for new fuel rods. In general, there is a decay period or "cooling time" between removal of the fuel from the reactor and the beginning of reprocessing. Several factors must be considered in setting the decay period. It might, for example, be about 150 days. During this period, the fuel-elements may be stored under water in a deep canal (at the power plant) from which they can be removed by remote handling techniques and transported in shielded containers to the reprocessing plant.

So far, most of the reprocessing has been done by the Federal Government (first the AEC, then ERDA, The Energy Research and Development Administration, and now DOE, The Department of Energy) at facilities designed to reprocess fuel elements from reactors used in the military program, e.g., naval reactors and reactors for producing plutonium. A privately owned reprocessing plant was built south of

Buffalo, New York adjacent to a radioactive waste disposal area. After about six years of operation, the plant was shut down for modifications in 1972. There is doubt that it will be reopened. A second plant was built in Illinois, but was never placed in operation. A third plant has been constructed in South Carolina, but has not yet been licensed to operate. Lack of privately owned reprocessing facilities for use by utility companies operating nuclear power reactors can be a serious problem if the capacity of existing spent fuel storage facilities is reached. Presumably DOE reprocessing facilities can be used, at least on an interim basis.

Because of the high levels of radioactivity, operation and maintenance require entirely different procedures from those normally used in a chemical plant. The problems encountered are severe. In terms of the environment, a fuel reprocessing plant is the place where even in normal operation the probability of release of radioactive materials into the environment is the highest in the whole fuel cycle (isotope enrichment, fabrication, use in a reactor, and reprocessing). Noble gases such as ^{85}Kr are released when the fuel is dissolved and are discharged through a tall stack. The spent fuel sent from a 1000 MW(el) power plant might contain about 10^5 Ci of ^{85}Kr per year. Although relatively little spent fuel from commercial nuclear power plants has been reprocessed, it has been estimated that the average concentration of ^{85}Kr in the troposphere was about 14 pCi/m^3 of standard air in 1973 and that the atmospheric content was about 55 megacuries at the end of 1973.

Most of the tritium in spent fuel would be discharged from a reprocessing plant as HTO in water vapor. About 10^4 Ci/yr would be sent in the fuel from a 1000 MW(el) plant. After the 150 day cooling period, most of the radioiodine in the fuel would have decayed. Only a small fraction of a curie would be released from the reprocessing plant under normal operating conditions. Some of the long-lived (1.59×10^7 yr) ^{129}I may be released in a volatile form. There is increasing concern about the environmental hazard of this radionuclide.

The nonvolatile, long-lived fission products and any plutonium or other trans-uranium elements such as curium that are not removed in the separation processes are stored for disposal. Low level liquid effluents from the reprocessing plant may contain a few curies per year of long-lived fission products. Radioactivity in the effluents must be less than the general MPC values already mentioned.

17.8.8 Radioactive Waste Disposal

Disposal of large amounts of radioactive waste has been studied for about 30 yr. Disposal really means storage in a manner such that the waste does not contaminate the environment. The waste cannot be degraded chemically into simple substances that are environmentally acceptable. The problem resides within the atom, not the molecular composition and structure. (It has been proposed that radioactive waste be recycled through power reactors to convert long-lived radionuclides to shorter-lived or stables ones. There are certain nuclides for which such transformation is a possibility, at least on paper. In a complex mixture, however, undesirable transformations could also occur). Fortunately, some of the shorter-lived radionuclides decay to stable materials in times that make short-term hold-up storage feasible, providing that such radionuclides have been separated from long-lived ones.

Radioactive wastes are generated at various stages in the processing and reprocessing of reactor fuel materials. Thus, the problem begins at the uranium mine. The waste of prime concern in terms of potential environmental impact is that which is produced in a fuel reprocessing plant. The risk arises from the possibility that some of the waste might find its way into the pathways between man and the environment. These pathways are discussed in Section 17.9.1.

The long-lived components in the waste make it necessary to think of storage times of thousands of centuries. The relative importance of the various radionuclides changes with time, of course. During the first 300-400 yr after production, the major hazard would be associated with the two β emitters, ^{90}Sr and ^{137}Cs, and the α emitter 17.9 yr ^{244}Cm. Later, the predominate radionuclides would be 9.5×10^5 yr ^{93}Zr, 2.13×10^5 yr ^{99}Tc, 2.3×10^6 yr ^{135}Cs, 24,390 yr ^{239}Pu, 6540 yr ^{240}Pu, 433 yr ^{241}Am and 8.5×10^3 yr ^{245}Cm.

At the present such wastes are commonly stored as aqueous solution in large tanks, which, as is characteristic of tanks, may develop slow leaks. Such leaks have been found at the Hanford facilities in Richland, Washington, for example. For long-term storage the waste must be converted to a solid form. As fuel from commercial nuclear power plants is reprocessed, the plan is to store the waste, probably as liquid, at the reprocessing site for about 5 yr during which time heat removal would be necessary. After the interim storage the waste would be solidified, placed in sealed containers, and transferred to a site for perpetual storage. The solidification process has not yet been selected. Many processes have been studied. Two with promise are (1) calcining to form oxides and (2) incorporating the waste into a heat-resistant glass. These processes closely fulfill the requirements that the solid be thermally stable and relatively inert chemically so that it will leach with difficulty if the integrity of the container (probably steel) is lost.

Various places have been proposed for the final deposition of the solidified waste. They include outer space, sealed trenches in deep ocean beds, the Antarctic ice cap, and caverns or mines in a geologic formation such as bedded salt. The last of these is probably the most advantageous. It is known that some of the bedded salt deposits (from ancient seas) have been stable for at least a hundred million years. Even so, it is difficult to find such a deposit completely free of some geologic imperfections that could conceivably provide a potential route for radioactive waste to reach the environment.

There is also a risk of a transportation accident while the waste is enroute to the storage site. About 10^7 Ci of waste requiring perpetual storage is generated per year by a 1000 MW(el) power plant. In solid form, this would probably have a volume of between 5 and 10 m^3. Obviously, the problem of disposal of radioactive waste cannot be dismissed as being trivial. On the other hand, the difficulties do not appear to be insurmountable, even though they have been under study for three decades.

17.8.9 Nuclear Safeguards

As the number of nuclear power plants increases, there is growing concern about the possibility of diversion of nuclear fuel materials from their intended use to use for the fabrication of explosive weapons or radioactivity contamination weapons. Such weapons might be utilized by nations at war or by terrorists and would spread radioactive contamination of the environment. Sabotage of a major nuclear facility could also release dangerous amounts of radioactivity.

Plutonium-239, rather than ^{235}U, is the likely material to be employed in any weapon because power reactors, in general, do not use highly enriched weapon-grade ^{235}U. Quite apart from supplying the energy of fission for a nuclear explosion, ^{239}Pu is extremely hazardous in its own right. It is made automatically in nuclear power plants without the need of a diffusion plant for isotope enrichment and less (about 10 kg) is required for an explosive weapon than for ^{235}U (about 30 kg). The required amount of each nuclide increases with increasing impurity content. In the case of plutonium the impurities are ^{240}Pu and ^{242}Pu.

The safeguard problem is not confined to the U.S. In 1974, there were 19 nations with reactors producing plutonium at the rate of about 200 kg/yr for each 1000 MW(el) LWR reactor. The number of nations is steadily increasing under assistance programs involving major nuclear powers such as the U.S. The Non-Proliferation Treaty of 1970 was intended to prevent the proliferation of nuclear weapons by maintaining control over the materials needed for weapons. Several nations with plutonium-producing capability have yet to sign or ratify the treaty. In 1974 doubts were raised about the feasibility of control when India detonated a nuclear explosive.

The potential for diversion of ^{239}Pu will be much greater if and when there are many breeder reactors operating on the plutonium fuel cycle. Diversion is most likely when the plutonium is easiest to handle (free of fission products) and is in a useful physical form. These conditions are met in fuel fabrication plants, fuel reprocessing plants, and during transportation of purified plutonium.

In order to avoid these conditions and thwart diversion of plutonium, the Civex process has been proposed for handling plutonium produced in fast breeder reactors. In this process the uranium (^{238}U) remaining in the irradiated fuel would be removed and purified, but the plutonium would be fabricated into new fuel rods without removal of the fission products. The impure highly radioactive plutonium could be used as fuel, in fast-breeder reactors, but could not be used readily for weapons.

Another proposed solution to the diversion problem is to use the ^{233}U-Th fuel cycle instead of the ^{239}Pu-U cycle for fast breeder reactors. The ^{233}U would be extracted for use as fuel in conventional LWR power plants after being denatured with ^{238}U to make it unsuitable for weapons.

There are two aspects to the nuclear safeguard problem. The first is that of reducing the diversion of nuclear materials to a level as close to zero as possible. The second is that of detecting any diversion, if it occurs. Until a satisfactory solution to the safeguard problem is found, major delays are likely in the development of an energy program depending upon breeder reactors using plutonium, especially in the U.S.

It should be recognized, however, that the proliferation of nuclear weapons cannot be completely prevented by simply avoiding a plutonium-based nuclear energy program. Plutonium can be made, albeit slowly, with small reactors, e.g., research reactors, and there are at least three processes for the enrichment of uranium that can be used for small production plants (the centrifuge, aerodynamic and laser enrichment techniques).

17.8.10 Natural Fission Reactors (the Oklo Phenomenon)

Until 1972 it was believed that the first nuclear reactor on the earth was the man-made one built under the University of Chicago stadium and successfully brought to criticality on December 2, 1942. It is now known that natural fission reactors existed on the earth about 1.8×10^9 yr ago in an uranium ore deposit at the Oklo mine in Gabon on the west coast of Africa. Six reactor zones or "fossil" reactors have been found to date. These zones are lense-shaped about one meter in thickness and several meters, e.g., ten, in length. They contain a relatively high uranium content of about 20% (compared to a normal content of about 0.5% for the area) as uraninite (UO_2), relatively free of neutron-absorbing impurities such as vanadium found in carnotite deposits. Depletion of the ^{235}U content (relative to that in normal uranium today) in the reactor zones led to the chance discovery of the Oklo phenomenon.

At the same time of operation, the ^{235}U content of the uranium was about 3 atom % and the water in the associated hydrated clay minerals most likely served as the neutron moderator. The chemical form of the fuel, i.e., UO_2, the ^{235}U content and the ratio of moderator to fuel are strikingly similar to those used today in power reactors of the LWR type. It is estimated that the reactors had an operational life of the order of 500,000 yr and a combined energy release about equivalent to 5 yr of operation of 3000 MW(th) reactor.

It will be years before study of the fossil reactors is completed and the data are fully interpreted. In the meantime, additional fossil reactors may be found elsewhere on the earth. The Oklo reactors have already provided information of value in connection with nuclear power reactors. For example, information has been obtained about the radiation damage of materials in and near the reactor zones and about long-term storage of ^{239}Pu and radioactive wastes. The ^{239}Pu was produced in the reactors. A fraction of it fissioned and the remainder decayed to ^{235}U, which also fissioned or decayed.

17.8.11 The Controversy Over Nuclear Fission Power Plants

During the period of preparation and revision of this chapter not only the public but also scientists and engineers in the U.S. have become strongly polarized into "pro-" and "anti-" nuclear groups relative to the expansion of nuclear power production. Nuclear power is not only uniquely different from that generated from fossil fuels because of the mechanisms of energy release, but also because in the minds of the public its history is associated with the "atom" bomb and its future may include reestablishment of that association. This, however, is but one of the reasons for the controversy.

In the preceding sections of this chapter we have attempted to present, from the viewpoint of nuclear chemistry, facts concerning nuclear fission, nuclear weapons, nuclear reactors, nuclear reactor safety, spent fuel reprocessing, radioactive waste storage, etc. It is a well-known political fact, however, that at the polling-place risk-benefit decisions, such as those affecting nuclear power development, are made to a great extent on an emotional rather than a factual basis. Facts can be questioned and disputed, but hopefully they will be used in the decision-making process. The extremes of the anti- and the pro-nuclear arguments owe some of their effectiveness to the emotional response they generate. Recognizing that it is not feasible to be completely objective while being a party to a public controversy, as we all are, it seems proper to set aside a section of this chapter to summarize the many aspects of the controversy. Hopefully, the chapter as a whole will be of value to the reader in evaluating the statements and arguments being made for and against nuclear energy.

The proponents of increasing the role of nuclear power in the U.S. argue that (1) the energy demands of the future cannot be met without nuclear power; (2) nuclear power from LWRs is safe, based on the safety record for about 2000 reactor years of experience (including naval reactors) and on the conclusions of the WASH-1400 report; (3) nuclear power is economically advantageous over fossil fuels in many areas of the U.S. despite the increasing cost of nuclear power plant construction (from about $180/kW of output in 1968 to about $450 today) and uranium (from about $9.50/lb in 1973 to over $40 currently); and (4) the problems of waste storage and safeguards will be solved. Further, the proponents of breeder reactors that produce and burn plutonium believe that (1) such reactors will be necessary to extend the uranium supply until fusion power is available; (2) plutonium can be handled safely on a large scale without danger to the public; and (3) the plutonium produced in power reactors could not be easily used for weapon fabrication because of the special procedures needed for working with plutonium and because the product contains significant amounts of an undesirable isotope ^{240}Pu.

The opponents of the construction of additional LWRs argue that the existing reactors are not as safe as claimed and cite the Browns Ferry accident (fire) as an alarming "near miss" of a serious accident despite the fact that the reactor was successfully shut down. The incident did weaken the impact of claims for completeness of safety planning for nuclear power reactors. At the base of the doubt is the possibility of human fallibility in (1) foreseeing all possible sources of accidents; (2) correctly designing, constructing, and testing of safety devices; and (3) reacting quickly and properly if a problem that could lead to a LOCA arises. The opponents feel that it has not been proven that the ECCS will function reliably because such systems have not been tested experimentally on a full scale and reduced-scale tests have not always been successful.

With respect to WASH-1400, the opponents consider the consequences of an accident to be underestimated. They question the validity of the event tree--fault tree methods and the resulting probabilities. The use of probabilities as a measure of risk is, of course, common, but probabilities generally have less than the hoped-for impact on the public, even when they are actuarial probabilities as in the case of death in an automobile accident or fire or in the case of lung cancer from smoking cigarettes. In the case of cigarette smoking, for example, the individual feels that he or she has the freedom to take the risk, but in the case of nuclear power, which is a societal as well as an individual issue, he or she feels that the freedom of choice has been taken away. Such does happen when the societal needs or those of the nation are given higher priority than those of the individual. In return, the individual does have the right to demand maximum safety.

Additional objections to the conclusions of WASH-1400 have arisen because of disagreement with some of the assumptions that had to be made and because the study did not deal with every conceivable accident. The latter objection again relates to the use of probabilities. Not all conceivable accidents were considered to have sufficient probability to warrant detailed analysis.

Some of the opponents allege that the proponents have not always been candid about the safety, economics, and environmental impact of nuclear power reactors. The result is, therefore, a credibility gap. In part this situation has been attributed to the Atomic Energy Act that charged the former AEC with both promoting the development and regulating the safety of nuclear power.

Finally, with respect to LWRs, many of the opponents of nuclear power argue that no additional plants should be built until (1) all questions of safety have been satisfactorily resolved, (2) proper waste storage facilities are available (assuming fuel reprocessing plants are also available), and (3) the power plants and related facilities are secure against sabotage and terrorist attack.

Efforts are being made to solve the problems associated with nuclear power. Included are experimental studies and new regulations related to safety. Funds are also being channeled to the development of other sources of energy, e.g., solar, to reduce the need for electricity, gas, and oil for heating and cooling buildings.

The breeder power reactor program in the U.S, and in particular the LMFBR, appears to be in trouble at this time. Serious objections have been raised by opponents with respect to reactor safety, the environmental hazards of a plutonium-based nuclear power economy, the diversion of plutonium for weapons, and the economic predictions for breeder reactors.

Without doubt, this is a critical time for the future of nuclear power in the U.S. However, as we look down the controversial road of our present LWR-based nuclear power industry, we see neither a green light nor a red light corresponding to the two extremes, but rather a yellow caution light in the use of one of the major sources of power available, albeit not without risk, for the benefit of mankind.

In the bibliography at the end of this chapter there is a section on the controversy containing references to articles and books that reflect the extremes and the middle-of-the-road attitudes. The collection of references given contain a large number of specific references on all aspects of the controversy. It should be noted that articles and books on the controversy are appearing at an overwhelming rate and government policies on nuclear power in the U.S. and elsewhere are changing rapidly.

Among the groups or organizations that have expressed opposition to the expansion of nuclear power are Business and Professional People for the Public Interest, The Committee on Nuclear Responsibility, Common Cause, The Friends of the Earth, Natural Resource Defense Council, The Sierra Club, The Union of Concerned Scientists, and The Wilderness Society. It is not claimed that this list is complete. There have been and are numerous groups concerned with nuclear power at the local level.

A partial listing of groups or organizations that are outspoken proponents of nuclear power are The American Assembly, The Atomic Industrial Forum, The Edison Electric Institute, and The Electric Power Research Institute. In addition, there are several electric utility companies and the suppliers of nuclear power plants.

17.8.12 Nuclear Fusion Power Plants

Attempts have been under way for over 25 yr to develop a device for carrying out controlled thermonuclear reactions, i.e., fusion reactions. Nuclear fusion has the potential of being an almost unlimited source of energy in the future.

There are several possible fusion reactions, four of which are

$$^{2}_{1}H + ^{2}_{1}H \rightarrow ^{3}_{2}He + ^{1}_{0}n \qquad (17\text{-}31)$$

$$^{2}_{1}H + ^{2}_{1}H \rightarrow ^{3}_{1}H + ^{1}_{1}H \qquad (17\text{-}32)$$

$$^{2}_{1}H + ^{3}_{1}H \rightarrow ^{4}_{2}He + ^{1}_{0}n \qquad (17\text{-}33)$$

$$^{2}_{1}H + ^{3}_{2}He \rightarrow ^{4}_{2}He + ^{1}_{1}H \qquad (17\text{-}34)$$

In the order given, these reactions are exothermic by 3.2, 4.0, 17.6, and 18.3 MeV. It is most likely that the D-T reaction given in reaction (17-33) will be used in the first generation of fusion power plants. (This reaction was introduced in Section 17.8.4 in connection with nuclear weapons.) Ultimately, the D-D reactions [reactions (17-31) and (17-32)] would be used because of the essentially "unlimited" amount of deuterium on the earth in the form of HDO and D_2O in normal water. Normal hydrogen has a deuterium content of 0.015 atom %. Isotopic enrichment of hydrogen with respect to deuterium is a relatively simple and inexpensive process when compared with that for enrichment of uranium with respect to ^{235}U. The products of the D-D reactions are also usable in reactions (17-33) and (17-34).

For a fusion reaction to occur, the reactants must have sufficient kinetic energy (about 10 keV) to overcome the Coulombic repulsion barrier. In order to obtain useful energy from fusion as in a power plant, it is necessary to use some of the energy released in the reaction to heat the reactants to the ignition temperature. The required temperature depends somewhat on the specific reaction, but it is about 10^8 °C. At these kinetic temperatures the electrons are stripped from the atoms and a plasma, a mixture of electrons and positive ions, is formed. Finding ways to attain such high kinetic temperatures is one of the major problems to be solved in developing controlled fusion power plants. The second major problem is confinement of the plasma for sufficient time to permit the fuel ions to react and release the heat of nuclear fusion.

There are two different approaches to solving the two problems. Both are being pursued in several laboratores in the U.S. and abroad. In the method that has been studied for the longer period of time, the plasma is contained in a vacuum chamber at a pressure in the range of 10^{-5} to 10^{-3} atm. Magnetic fields can be used in a number of ways to confine the plasma within the vessel so that it does not contact the wall. The tokamak is an example of a device that uses magnetic confinement. The initial heating can be achieved in several ways, namely, ohmic (resistance) heating of the fuel, magnetic compression, and injection of a high energy "neutral" beam containing an electrically neutral mixture of ions and electrons. Confinement times required are in the range of a few hundredths of a second to seconds.

In the second approach, large amounts of energy are delivered to the fuel mixture, ^{2}H and ^{3}H in molecular form, in pulses of perhaps a microsecond to a nanosecond duration. The energy is supplied by high-powered lasers or sources of electrons or positive ions. Confinement, called inertial confinement, is achieved by sealing the fuel in small pellets or micro spheres that are about 1 mm or less in diameter and are made of glass, for example. The beam of light or electrons or ions of high energy rapidly vaporizes the confining shell of the pellet creating an implosion that increases the density of the fuel mixture and heats it to the ignition temperature.

In the D-T reaction, ^{4}He is formed and a high energy neutron is released. Neutrons so released will produce radioisotopes in the containment vessel and other materials such as ceramics in the reactor. In a fusion reactor, as many of the neutrons as possible would be used to produce the tritium needed for the D-T reaction. In order to consider this point, we must discuss some of the fusion reactor concepts that have been proposed. These drawing-board reactors present some difficult challenges quite apart from the ignition and confinement problems associated with the plasma in the core. The most commonly proposed method of producing or even "breeding" tritium within the reactor while burning it in the core is by surrounding the plasma containment vessel with a blanket containing lithium. Both isotopes of normal lithium, ^{6}Li and ^{7}Li react with neutrons to form tritium. The reactions are $^{6}Li(n,{}^{3}H)^{4}He$ and $^{7}Li(n,n{}^{3}H)^{4}He$. In the ^{6}Li reaction 4.8 MeV are released; in the ^{7}Li reaction 2.5 MeV are required. Materials proposed for the blanket include molten lithium, a molten mixture of lithium fluoride and beryllium fluoride, and solid lithium compounds or alloys.

Heat could be removed to generate steam by using the molten lithium or molten salt mixture itself as a heat transfer medium or by passing helium gas through the blanket for the case of solid breeding material. Fusion reactors would operate at higher temperatures than fission reactors and could, therefore, generate electricity with higher thermodynamic efficiency. The conceptual designs assume an efficiency of about 50%.

Tritium would be removed from the molten lithium or molten salt mixture or from the helium gas while it is outside the reactor in the heat exchanger system. A molten lithium blanket might contain perhaps 1-10 kg of tritium. It is to be expected that there will be some leakage of tritium from the system by diffusion of the $^{3}H_2$ through metal pipes and vessels quite apart from any accident. Molten lithium is itself a source of problems. These include the need to develop container technology and the hazards of fire and explosion when this active metal comes into contact with water.

In terms of the environment, tritium is the only radioactive substance that might be released by leakage during normal operation of a fusion reactor apart from any activity induced in air in the reactor building. The total radioactivity in a reactor depends on the type of refractory metal alloy used for the containment vessel, and, therefore subjected to a high neutron flux. One alloy proposed contains about 99% Nb and 1% Zr. Neutron-induced activities would include ^{89}Zr(78.5 hr), ^{93}Zr(9.5×10^5 yr), ^{95}Zr(65.5 day), ^{97}Zr(16.8 hr), ^{92}Nb($\sim 2 \times 10^7$ yr), ^{93}Nb(12 yr), ^{94}Nb(2.0×10^4 yr) and short-lived radiostopes of yttrium. The radioactivity of the alloy at any time would depend on the number of watts(th) of power generated up to that time. For a fusion reactor containing the niobium alloy, the radioactivity in curies per watt of thermal power generated might be comparable to that of fission products in a fission reactor for as long as about one year after shutdown of the two types of reactors following about five years of operation. Beyond one year after shutdown, the fission product activity would be greater. Any hazard to the public would arise from an accident causing release of the induced activity (and the tritium) into the atmosphere. Statements sometimes made to the effect that fusion power reactors would have no radioactivity hazard are clearly misleading. There are, in fact, concepts of hybrid fusion-fission reactors in which uranium would be mixed with the lithium-containing material in the blanket in order to use some of the fast neutrons produced in the D-T reaction to produce plutonium. These hybrid reactors would then be coupled with fission breeder reactors such as the LMFBR in a plutonium fuel cycle.

In the absence of uranium, the main potential hazards of fusion reactors foreseen at this time would be associated with failure of tritium containment, biological effects of exposure to magnetic fields, and reactor accidents.

17.9 POPULATION EXPOSURE TO IONIZING RADIATION

17.9.1 Pathways to Exposure

We have seen that man is exposed to numerous external and internal sources of ionizing radiation. In fact, he has evolved in an environment where naturally radioactive materials are present in the atmosphere, soil, water supplies, and food of all types. When man's technology leads to the discharge of radioactive material (naturally occurring or man-made) as gases into the atmosphere or as liquids into soil or water supplies (surface or ground), the pathways to man must be examined in detail before an estimate of the hazard can be made. Figure 17-13 shows the major pathways of radioactive materials released to the atmosphere, to surface or ground water and to soil.

Let us consider a specific example of a pathway that happens to be only partially shown in Fig. 17-13. There is experimental evidence that radioactivity may be one of the carcinogens in smoke that cause higher incidence of lung cancer among cigarette smokers. The radionuclide directly involved is the α emitter ^{210}Po, which is the last radioactive member of the uranium series (see Fig. 17-3 and Table 17-2). This isotope is contained in cigarette smoke and it is formed by decay of its long-lived (22 yr) grandparent ^{210}Pb, which is also contained in the smoke. The ^{210}Po that is carried directly is water soluble and very likely eliminated from the lungs. Atoms of ^{210}Pb, however, are incorporated in insoluble particles in the smoke and remain in the lung. Tissue near the particle is irradiated with alpha particles (high LET) from the ^{210}Po daughter year after year, since the ^{210}Po decays with the half-life of the ^{210}Pb after secular equilibrium is attained. The proposed pathway for most of the ^{210}Pb in tobacco starts with ^{226}Ra in the soil or in phosphate fertilizer produced from uranium-bearing phosphate rock. The daughter ^{222}Rn is released as a gas below the tobacco plant. Atoms of the radon daughters attach themselves to particles in air and some of these are collected on the surface of the leaf of the plant. They are retained by the leaf and in this way ^{210}Pb is incorporated in the cigarette. Unfortunately for the smoker, the specific activity (e.g., pCi/gm) of ^{210}Pb is about 10^4 times greater in smoke particles than in dried tobacco.

Fission products such as ^{131}I, ^{137}Cs, and ^{90}Sr in the atmosphere or in the soil can contaminate milk by first contaminating plants that serve as food for cows. Although ^{131}I has a relatively short radioactivity half-life, it is

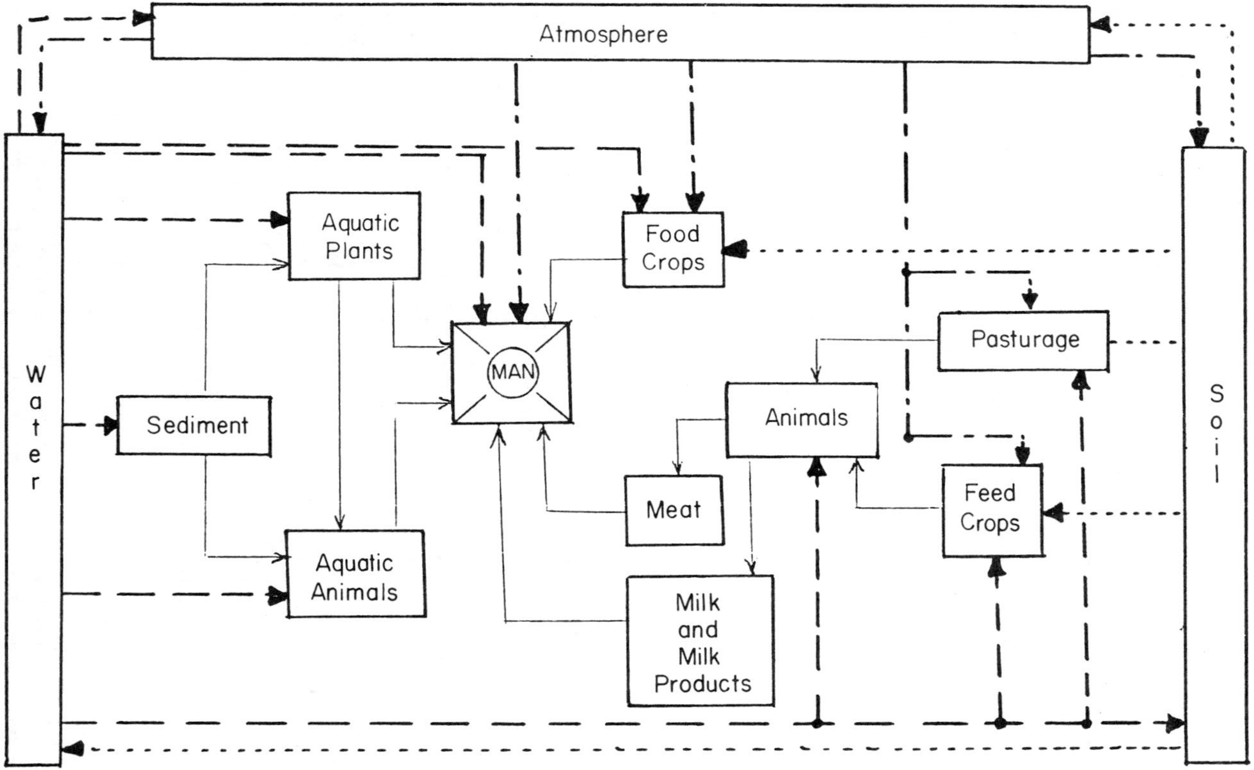

FIGURE 17-13. Major pathways for the uptake by man of radioactive material from the environment.

concentrated in the body after ingestion and is retained in the thyroid where, as we have seen (Section 17.6.3), it has a relatively long biological half-life. Strontium-90 concentrates in the bone and ^{137}Cs becomes rather widely distributed in the body.

Nature provides ways for reducing man's uptake of certain radionuclides. One way is by dilution with stable isotopes. Such dilution is effective for tritium and sodium, for example, but not significantly so for ^{131}I, ^{90}Sr, and ^{137}Cs. A second way is through the ability of biological systems to discriminate against certain elements. The effectiveness can be expressed in various ways. Consider the case of ^{90}Sr in milk. We can calculate a ratio of the ^{90}Sr concentration in milk to that in forage. By taking into account the chemical similarity between Sr and Ca, we can express the concentration of ^{90}Sr in the same unit, the "strontium unit" for all materials, solid or liquid. One Sr unit is equal to 1 pCi of ^{90}Sr/gm of calcium. The observed ratio (OR) for milk relative to forage is

$$\mathrm{OR} = \frac{^{90}\mathrm{Sr/Ca\ (in\ milk)}}{^{90}\mathrm{Sr/Ca\ (in\ forage)}} \qquad (17\text{-}35)$$

Under steady-state conditions, the OR values are in the range of 0.1-0.2.

Biological systems may also concentrate certain elements. Examples of such systems are plants growing in contaminated soil and organisms growing in a marine or a freshwater environment. A concentration factor (CF) for aquatic organisms, for example, is defined as C_o/C_w, where C_o is the concentration of the nuclear species of interest in the organism and C_w is the concentration of the same species in the ambient water. Equilibrium conditions are assumed. For the element zinc, as an example, CF values for plants, molluscs, crustacea, and fish in freshwater may be between 10 and 15,000. For similar marine organisms, the CF values for zinc may be between 100 and 300,000. For a given organism and a given element, the values vary widely with the local conditions. There is, then, a need for caution in drawing conclusions about the significance of CF.

17.9.2 Annual Dose Rate

Table 17-10 contains estimated current annual average whole-body dose rates from all sources in the U.S. The average is over the whole population. Note that almost half of

TABLE 17-10

Estimated Annual Average Whole-body Dose Rates from All Sources of Ionizing Radiations in the United States

Source	*Dose rate (mrem/yr)*
Natural background	
Cosmic rays	*45*
Terrestrial radiation	
External	*40*
Internal (^{40}K)	*17*
Fallout	*4*
Nuclear power	*0.003-0.01*
Medical	*90*
Occupational	*1*
Miscellaneous[a]	*3*
Total	*200*

[a] *Color television receivers, air travel, etc.*

the dose is from medical exposure and about one-eleventh is from ^{40}K in the body.

Several professional groups are continually evaluating the recommended maximum permissible occupational and population exposure to ionizing radiation. These groups include the National Council on Radiation Protection and Measurement (NCRP) and the International Commission on Radiological Protection (ICRP). The ICRP is the organization that sets levels of maximum permissible exposure to ionizing radiation for use as international standards. The NCRP has an analogous role in the U.S. and is affiliated with the National Bureau of Standards. During the period 1959-1970, the Federal Radiation Council (FRC) advised the President with respect to radiation matters directly or indirectly affecting health. The FRC issued reports that are referenced in journal articles and in some books on radiological health. It was abolished in 1970 as part of a major reorganization of governmental health agencies and its functions were transferred to the Environmental Protection Agency (EPA).

The maximum permissible dose (MPD) set for the general public for whole-body exposure for an *individual* is 500 mrem/yr. The maximum whole-body exposure as an *average value for the population* is 170 mrem/yr. The latter is consistent with a maximum genetic dose (to the gonads) of 5000 mrem in 30 yr.

Examples of the maximum permissible concentration (MPC) of radionuclides in water and air set by the NCRP are given in Table 17-11. Note that the values are for occupational exposure in a controlled area where normal operations could involve sources of ionizing radiation. Values for 40-hr exposure are also given in the reference for Table 17-11. For persons outside a controlled area the NCRP recommendations are one-tenth the MPBB and MPC values given in Table 17-11.

During the last three decades the maximum permissible levels of exposure have been reduced several times as more information has been obtained about the long-term health effects of ionizing radiation. At what level can the exposure (or more correctly, the dose) be considered safe? Because it has been difficult to obtain good data on the effects at low level, the usual practice has been to extrapolate to low levels the data obtained at higher levels where the effects are measurable. The type of extrapolation used has been the subject of an ongoing controversy based on disagreement over the existence of a threshold. A threshold for an effect such as leukemia implies that there is a level below which no effect occurs. The absence of a threshold then implies that the effect can occur at any level. There is increasing evidence that there is no threshold for long-term effects.

TABLE 17-11

Maximum Permissible Body Burdens (MPBB) and Maximum Permissible Concentrations (MPC) in Water and Air for Selected Radionuclides for Occupational Exposure[a]

Radionuclide	Form[b]	Location[c]	MPBB (μCi) (Total body)	MPC (μCi/ml) (168 hr-week) Water[d]	Air[d]
^{3}H	S	Body tissue	10^{3}	0.03	2×10^{-6}
	S	Total body	2×10^{3}	0.05	3×10^{-6}
^{14}C	S	Fat	300	8×10^{-3}	10^{-6}
	S	Total body	400	0.01	2×10^{-6}
^{85}Kr	See note[b]	Total body	-	-	3×10^{-6}
^{90}Sr	S	Bone	2	10^{-6}	10^{-10}
	S	Total body	20	4×10^{-6}	3×10^{-10}
	I	Lung	-	-	2×10^{-9}
	I	GI	10^{-3}	4×10^{-4}	6×10^{-8}
^{131}I	S	Thyroid	0.7	2×10^{-5}	3×10^{-9}
	S	Total body	50	2×10^{-3}	3×10^{-7}
	I	GI	-	6×10^{-4}	10^{-7}
	I	Lung	-	-	10^{-7}
^{137}Cs	S	Total body	30	2×10^{-4}	2×10^{-8}
	S	Liver	40	2×10^{-4}	3×10^{-8}
	S	Spleen	50	2×10^{-4}	3×10^{-8}
	S	Muscle	50	2×10^{-4}	4×10^{-8}
	I	Lung	-	-	5×10^{-9}
	I	GI	-	4×10^{-4}	8×10^{-8}

TABLE 17-11 (Continued)

Radionuclide	Form[b]	Location[c]	MPBB (μCi) (Total body)	MPC (μCi/ml) (168 hr-week) Water[d]	Air[d]
^{210}Po	S	Spleen	0.03	7×10^{-6}	2×10^{-10}
	S	Kidney	0.04	8×10^{-6}	2×10^{-10}
	S	Total body	0.4	8×10^{-5}	2×10^{-9}
^{222}Rn	–	Lung	–	–	10^{-8}
^{226}Ra	S	Bone	0.1	10^{-7}	10^{-11}
	S	Total body	0.2	2×10^{-7}	2×10^{-11}
	I	Lung	–	–	2×10^{-11}
	I	GI	–	3×10^{-4}	6×10^{-8}
^{238}U	S	GI	–	4×10^{-4}	8×10^{-8}
	S	Kidney	5×10^{-3}	6×10^{-4}	3×10^{-11}
	S	Total body	0.5	0.01	6×10^{-10}
	I	Lung	–	–	5×10^{-11}
	I	GI	–	4×10^{-4}	6×10^{-8}
^{239}Pu	S	Bone	0.04	5×10^{-5}	6×10^{-13}
	S	Total body	0.4	3×10^{-4}	5×10^{-12}
	I	Lung	–	–	10^{-11}
	I	GI	–	3×10^{-4}	5×10^{-8}

[a] From "Maximum Permissible Body Burdens and Maximum Permissible Concentrations of Radionuclides in Air and in Water for Occupational Exposure." National Bureau of Standards Handbook 69, 1963, U.S. Government Printing Office, Washington, D.C.

[b] S means soluble form; I means insoluble form; for ^{85}Kr, immersion in air is assumed.

[c] Where a specific organ is given, it is considered to be a critical organ. GI means gastrointestinal tract and the entries in this table refer specifically to the lower large intestine. For ^{137}Cs the total body is in the critical organ category.

[d] Values are for continuous exposure for 50 yr, 50 weeks/yr.

REFERENCES

Principles of Nuclear Chemistry

R. D. Evans, "The Atomic Nucleus." McGraw-Hill, New York, 1955.

G. Friedlander, J. W. Kennedy, and J. M. Miller, "Nuclear and Radiochemistry," 2nd ed. Wiley, New York, 1964.

S. Glasstone, "Sourcebook on Atomic Energy," 3rd ed. Van Nostrand, Princeton, New Jersey, 1967.

B. G. Harvey, "Introduction to Nuclear Physics and Chemistry," 2nd ed. Prentice-Hall, Englewood Cliffs, New Jersey, 1969.

Nuclear Data

"Chart of the Nuclides," The General Electric Co., Knolls Atomic Power Laboratory, Schenectady, New York, 1977.

"Handbook of Chemistry and Physics." The Chemical Rubber Co., Cleveland, Ohio.

"Lange's Handbook of Chemistry." McGraw-Hill, New York, 1973.

C. M. Lederer, J. M. Hollander, and I. Perlman, "Table of Isotopes," 6th ed. Wiley, New York, 1967(7th ed in Press).

"Nuclear Data Sheets" (published monthly). Academic Press, New York.

"Radiological Health Handbook," Revised ed. U.S. Dept. of Health, Education, and Welfare, Public Health Service, Washington, D.C., 1960.

Y. Wang, Ed., "Handbook of Radioactive Nuclides." The Chemical Rubber Co., Cleveland, Ohio, 1969.

Radiation Chemistry

J. H. O'Donnell and D. F. Sangster, "Principles of Radiation Chemistry." American Elsevier, New York, 1970.

J. W. T. Spinks and R. J. Woods, "An Introduction to Radiation Chemistry." Wiley, New York, 1964.

Radiological Health (Dosimetry, Effects)

"Basic Radiation Protection Criteria." NCRP Report 39, National Council on Radiation Protection and Measurements, Washington, D.C., 1971.

H. Blatz, "Introduction to Radiological Health." McGraw-Hill, New York, 1964.

H. Cember, "Introduction to Health Physics," 2nd ed. Pergamon Press, Elmsford, New York, 1976.

"The Effects on Populations of Exposure to Low Levels of Ionizing Radiation" (The BEIR Report). National Academy of Sciences - National Research Council, Washington, D.C., 1972.

G. G. Eichholz, "Environmental Aspects of Nuclear Power." Ann Arbor Science Publishers, Ann Arbor, Michigan, 1976.

M. Eisenbud, "Environmental Radioactivity," 2nd ed. Academic Press, New York, 1973.

S. Glasstone and A. Sesonske, "Nuclear Reactor Engineering." Van Nostrand, Princeton, New Jersey, 1967.

E.F. Gloyna and J. O. Ledbetter, "Principles of Radiological Health." Marcel Dekker, New York, 1969.

W. C. Hanson, Ecological considerations of the behavior of plutonium in the environment, *Health Phys. 28,* 529 (1975).

H. F. Henry, "Fundamentals of Radiation Protection." Wiley (Interscience), New York, 1969.

K. Z. Morgan and J. E. Turner, Eds., "Principles of Radiation Protection." Wiley, New York, 1967.

W. D. Norwood, "Health Protection of Radiation Workers." C. C. Thomas, Springfield, Illinois, 1975.

"Radioactivity in the Marine Environment." National Academy of Sciences, Washington, D.C., 1971.

"Radiological Health Handbook," Revised ed. U.S. Dept. of Health, Education and Welfare, Public Health Service, Washington, D.C., 1960.

"Rules and Regulations: Standards for Protection Against Radiation," Title 10, Chapter 1. Code of Federal Regulations, Part 20, U.S. Regulatory Commission, Washington, D.C., Dec. 26, 1975.

J. Schubert, Radioactive poisons, in "Chemistry in the Environment, Readings from Scientific American." W. H. Freeman and Co., San Francisco, California, 1973.

Y. Wang, Ed., "Handbook of Radioactive Nuclides." The Chemical Rubber Co., Cleveland, Ohio, 1969.

Natural Background of Ionizing Radiation

J. A. S. Adams and W. H. Lowder, Eds., "The Natural Radiation Environment." The University of Chicago Press, Chicago, Illinois, 1964.

"The Effects on Populations of Exposure to Low Levels of Ionizing Radiation" (The BEIR Report). National Academy of Sciences - National Research Council, Washington, D.C., 1972.

M. Eisenbud, "Environmental Radioactivity," 2nd ed. Academic Press, New York, 1973.

M. Eisenbud and H. G. Petrow, Radioactivity in the atmospheric effluents of power plants that use fossil fuels, *Science 144*, 288 (1974).

T. F. Gesell and J. A. S. Adams, Geothermal power plants: Environmental Impact, *Science 189*, 328 (1975).

T. F. Gesell and H. M. Prichard, The technologically enhanced natural radiation environment, *Health Phys. 28*, 361 (1975).

R. H. Johnson, Jr., D. E. Bernhardt, N. S. Nelson, and H. W. Calley, Jr., "Assessment of Potential Radiological Health Effects from Radon in Natural Gas." Report EPA-520/1-73-004, U.S. Environmental Protection Agency, Washington, D.C., 1973.

D. Lal and H. E. Suess, The Radioactivity of the atmosphere and hydrosphere, *Annu. Rev. Nucl. Sci. 18*, 407 (1968).

W. S. Osborn, "Primordial radionuclides: Their distribution, movement, and possible effect within terrestial ecosystems, *Health Phys. 11*, 1275 (1965).

R. W. Perkins and J. M. Nielsen, Cosmic-ray produced radionuclides in the environment, *Health Phys. 11,* 1297 (1965).

"Radioactivity in the Marine Environment." National Academy of Sciences, Washington, D.C., 1971.

"Reports of the United Nations Scientific Committee on the Effects of Atomic Radiation" (the "UNSCEAR" Reports). United Nations, New York, 1958, 1962, 1964, 1966, 1969, and 1972.

J. V. Rouse, Radioactive slime ponds, *Environ. Sci. Technol. 8,* 876 (1974).

H. J. Schaefer, Radiation exposure in air travel, *Science 173,* 780 (1971).

H. E. Suess, The radioactivity of the atmosphere and hydrosphere, *Annu. Rev. Nucl. Sci. 8,* 243 (1958)

Radioisotope Power Generation

W. R. Corliss and D. G. Harvey, "Radioisotope Power Generation." Prentice-Hall, Englewood Cliffs, New Jersey, 1964.

Nuclear Fission

G. Friedlander, J. W. Kennedy, and J. M. Miller, "Nuclear and Radiochemistry," 2nd ed. Wiley, New York, 1964.

S. Glasstone, "Sourcebook on Atomic Energy," 3rd ed. Van nostrand, Princeton, New Jersey, 1967.

S. Glasstone and A. Sesonske, "Nuclear Reactor Engineering." Van Nostrand, Princeton, New Jersey, 1967.

B. G. Harvey, "Introduction to Nuclear Physics and Chemistry," 2nd ed. Prentice-Hall, Englewood Cliffs, New Jersey, 1969.

Nuclear Weapons

S. D. Drell and F. von Hippel, Limited nuclear war, *Sci. Am. 235,* No. 5, 27 (1976).

W. Epstein, Nuclear-free zones, *Sci. Am. 233,* No. 5, 25 (1975).

W. Epstein, The proliferation of nuclear weapons, *Sci. Am. 232,* No. 4, 18 (1975).

B. T. Feld, Nuclear proliferation - thirty years after Hiroshima, *Phys. Today 28* , No. 7, 23 (1975).

S. Glasstone, Ed., "The Effects of Nuclear Weapons." U.S. Atomic Energy Commission, U.S. Govt. Printing Office, Washington, D.C., 1962.

H. F. Henry, "Fundamentals of Radiation Protection." Wiley (Interscience), New York, 1969.

D. R. Inglis, "Nuclear Energy: Its Physics and Its Social Challenge." Addison-Wesley, Reading, Massachusetts, 1973.

F. M. Kaplan, Enhanced-radiation weapons, *Sci. Am. 238,* No. 5, 44 (1978).

A. S. Krass, Laser enrichment of uranium: the proliferation connection, *Science 196,* 721 (1977).

J. C. Mark, Global consequences of nuclear weaponry, *Annu. Rev. Nucl. Sci. 26,* 51 (1976).

J. McPhee, "The Curve of Binding Energy." Ballantine Books, New York, 1975.

D. J. Rose and R. K. Lester, Nuclear power, nuclear weapons and international stability, *Sci. Am. 238,* No. 4, 45 (1978).

E. J. Moniz and T. L. Neff, Nuclear power and nuclear-weapons proliferation, *Phys. Today 31,* No. 4, 42 (1978).

T. B. Taylor, Nuclear safeguards, *Annu. Rev. Nucl. Sci. 25,* 407 (1975).

Fallout (see also Nuclear Weapons)

J. R. Arnold and E. A. Martell, The circulation of radioactive isotopes, in "Chemistry in the Environment, Readings from Scientific American." W. H. Freeman and Co., San Francisco, California, 1973.

C. L. Comar, Movement of fallout radionuclides through the biosphere and man, *Annu. Rev. Nucl. Sci. 15,* 175 (1965).

M. Eisenbud, "Environmental Radioactivity," 2nd ed. Academic Press, New York, 1973.

E. G. Fowler, Ed., "Radioactive Fallout, Soils, Plants, Foods, Man." Elsevier Publishing Co., New York, 1965.

S. Glasstone, Ed., "The Effects of Nuclear Weapons." U.S. Atomic Energy Commission, U.S. Govt. Printing Office, Washington, D.C., 1962.

D. Lal and H. E. Suess, The radioactivity of the atmosphere and hydrosphere, *Annu. Rev. Nucl. Sci. 18,* 407 (1968).

"Radioactivity in the Marine Environment." National Academy of Science, Washington, D.C., 1971.

"Radionuclides in the Environment." Advances in Chemistry Series No. 93, American Chemical Society, Washington, D.C., 1970.

H. E.Suess, The radioactivity of the atmosphere and hydrosphere, *Annu. Rev. Nucl. Sci. 8,* 243 (1958).

Nuclear Fission Power Plants (Types, Safety, etc.)

"Advanced Nuclear Reactors: An Introduction." ERDA-76-107, U.S. Govt. Printing Office, Washington, D.C., 1976.

T. B. Cochran, "The Liquid Metal Fast Breeder Reactor." Resources for the Future, Inc. (distributed by the Johns Hopkins Univ. Press, Baltimore, Maryland), 1974.

Breeder reactors: Power for the future, *Sience 174,* 807 (1971).

G. G. Eichholz, "Environmental Aspects of Nuclear Power." Ann Arbor Science Publishers, Inc., Ann Arbor, Michigan, 1976.

M. M. El-Wakil, "Nuclear Energy Conversion." Intertext Educational Publishers, Scranton, Pennsylvania, 1971.

"Environmental Radioactivity Surveillance Guide." Report ORP/SID 72-2, U.S. Environmental Protection Agency, Washington, D.C., 1972.

EPA's environmental radiation-assessment program, *Nucl. Saf. 16,* 667 (1975).

European breeders, *Science 190,* 1279 (1975); *191*, 368 (1976); *191*, 551 (1976).

F. R. Farmer, Ed., "Nuclear Reactor Safety." Academic Press, New York, 1977.

M. Fogiel, Ed., "Modern Energy Technology," Vol. I. Research and Education Association, New York, 1975.

S. Glasstone, "Sourcebook on Atomic Energy," 3rd ed. Van Nostrand, Princeton, New Jersey, 1967.

S. Glasstone and A. Sesonske, "Nuclear Reactor Engineering." Van Nostrand, Princeton, New Jersey, 1967.

J. Graham, "Fast Reactor Safety." Academic Press, New York, 1971.

W. Häfele, D. Faude, E. A. Fischer, and H. J. Love, Fast breeder reactors, *Annu. Rev. Nucl. Sci. 20,* 393 (1970).

R. P. Hammond, Nuclear power risks, *Am. Sci 62,* 155 (1974).

A. L. Hammond, W. D. Metz, and T. H. Maugh II, "Energy and the Future." American Association for the Advancement of Science, Washington, D.C., 1973.

H. C. Hottel and J. B. Howard, "New Energy Technology: Some Facts and Assessments." MIT Press, Cambridge, Massachusetts, 1971.

D. R. Inglis, "Nuclear Energy: Its Physics and Its Social Challenge." Addison-Wesley, Reading, Massachusetts, 1973.

C. K. Leeper, How safe are reactor emergency cooling systems? *Phys. Today 26,* No. 8, 30 (1973).

H. C. McIntyre, Natural-uranium heavy water reactors, *Sci. Am. 233,* No. 4, 17 (1974).

Nuclear reactor safety - the APS submits its report, *Phys. Today 28,* No. 7, 38 (1975).

T. H. Pigford, Environmental aspects of nuclear energy production, *Annu. Rev. Nucl. Sci. 24,* 515 (1974).

"Reactor Safety Study, An Assessment of Accident Risks in U.S. Commercial Nuclear Power Plants." U.S. Nuclear Regulatory Commission, WASH-1400 (NUREG 75/014), NTIS, Springfield, Virginia, 1975.

Report to the American Physical Society by the study group on light-water reactor safety, *Rev. Mod. Phys. 47,* Suppl. No. 1, (1975).

D. J. Rose, Nuclear electric power, *Science 184,* 351 (1974); and in "Energy, Use, Conservation and Supply." American Association for the Development of Science, Washington, D.C., 1974.

"Rules and Regulations: Licensing of Production and Utilization Facilities," Title 10, Chapter 1. Code of Federal Regulations, Part 50, U.S. Nuclear Regulatory Commission, Washington, D.C., June 20, 1975.

G. T. Seaborg and J. L. Bloom, Fast breeder reactors, in "Chemistry in the Environment, Readings from Scientific American." W. H. Freeman and Co., San Francisco, California, 1973.

A. Sesonske, "Nuclear Power Plant Design Analysis." TID-26241, National Technical Information Service, Springfield, Virginia, 1973.

G. A. Vendryes, Superphenix: A full-scale breeder reactor, *Sci. Am. 236,* No. 3, 26 (1977).

Nuclear Fuel Reprocessing

W. P. Bebbington, The reprocessing of nuclear fuels, *Sci. Am. 235,* No. 6, 30 (1976).

M. Benedict and T. H. Pigford, "Nuclear Chemical Engineering." McGraw-Hill, New York, 1957.

G. G. Eichholz, "Environmental Aspects of Nuclear Power." Ann Arbor Science Publishers, Inc., Ann Arbor, Michigan, 1976.

S. Glasstone and A. Sesonske, "Nuclear Reactor Engineering." Van Nostrand, Princeton, New Jersey, 1967.

J. M. Palms, V. R. Veluri, and F. W. Boone, The environmental impact of ^{129}I released by a nuclear fuel-reprocessing plant, *Nucl. Saf. 16,* 593 (1975).

T. H. Pigford, Environmental aspects of nuclear energy production, *Annu. Rev. Nucl. Sci. 24,* 515 (1974).

K. Telegadas and G. J. Ferber, Atmospheric concentrations and inventory of Krypton-85 in 1973, *Science 190,* 882 (1975).

The nuclear fuel cycle: an appraisal, Summary of report of the study group on nuclear fuel cycles and waste management, American Physical Society, *Phys. Today 30,* No. 10, 32 (1977).

Radioactive Waste Disposal

E. E. Angino, High-level and long-lived radioactive waste disposal, *Science 198,* 885 (1977).

J. O. Blomeke, J. P. Nichols, and W. C. McClain, Managing radioactive wastes, *Phys. Today 26,* No. 8, 36 (1973).

M. H. Campbell, Ed., "High Level Radioactive Waste Management." Advances in Chemistry Series No. 153, American Chemical Society, Washington, D.C., 1976.

B. L. Cohen, High-level radioactive waste from light-water reactors, *Rev. Mod. Phys. 49,* 1 (1977).

G. G. Eichholz, "Environmental Aspects of Nuclear Power." Ann Arbor Science Publishers, Inc., Ann Arbor, Michigan, 1976.

M. Eisenbud, "Environmental Radioactivity," 2nd ed. Academic Press, New York, 1973.

Environmental hazards in radioactive waste disposal, *Phys. Today 29,* No. 1, 9 (1976).

"An Evaluation of the Concept of Storing Radioactive Wastes in Bedrock Below the Savannah River Plant Site." National Academy of Sciences, Washington, D.C., 1972.

E. F. Gloyna and J. O. Ledbetter, "Principles of Radiological Health." Marcel Dekker, New York, 1969.

"Interim Storage of Solidified High-Level Radioactive Wastes." National Academy of Sciences, Washington, D.C., 1975.

G. de Marsily, E. Ledoux, A. Barbreau and J. Margat, Nuclear waste disposal: can the geologist guarantee isolation? *Science 197,* 519 (1977).

T. H. Pigford, Environmental aspects of nuclear energy production, *Annu. Rev. Nucl. Sci. 24,* 515 (1974).

G. I. Rochlin, Nuclear waste disposal: Two social criteria *Science 195,* 23 (1977).

The nuclear fuel cycle: an appraisal, Summary of report of the study group on nuclear fuel cycles and waste management, American Physical Society, *Phys. Today 30,* No. 10, 32 (1977).

Natural Fission Reactors

G. A. Cowan, A natural fission reactor, *Sci. Am. 235,* No. 1, 36 (1976).

M. Maurette, Fossil nuclear reactors, *Annu. Rev. Nucl. Sci. 26,* 319 (1976).

"The Oklo Phenomenon." IAEA Proceedings, STI/PUB/405, Vienna, 1975.

R. West, Natural nuclear reactors: The Oklo phenomenon, *J. Chem. Educ. 53,* 336 (1976).

The Controversy Over Nuclear Fission Power Plants

E. V. Anderson, Nuclear energy: A key role despite problems, *Chem. Eng. News 55,* No. 10, 8 (1977).

J. J. Berger, "Nuclear Power: The Unviable Option." Ramparts Press, Palo Alto, California, 1976.

H. A. Bethe, The necessity of fission power, *Sci. Am. 234,* No. 1, 21 (1976).

P. M. Boffey, Reactor safety: Congress hears critics of Rasmussen report, *Science 192,* 1312 (1976).

B. G. Chow, The economic issues of the fast breeder reactor program, *Science 195,* 551 (1977).

B. Cohen, Impacts of the nuclear energy industry on human health and safety, *Am. Sci. 64,* 550 (1976).

B. Commoner, "The Poverty of Power." A. A. Knopf, New York, 1976.

J. J. Di Certo, "The Electric Wishing Well." Macmillan, New York, 1976.

S. Ebbin and R. Kasper, "Citizens Groups and the Nuclear Power Controversy." MIT Press, Cambridge, Massachusetts, 1974.

J. G. Fuller, "We Almost Lost Detroit." Ballantine Books, New York, 1976.

C. Hohenemser, R. Kasperson, and R. Kates, The distrust of nuclear power, *Science 196,* 25 (1977).

A. Legault and G. Lindsey, "The Dynamics of Nuclear Balance." Cornell Univ. Press, Ithaca, New York, 1976.

A. Lovins and J. Price, "Non-Nuclear Futures: The Case for an Ethical Energy Strategy." Ballinger, Cambridge, Massachusetts, 1975.

W. W. Lowrance, "Of Acceptable Risk: Science and the Determination of Safety." W. Kaufman, Los Altos, California, 1976.

J. McPhee, "The Curve of Binding Energy." Ballantine Books, New York, 1975.

E. J. Moniz and T. L. Neff, Nuclear power and nuclear-weapons proliferation, *Phys. Today 31,* No. 4, 42 (1978).

A. W. Murphy, "The Nuclear Power Controversy." Prentice-Hall Englewood Cliffs, New Jersey, 1976.

J. Neyman, Public health hazards from electricity-producing plants, *Science 195,* 754 (1977).

S. Novick, "The Electric War: The Fight Over Nuclear Power." Sierra Club Books, San Francisco, 1976.

"Nuclear Energy." Report of the Fiftieth American Assembly, American Assembly, Columbia Univ., New York, 1976.

Nuclear energy: How bright a future? *Environ. Sci. Technol. 11,* 128 (1977).

"The Nuclear Fuel Cycle. A Survey of the Public Health, Environmental, and National Security Effects of Nuclear Power." Union of Concerned Scientists, MIT Press, Cambridge, Massachusetts, 1975.

M. C. Olson, "Unacceptable Risk: The Nuclear Power Controversy." Bantam Books, New York, 1976.

E. M. Page, "We Did Not Almost Lose Detroit." Detroit Edison, Detroit, Michigan, 1976.

W. L. Rankin and S. M. Nealy, Attitudes of the public about nuclear wastes, *Nucl. News 21*, No. 8, 112 (1978).

"Reactor Safety Study: An Assessment of Accident Risks in U.S. Commercial Nuclear Power Plants." Report: WASH-1400 (NUREG 75/014), U.S. Nuclear Regulatory Commission, Washington, D.C., 1975.

D. J. Rose, P. W. Walsh, and L. L. Leskovjan, Nuclear power - compared to what?, *Am. Sci. 64*, 291 (1976).

D. J. Rose and R. K. Lester, Nuclear power, nuclear weapons and international stability, *Sci. Am. 238*, No. 4, 45 (1978).

F. H. Schmidt and D. Bodansky, "The Fight over Nuclear Power." Albion, San Francisco, 1976.

D. Shapley, Nuclear power plants: Why do some work better than others?, *Science 195*, 1311 (1977).

T. B. Taylor, Nuclear safeguards, *Annu. Rev. Nucl. Sci. 25*, 407 (1975).

R. E. Webb, "The Accident Hazards of Nuclear Power Plants." The Univ. of Massachusetts Press, Amherst, Massachusetts, 1976.

A. M. Weinberg, Social institutions and nuclear energy, *Science 177*, 27 (1972).

A. M. Weinberg, The maturity and future of nuclear energy, *Am. Sci. 64*, 16 (1976).

Nuclear Fusion Power Plants

M. M. El-Wakil, "Nuclear Energy Conversion." Intertext Educational Publishers, Scranton, Pennsylvania, 1971.

M. Fogiel, Ed., "Modern Energy Technology," Vol. I. Research and Education Association, New York, 1975.

W. C. Gough and B. J. Eastlund, The prospects of fusion power, in "Chemistry in the Environment, Readings from Scientific American." W. H. Freeman and Co., San Francisco, California, 1973.

D. M. Gruen, Ed., "The Chemistry of Fusion Technology." Plenum, New York, 1972.

A. L. Hammond, W. D. Metz, and T. H. Maugh II, "Energy and the Future." American Association for the Advancement of Science, Washington, D.C., 1973.

R. L. Hirsch, Status and future directions of the world program in fusion research and development, *Annu. Rev. Nucl. Sci. 25,* 79 (1975).

J. P. Holdren, Fusion energy in context: Its fitness for the long term, *Science 200,* 168 (1978).

B. B. Kadomtsev and T. K. Fowler, Fusion reactors (USSR), *Phys. Today 28,* No. 11, 36 (1975).

J. W. Landis, Fusion power, *J. Chem. Educ. 50,* 658 (1973).

J. Nuckolls, J. Emmett, and L. Wood, Laser-induced thermonuclear fusion, *Phys. Today 26,* No. 8, 46 (1973).

W. E. Parkins, Engineering limitations of fusion power plants, *Science 199,* 1403 (1978).

R. F. Post, Prospects for fusion power, *Phys. Today 26,* No. 4, 30 (1973).

R. F. Post and F. L. Ribe, Fusion reactors as future energy sources, *Science 186,* 397 (1974).

D. Steiner and J. F. Clarke, The tokamak: Model T fusion reactor, *Science 199,* 1395 (1978).

Population Exposure to Ionizing Radiation

J. R. Arnold and E. A. Martell, The circulation of radioactive isotopes, in "Chemistry in the Environment, Readings from Scientific American." W. H. Freeman and Co., San Francisco, California, 1973.

C. L. Comar, Movement of fallout radionuclides through the biosphere and man, *Annu. Rev. Nucl. Sci. 15,* 175 (1965).

"The Effects on Populations of Exposure to Low Levels of Ionizing Radiation (The BEIR Report). National Academy of Sciences - National Research Council, Washington, D.C., 1972.

G. G. Eichholz, "Environmental Aspects of Nuclear Power." Ann Arbor Science Publishers, Ann Arbor, Michigan, 1976.

M. Eisenbud, "Environmental Radioactivity," 2nd ed. Academic Press, New York, 1973.

J. W. Elwood, Ecological aspects of tritium behavior in the environment, *Nucl. Saf. 12,* 236 (1971).

E. G. Fowler, Ed., "Radioactive Fallout, Soils, Plants, Foods, Man." Elsevier, New York, 1965.

D. R. Inglis, "Nuclear Energy: Its Physics and Its Social Challenge." Addison-Wesley, Reading, Massachusetts, 1973.

S. M. Jinks and M. Eisenbud, Concentration factors in the aquatic environment, *Radiat. Data Rep. 13,* 243 (1972).

R. H. Johnson, D. E. Bernhardt, N. S. Nelson, and H. W. Calley, Jr., Assessment of potential radiological health effects from radon in natural gas, Report EPA-520/1-73-004, U.S. Environmental Protection Agency, Washington, D.C., 1973.

J. A. Lieberman, Ionizing-radiation standards for population exposure, *Phys. Today 24,* No. 11, 32 (1971).

J. B. Little, A. R. Kennedy, and R. B. McGandy, Lung cancer induced in hamsters by low doses of alpha radiation from polonium-210, *Science 188,* 737 (1975).

E. A. Martell, Tobacco radioactivity and cancer in smokers, *Am. Sci. 63,* 404 (1975).

"Maximum Permissible Body Burdens and Maximum Permissible Concentrations of Radionuclides in Air and in Water for Occupational Exposure." National Bureau of Standards Handbook 69, U.S. Govt. Printing Office, Washington, D.C., 1963.

K. Z. Morgan and J. E. Turner, Eds., "Principles of Radiation Protection." Wiley, New York, 1967.

R. S. Morse and G. A. Walford, Dietary intake of ^{210}Pb, *Health Phys. 21,* 53 (1971).

D. Nelkin, "Nuclear Power and Its Critics." Cornell Univ. Press, Ithaca, New York, 1971.

Population exposure to X-rays - U.S. 1970, *Nucl. Saf. 15,* 453 (1974).

"Principles of Environmental Monitoring Related to the Handling of Radioactive Materials." The International Commission on Radiological Protection, ICRP Publication 7, Pergamon Press, Elmsford, New York, 1966.

"Proceedings of the Third National Symposium on Radioecology," held at Oak Ridge, Tennessee, 2 Vols., NTIS, Springfield, Virginia, 1971.

"Radioactivity in the Marine Environment." National Academy of Sciences, Washington, D.C., 1971.

"Reports of the United Nations Scientific Committee on the Effects of Atomic Radiation" (The "UNSCEAR" Reports). United Nations, New York, 1958, 1962, 1964, 1966, 1969, and 1972.

"Rules and Regulations: Standards for Protection Against Radiation," Title 10, Chapter 1, Code of Federal Regulations, Part 20, U.S. Regulatory Commission, Washington, D.C., Dec. 26, 1975.

R. Scott-Russell, Ed., "Radioactivity and Human Diet." Pergamon Press, Oxford, 1966.

INDEX

B

F

G

H

O

P

Q

R

S

U

V

W

X

Z